PEM FUEL CELL DIAGNOSTIC TOOLS

PEM FUEL CELL DURABILITY HANDBOOK

PEM Fuel Cell Failure Mode Analysis

PEM Fuel Cell Diagnostic Tools

PEM FUEL CELL DIAGNOSTIC TOOLS

EDITED BY
HAIJIANG WANG
XIAO-ZI YUAN
HUI LI

CRC Press
Taylor & Francis Group
Boca Raton London New York

CRC Press is an imprint of the
Taylor & Francis Group, an **informa** business

CRC Press
Taylor & Francis Group
6000 Broken Sound Parkway NW, Suite 300
Boca Raton, FL 33487-2742

First issued in paperback 2017

© 2012 by Taylor & Francis Group, LLC
CRC Press is an imprint of Taylor & Francis Group, an Informa business

No claim to original U.S. Government works

Version Date: 20110707

ISBN-13: 978-1-4398-3919-5 (hbk)
ISBN-13: 978-1-138-11334-3 (pbk)

Library of Congress Cataloging-in-Publication Data

PEM fuel cell diagnostic tools / editors, Haijiang Wang, Xian-Zi Yuan, and Hui Li.
    p. cm.
    Summary: "Compared to other electrochemical power devices such as the battery, the PEM fuel cell is much more
    Includes bibliographical references and index.
    ISBN 978-1-4398-3919-5 (hardback)
    1. Proton exchange membrane fuel cells--Testing. 2. Proton exchange membrane fuel cells--Testing--Equipment and supplies. I. Wang, Haijiang Henry. II. Yuan, Xiao-Zi. III. Li, Hui, 1964- IV. Title.

TK2933.P76P46 2011
621.31'24290287--dc22

2011007915

Visit the Taylor & Francis Web site at
http://www.taylorandfrancis.com

and the CRC Press Web site at
http://www.crcpress.com

Contents

PART I *In Situ* Diagnostic Tools

PART II *Ex Situ* Diagnostic Tools

Preface

Compared to other electrochemical power devices such as the battery, the polymer electrolyte membrane (PEM) fuel cell is much more complicated. Its complexity derives from the following aspects: (1) Most of the components are composite materials. (2) Porous materials must be used for gas and water transport. (3) Nanomaterials have to be used to achieve high electrochemical activity. (4) Complicated processes take place within the fuel cell in addition to the electrochemical reactions, such as the transport of electrons, protons, reactant gases, product water and vapor, and heat. (5) The electrode reaction occurs at a multiphase boundary and transport may occur across multiple boundaries. (6) Multiphase flow happens in flow field channels and porous media. (7) The scale at which researchers have to look ranges from nanometers to meters. (8) Three-dimensional architecture is vitally important to performance and durability, due to the large size of PEM fuel cell stacks. (9) Local performance can seriously affect the system's performance and durability. (10) There are complicated operating conditions, such as load, temperature, pressure, gas flow, and humidification.

To study such a multifaceted system, different types of tools are needed. By adopting some existing tools and developing new ones, fuel cell researchers now have a handful of methods available for carrying out PEM fuel cell performance and durability diagnosis. The research focus in PEM fuel cell diagnostics is mainly on uncovering the detailed relationships between the performance and durability of a fuel cell and the cell's structure, design, and operation, in order to use this knowledge to design fuel cell systems that exhibit improved performance and longevity.

Owing to the electrochemical nature of PEM fuel cells, the electrode process plays a key role in fuel cell performance. Common electrochemical methods, such as electrochemical impedance spectroscopy, cyclic voltammetry, linear sweep voltammetry, CO stripping voltammetry, current stepping, voltage stepping, coulometry, and rotating disc electrode, have already become routine techniques in fuel cell research labs. These techniques can enable measurements for the study of various electrode processes, according to different research objectives. Commercial electrochemical equipment can usually handle only very limited power, so certain modifications are usually necessary to use such equipment to carry out fuel cell measurements.

As mentioned, the fuel cell system is highly complex. The application of electrochemical methods is therefore subject to limitations, and other types of tools are also necessary. For example, multiphase flow and water management are very important aspects of PEM fuel cells, so tools that can help with studying water transport inside the fuel cell are essential for understanding these phenomena. Fuel cell researchers have achieved a degree of success using transparent cells, magnetic resonance imaging (MRI), neutron scattering, and other techniques to study water transport in fuel cells.

Tools for material characterization, especially ones that can reveal the micro- or nanostructures of materials (such as SEM, TEM, and x-ray CT) are a very important category of diagnostic tools for PEM fuel cell research. Another important category consists of tools for mapping characteristics such as temperature, current, voltage, gas pressure, oxygen concentration, and liquid water profile. Segmented cell and embedded sensor techniques are frequently used methods to achieve mapping of local properties.

Most diagnostic tools can have several functions, while some may have a unique function. Up to now, fuel cell researchers have been using these tools to perform various PEM fuel cell diagnostic measurements. Fundamental knowledge of fuel cell performance and durability in relation to component material structure, fuel cell architecture, and operating conditions has been accumulated, and this knowledge has been important in the design of new fuel cell products. In this book we bring together the different types of diagnostic tools that fuel cell researchers have been using in PEM fuel cell research. We would like this book to be the manual for your tool box.

Editors

Haijiang Wang is a senior research officer, project manager of multiprojects, and the core competency leader of the Unit Fuel Cell Team in the National Research Council of Canada Institute for Fuel Cell Innovation (NRC-IFCI). He is currently leading a team of over 10 scientists to carry out research and development on novel fuel cell design and materials, as well as fuel cell diagnosis and durability. Dr. Wang received his PhD in electrochemistry from the University of Copenhagen, Denmark, in 1993. He then joined Dr. Vernon Parker's research group at Utah State University as a postdoctoral researcher to study electrochemically generated anion and cation radicals. In 1997, he began working with Natural Resources Canada as a research scientist to carry out research on fuel cell technology. In 1999, he joined Ballard Power Systems as a senior research scientist to continue his investigations. After spending five years with Ballard Power Systems, he joined NRC-IFCI in 2004. He is currently adjunct professor at five universities, including the University of British Columbia and the University of Waterloo. Dr. Wang has 25 years of professional research experience in electrochemistry and fuel cell technology. To date, he has published 115 journal papers, 3 books, 40 industrial reports, and 30 conference papers or presentations, and has been issued three patents.

Xiao-Zi Yuan is a research officer and project leader of the Unit Cell Team at the Institute for Fuel Cell Innovation, National Research Council of Canada (NRC-IFCI). Dr. Yuan received her BS and MSc in electrochemical engineering from Nanjing University of Technology in 1991 and 1994, respectively, under the supervision of Professor Baoming Wei, and her PhD in material science from Shanghai Jiaotong University in 2003, under the supervision of Professor Zi-Feng Ma. After graduating in MSc, she held a lecturer position at Nantong University for six years, and on completing her PhD she was an associate professor in the same university for one year. Beginning in 2005, she carried out a three-year postdoctoral research program at NRC-IFCI with Dr. Haijiang Wang. Dr. Yuan has over 16 years of R&D experience in applied electrochemistry, including over 10 years of fuel cell R&D (among these three years at Shanghai Jiaotong University, one year at Fachhochschule Mannheim, and six years, to date, at NRC-IFCI). Currently, her research focuses on PEM fuel cell design, testing, diagnosis, and durability. Dr. Yuan has published more than 50 research papers in refereed journals, produced two books and five book chapters, presented more than 30 conference papers or presentations, and holds five China patents.

Hui Li is a research council officer and project technical leader both for the PEM Fuel Cell Failure Mode Analysis project and PEM Fuel Cell Contamination Consortium at the National Research Council of Canada Institute for Fuel Cell Innovation (NRC-IFCI). Dr. Li received her BS and MSc in chemical engineering from Tsinghu University in 1987 and 1990, respectively. After completing her MSc, she joined Kunming Metallurgical Institute as a research engineer for four years and then took a position as an associate professor at Sunwen University (then a branch of Zhongshan University) for eight years. In 2002, she started her PhD program in electrochemical engineering at the University of British Columbia

(UBC) under the supervision of Professor Colin Oloman. After obtaining her PhD in 2006, she carried out one term of postdoctoral research at the Clean Energy Research Centre (CERC) at UBC with Professor Colin Oloman and Professor David Wilkinson. In 2007, she joined the Low-temperature PEM Fuel Cell Group at NRC-IFCI, working on PEM fuel cell contamination and durability. Dr. Li has years of research and development experience in theoretical and applied electrochemistry and in electrochemical engineering. Her research is based on PEM fuel cell contamination and durability testing; preparation and development of electrochemical catalysts with long-term stability; catalyst layer/cathode structure; and catalyst layer characterization and electrochemical evaluation. Dr. Li has coauthored more than 30 research papers published in refereed journals and has one technology licensed to the Mantra Energy Group. She has also produced many industrial technical reports.

Contributors

Robert Alink
Fraunhofer Institute for Solar Energy
   Systems
Freiburg, Germany

Bruce J. Balcom
Department of Physics
University of New Brunswick
Fredericton, New Brunswick, Canada

Dmitri Bessarabov
HySA Infrastructure Center of Competence
North-West University
Potchefstroom, South Africa

Xiaotao Bi
Department of Chemical and Biological
   Engineering
Clean Energy Research Centre
University of British Columbia,
Vancouver, British Columbia, Canada

Pierre Boillat
Paul Scherrer Institut
Villigen, Switzerland

Xuan Cheng
Department of Materials Science and
   Engineering
and
Fujian Key Laboratory of Advanced
   Materials
Xiamen University
Xiamen Fujian, China

Sebastian Dormido
Departamento de Informática y Automática
Universidad Nacional de Educacian a Distancia
Madrid, Spain

Khalid Fatih
National Research Council
Institute for Fuel Cell Innovation
Vancouver, British Columbia, Canada

Michael W. Fowler
Department of Chemical Engineering
University of Waterloo
Waterloo, Ontario, Canada

K. Andreas Friedrich
German Aerospace Center
Institute for Technical Thermodynamics
Stuttgart, Germany

Dietmar Gerteisen
Fraunhofer Institute for Solar Energy
   Systems
Freiburg, Germany

Jeff T. Gostick
Department of Chemical Engineering
McGill University
Montreal, Quebec, Canada

Herwig Robert Haas
Automotive Fuel Cell Cooperation Corporation
Burnaby, British Columbia, Canada

Andrea Haug
German Aerospace Center
Institute for Technical Thermodynamics
Stuttgart, Germany

Renate Hiesgen
Department of Basic Science
University of Applied Sciences
Esslingen, Germany

Chun-Ying Hsu
Mechanical Engineering Department
Fuel Cell Center
Yuan-Ze University
Chung-Li, Taiwan, Republic of China

Marios A. Ioannidis
Department of Chemical Engineering
University of Waterloo
Waterloo, Ontario, Canada

Hengyi Li
Department of Materials Science and
    Engineering
Xiamen University
Xiamen Fujian, China

Hui Li
National Research Council
Institute for Fuel Cell Innovation
Vancouver, British Columbia, Canada

Rui Lin
Clean Energy Automotive Engineering
    Center
School of Automotive Studies
Tongji University
Shanghai, China

Jian-Xin Ma
Clean Energy Automotive Engineering
    Center
School of Automotive Studies
Tongji University
Shanghai, China

Zi-Feng Ma
Department of Chemical Engineering
Shanghai Jiao Tong University
Shanghai, China

Jonathan J. Martin
National Research Council
Institute for Fuel Cell Innovation
Vancouver, British Columbia, Canada

Dilip Natarajan
Experimental Station
Du Pont de Nemours
Wilmington, Delaware

Trung Van Nguyen
Department of Chemical and Petroleum
    Engineering
The University of Kansas
Lawrence, Kansas

Mark D. Pritzker
Department of Chemical Engineering
University of Waterloo
Waterloo, Ontario, Canada

Justin Roller
National Research Council
Institute for Fuel Cell Innovation
Vancouver, British Columbia, Canada

Miguel A. Rubio
Departamento de Inform'atica y Autom'atica
Universidad Nacional de Educacian a Distancia
Madrid, Spain

Pierre Sauriol
Department of Chemical and Biological
    Engineering
Clean Energy Research Centre
University of British Columbia,
Vancouver, British Columbia, Canada

Günther G. Scherer
Paul Scherrer Institut
Villigen, Switzerland

Günter Schiller
German Aerospace Center
Institute for Technical Thermodynamics
Stuttgart, Germany

Mathias Schulze
German Aerospace Center
Institute for Technical Thermodynamics
Stuttgart, Germany

Jun Shen
National Research Council
Institute for Fuel Cell Innovation
Vancouver, British Columbia, Canada

Jürgen Stumper
Automotive Fuel Cell Cooperation Corporation
Burnaby, British Columbia, Canada

Alfonso Urquia
Departamento de Inform´atica y Autom´atica
Universidad Nacional de Educacian a Distancia
Madrid, Spain

Norbert Wagner
German Aerospace Center
Institute for Technical Thermodynamics
Stuttgart, Germany

Haijiang Wang
National Research Council
Institute for Fuel Cell Innovation
Vancouver, British Columbia, Canada

Fang-Bor Weng
Mechanical Engineering Department
Fuel Cell Center
Yuan-Ze University
Chung-Li, Taiwan, Republic of China

Jinfeng Wu
National Research Council
Institute for Fuel Cell Innovation
Vancouver, British Columbia, Canada

Tak Cheung Yau
Department of Chemical and Biological
    Engineering
Clean Energy Research Centre
University of British Columbia,
Vancouver, British Columbia, Canada

Xianxia Yuan
Department of Chemical Engineering
Shanghai Jiao Tong University
Shanghai, China

Xiao-Zi Yuan
National Research Council
Institute for Fuel Cell Innovation
Vancouver, British Columbia, Canada

Dong-Yun Zhang
Department of Material Science and
    Engineering
Shanghai Institute of Technology
Shanghai, China

Shengsheng Zhang
National Research Council
Institute for Fuel Cell Innovation
Vancouver, British Columbia, Canada

Ziheng Zhang
Department of Diagnostic Radiology
Yale University School of Medicine
New Haven, Connecticut

Junsheng Zheng
Clean Energy Automotive Engineering
    Center
School of Automotive Studies
Tongji University
Shanghai, China

Qiaoming Zheng
Department of Materials Science and
    Engineering
Xiamen University
Xiamen Fujian, China

Introduction

Xiao Zi Yuan, Haijiang Wang, and Hui Li

The Proton Exchange Membrane Fuel Cell and Its Durability

Different technological solutions to deliver cleaner and greener energy are being investigated worldwide, and important research programs have been launched to develop new electrochemical devices, such as fuel cell (FC) generators. Owing to their high-energy efficiency, convenient operation, and environmentally friendly characteristics, low-temperature FCs, especially polymer electrolyte membrane fuel cells (PEMFCs) and stacks, are considered one of the most promising technologies for both stationary and mobile applications. Significant progress has been achieved over the past few decades, especially in the areas of increasing volumetric and/or gravimetric specific power density and more effective materials utilization. However, the commercialization of this technology has been held up due to technical challenges, including the FC system itself, as well as problems of on-board storage and the need of an infrastructure for hydrogen fuel. The U.S. Department of Energy (DOE) has identified durability and cost as the top two issues in PEMFC technology and recognized that only when FC costs are dramatically reduced to the DOE target of $50 kW^{-1} will FCs be competitive for virtually every type of power application. The success of PEMFCs depends ultimately on their performance, durability, and cost competitiveness.

Durability is the ability of a PEMFC or stack to resist permanent change in performance over time. Durability decay does not lead to catastrophic failure but simply to a decrease in performance that is not recoverable or reversible (i.e., due to loss of electrochemical surface area, carbon corrosion, etc.). Depending upon the application, the requirements for FC lifetime vary significantly, ranging from 5000 h for cars to 20,000 h for buses, and 40,000 h of continuous operation for stationary applications. Although the life targets for automobiles are much lower than those for stationary applications, operating conditions such as dynamic load cycling, startup/shutdown, and freezing/thawing make this goal very challenging for current FC technologies. Unfortunately, at present most PEMFC stacks provided by manufacturers and research institutes cannot achieve these goals. Many internal and external factors affect the performance of a PEMFC or stack, such as FC design and assembly, materials degradation, operational conditions, and impurities or contaminants. Performance degradation is unavoidable, but the rate can be minimized through a comprehensive understanding of degradation and failure mechanisms. For example, normal degradation targets require less than 10% loss in the efficiency of the FC system by the end of its life, and a degradation rate of 2–10 $\mu V\ h^{-1}$ is commonly accepted for most applications (Wu et al., 2008a).

Importance of Diagnostic Tools

To improve FC durability and thus reduce the cost of FC devices, diagnostic procedures and tools are needed. To date, a wide range of experimental diagnostic tools for the accurate analysis of PEMFCs and stacks have been developed to help gain a fundamental understanding of FC dynamics, to diagnose failure modes and degradation mechanisms, to mitigate performance losses, and to provide benchmark-quality data for modeling research. A better understanding of all the operating parameters influencing the entire cell's function and of the phenomena occurring in the electrodes, especially the performance- and efficiency-limiting processes, is essential for advancing this promising technology (Wu et al., 2008b).

FC science and technology cut across multiple disciplines, including materials science, interfacial science, transport phenomena, electrochemistry, and catalysis. It is always a major challenge to fully understand the thermodynamics, fluid mechanics, FC dynamics, and electrochemical processes within a FC. Numerous researchers are currently focusing on experimental diagnostics and mathematical modeling. On the one hand, diagnostic tools can help distinguish the structure–property–performance relationships between a FC and its components. On the other hand, results obtained from experimental diagnostics also provide benchmark-quality data for fundamental models, which further benefit in the prediction, control, and optimization of various transport and electrochemical processes occurring within FCs.

Owing to the complexity of the heat and mass transport processes in FCs, there are typically a multitude of parameters to be determined. A number of issues, including *in situ* water distribution (Men et al., 2003; Dong et al., 2005), ohmic voltage losses (Mennola et al., 2002), ionic conductivity of PEMFC electrodes (Li and Pickup, 2003), FC operating conditions (Yuan et al., 2006), current density distribution in PEMFCs (Cleghorn et al., 1998; Stumper et al., 1998), temperature variation in a single cell (Wilkinson et al., 2006), and flow visualization within the FC (Tüber et al., 2003; Ma et al., 2006), have been tackled using various tools. It is important to examine the operation of PEMFCs or stacks with suitable techniques that allow for separate evaluation of these parameters and can determine the influence of each on overall FC performance.

A few review papers have been published regarding prior efforts in PEMFC diagnostics. In a recent review of fundamental models for FC engineering, Wang (2004) briefly summarized some of the diagnostic techniques that were particularly pertinent to the modeling of PEMFCs. Hinds (2004) provided a good review of the literature on experimental techniques employed in the characterization of FC performance and durability, but limited the review to *ex situ* tools and applications. The objective of this book is to provide a comprehensive and detailed review of both *in situ* and *ex situ* diagnostic tools presently used in PEMFC research, with an attempt to incorporate the most recent technical advances in PEMFC diagnosis.

Brief Chapter Introduction

Diagnostic tools can help to understand the physical and chemical phenomena involved in PEMFCs. The tools introduced in this book include the most commonly used conventional tools, such as cyclic voltammetry (CV), electrochemical impedance spectroscopy (EIS), scanning electron microscopy (SEM), and transmission electron microscopy (TEM), special tools developed for PEMFCs, such as transparent cells, cathode discharge, and current mapping, and the most recent advanced tools for PEMFC diagnosis, such as magnetic resonance imaging (MRI) and atomic force microscopy (AFM). To better understand these tools, PEMFC testing is also included (Chapter 1).

These diagnostic methodologies can be classified according to various characteristics (real time, *ex situ*, capability, accuracy, cost, etc.). For perspicuity, various diagnostic tools employed in the characterization and determination of FC performance are summarized into two general categories—Part I: *In Situ* Diagnostic Tools, and Part II: *Ex Situ* Diagnostic Tools. However, some measurements for a specific purpose can be done in both situations; for example, measurement of gas permeability includes both *in situ* and *ex situ* methods, but with a focus on *ex situ* tools. Chapter 21, on gas permeability, has thus been placed in the category of *ex situ* tools.

Each chapter of Part I and Part II describes essentially one diagnostic tool or several tools on one topic. For example, Chapter 22, on species detection, covers the most frequently used tools to detect fluorine-containing species, including ion-selective electrode, ion-exchange chromatography, and nuclear magnetic resonance (NMR). Generally, each chapter is independent from the others in terms of context, and attempts to cover a wide range of material for the topic under discussion. To avoid any overlap in chapter content, cross-references to other chapters are included. Typically, each diagnostic chapter contains the following sections: introduction, principle, instruments and measurements, applications, literature review/recent advances, advantages and limitations, and outlook, with a focus on both fundamentals and applications, especially applications in durability studies.

To give the readers an overview of all the tools in this book, Tables 1 and 2 compare, respectively, the main features of the *in situ* and *ex situ* diagnostic tools for each chapter.

TABLE 1   Comparison of *In Situ* Diagnostic Tools

Chapters	Other Tools Included	Typical Equipment	Major Applications in PEMFC Diagnosis
Chapter 1 PEM Fuel Cell Testing	—	FC test station	Performance assessment
Chapter 2 Polarization Curve	—	FC test station	Performance assessment; performance loss diagnosis
Chapter 3 Electrochemical Impedance Spectroscopy	—	Impedance analyzer; potentiostat/galvanostat	Measurement of cell impedance; electrode, membrane, and cell operation management-related failure diagnosis
Chapter 4 Cyclic Voltammetry	CO stripping	Potentiostat	Determination of electrochemical surface area
Chapter 5 Linear Sweep Voltammetry	Chronocoulometry	Potentiostat	Measurement of gas crossover; membrane-related failure diagnosis
Chapter 6 Current Interruption	—	Fast electric switch	Determination of ohmic resistance; estimation of impedance models
Chapter 7 Cathode Discharge	—	Load bank; multichannel oscilloscope or equivalent; Hall effect current transducers	Performance loss diagnostics; performance model optimization; detection of error states in cells and stacks
Chapter 8 Water Transfer Factor Measurement	—	Electronic balance; RH sensors; thermoconductivity detector; infrared sensor	Measurement of water crossover; water management diagnosis
Chapter 9 Current Mapping	—	Potentiostat/galvanostat; electrochemical impedance analyzer	Measurement of current distribution over the electrode surface, mainly along the flow direction
Chapter 10 Transparent Cell	—	FC test platform; CCD (charge-coupled device) camera	Study of the connection between water flooding and performance
Chapter 11 Magnetic Resonance Imaging	—	Superconducting magnet	Water content determination in space and time within the Nafion® membrane
Chapter 12 Neutron Imaging	—	Neutron sources; detecting system (e.g., scintillator screen with optical detectors)	Qualitative and quantitative determination of liquid water content and distribution with sufficient spatial and temporal resolution

TABLE 2 Comparison of *Ex Situ* Diagnostic Tools

Chapters	Other Tools Included	Typical Equipment	Major Applications in PEMFC Diagnosis
Chapter 13 X-Ray Diffraction	—	X-ray diffractometer	Measurement of particle size; degree of alloying; phase identification
Chapter 14 Scanning Electron Microscopy	Energy-dispersive x-ray; environmental scanning electron microscopy (ESEM)	Scanning electron microscope; cryotable (in ESEM)	Qualitative examination of physical electrode/gas diffusion layer degradation, catalyst/element dissolution, change in wetting properties
Chapter 15 Transmission Electron Microscopy	Energy-dispersive x-ray	Transmission electron microscope	Diagnosis of electrocatalyst degradation
Chapter 16 Infrared Imaging	Perforation detection methods	Infrared camera	Membrane thinning and pinhole detection
Chapter 17 Fourier Transform Infrared Spectroscopy	Attenuated total reflectance (ATR)	Fourier transform infrared spectrometer	Investigation of polymer alterations induced by degradation; investigation of chemical reactions on catalyst surfaces
Chapter 18 X-Ray Photoelectron Spectroscopy	—	X-ray photoelectron spectrometer	Determination of surface composition
Chapter 19 Atomic Force Microscopy	—	Atomic force microscope	Measurement of surface topography with local mechanical, chemical, and electrical properties of all FC components
Chapter 20 Gas Diffusion	—	Loschmidt diffusion cell	Measurement of effective gas diffusion coefficients of catalyst layers and gas diffusion layers
Chapter 21 Gas Permeability of Proton Exchange Membrane	Steady-state and transient gas permeation methods based on direct gas concentration measurements	Membrane permeation cell and gas chromatograph	Direct measurements of gas permeability constant and gas diffusion coefficients with a large variety of feed gases (N_2, O_2, CO_2, He, O_2, etc.); membrane-related failure diagnostics
Chapter 22 Species Detection	Fluoride-selective electrode; ion-exchange chromatography; 19F nuclear magnetic resonance (NMR)	Ion-selective electrode and multimeter; ion-exchange chromatography; NMR	Measurement of membrane and ionomer degradation through analysis of dissolved fluoride ions; measurement of degradation products through analysis of dissolved ions; measurement of degradation products containing fluorine
Chapter 23 Rotating Disk Electrode/ Rotating Ring-Disk Electrode	—	Potentiostat; rotator	Evaluation of catalyst activity; fuel contamination diagnosis; kinetics of hydrogen oxidation reaction and oxygen reduction reaction
Chapter 24 Porosimetry and Characterization of the Capillary Properties of Gas Diffusion Media	Capillary pressure measurement; contact angle measurement	Mercury porosimetry equipment; sessile drop imaging equipment; air–water capillary pressure rigs	Porosity of gas diffusion layer; relative permeability; saturation of gas diffusion layer

References

Cleghorn, S. J. C., Derouin, C. R., Wilson, M. S., and Gottesfeld, S. 1998. A printed circuit board approach to measuring current distribution in a fuel cell. *J. Appl. Electrochem.* 28: 663–672.

Dong, Q., Kull, J., and Mench, M. M. 2005. *In situ* water distribution measurements in a polymer electrolyte fuel cell. *J. Power Sources* 139: 106–114.

Hinds, G. 2004. Performance and durability of PEM fuel cells. NPL Report DEPC-MPE 002, National Physical Laboratory, Teddington, UK.

Li, G. and Pickup, P. G. 2003. Ionic conductivity of PEMFC electrodes. *J. Electrochem. Soc.* 150: C745–C752.

Ma, H. P., Zhang, H. M., Hu, J., Cai, Y. H., and Yi, B. L. 2006. Diagnostic tool to detect liquid water removal in the cathode channels of proton exchange membrane fuel cells. *J. Power Sources* 162: 469–473.

Mench, M. M., Dong, Q. L., and Wang, C. Y. 2003. *In situ* water distribution measurements in a polymer electrolyte fuel cell. *J. Power Sources* 124: 90–98.

Mennola, T., Mikkola, M., and Noponen, M. 2002. Measurement of ohmic voltage losses in individual cells of a PEMFC stack. *J. Power Sources* 112: 261–272.

Stumper, J., Campbell, S. A., Wilkinson, D. P., Johnson, M. C., and Davis, M. 1998. *In-situ* methods for the determination of current distributions in PEM fuel cells. *Electrochim. Acta.* 43: 3773–3783.

Tüber, K., Pócza, D., and Hebling, C. 2003. Visualization of water buildup in the cathode of a transparent PEM fuel cell. *J. Power Sources* 124: 403–414.

Wang, C. Y. 2004. Fundamental models for fuel cell engineering. *Chem. Rev.* 104: 4727–4766.

Wilkinson, M., Blanco, M., Gu, E., Martin, J. J., Wilkinson, D. P., Zhang, J. et al. 2006. *In situ* experimental technique for measurement of temperature and current distribution in proton exchange membrane fuel cells. *Electrochem. Solid-State Lett.* 9: A507–A511.

Wu, J., Yuan, X. Z., Martin, J. J., Wang, H., Zhang, J., Shen, J. et al. 2008a. A review of PEM fuel cell durability: Degradation mechanisms and mitigation strategies. *J. Power Sources* 108: 104–119.

Wu, J., Yuan, X. Z., Wang, H., Blanco, M., Martin, J. J., and Zhang, J. 2008b. Diagnostic tools for PEM fuel cells, Part I, Electrochemical techniques. *Int. J. Hydrogen Energy* 33: 1735–1746.

Yuan, X. Z., Sun, J. C., Blanco, M., Wang, H., Zhang, J., and Wilkinson, D. P. 2006. AC impedance diagnosis of a 500W PEM fuel cell stack: Part I: stack impedance. *J. Power Sources* 161: 920–928.

I

In Situ Diagnostic Tools

1

Proton Exchange Membrane Fuel Cell Testing

Jonathan J. Martin
Institute for Fuel Cell Innovation

1.1 General Overview

PEM fuel cell testing is an integral part of any fuel cell research. A small sample of its uses includes: evaluating new materials, validating new designs, assessing new experimental techniques, and optimizing operating procedures. It is an important companion to any set of diagnostic tools, so that the results can be viewed in the proper context.

It is also necessary for all fuel cell tests to be conducted in a way that not only complies with the standards of the particular organization conducting the testing, but also meets the standards of the international community. Organizations with interest in testing standards include the American Society of Mechanical Engineers (ASME), Japan Automobile Research Institute (JARI), Joint Research Council (JRC), New Energy and Industrial Technology Development Organization (NEDO), Society of Automotive Engineers (SAE), and US Fuel Cell Council (USFCC).

According to the USFCC Joint Hydrogen Quality Task Force (JHQTF), a test protocol is a methodology noting the key parameters such as pressure, temperature, contamination concentration, contamination exposure, test duration, and so on, which should be measured and reported to ensure that the data generated and supplied are (USFCC JHQTF, 2006a):

- Applicable—test results are relevant to the needs of the industry.
- Repeatable—test results could be duplicated under the same conditions, with the same type of equipment.

3

- Reproducible—test results could be duplicated under the same conditions, with a different type of equipment.
- Scalable and serviceable—sufficient information is provided such that single cell test data can be scaled up to represent a full-size stack.

1.2 Pretest Evaluation Procedures

The USFCC (JHQTF, 2006b) and JRC (Malkow et al., 2010) outline several steps that should be conducted at the outset of testing. Pretest evaluation procedures are an important part of ensuring the safety of the test operator and the quality of the test results. The most common procedures include verification of the cell assembly, preparation of the cell, and noninvasive operating diagnostic procedures.

1.2.1 Verification of the Assembly

Before the initiation of any test, it is imperative to verify proper assembly of the cell. As outlined in USFCC 04-003A (JHQTF, 2006b), cell testing can expose the investigator to a number of potential hazards; voltage (multicell stacks), current, water, flammable gases, poisonous gases, and gases causing asphyxia. A properly designed facility in combination with proper pretest procedures (such as a visual inspection, a pressure and leakage test, an electrical isolation test, and an instrumentation checkout) can minimize the potential hazards and increase the reliability of results (USFCC JHQTF, 2006b).

1.2.1.1 Visual Inspection

A visual inspection is a simple test to perform and can discover a variety of assembly errors. The inspection can be part of the assembly procedure and should include confirmation that all fluid lines and cell connecting hardware are securely fastened and all necessary components have been assembled in proper order.

1.2.1.2 Pressure and Leakage Test

The purposes of the pressure and leak tests are to verify proper hardware sealing, determine a gross crossover leak rate, and electrochemically determine hydrogen crossover (Rockward, 2004). If a coolant is used, the coolant circuit should also be tested for leaks by circulating the fluid. The USFCC Single Cell Testing Task Force (SCTTF) outlines a leak tightness test procedure that should be performed separately on both the fuel and air circuits (USFCC SCTTF, 2004). This procedure involves closing off the exhaust valve, pressurizing the line with an inert gas, monitoring the gas pressure over a period of time, and using soapy water to identify the source of any leaks for repair.

An alternate procedure from the USFCC (JHQTF, 2006c) submerses the entire pressurized stack under water to diagnose and locate leaks. The procedure, which is similar to the one previously mentioned, also uses room-temperature N_2 as the working gas.

1.2.1.3 Electrical Isolation Test

In addition to safety considerations, electrical isolation tests are necessary to verify cell and electrical bus isolation from ground. The USFCC (JHQTF, 2006c) suggests a procedure where the operator ensures electrical isolation of all tie rods from the pressure plate, the reactant manifolds from the pressure plate, and the pressure plate to ground.

1.2.1.4 Axial Load Verification

Axial load verification ensures that improper cell fabrication is not influencing the performance issues being investigated. The procedure used for axial load verification depends on the loading method employed in the cell or stack. A bladder system utilizing pneumatic loading can be verified through a modified leak tightness test and by monitoring the compression pressure. For a torque-tightened system, the use of a torque wrench is a valid verification procedure. Additionally, stacking heights can be measured before and after compression, yielding the spring load. In both systems it is highly recommended to verify the magnitude and proper distribution of the compression load through the use of pressure-sensitive paper (such as Pressurex® from Sensor Products Inc).

1.2.1.5 Instrumentation Checkout

As outlined by the USFCC (JHQTF, 2006b), an instrument checkout is to verify the instrumentation is properly installed, is correctly routed to the data-acquisition system, and generates accurate and expected values. Rework of some instrumentation may be required if the performance does not meet the required standards. Additionally, the USFCC (SCTTF, 2004) outlines standard fuel cell test station requirements and procedures. While a variety of procedures can be used, it is important for all equipment and instrumentation to be regularly calibrated.

1.2.2 Cell Preparation

For a fuel cell to perform adequately, the membrane electrode assembly (MEA) must be properly prepared. Typically, preparation of the fuel cell is accomplished through operation at set conditions, which humidifies the membrane and increases the temperature of the cell. Several benefits of the temperature increase can include decreasing contact resistances in the cell, improved catalyst activity, and decreased internal resistances. Two similar procedures that are typically required for fuel cell preparation, depending on the state of the cell, are break-in/start-up and conditioning.

1.2.2.1 Break-In/Start-Up

As stated by the USFCC (SCTTF, 2006), the main consideration is to perform a repeatable break-in procedure, which brings the cell materials to a stable level of performance for subsequent testing. Since the break-in procedure has a lasting effect on the fuel cell materials, it is also possible to permanently bias test results with inappropriate conditions early in the fuel cell life. Typically, the break-in protocols are defined by: a material or component supplier in consideration of unique materials properties, a fuel cell developer in order to maintain conditions consistent with other program testing, a research organization that has defined a common practice, or a respected international organization.

The USFCC (SCTTF, 2006) developed its own clearly defined break-in protocol. The protocol, spanning 19 h, consists of one cycle at 0.60 V for 60 min, nine cycles between 0.70 and 0.50 V for 20 min at each set point (spanning 7 h), and constant current operation at 200 mA cm^{-2} for 12 h.

The JRC developed two working groups to develop standardized fuel cell testing procedures: the Fuel Cells Testing & Standardisation NETwork (FCTestNet) and the Fuel Cell Systems Testing, Safety and Quality Assurance (FCTes^{QA}) groups. The procedure developed, consists of increasing the current density of the cell by 100 mA cm^{-2} intervals while keeping the cell voltage above 0.5 V. They recommended conducting the conditioning for at least 24 h, until the cell voltage variation is less than ±5 mV for at least 1 h (FCTestNet/FCTes^{QA}, 2010a).

Partially funded by the European Union, the Hydrogen System Laboratory (HySyLab) developed a conditioning procedure (Cuccaro et al., 2008). Their procedure involved 10 cycles between 0.6 and 0.4 V for 30 min at each set point. After each cycle, the cell is held at OCV for at least 1 min and a polarization curve is obtained. Conditioning is considered complete when the maximum deviation between successive polarization curves is less than 1.7%.

Break-in protocols are typically applied only to a newly assembled cell; however, reconditioning procedures may also be used upon restarting a cell before each round of testing. Such procedures involve a stabilization period at a specified load and cell conditions in order to bring the cell to a predictable state of hydration before testing is resumed (USFCC SCTTF, 2006).

1.2.2.2 Conditioning

Cell conditioning is necessary to rehumidify a previously used MEA and should be performed prior to any testing. Typically, cell conditioning is similar in nature to a break-in procedure but, depending on the technique, can be a significantly shorter process. The cell conditioning can be considered complete when the stability criteria outlined in the break-in procedure are reached.

It is important that all conditioning is performed at operation conditions in the ohmic losses region of the polarization curve. If operation in the suggested range would push the cell into the mass transport region, the applied load should be reduced until cell potential is greater than 0.6 V. Any conditioning procedure will be affected by fuel cell size and targeted operating conditions.

1.2.3 Noninvasive Operating Diagnostic Procedures

At this step, the intent is to evaluate the status of the cell without performing actions that may alter the future performance of the cell. These procedures, which may include use of a polarization curve, hydrogen crossover, the current-interrupt method, and electrochemical impedance spectroscopy, are described fully in subsequent chapters.

1.3 Testing Procedures

As stated by the USFCC (JHQTF, 2006a), a test plan should contain the range of concentrations, temperatures, pressures, current densities, and stoichiometries to be tested, the types of testing to be conducted, the order of the tests to be conducted, and the posttest analysis desired. A well-conceived test plan should accomplish these aims while keeping a clear record that can be used for guiding future research. For any test plan to be successful, it is important for specific parameters to be measured and recorded. The following sections describe various established fuel cell evaluation procedures.

1.3.1 Testing Profiles

For any test plan, the specific tests to be performed will be dependent on the goals of the experiment, the characteristics of the fuel cell, and the parameters of interest. Some examples of specific test protocols outlined by USFCC (SCTTF, 2006) are: steady-state testing at singular, critical operating points; parametric sensitivity testing centred on key operating conditions; very low-power stability; and high-current density operation. This section describes typical methods used to characterize the performance of a fuel cell. Depending on the situation it may be desirable to implement various combinations of the test procedures. When designing a successful test plan, the following should be considered:

- When increasing load, first set reactant flows to the higher flow rates. When decreasing load, drop the load first, then adjust reactant flows to the lower flow rates.
- Choose a stabilization time at each test point that includes the equilibrium time for the test stand plus the fuel cell under test. Typical fuel cell test stations will require 5–10 min to stabilize and the fuel cell will require a period of time to equilibrate after the test conditions are stable. The time required for the cell to equilibrate will highly depend on the cell size, components, and composition.
- Choose a data acquisition rate that is fast enough to observe trends in test conditions and cell voltage.

A highly detailed resource covering fuel cell testing procedures, specifically for automotive applications, is SAE J2615 (SAE, 2005). The recommended practice defines the components of a fuel cell system and various reference conditions, testing description specifications, and data analysis procedures. SAE J2615 is a resource that should be consulted by any personnel responsible for the development of a test plan.

An additional resource for performance testing by ASME is PTC 50-2002 (ASME, 2002). It outlines all the test parameters that should be measured, presents which parameters should be calculated from the measurements, describes how to perform the calculations, defines standard conditions, provides conversion factors, and provides values for constants. Since it would be unwieldy to provide all the data from the test code in this chapter, it is recommended reading for any individuals developing test plans. Another recommended resource from ASME is PTC 1-2004 (ASME, 2004). While it does not explicitly apply to fuel cell testing, it does explain in detail some of the concepts suggested in PTC 50-2002.

1.3.1.1 Polarization Curve

A polarization curve is a standard technique, where the test results are plotted as the cell voltage versus the current density. It is a quasi-steady-state test in which the current is held at specified points until the voltage stabilizes. Polarization curves can be performed such that the set current increases for ensuing points, decreases for ensuing points, or is random. Typically, the test is performed with increasing points then decreasing points and the results are averaged. Polarization curves are useful in that they can quickly evaluate the performance of a particular cell composition or the effects of particular operating parameters and can also be used to distinguish between kinetic, ohmic, and mass transport losses.

The JRC developed detailed testing procedures for obtaining polarization curves for a single cell, TM PEFC SC 5-2 (FCTestNet/FCTes^{QA} 2010a), and a stack, TM PEFC ST 5-3 (Malkow et al., 2010). Additionally, the stack testing procedure is validated with a controlled series of 8 round robin tests (referred to as Test No. 1a, 1b, 2a, 2b, 3a, 3b, 4a, and 4b). Further, the JRC also developed testing procedures for examining the effects of pressure (FCTestNet/FCTes^{QA}, 2010b) and stoichiometry (FCTestNet/FCTes^{QA}, 2010c) on the voltage, power, and efficiency of a water-cooled PEM fuel cell stack.

1.3.1.2 Dynamic/Transient Performance

Bucci et al. (2004) conducted four different transient performance tests on a self-humidified stack with a 64 cm^2 active area, including: a static run-up or run-down procedure to examine the fuel cells response to a constant load change; increasing and decreasing steps, to examine the response to a series abrupt load changes; increasing and decreasing current pulses, to examine the response to a brief increase or decrease in the load; and hydrogen flow rate variation, to examine the behavior of the cells in the case of a fault in the H$_2$ control system.

Tang et al. (2010) investigated a variety of transient responses in a kW-class PEM fuel cell stack by utilizing dynamic performance profiles. They examined responses during start-up, shutdown, increasing constant step load variation, combined regularly increasing and decreasing load variation, and combined irregularly increasing and decreasing load variation (Figure 1.1).

1.3.2 Durability Testing

An additional and important category of testing is durability testing. Rather than focusing on the short-term response to various phenomena, durability testing investigates the long-term response to typical operating conditions. The three topics discussed are a dynamic testing profile, standard operating conditions, and accelerated testing conditions.

1.3.2.1 Dynamic Profiles

When testing fuel cells for automotive applications, it may be necessary to use the dynamic testing profile (DTP) developed by the USFCC (JHQTF, 2006d). The DTP (Figure 1.2) is derived from the Dynamic Stress Test (DST) for batteries, which in turn is based on the *simplified* version of the Federal Urban Driving Schedule (SFUDS). The main changes made for the DTP cycle, when modifying the DST profile, involved elimination of the battery charging points and extension of the higher power points.

An alternate dynamic testing profile was developed by JRC and is detailed in TM PEFC ST 5-7 (FCTest Net/FCTes^{QA}, 2010d). The testing profile, which was not developed for automotive applications, consists of

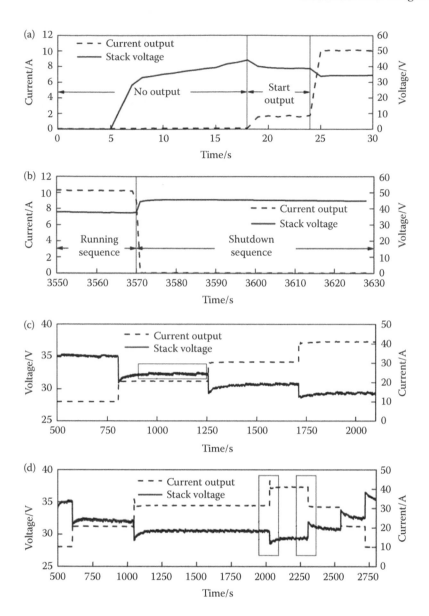

FIGURE 1.1 Responses of a kW-class PEM fuel cell stack during (a) start-up, (b) shutdown, (c) increasing constant step load variation, and (d) combined regularly increasing and decreasing load variation. (Reprinted from *Applied Energy*, 87, Yuan, W. et al., Experimental investigation of dynamic performance and transient responses of a kW-class PEM fuel cell stack under various load changes, 1410–1417, Copyright (2010), with permission from Elsevier.)

repeated cycles of 40 seconds at 20% of the maximum power during normal operation and 20 s at 100% of the maximum power during normal operation, with periodic polarization curves to quantify degradation.

1.3.2.2 Standard Operating Conditions

Durability testing is vital to the design of a fuel cell for stationary applications. Unlike automotive applications, stationary power requirements typically require the fuel cell to run at constant loads. Thus, a typical durability test for stationary power applications will run the fuel cell at a single design load for

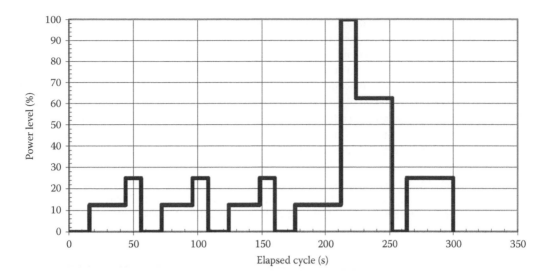

FIGURE 1.2 DTP cycle developed for automotive applications by the USFCC (JHQTF, 2006d). (Figure courtesy of US Fuel Cell Council, www.usfcc.com.)

an extended period of time and performing periodic noninvasive diagnostic procedures. Table 1.1 includes the parameters from the residential fuel cell test protocol of W. L. Gore & Associates (Liu et al., 2001) and the JRC procedures for a long-term durability test for both single cells, TM PEFC SC 5-6 (FCTestNet/FCTes^{QA}, 2010e), and for a stack, TM PEFC ST 5-6 (FCTestNet/FCTes^{QA}, 2010f).

1.3.2.3 Accelerated Testing Conditions

Interest in accelerated testing procedures is increasing, as confirmation of fuel cell durability is necessary. Particularly in stationary applications with lifetimes of 50,000 h, it is not feasible to operate a fuel cell at standard conditions until failure. Even in other applications such as transportation, with lifetimes of 5000 h, it is desirable to verify durability on a shorter timescale. However, as stated by Dillard et al. (2004), it is difficult at this time to propose an accelerated protocol without performing a detailed examination of the existing fuel cell performance and diagnostics data such as ionic resistance, polarization curves, lifetime tests, fuel crossover rates, and effluent concentrations. Regardless of the accelerated protocol developed, care must be taken to perform accelerated testing under conditions, which do not introduce new failure mechanisms that are unrealistic in fuel cell environments.

Despite the difficulties, several studies have examined candidate conditions for accelerated testing. LaConti et al. (2003) describe multiple methods of accelerating degradation in membranes. It is possible to simulate the degradation of PEMs using an accelerated test medium such as Fenton's reagent (small amounts of hydrogen peroxide [~3%] and Fe^{2+} ions [4 ppm] in solution). Additionally, they describe accelerated testing conditions for a fuel cell of: $T_{cell} = 100°C$ and $P_{H2} = P_{Air} \sim 65$ psig (448 kPa). Meanwhile, Liu and Crum (2006) utilized OCV testing to accelerate chemical degradation of the ionomer.

TABLE 1.1 Example Standard Operating Conditions

Reference	T_{cell}	RH_a, RH_c	$Flow_{H2}$	$Flow_{Air}$	P_{H2}, P_{Air}
Liu et al. (2001)	70°C	100%	1.2	2.0	0 psig
FCTestNet/FCTes^{QA} 2010e	80°C	—	1.1–2.0	2.0–3.0	0–300 kPa

W. L. Gore & Associates have conducted extensive investigations into durability testing and accelerated testing procedures. They observed an approximately 10× relationship for membrane life between their standard and accelerated residential test procedures, while H_2 crossover (monitored regularly for both standard and accelerated testing) gradually increased over a period of time. They concluded that the accelerated fuel cell life test protocol (compared to the standard protocol in Table 1.2) was a valid testing condition to evaluate membrane durability for residential fuel cell applications (Liu et al., 2001).

As demonstrated by Crum and Liu (2006), RH cycling can be used to examine mechanical durability in isolation, or to examine combined mechanical and chemical durability. In an inert nitrogen atmosphere with RH cycling, the mechanical durability associated with expansion and contraction of the ionomer under the flow-field channels is isolated. When the same RH cycling is employed with air and hydrogen feed gases, the mechanical and chemical durability are combined.

The DOE (Garland et al., 2007) produced a comprehensive set of accelerated degradation protocols for single-cell testing. The accelerated electrocatalyst test procedure involved 30,000 cycles between 0.7 and 0.9 V (30 s at each step), with fully humidified hydrogen at the anode and fully humidified nitrogen at the cathode. Their accelerated catalyst support test utilized a steady 1.2 V potential for 200 h (diagnostics every 24 h), at an elevated temperature of 95°C, with 80% RH hydrogen at the anode, and with 80%RH nitrogen at the cathode. Their accelerated MEA chemical stability test procedure utilized OCV conditions for 200 h, an elevated temperature of 90°C, with a stoichiometry of 10, and an RH of 30% for both the hydrogen anode and the air cathode. Their accelerated membrane mechanical durability test utilized RH cycling from 0% to 90% every 2 min, with air at both the anode and cathode. They also describe a combined humidity/load cycle developed by Dupont. The RH is cycled between 1% and 100% at a fixed load for 24 h followed by a load cycle between 10 and 800 mA cm^{-2} at fixed RH for 24 h. While effective, this accelerated test does not isolate the degradation mechanisms and effects.

Detailed accelerated stress test procedures for the various components of the MEA were also examined by Zhang et al. (2009). Accelerated membrane tests can include: undesirable temperature and RH, OCV, load cycling, Fenton's test, freeze/thaw cycling, stress cycling under tension, high-pressure testing, and high-temperature exposure. Electrocatalyst degradation can be accelerated by potential control, undesirable temperatures and humidities, contaminants, and load cycling. Corrosion of the carbon support can typically be accelerated with fuel starvation, start-up/shutdown cycling, cold start-up, and control of the temperature, potential, and/or RH. Gas diffusion layer accelerated testing procedures can include elevated temperatures, freeze/thaw cycling, fuel starvation, elevated potentials, and high flow rates.

Accelerated testing procedures for both a single cell and a stack were also developed by the JRC. For the single-cell test module TM PEFC SC 5-4 (FCTestNet/FCTes^{QA}, 2010g), they utilized on/off cycling to accelerate aging. However, for the stationary fuel cell system test module (FCTestNet, 2006), they used four different subtests (normal efficiency test, thermal load cycling, electrical load cycling, and start-up/shutdown) in sequence.

1.4 Posttest Evaluation Procedures

This section provides a general description of the posttest evaluation procedures that should be conducted at the conclusion of testing. These steps, as partially outlined by the USFCC (JHQTF, 2006b), include noninvasive operating diagnostic procedures, cleanup of the cell, invasive operating diagnostic procedures, verification of the assembly, disassembly of the cell, and destructive postmortem.

TABLE 1.2   Accelerated versus Standard Operating Conditions

Type	T_{cell}	DP_a, DP_c	Flow$_{H2}$	Flow$_{Air}$	P_{H2}, P_{Air}
Accelerated	90°C	83°C	1.2	2.0	5, 15 psig
Standard	70°C	70°C	1.2	2.0	0 psig

1.4.1 Noninvasive Diagnostic Procedures

Noninvasive operating diagnostic procedures should be the first step in posttest evaluation. It is acceptable to utilize the techniques at this stage as they will not cause irreversible damage to the materials and components. These procedures may include use of a polarization curve, hydrogen crossover, the current-interrupt method, and electrochemical impedance spectroscopy.

1.4.2 Cleanup of the Cell

Cleanup of the tested cell is similar to invasive diagnostic procedures, in that they both must be conducted after all relevant data have been obtained from the unaltered cell. The main difference between the two is that alteration of the cell is the goal of a cleanup (i.e., removing contaminants), while it is an effect of invasive diagnostic procedures. Cleanup of the cell is typically required when contaminants were used during the investigation. For directions on cell cleanup (load conditions and gases required), with diagnostics conducted between each option, consult USFCC 04-003A (JHQTF, 2006b).

1.4.3 Invasive Diagnostic Procedures

Invasive diagnostic procedures may complicate the interpretation of test data by stressing or adversely affecting the cell. They should therefore only be used after as much relevant data have been collected through noninvasive means. At this point it is acceptable to conduct these procedures, as there is no longer a need to leave the components unaltered. Three typical invasive procedures used are: stoichiometry sweeps, limiting current, and cyclic voltammetry tests.

1.4.4 Verification of the Assembly

A verification of the assembly should be the next step in the posttest evaluation. Available techniques are outlined in detail in Section 1.2.

1.4.5 Disassembly of the Cell

Disassembly protocols are a necessary precursor to failure analysis to check for problems such as crushed gas diffusion layers, adherence of seal/gaskets to PEM, deterioration of gaskets, electrode indentations, extruded sections of MEA, and perforations near seal and electrode edges where stress concentrations can be the highest (Dillard et al., 2004).

LaConti et al. (2003) also detail a disassembly procedure consisting of visual checks of the components and attention to signs of leaks, electrical shorts, and foreign debris and particles.

1.4.6 Destructive Postmortem

At this stage it is possible to conduct a test that will destroy the electrodes, providing it will not be needed for subsequent tests. Careful planning is required for such tests, as they most likely cannot be repeated. The USFCC (JHQTF, 2006b) found techniques such as electron microscopy, elemental analysis, and electrochemical analysis yield valuable information on MEAs and electrodes.

1.5 Summary

To maximize the value of the work, it is necessary to conduct fuel cell research according to standards that comply with other international organizations. Doing so ensures that the results are relevant to a wide audience. Test plans must include pretest evaluation procedures, testing profiles, and posttest evaluation procedures that are accepted by the stakeholders in the PEM fuel cell industry. Data must be acquired such that they are applicable, repeatable, reproducible, and scalable and serviceable. With the standards referenced in this chapter, developed by various organizations, universities, and corporations involved in fuel cell research, researchers can comply with international testing standards.

1.6 Nomenclature

DP_a	Dew point temperature of the humidifier at the anode	°C
DP_c	Dew point temperature of the humidifier at the cathode	°C
$Flow_{H2}$	Stoichiometry of the hydrogen flow	unitless
$Flow_{Air}$	Stoichiometry of the air flow	unitless
P_{H2}	Backpressure at the anode	psig/kPa
P_{Air}	Backpressure at the cathode	psig/kPa
RH_a	Relative humidity of the reactant gas at the anode	°C
RH_c	Relative humidity of the reactant gas at the cathode	°C
T_{cell}	Cell operating temperature	°C

References

ASME. 2002. *Fuel Cell Power Systems Performance*. ASME PTC 50-2002. ASME. Performance Test Codes.

ASME. 2004. *General Instructions*. ASME PTC 1-2004. ASME. Performance Test Codes.

Bucci, G., Ciancetta, F., and Fiorucci, E. 2004. An automatic test system for the dynamic characterization of PEM fuel cells. *IMTC* 2004: 675–680.

Crum, M. and Liu, W. 2006. Effective testing matrix for studying membrane durability in PEM fuel cells: Part 2. Mechanical durability and combined mechanical and chemical durability. *ECS Trans* 3: 541–550.

Cuccaro, R., Lucariello, M., Battaglia, A., and Graizzaro, A. 2008. Research of a HySyLab internal standard procedure for single PEMFC. *International Journal of Hydrogen Energy* 33: 3159–3166.

Dillard, D. A., Gao, S., Ellis, M. W., et al. 2004. Seals and sealants in PEM fuel cell environments: Material, design, and durability challenges. *Conference on Fuel Cell Science, Engineering and Technology* 2004: 553–560.

FCTestNet. 2006. *WP2 Stationary Fuel Cell Systems: Test Programs and Test Modules*.

FCTestNet/FCTes[QA]. 2010a. Testing the voltage and the power as a function of the current density (Polarisation curve for a PEFC single cell). Test Module PEFC SC 5-2.

FCTestNet/FCTes[QA]. 2010b. PEFC power stack performance testing procedure: Measuring voltage, power and efficiency as a function of pressure for a water-cooled PEFC stack. Test Module PEFC ST 5-1.

FCTestNet/FCTes[QA]. 2010c. PEFC power stack performance testing procedure: Measuring voltage, power and efficiency as a function of reactant stoichiometry for a water-cooled PEFC stack. Test Module PEFC ST 5-2.

FCTestNet/FCTes[QA]. 2010d. PEFC power stack performance testing procedure: Measuring voltage, power and efficiency as a function of current density following a dynamic profile versus time. Test Module PEFC ST 5-7.

FCTestNet/FCTes[QA]. 2010e. Testing the voltage and the power as a function of time at a fixed current density (Long term durability steady test for a PEFC single cell). Test Module PEFC SC 5-6.

FCTestNet/FCTes[QA]. 2010f. PEFC power stack performance testing procedure: Measuring voltage and power as a function of time and current density, long term durability steady test. Test Module PEFC ST 5-6.

FCTestNet/FCTes[QA]. 2010g. Testing the voltage and the power as a function of the current density following an on/off profile versus time (Accelerated ageing on/off cycling test for a PEFC single cell). Test Module PEFC SC 5-4.

Garland, N. L., Benjamin, T. G., and Kopasz, J. P. 2007. DOE fuel cell program: Durability technical targets and testing protocols. *ECS Transactions* 11: 923–931.

LaConti, A. B., Hamdan, M., and McDonald, R. C. 2003. Mechanisms of membrane degradation. In *Handbook of Fuel Cells—Fundamentals, Technology and Applications. Volume 3*, eds. W. Vielstich, H. A. Gasteiger, and A. Lamm. New York, NY: John Wiley & Sons.

Liu, W., Ruth, K., and Rusch, G. 2001. Membrane durability in PEM fuel cells. *Journal of New Materials for Electrochemical Systems* 4: 227–232.

Liu, W. and Crum, M. 2006. Effective testing matrix for studying membrane durability in PEM fuel cells: Part 1. Chemical durability. *ECS Transactions* 3: 531–540.

Malkow, T., De Marco, G., Pilenga, A., et al. 2010. PEFC power stack performance testing procedure: Measuring voltage and power as a function of current density, polarization curve method. Test Module PEFC ST 5-3.

Rockward, T. Q. T. 2004. Presentation on establishing a standardized single cell testing procedure through industry participation, consensus and experimentation. *2004 Fuel Cell Seminar: Technology, Markets, and Commercialization*

SAE. 2005. Testing performance of fuel cell systems for automotive applications. J2615. SAE International. Surface Vehicle Recommended Practice.

Tang, Y., Yuan, W., Pan, M., Li, Z., Chen, G., and Li, Y. 2010. Experimental investigation of dynamic performance and transient responses of a kW-class PEM fuel cell stack under various load changes. *Applied Energy* 87: 1410–1417.

USFCC Joint Hydrogen Quality Task Force. 2006a. Protocol on fuel cell component testing: Types of measurements necessary for industry to understand and apply the test data generated. USFCC 04-007A.

USFCC Joint Hydrogen Quality Task Force. 2006b. Protocol on fuel cell component testing: Primer for generating test plans. USFCC 04-003A.

USFCC Joint Hydrogen Quality Task Force. 2006c. Protocol on fuel cell component testing: Suggested test plan template. USFCC 05-002A.

USFCC Joint Hydrogen Quality Task Force. 2006d. Protocol on fuel cell component testing: Suggested dynamic testing profile (DTP). USFCC 04-068A.

USFCC Single Cell Testing Task Force. 2004. Fuel cell test station requirements and verification procedure. USFCC 04-011.

USFCC Single Cell Testing Task Force. 2006. Single cell test protocol. USFCC05-014B.2.

Zhang, S., Yuan, X. Z., Wang, H., et al. 2009. A review of accelerated stress tests of MEA durability in PEM fuel cells. *International Journal of Hydrogen Energy* 34: 388–404.

2

Polarization Curve

Dong-Yun Zhang
Shanghai Institute of Technology

Xianxia Yuan
Shanghai Jiao Tong University

Zi-Feng Ma
Shanghai Jiao Tong University

2.1 Introduction

A fuel cell is an electrochemical device that converts the chemical energy of fuels directly to electricity. A popular way of evaluating a fuel cell is to measure its polarization curve, which is a plot showing the cell voltage change with current or current density. A good fuel cell should display a polarization curve with high-current density at high cell voltage, indicating high-power output. Figure 2.1 shows typical polarization curves for a proton exchange membrane fuel cell (PEMFC) (Lee et al., 1998).

The polarization curve is the standard electrochemical technique for characterizing the performance of fuel cells (both single cells and stacks), yielding information on performance losses in the cell or stack under operating conditions. By measuring polarization curves, the effects that certain parameters—such as composition, flow rate, temperature, and relative humidity of the reactant gases—have on fuel cell performance can be characterized and compared systematically.

Polarization curves provide information on the performance of the cell or stack as a whole, but although they are useful indicators of overall performance under specific operating conditions, they fail to produce much information about the performance of individual components within the cell. They cannot be obtained during normal operation of a fuel cell, and take significant time to complete. In addition, they fail to differentiate mechanisms from each other; for example, flooding and drying inside a fuel cell cannot be distinguished in a single polarization curve. They are also incapable of resolving time-dependent processes occurring in the fuel cell or in the stack.

2.2 Fundamentals of Polarization Curves

2.2.1 Nernst Equation

The theoretical voltage of an electrochemical reaction depends on its Gibbs free energy, ΔG_r:

$$\Delta G_r = -nFE \tag{2.1}$$

15

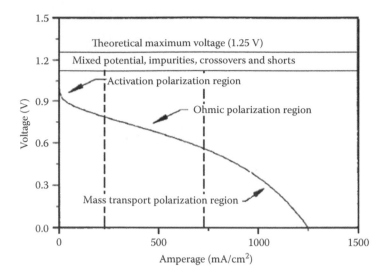

FIGURE 2.1 A typical polarization curve for a PEMFC. (Reproduced from *Journal of Power Sources*, 70, Lee, J. H., Lalk, T. R., and Appleby, A. J. Modeling electrochemical performance in large scale proton exchange membrane fuel cell stacks. 258–268, Copyright (1998), with permission from Elsevier.)

where n is the electron transfer number in the reaction, F is the Faraday constant, E is the theoretical voltage, and nFE is the electrical work done by an electrochemical reaction.

The reaction's Gibbs free energy is obtained by the following equation:

$$\Delta G_r = \Delta G_r^0 + RT \ln \frac{\Pi \gamma_{Products}^{v_i}}{\Pi \gamma_{Reactants}^{v_i}} \tag{2.2}$$

where ΔG_r^0 is the standard reaction Gibbs free energy, R is the gas constant, T is the absolute temperature, γ is the activity of the product, and v_i is the stoichiometric coefficient of species i. Thus,

$$E = E^0 - \frac{RT}{nF} \ln \frac{\Pi \gamma_{Products}^{v_i}}{\Pi \gamma_{Reactants}^{v_i}} \tag{2.3}$$

where E^0 is the standard thermodynamic voltage at 25°C and the activities of both reactant and product are 1. Formula 2.3 is the Nernst equation.

For an H_2/O_2 PEMFC with the electrochemical reaction

$$H_2(g) + \frac{1}{2}O_2(g) \Leftrightarrow H_2O \tag{2.4}$$

the Nernst equation takes the form

$$E = E^0 - \frac{RT}{2F} \ln \left[\frac{\gamma_{H_2O}}{p_{H_2} \sqrt{p_{O_2}}} \right] \tag{2.5}$$

where E^0 is 1.229 V under standard conditions (at 25°C, if the pressures of both H_2 and O_2 are 1 atm), the produced water is assumed to be in the liquid form, and p_i is the activity of the species i, expressed as partial pressure in atmospheres with the standard state of 1 atm. Equation 2.5 shows that E is temperature dependent, and so an increase in temperature leads to a decrease in theoretical voltage.

2.2.2 Open-Circuit Voltage

The open-circuit voltage (OCV) is the voltage provided by a single fuel cell under the condition of no electrical load. When the discharge current approaches zero, true electrochemical equilibrium can be achieved, which represents the thermodynamic voltage E. In practice, the measured OCV is always lower than the theoretical OCV, as shown in Figure 2.1. This difference results mainly from fuel crossover from the anode to the cathode. Sometimes, impurities or current leakage exist, which will also lower the measured OCV. The difference between the measured OCV and the theoretical voltage is called the irreversible voltage loss.

2.2.3 Working Potential

The OCV is of less practical use because no power output occurs under this condition. It is much more useful to record the working potential as a function of the current density changes—that is, the polarization curve—that occur with power release. The working potential is always lower than the OCV, and decreases as the current density increases. This is the most common way to evaluate a fuel cell's performance. The polarization curve provides information on polarization loss, which includes kinetic-related activation losses, cell resistance losses, and mass transport losses. A polarization curve can be divided into three regions, as shown in Figure 2.1, corresponding to activation, ohmic, and mass transport-dominated voltage losses. Through simulating and analyzing polarization curves, we can optimize cell structures and operating conditions, and thereby improve cell performance.

2.2.3.1 Activation Losses

In the low current density region the working potential drops, mainly due to the charge-transfer kinetics, that is, the O_2 reduction and H_2 oxidation rates at the electrode surfaces, which are dominated by the Butler–Volmer equation:

$$i = i_0 \left\{ \exp\left(\frac{\alpha n F \eta_{\text{act}}}{RT} \right) - \exp\left(-\frac{\beta n F \eta_{\text{act}}}{RT} \right) \right\} \tag{2.6}$$

where η_{act} denotes the activation losses; α and β are the transfer coefficients and $\alpha + \beta = 1$; n is the apparent electron number involved in the rate-determining step of a multielectron reaction, which is different from the reaction's total electron transfer number; and i_0 is the exchange current density.

Taking Equation 2.6, when η_{act} is large enough that the other factors can be neglected, the equation can be simplified and rearranged as

$$\eta_{\text{act}} = a + b \log i \tag{2.7}$$

where $a = -(2.303RT/\alpha nF)\log i_0$, and the slope $b = 2.303RT/\alpha nF$. As shown in Equation 2.7, the activation losses are linearly dependent on the logarithmic of current density; this is the so-called Tafel equation, which expresses the relationship between the current density of an electrochemical reaction and the activation losses.

When a H_2/O_2 fuel cell is under operation, the electrode potentials of the anode and cathode sides deviate from their equilibrium and Equation 2.6 works for both sides. Thus, at the anode side,

$$i_a = i_0^H \left\{ \exp\left(\frac{\alpha_H n_H F \eta_a}{RT} \right) - \exp\left[-\frac{\beta_H n_H F \eta_a}{RT} \right] \right\} \tag{2.8}$$

where i_0^H is the exchange current density of the hydrogen oxidation reaction (HOR), n_H is the electron number at the rate-determining step, α_H and β_H are the electron transfer coefficients, and η_a is the anode overpotential for the HOR. The overpotential η_c for the oxygen reduction reaction (ORR) at the cathode side can be similarly expressed.

The working potential drop at low current densities due to polarization losses both at anode and cathode sides can be denoted as the activation losses, $\eta_{act} = \eta_a + \eta_c$.

2.2.3.2 Ohmic Losses

In the intermediate current density range of a polarization curve, the working potential drop is mainly caused by internal resistance, including the electric contact resistance among the fuel cell components, and the proton resistance of the proton-conducting membrane. This internal resistance is called ohmic resistance, R_{ohmic}. The working potential drop due to ohmic resistance is denoted as ohmic losses, η_{ohmic}, defined as $\eta_{ohmic} = iR_{ohmic}$. In a PEMFC, the proton resistance of the polymer electrolyte membrane contributes most to the total ohmic resistance. In this region, the working potential is linearly dependent on current density.

2.2.3.3 Mass Transport Losses

In the high-current density range, the working potential decreases rapidly because the transfer speed of reactants and products is slower than the reaction rate. The transfer of reactants and products in fuel cells is called mass transport, and consists of convection and diffusion. Convection is liquid bulk movement under mechanical force, mainly in a flow-field channel. Mass transport due to a concentration gradient in a gas diffusion layer is called diffusion, and affects fuel cell performance in two ways. The boundary between convection and diffusion always appears at the interface of a flow channel and a porous electrode, as shown in Figure 2.2. In practical terms, the boundaries of convection and diffusion are often vague—for example, at low velocity, the diffusion layer may extend to the flow channel, while conversely at a high flow rate convection may penetrate to the electrode. In any case, only the reactants arriving at the catalyst layer can react, and so the theoretical voltage should be calculated based on the concentration of reactants in the catalyst layer, which is much lower than that in flow-field plates.

The theoretical voltage losses resulting from mass diffusion are called Nernst losses (denoted as η_{conc}^l), which can be calculated according to

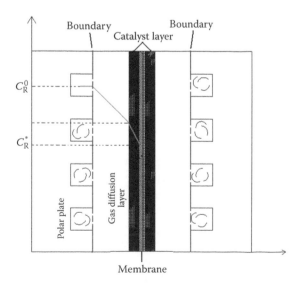

FIGURE 2.2   Schematic diagram of mass transport in a PEMFC.

$$\eta_{conc}^1 = E - E^* = \left(E^0 - \frac{RT}{nF} \ln \frac{1}{C_R^0} \right) - \left(E^0 - \frac{RT}{nF} \ln \frac{1}{C_R^*} \right) = \frac{RT}{nF} \ln \frac{C_R^0}{C_R^*} \tag{2.9}$$

where E^* is the real voltage, C_R^0 is the bulk concentration of the reactant, C_R^* is the reactant concentration on the catalyst layer, and the accumulation of product is ignored.

It must be noted that Equation 2.6 does not consider the concentration changes in reactant and product that occur as a result of the reaction, yet these changes also affect the activation losses. Therefore, we have to modify the Butler–Volmer equation as follows:

$$i = i_0' \left\{ \frac{C_R^*}{C_R'} \exp\left(\frac{\alpha n F \eta_{act}^*}{RT} \right) - \frac{C_R^*}{C_R'} \exp\left(-\frac{\beta n F \eta_{act}^*}{RT} \right) \right\} \tag{2.10}$$

where i_0' is the exchange current density obtained when the concentrations of reactant and product are C_R' and C_P', respectively, and η_{act}^* is the overpotential corresponding to E^*. Under high-current density, the second term of Equation 2.10 can be ignored, thus simplifying it to

$$i = i_0' \frac{C_R^*}{C_R'} \exp\left(\frac{\alpha n F \eta_{act}^*}{RT} \right) \tag{2.11}$$

$$\eta_{act}^* = \frac{RT}{\alpha n F} \ln \frac{i C_R'}{i_0' C_R^*} \tag{2.12}$$

The increase in activation losses resulting from the transfer of reactants and products is denoted as η_{conc}^2, which can be calculated by

$$\eta_{conc}^2 = \eta_{act}^* - \eta_{act} = \left(\frac{RT}{\alpha n F} \ln \frac{i C_R'}{i_R' C_R^*} \right) - \left(\frac{RT}{\alpha n F} \ln \frac{i C_R'}{i_0' C_R^0} \right) = \frac{RT}{\alpha n F} \ln \frac{C_R^0}{C_R^*} \tag{2.13}$$

The total voltage losses caused by mass diffusion are the sum of η_{conc}^1 and η_{conc}^2,

$$\eta_{conc} = \left(1 + \frac{1}{\alpha} \right) \frac{RT}{nF} \ln \frac{C_R^0}{C_R^*} \tag{2.14}$$

Since the reactant is consumed immediately upon arrival at the catalyst layer, the current density depends on the diffusion mass flow rate, μ_{diff}, in the catalyst layer. Thus,

$$i = n F \mu_{diff} \tag{2.15}$$

According to Fick's diffusion laws, diffusion mass flow rate on arrival at the catalyst layer can be expressed as

$$\mu_{diff} = -D^{eff} \frac{C_R^* - C_R^0}{\delta} \tag{2.16}$$

where D^{eff} is the effective diffusivity and δ is the electrode thickness. When C_R^* decreases to zero, the current density achieves its maximum value, called the limiting current density, i_L:

$$i_L = nFD^{eff}\frac{C_R^0}{\delta} \tag{2.17}$$

Therefore, Equation 2.14 becomes

$$\eta_{conc} = \left(1+\frac{1}{\alpha}\right)\frac{RT}{nF}\ln\frac{i_L}{i_L-i} \tag{2.18}$$

Increasing i_L will thus help to reduce mass transport losses. According to Equation 2.17, there are two methods to increase i_L. One is optimizing the working conditions, electrode structure, and diffusion layer thickness to raise D^{eff} and lower δ. The other is optimizing the flow field design to enable uniform reactant distribution, which can ensure a high C_R^0 value.

However, the reactant input concentration is not equal to C_R^0, due to mass convection from the flow field channel to the electrode, according to

$$\omega_{C,i} = h_m\left(\rho_{i,s}-\overline{\rho_i}\right) \tag{2.19}$$

where $\omega_{C,i}$ is the convective mass flow rate (kg m^{-2} s^{-1}), $\rho_{i,s}$ is the density of species i on the electrode surface, $\overline{\rho_i}$ is the average density of species i in the flow channel, and h_m is the convection mass transfer coefficient, whose value depends on the flow channel geometry and the physical properties of species i and j. Usually, h_m can be calculated as follows:

$$h_m = Sh\frac{D_{ij}}{d_h} \tag{2.20}$$

where Sh is the Sherwood number, which is a dimensionless value, d_h is the hydraulic diameter, and D_{ij} is the two-phase diffusion coefficient of species i and j.

The equations for η_{conc}, current density, and limiting current density therefore should be correspondingly modified.

2.3 Measurement of Polarization Curves

When we run PEMFCs, purging is both the first and the last step. Initially, the reactant gases are used to purge away the air in fuel cells, while the temperature gradually rises to a set value. At the end of the operation, similarly, we use dry inert gas to purge away the residual reactant gases and water in the fuel cells while they cool down to room temperature.

Before testing, PEMFCs have to be conditioned under the same operating conditions as they will undergo during testing (Bashyam and Zelenay, 2006). Conditioning is continued at a constant voltage until the current density reaches a constant level, or at a constant current density until the voltage reaches a constant level.

The polarization curve of PEMFCs can usually be measured using either steady-state or transient-state techniques. A steady-state polarization curve can be obtained by recording the cell potential as the cell current changes; the cell potential should be recorded when the current change rate is less than a set value. The steady-state cell potential here is an approximation. Conversely, a nonsteady-state polarization curve can be obtained using a slow current sweep rate.

2.4 Empirical Modeling of Polarization Curves

In order to understand the electrochemical behavior of PEMFCs, several types of empirical modeling have been introduced to mimic polarization curves.

Srinivasan et al. (1988) developed the following equation to describe the relationship between working potential, V_{cell}, and current density, i, in the low- and mid-current density ranges, where the cell overpotential is dominated by activation and ohmic losses:

$$V_{cell} = E + b\log(i_0) - b\log(i) - Ri \tag{2.21}$$

In Equation 2.21, i_0 and b are the exchange current density and the Tafel slope for oxygen reduction, respectively, since anode polarization is negligible. The third term in this equation is predominant at low-current densities and describes the cathodic activation overpotential. The last term, R, is the cell resistance, which causes the working potential to decrease linearly as the current density increases, and predominates in the intermediate current density region. As shown in Figure 2.3 (Kim et al., 1995), this model is not a good fit for the entire polarization curve, which deviates dramatically from the fitting line at high-current densities.

Kim et al. (1995) expanded Equation 2.21 by introducing an exponential term in order to fit the high-current density region of polarization curves, which is predominated by mass transport losses:

$$V_{cell} = E + b\log(i_0) - b\log(i) - Ri - m\exp(\xi i) \tag{2.22}$$

where m and ξ are the parameters related to mass transport limitations. The empirical equation 2.22 excellently fits the polarization curve over the entire current density range, as shown in Figure 2.4 (Kim et al., 1995). However, it lacks a theoretical electrode kinetic interpretation of the exponential term. In

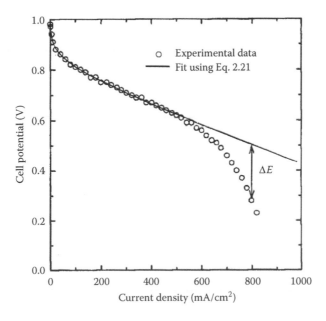

FIGURE 2.3 Modeling of E versus i curves using Equation 2.21. (Reproduced from Kim, J., Lee, S., and Srinivasan, S. 1995. *Journal of Electrochemical Society* 142: 2670–2674. With permission from the Electrochemical Society, Inc.)

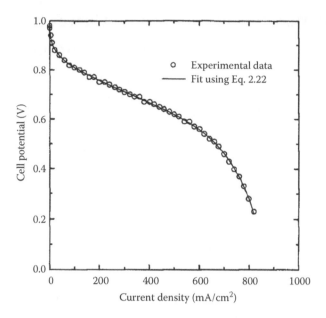

FIGURE 2.4 Modeling of E versus i curves using Equation 2.22. (Reprinted from Kim, J., Lee, S., and Srinivasan, S. 1995. *Journal of Electrochemical Society* 142: 2670–2674. With permission from the Electrochemical Society, Inc.)

the one-dimensional model developed by Bevers et al. (1997), ξ and m correlate, respectively, to the electrolyte conductivity and the parameters determining the position of the diffusion controlled region. Their model fits the real data well, with a changing Tafel slope at about 800 mV, as shown in Figure 2.5 (Bevers et al., 1997). Only in the very high-current density region is there a slight deviation from the experimental data, due to reduction of the active surface resulting from pore flooding.

Squadrito et al. (1999) developed a logarithmic formation based on mechanistic analysis to express the concentration polarization, which replaced the exponential empirical term in Equation 2.22:

$$V_{cell} = E + b\log(i_0) - b\log(i) - Ri - \alpha i^k \ln(1 - yi) \tag{2.23}$$

where αi^k is the prelogarithmic term attributed to the different contributions and acts as an "amplification term," expressed in potential units; k is a dimensionless number; and y is the inverse of the limiting current density. Equation 2.23 was claimed to provide a more accurate prediction of behavior at high-current densities, since k mainly influences the point at which there is a departure from the linear behavior, and α determines the shape of the curve at high-current densities. As shown in Figure 2.6 (Squadrito et al., 1999), the experimental data (symbols) are an excellent fit with Equation 2.23. An attempt to correlate α to the cell and electrode properties was proposed through dimensional analysis. However, there is insufficient evidence to define the correct mathematical correlation between this parameter and the physical properties of the system.

A semiempirical equation was developed by Pisani et al. (2002) by attributing the nonlinear drop in the high-current density range to the interface phenomena occurring in the cathode reaction region:

$$V_{cell} = E + b\log(i_0) - f\log(i) - Ri + t\ln\left(1 - \frac{i}{i_L}S^{-\mu(1 - i/i_L)}\right) \tag{2.24}$$

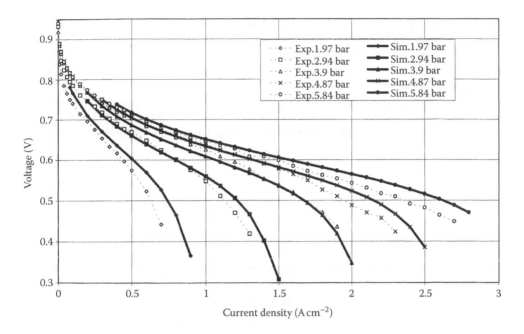

FIGURE 2.5 Fit of complete current voltage curves (pressures: 1.97, 2.94, 3.9, 4.87, and 5.48 bar, $\varepsilon_{g,diff/rea}$ = 22.5%/2.3%, $\gamma_{p,diff/rea}$ = 20 nm/20 nm, $d_{diff/rea}$ = 170 μm/60 μm, b = 120 mV, α_{act} = 2.3 × 10⁷, $\kappa_{el} \times \varepsilon_{el}$ = 0.8 Ω^{-1} m⁻¹, U_o = 946 mV, $p°$ = 4.87 bar, and i_o = 14.3 × 10⁻² A m⁻²). (Reprinted from Bevers, D. et al. 1997. *Journal of Applied Electrochemistry* 27: 1254–1264. With permission from Kluwer Academic Publishers.)

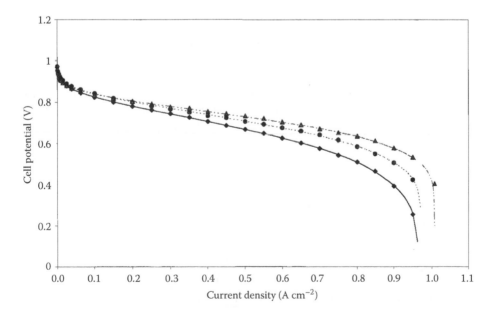

FIGURE 2.6 Comparison between experimental (bullets) and simulated (lines) data at 80°C and gas pressures of 3/5 bar (absolute) for H₂/air, respectively, varying the Nafion® membrane: NF112 (▲), NF115 (●), NF117 (◆). (Reprinted from Squadrito, G. et al. 1999. *Journal of Applied Electrochemistry* 29: 1449–1455. With permission from Kluwer Academic Publishers.)

where S is a flooding parameter and μ is an empirical constant. The prelogarithmic terms f and t are functions of the parameter N_d, which characterizes the gas pore structure inside the cathode reactive region. The simulated polarization curves agree well with those obtained from Equations 2.23 and 2.24, as shown in Figure 2.7 (Pisani et al., 2002), but with the advantage of having coefficients with a precise physical origin and a precise physical meaning.

Sena et al. (1999) treated the catalyst layer as a thin-film/flooded agglomerate, whereby the gas diffusion electrode is assumed to be composed of an assembly of flooded zones (catalytic zones) and empty zones (no catalyst present). The final equation related to oxygen diffusion effects in a gas diffusion electrode becomes

$$V_{cell} = E + b\log\left(i_0\right) - b\log\left(i\right) - Ri + b\log\left(1 - \frac{i}{i_L^{O_2}}\right) \tag{2.25}$$

where $i_k^{O_2}$ is the limiting current density due to a limiting oxygen diffusion effect. This model leads to good fittings between simulated results and experimental polarization curves for a PEMFC working with dry hydrogen and various membranes, because it takes into account the influence of water transport in the membrane.

A step-by-step technique was developed by Williams et al. (2005) to evaluate six sources of polarization, including nonelectrode ohmic overpotential, electrode ohmic overpotential, nonelectrode concentration overpotential, electrode concentration overpotential, activation overpotential from the Tafel slope, and activation overpotential from catalyst activity, mainly associated with the cathode side of a PEMFC. Applying this technique to understanding polarization curves for differently constructed PEMFCs allows systematic comparison of the polarization sources associated with various operating conditions and membrane electrode assembly (MEA) structures.

The influence of pressure parameters on the concentration polarization in PEMFC stack models was taken into account by Lee et al. (1998), who obtained the following equation:

$$V_{cell} = E + b\log\left(i_0\right) - b\log\left(i\right) - Ri - m\exp\left(\xi i\right) - z\log\left(\frac{p}{p_{O_2}}\right) \tag{2.26}$$

where p is the total pressure, p_{O_2} is the partial pressure of oxygen, and z is a constant. The pressure ratio logarithm is a form of the Nernst equation, primarily describing potential changes in the cathode. This modeling technique is a useful tool for investigating fuel cell power systems. The model can be used to

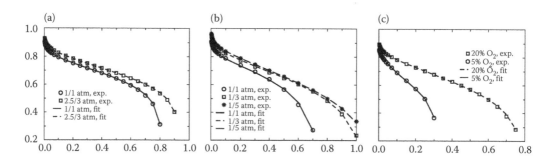

FIGURE 2.7 Comparison between experimental data and theoretical results. Experimental data are from simulation results of Equation 2.23 (a) and Equation 2.22 (b and c). (Reprinted from *Journal of Power Sources*, 108, Pisani, L. et al. A new semiempirical approach to performance curves of polymer electrolyte fuel cells, 192–203, Copyright (2002), with permission from Elsevier.)

investigate various MEA designs and assembly methods by changing the coefficients of the polarization equations used to describe the electrochemical performance of the PEMFC. These coefficients would be determined from the results of laboratory-scaled, single-cell experiments.

Recently, more complex models based on computational fluid dynamics code have been developed. Here, we present a brief introduction without expanding on the specific theoretical formula because of the complexity of these models. As discussed above, the most complicated part of polarization curves—the mass transport dominant, high-current density region—is primarily influenced by the reactant concentrations at the reactive sites and by water transport in the porous electrode. Therefore, the convection and diffusion in flow channels cannot be ignored if one wishes to accurately simulate mass transport in the porous electrode. Thus, two-dimensional (2D) and three-dimensional (3D) models have been developed (Hsing and Futerko, 2000; Kazim et al., 1999; Kulikovsky et al., 1999; Liu et al., 2010; Nguyen and White, 1993; Singh et al., 1999; Sivertsen and Djilali, 2005; Su et al., 2010). These models can give a more precise picture of the variations in temperature and reactants in typical 3D PEMFCs as well as a better understanding of how the actual fuel cell performs, and they are generally used to optimize PEMFC structure and operating parameters.

2.5 Applications of Polarization Curves in PEM Fuel Cells

A polarization curve is generally used to express the characteristics of a PEMFC system. The curve reflects an integral effect of all parameters of cell structure and operating conditions. The behavior of a PEMFC is highly nonlinear, and so it is important to incorporate process nonlinearity for control system design and process optimization. The correlation between a polarization curve and the characteristics of a PEMFC can be defined using experimental techniques and modeling, and this correlation is always used to optimize PEMFC structure and operating conditions.

2.5.1 Optimizing Cell Structure

Catalyst layer (CL). Das et al. (2007) developed a parametric model using a combination of analytical solutions and empirical formulae to optimize the cathode catalyst layer structure. The optimum catalyst loading was found to be 0.19–0.20 mg cm^{-2} when the cell output was set to 0.8 V, as shown in Figure 2.8 (Das et al., 2007). The higher the membrane content in the catalyst layer, the better the cell performance in terms of activation overpotential, as shown in Figure 2.9 (Das et al., 2007). The optimum catalyst layer thickness was 9–11 μm for a platinum loading of 0.20 mg cm^{-2} at a cell voltage of 0.8 V, as shown in Figure 2.10 (Das et al., 2007).

Gas diffusion layer (GDL). Beuscher (2006) developed a very quick and simple means to measure the average mass transport resistance *in situ* by extrapolating the polarization curve to obtain the limiting current (using Equation 2.25), then using a resistance-in-series approach to separate the mass transport resistance of the gas diffusion media from that of the electrode. The mass transport resistance of CARBEL CL gas diffusion media in their studied system accounted for only 26% of the total mass transport resistance. Gerteisen et al. (2010) investigated the performance of a 6-cell PEMFC stack with various GDLs (Toray TGP-H-060 untreated and laser perforated). Performance analysis and transient analysis showed lower overpotential and higher stability in the medium and high-current density range with perforated GDLs (as shown in Figure 2.11 (Gerteisen et al., 2010)), results that could be attributable to pore flooding being minimized.

Flow pattern. Weng et al. (2005) investigated commonly used flow patterns—parallel straight, serpentine, and interdigitated (see Figure 2.12 for each configuration). Performance was simulated using a rigorous 3D mathematical model, employing the CFDRC commercial code, as shown in Figure 2.13 (Weng et al., 2005). The results revealed that serpentine and interdigitated flow patterns are preferable for strong convection and high mass transfer. However, they also experience higher pressure loss. Miansari et al. (2009) experimentally studied the effect of channel depth on PEMFC performance, and

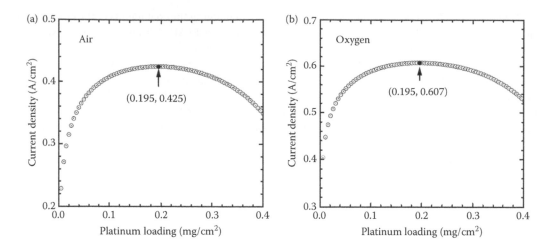

FIGURE 2.8 Optimum platinum loading as a function of the current density at a given cell potential of 0.8 V for (a) air as the cathode gas, and (b) oxygen as the cathode gas. (Reprinted from *Journal of Electroanalytical Chemistry*, 604, Das, P. K., Li, X., and Liu, Z. Analytical approach to polymer electrolyte membrane fuel cell performance and optimization, 72–90, Copyright (2007), with permission from Elsevier.)

observed high performance at a depth of 1.5 mm for the anode and 1 mm for the cathode, as shown in Figure 2.14. Two bio-inspired flow channel patterns—leaf and lung (as shown in Figure 2.15)—were designed and investigated by Kloess et al. (2009). The same group also numerically simulated the performance of PEMFCs with various flow channel designs in bipolar plates, as shown in Figure 2.16 (Kloess et al., 2009). The results indicated a lower pressure drop from the inlet to the outlet in the leaf or lung design as compared with existing serpentine or interdigitated flow patterns. Flow diffusion to the GDL was found to be more uniform with the new flow channel patterns.

Membrane. Squadrito (1999) used Equation 2.23 to simulate the performance of PEMFCs with various Nafion membranes. The results revealed that cell resistance (R) is clearly influenced by membrane

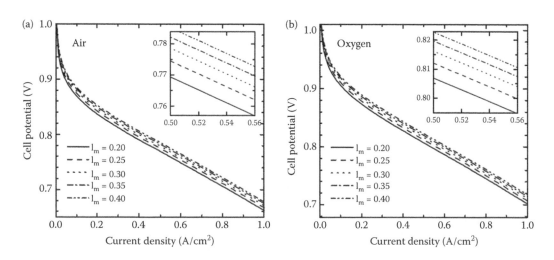

FIGURE 2.9 Effect of membrane content in the catalyst layer on cell potential for (a) air as the cathode gas, and (b) oxygen as the cathode gas with a platinum loading of 0.2 mg cm^{-2}. (Reprinted from *Journal of Electroanalytical Chemistry*, 604, Das, P. K., Li, X., and Liu, Z. Analytical approach to polymer electrolyte membrane fuel cell performance and optimization, 72–90, Copyright (2007), with permission from Elsevier.)

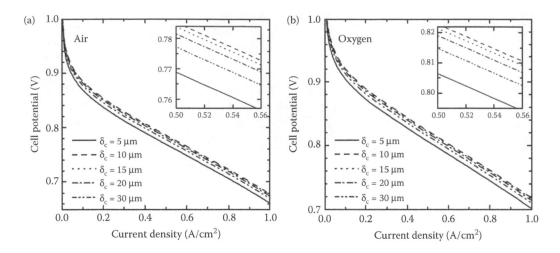

FIGURE 2.10 Variation of cell potential with current density for (a) air as the cathode gas, and (b) oxygen as the cathode gas with a platinum loading of 0.2 mg cm^{-2} for different catalyst layer thicknesses, as indicated in the legend. (Reprinted from *Journal of Electroanalytical Chemistry*, 604, Das, P. K., Li, X., and Liu, Z. Analytical approach to polymer electrolyte membrane fuel cell performance and optimization, 72–90, Copyright (2007), with permission from Elsevier.)

thickness; under the same cell conditions, a decrease in R was observed from Nafion® 117 to Nafion® 115 to Nafion® 112. According to Sena's study (1999), different limiting effects were found in PEMFCs with dry hydrogen and various membranes. With Nafion 117 and Nafion 115, limiting effects due to oxygen diffusion were minimal and water transport became a limiting effect, while with Nafion 112 oxygen diffusion effects dominated.

2.5.2 Optimizing Operating Conditions

The operating parameters for a PEMFC include temperature, pressure, gas composition, humidification, and stoichiometrics. Some of these can be further subdivided—for example, temperature can be

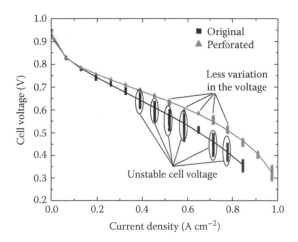

FIGURE 2.11 A comparison of polarization curves of cell #1 between Stack$_{org}$ and Stack$_{perf}$. (Reprinted from *Journal of Power Sources*, 195, Gerteisen, D. and Sadeler, C., Stability and performance improvement of a PEM fuel cell stack by laser perforation of gas diffusion layers, 5252–5257, Copyright (2010), with permission from Elsevier.)

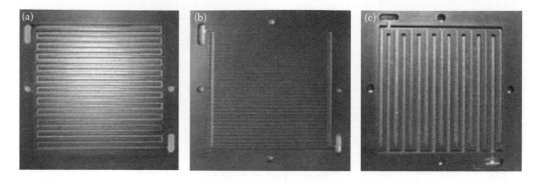

FIGURE 2.12 (a) Parallel straight flow pattern, (b) serpentine flow pattern, and (c) interdigitated flow pattern. (Reprinted from *Journal of Power Sources*, 145, Weng, F. et al. Numerical prediction of concentration and current distributions in PEMFC, 546–554, Copyright (2005), with permission from Elsevier.)

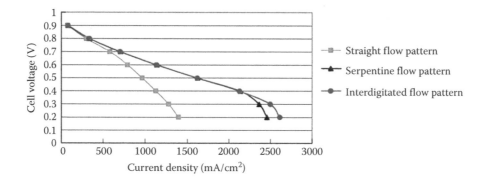

FIGURE 2.13 Polarization curves of straight flow, serpentine flow, and interdigitated flow patterns under the same operating conditions (numerical simulation). (Reprinted from *Journal of Power Sources*, 145, Weng, F. et al. Numerical prediction of concentration and current distributions in PEMFC, 546–554, Copyright (2005), with permission from Elsevier.)

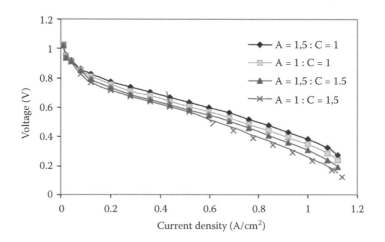

FIGURE 2.14 The effect of channel depth on cell performance. (Reprinted from *Journal of Power Sources*, 190, Miansari, M. et al. Experimental and thermodynamic approach on proton exchange membrane fuel cell performance, 356–361, Copyright (2009), with permission from Elsevier.)

(a)

(b)

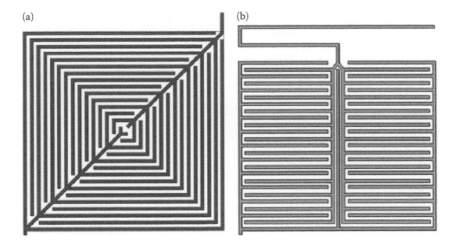

FIGURE 2.15 (a) New bio-inspired leaf flow pattern. (b) New bio-inspired lung flow pattern. (Reprinted from *Journal of Power Sources*, 188, Kloess, J. P. et al. Investigation of bio-inspired flow channel designs for bipolar plates in proton exchange membrane fuel cells, 132–140, Copyright (2009), with permission from Elsevier.)

subdivided into cell temperature, anode gas temperature, and cathode gas temperature. Considerable work has been done to investigate the influence of these operating parameters, some examples of which are described in the following paragraphs.

Santarelli et al. (2006) developed a regression model of the polarization curve (Figure 2.17a), and analyzed the effects of temperature on the behavior of the curve's main parameters (Figure 2.17b,c,d). As shown in Figure 2.17b, the cathode exchange current density, $i_{0,C}$, linearly increases from 50°C to 80°C with a slope of 10^{-4} A cm^{-2}. The trend of parameter C_1 (linked to cell resistance) is typical of electrolytes, with a fairly linear decrease in temperature, as shown in Figure 2.17c. The internal current density, i_n, increases significantly as the temperature increases in the same range, as shown in Figure 2.17d. However, it is impossible to propose general analytic correlations linking the parameters to the operating temperature.

A one-dimensional water and thermal management model was developed and adopted by Hung et al. (2007) to generate a polarization curve. As shown in Figure 2.18 (Hung et al., 2007), this model is able to describe fuel cell behavior while operating conditions change (i.e., fuel cell temperature, anode pressure, cathode pressure, hydrogen stoichiometric ratio, air stoichiometric ratio, hydrogen humidification temperature, and air humidification temperature). The analytic results illustrate that fuel cell temperature,

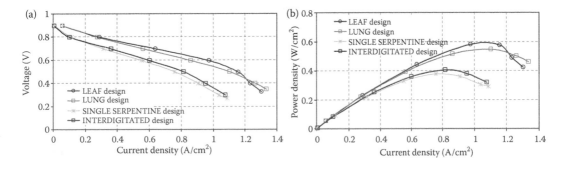

FIGURE 2.16 (a) Polarization curve comparison for various flow pattern designs: $P = 2$ atm, $T = 75$°C, and RH = 100%. (b) Power density comparison for various flow pattern designs: $P = 2$ atm, $T = 75$°C, and RH = 100%. (Reprinted from *Journal of Power Sources*, 188, Kloess, J. P. et al. Investigation of bio-inspired flow channel designs for bipolar plates in proton exchange membrane fuel cells, 132–140, Copyright (2009), with permission from Elsevier.)

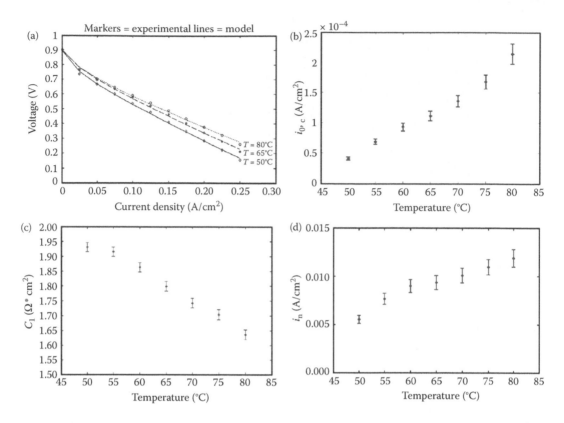

FIGURE 2.17 (a) Regression curves obtained at different values across a temperature range. 50°C: rhombus experimental, solid line model; 65°C: asterisks experimental, dashed line model; and 80°C: circles experimental, dotted line model. Maximum voltage uncertainty less than 3%. Values of the parameter (b) $i_{0,c}$, (c) C_l, and (d) i_n versus cell operating temperature. (Reprinted from *Energy Conversion and Management*, 48, Santarelli, M. G. and Torchio, M. F. Experimental analysis of the effects of the operating variables on the performance of a single PEMFC, 40–51, Copyright (2007), with permission from Elsevier.)

cathode pressure, humidification temperature at the anode, and stoichiometric ratios of air are the dominant process variables in a PEMFC system. The correlation between dominant process variables and internal variables was investigated, and the results showed that cell temperature has the greatest effect on the transfer coefficient, followed by cathode pressure, whereas the humidification temperature at the anode and the stoichiometric ratios of air do not significantly affect the transfer coefficient. Weng et al. (2005) studied the effects of operating temperature and relative humidity using a 3D model and obtained a similar result as Santarelli et al. (2007), that cell performance increases with temperature (30–70°C) due to high membrane conductivity. However, the membrane will dry out at a high cell temperature.

2.5.3 Diagnosing Degradation of PEM Fuel Cells

Durability is one of the biggest challenges in the commercialization of PEMFCs. Cell performance is prone to material degradation and affected by operating conditions. Cleghorn et al. (2006) conducted a 26,300 h single-cell life test under conditions relevant to stationary fuel cell applications. Performance degraded at a rate of between 4 and 6 μV h^{-1} at an operating current density of 800 mA cm^{-2}, corresponding to a maximum total voltage loss of less than 150 mV over the length of the test, as shown in Figure 2.19.

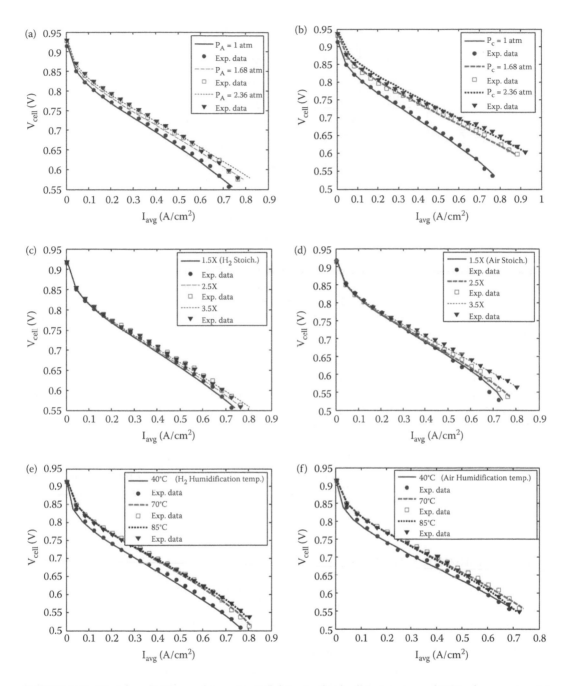

FIGURE 2.18 Model predictions and experimental data at a fixed cell temperature of 60°C when varying: (a) anode pressure, (b) cathode pressure, (c) hydrogen stoichiometric ratio, (d) air stoichiometric ratio, (e) hydrogen humidification temperature, and (f) air humidification temperature. (Reprinted from *Journal of Power Sources*, 171, Hung, A. et al., Operation-relevant modeling of an experimental proton exchange membrane fuel cell, 728–737, Copyright (2007), with permission from Elsevier.)

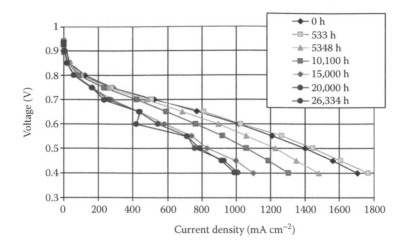

FIGURE 2.19 Polarization curves at various periods in time (0, 500, 5348, 10,100, 15,000, 20,000, and 26,330 h) during life test. Cell temperature at 70°C. Air: 2.0 × stoichiometry, ambient pressure, 100% RH. Hydrogen: 1.2 × stoichiometry, ambient pressure, 100% RH. (Reprinted from *Journal of Power Sources*, 158, Cleghorn, S. J. C. et al. A polymer electrolyte fuel cell life test 3 years of continuous operation, 446–454, Copyright (2006), with permission from Elsevier.)

Lin et al. (2009) studied PEMFC performance degradation in a dynamic load cycling situation to simulating automotive working conditions. The polarization curves, measured at regular intervals, revealed rapid performance degradation after 280 h of operation, as shown in Figure 2.20.

Wu et al. (2010) conducted a 1200 h lifetime test of a six-cell PEMFC stack under close to open-circuit conditions. The overall cell degradation rate under this accelerated stress testing was approximately 0.128 mV h^{-1}. As Figure 2.21 shows, the degradation rate after 800 h was obviously higher, due to the dominance of different degradation mechanisms. For the first period, degradation was mainly attributable to catalyst decay, while the subsequent dramatic degradation was likely caused by membrane failure.

2.6 Concluding Remarks

The polarization curve reflects an integrated effect of all variables, including the structure of cell components and the cell operating conditions, on PEMFC performance. Two aspects of polarization curve studies require further work. One is separating the influence of each variable on performance and establishing a quantitative relationship between macroscopic variables and modeling parameters, which requires establishing a correlation between the parameters and the physical properties. The other is developing an accurate and quick model to predict the performance of a PEMFC. Most one-dimensional models are based on empirical or semiempirical equations, and cannot be adapted to various types of PEMFCs. 2D and 3D two-phase models should be more promising for correlating the parameters and physical properties of PEMFCs. However, two bottlenecks are impeding progress. One is how to determine the values of the parameters, and the other is how to verify their accuracy. In other words, 2D and 3D models are difficult to verify, especially those simulating the spatial distribution of the variables. In summary, the development of *in situ* diagnostic technology lags far behind the development of computing technology in investigating the polarization curves of PEMFCs.

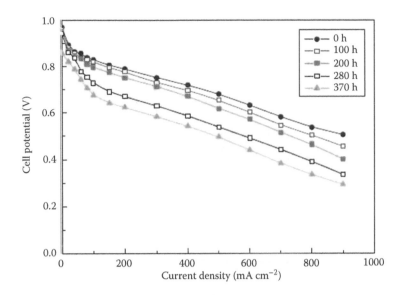

FIGURE 2.20 Polarization curves of initial MEA and MEAs measured at time intervals (after 100, 200, 280, and 370 h of running) under driving cycles. (Reprinted from *International Journal of Hydrogen Energy*, 34, Lin, R. et al. Investigation of dynamic driving cycle effect on performance degradation and micro-structure change of PEM fuel cell, 2369–2376, Copyright (2009), with permission from Elsevier.)

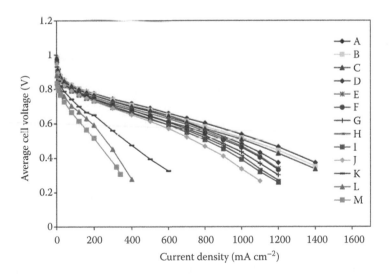

FIGURE 2.21 Average performance change in a PEMFC stack with time: (A) BOL, (B) 100 h, (C) 200 h, (D) 300 h, (E) 400 h, (F) 500 h, (G) 600 h, (H) 700 h, (I) 800 h, (J) 900 h, (K) 1000 h, (L) 1100 h, and (M) 1200 h. Cell temperature at 70°C; humidifier temperature of air/hydrogen at 70/70°C; gas pressure of air/hydrogen at atmosphere; stoichiometries of air/hydrogen at 2.5/1.5. (Reprinted from *Journal of Power Sources*, 195, Wu, J. et al. Proton exchange membrane fuel cell degradation under close to open-circuit conditions Part I *In situ* diagnosis, 1171–1176, Copyright (2010), with permission from Elsevier.)

References

Bashyam, R. and Zelenay, P. 2006. A class of non-precious metal composite catalysts. *Nature* 443: 63–66.

Beuscher, U. 2006. Experimental method to determine the mass transport resistance of a polymer electrolyte fuel cell. *Journal of the Electrochemical Society* 153: A1788–A1793.

Bevers, D., Wöhr, M., Yasuda, K. et al. 1997. Simulation of a polymer electrolyte fuel cell electrode. *Journal of Applied Electrochemistry* 27: 1254–1264.

Cleghorn, S. J. C., Mayfield, D. K., Moore, D. A. et al. 2006. A polymer electrolyte fuel cell life test 3 years of continuous operation. *Journal of Power Sources* 158: 446–454.

Das, P. K., Li, X., and Liu, Z. 2007. Analytical approach to polymer electrolyte membrane fuel cell performance and optimization. *Journal of Electroanalytical Chemistry* 604: 72–90.

Gerteisen, D. and Sadeler, C. 2010. Stability and performance improvement of a PEM fuel cell stack by laser perforation of gas diffusion layers. *Journal of Power Sources* 195: 5252–5257.

Hsing, I. M. and Futerko, P. 2000. Two-dimensional simulation of water transport in polymer electrolyte fuel cells. *Chemical Engineering Science* 55: 4209–4218.

Hung, A., Sung, L., Chen, Y. et al. 2007. Operation-relevant modeling of an experimental proton exchange membrane fuel cell. *Journal of Power Sources* 171: 728–737.

Kazim, A., Liu, H. T., and Forges, P. 1999. Modeling of performance of PEM fuel cells with conventional and interdigitated flow fields. *Journal of Applied Electrochemistry* 29: 1409–1416.

Kim, J., Lee, S., and Srinivasan, S. 1995. Modeling of proton exchange membrane fuel cell performance with an empirical equation. *Journal of Electrochemical Society* 142: 2670–2674.

Kloess, J. P., Wang, X., Liu, J. et al. 2009. Investigation of bio-inspired flow channel designs for bipolar plates in proton exchange membrane fuel cells. *Journal of Power Sources* 188: 132–140.

Kulikovsky, A. A., Divisek, J., and Kornyshev, A. A. 1999. Modeling the cathode compartment of polymer electrolyte fuel cells dead and active reaction zones. *Journal of Electrochemical Society* 146: 3981–3991.

Lee, J. H., Lalk, T. R., and Appleby, A. J. 1998. Modeling electrochemical performance in large scale proton exchange membrane fuel cell stacks. *Journal of Power Sources* 70: 258–268.

Lin, R., Li, B., Hou, Y. P. et al. 2009. Investigation of dynamic driving cycle effect on performance degradation and micro-structure change of PEM fuel cell. *International Journal of Hydrogen Energy* 34: 2369–2376.

Liu, X., Lou, G., and Wen, Z. 2010. Three-dimensional two-phase flow model of proton exchange membrane fuel cell with parallel gas distributors. *Journal of Power Sources* 195: 2764–2773.

Miansari, M., Sedighi, K., Amidpour, M. et al. 2009. Experimental and thermodynamic approach on proton exchange membrane fuel cell performance. *Journal of Power Sources* 190: 356–361.

Nguyen, T. V. and White, R. E. 1993. A water and heat management model for proton exchange membrane fuel cells. *Journal of Electrochemical Society* 140: 2178–2186.

Pisani, L., Murgia, G., Valentini, M. et al. 2002. A new semiempirical approach to performance curves of polymer electrolyte fuel cells. *Journal of Power Sources* 108: 192–203.

Santarelli, M. G. and Torchio, M. F. 2007. Experimental analysis of the effects of the operating variables on the performance of a single PEMFC. *Energy Conversion and Management* 48: 40–51.

Santarelli, M. G., Torchio, M. F., and Cochis, P. 2006. Parameters estimation of a PEM fuel cell polarization curve and analysis of their behavior with temperature. *Journal of Power Sources* 159: 824–835.

Sena, D. R., Ticianelli, E. A., Paganin, V. A. et al. 1999. Effect of water transport in a PEFC at low temperatures operating with dry hydrogen. *Journal of Electroanalytical Chemistry* 477: 164–170.

Singh, D., Lu, D. M., and Djilali, N. 1999. A two-dimensional analysis of mass transport in proton exchange membrane fuel cells. *International Journal of Engineering Science* 37: 431–452.

Sivertsen, B. R. and Djilali, N. 2005. CFD-based modelling of proton exchange membrane fuel cells. *Journal of Power Sources* 141: 65–78.

Squadrito, G., Maggio, G., Passalacqua, E. et al. 1999. An empirical equation for polymer electrolyte fuel cell (PEFC) behaviour. *Journal of Applied Electrochemistry* 29: 1449–1455.

Srinivasan, S., Ticianelli, E. A., Derouin, C. R. et al. 1988. Advances in solid polymer electrolyte fuel cell technology with low platinum loading electrodes. *Journal of Power Sources* 22: 359–375.

Su, A., Ferng, Y. M., and Shih, J. C. 2010. CFD investigating the effects of different operating conditions on the performance and the characteristics of a HT-PEMFC. *Energy* 35: 16–27.

Weng, F., Su, A., Jung, G. et al. 2005. Numerical prediction of concentration and current distributions in PEMFC. *Journal of Power Sources* 145: 546–554.

Williams, M. V., Kunz, H. R., and Fenton, J. M. 2005. Analysis of polarization curves to evaluate polarization sources in hydrogen-air PEM fuel cells. *Journal of Electrochemical Society* 152: A635–A644.

Wu, J., Yuan, X., Martin, J. J. et al. 2010. Proton exchange membrane fuel cell degradation under close to open-circuit conditions Part I *In situ* diagnosis. *Journal of Power Sources* 195: 1171–1176.

3

Electrochemical Impedance Spectroscopy

Norbert Wagner
Institute for Technical
Thermodynamics

3.1 Introduction

The need for an efficient, nonpolluting power source for vehicles in urban environments, emphasized by legislative initiatives, has resulted in increased attention to the option of fuel-cell-powered vehicles of high efficiency and low emissions. Recently, fuel cells are also used in a variety of new applications like portable devices (e.g., radio communications, mobiles, laptop, etc.), residential applications (combined heat and power generation), and other transport applications (e.g., submarines, ships, rail-guided vehicles, etc.).

Fuel cells can continuously convert the chemical energy of a fuel (hydrogen, methanol, methane, etc.) and an oxidant into electrical energy at up to 83% efficiency with very low pollutant emissions. Depending on the type of electrolyte used in a fuel cell, one can distinguish the following six main types:

1. Alkaline fuel cell (AFC)
2. Proton exchange membrane fuel cell (PEMFC)—also called polymer electrolyte fuel cell (PEFC) or solid polymer fuel cell (SPFC)
3. Direct methanol fuel cell (DMFC)
4. Phosphoric acid fuel cell (PAFC)
5. Molten carbonate fuel cell (MCFC)
6. Solid oxide fuel cell (SOFC)

The first four types of fuel cells operate at temperatures below 200°C, MCFC and SOFC operate at temperatures higher than 650°C.

In order to increase the reaction area (triple phase boundary) electrodes used in fuel cells are porous electrodes with highly dispersed catalysts grains (agglomerates) linked to an ionic conductor (electrolyte). The key technical problems common to all types of fuel cell are the optimal design, stability, and durability of such porous electrodes. Other R&D challenges are the investigation of the reaction mechanism and kinetic at each electrode/electrolyte interface, determination of degradation (poisoning) mechanism, production of cheap and efficient electrodes with low catalyst loading, development of suitable catalysts, long-term stability, and so on. In order to solve these problems, a better understanding of the electrochemical reactions and mass transport in the fuel cell is essential. Moreover, quality control and understanding of degradation require new nondestructive methods and a better understanding of experimental results based on modeling and simulation. However, EIS is such an *in situ* investigation method that is increasingly applied in fuel cell research and development.

3.2 Principle of EIS

Impedance analysis is a very popular, nondestructive measurement technique that provides detailed diagnostic information about a wide range of electrochemical phenomena including charge transfer reaction at the interface electrode/electrolyte, reaction mechanisms, state of charge of batteries, electrode material properties, and state of health of fuel cells, and so on.

The technique involves applying a low-level alternative current (AC) waveform to the electrochemical system (half cell, single cell, stack, etc.) under investigation and measuring the response of the cell to this stimulus (the AC voltage across the cell and the AC current through the cell). The impedance of the cell is obtained by taking the ratio of AC voltage/AC current (Figure 3.1). Since a low-level AC voltage or AC current excitation signal (stimulus) is used, no damage or change is suffered by the cell. Typically, a swept frequency sinewave is used as the stimulus so that the impedance can be evaluated across the frequency range of interest, in most cases from 1 mHz to 100 kHz. The AC waveform is usually applied to the cell via a potentiostat that provides signal conditioning, amplification and DC/AC level control so that the programmed voltage or current is correctly applied. Connections are made from the cell to the potentiostat so that the potential across the cell and the current through the cell may be monitored. The potentiostat provides buffering of the voltage input signals (reference inputs) so that voltage measurements may be performed with minimal disturbance to the cell. In addition, the potentiostat provides current to voltage conversion circuitry to convert the AC current passing through the cell into an AC voltage waveform that can be measured by a frequency response analyzer (FRA) to

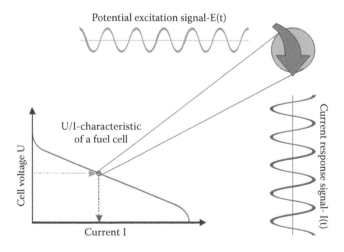

FIGURE 3.1 Schematic representation of EIS applied to fuel cell characterization.

provide impedance analysis. EIS, initially applied by Graham (1947) and later by Delahay (1965) to study the double layer capacitance and also applied in AC polarography (Schmidt and Stackelberg, 1963) to separate the double-layer charging current and Faradaic current from the total current, is now applied to characterize electrode processes and complex interfaces. The analysis of the system response yields information about the reactivity and structure of the interface, also about the electrochemical reactions and mass transport limitations taking place there. Its importance recently increased due to the modern computer-controlled devices and the corresponding analytical software. Accordingly, the fields of application and therefore the number of publications and recommendable textbooks (Lasia, 1999; Bard and Faulkner, 2001; Wagner, 2005, 2009; Orazem and Tribollet, 2008; Yuan et al., 2010) increased in the last 10 years.

First impedance measurements and interpretation of EIS performed to characterize PEMFC were reported in Srinivasan et al. (1988), Fletcher (1992), Wilson et al. (1993), and Poltarzewski et al. (1992).

3.2.1 Impedance Elements and Equivalent Circuits Used for PEMFC Characterization

During oxidation of the fuel at the anode and reduction of the oxygen at the cathode different chemical and electrochemical reactions can take place in front or on the electrode surface. In addition, concentration gradient or mass transport hindrance has an impact on the fuel cell performance. A schematic representation of the different steps and their location during the electrochemical reactions as a function of distance from the electrode surface is given in Figure 3.2.

Fortunately, these different reaction steps proceed with different time constants ranging from microsecond to month, as schematically shown in Figure 3.3 and described by Wagner and Friedrich (2009).

After transformation of the time constants into the frequency domain the different electrochemical processes can be distinguished from each other. In the high-frequency range of measured EIS at the PEMFC electrolyte (R_Ω) and contact resistances can be determined. At lower frequencies mainly the double-layer charging and charge transfer reactions can be determined and in the lowest frequency range slow processes such as diffusion processes are observed. Given by the different reaction rates of the involved charge transfer reactions hydrogen oxidation and oxygen reduction, as well as the time

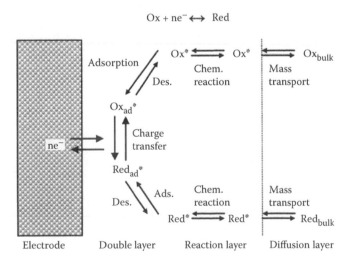

FIGURE 3.2 Schematic representation of the different steps and their location during the electrochemical reactions as a function of distance from the electrode surface.

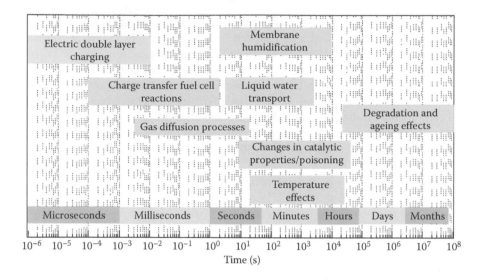

FIGURE 3.3 Overview of the wide range of dynamic processes in fuel cells.

constant of these reactions, will be different and in most cases are distinguishable from each other as schematically shown in Figure 3.4.

The most common representations of electrochemical impedance spectra are the Bode plots, where the logarithm of impedance magnitude $|Z|$ and the phase-shift (α) are plotted versus the logarithm of the frequency (f) and Nyquist plots, where the imaginary part of the impedance is plotted versus the real part of the impedance. Both plots are useful. The Bode plot is used when the impedance magnitude covers a large range of sizes. From the shape of the Nyquist plot one can

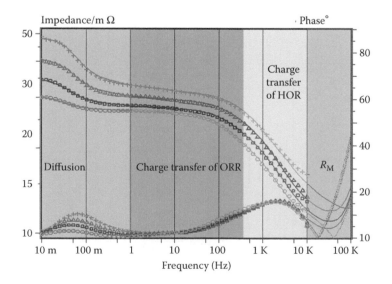

FIGURE 3.4 Bode plot of EIS measured at different current densities, PEMFC operated at 80°C with H_2 and O_2 at 2 bar.

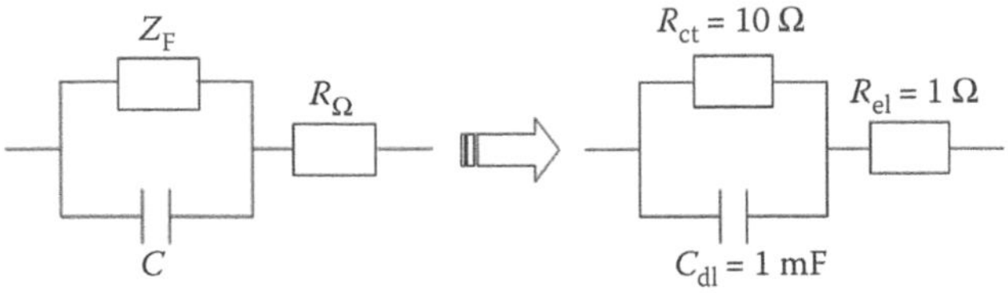

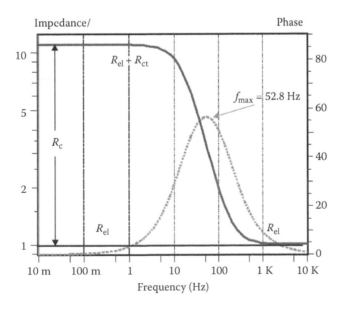

FIGURE 3.5 Common equivalent circuit describing the frequency response of a fuel cell (left) and simplest case of an equivalent circuit with charge transfer resistance parallel to the double-layer capacity.

distinguish between different diffusion processes and evaluate the deviation of the double-layer capacity from ideal behavior, that is, constant phase element behavior. The electrode/electrolyte interface can be represented by an equivalent circuit which contains various impedance elements representing the involved reaction steps. These elements are generally represented as ohmic, capacitive or inductive components with particular dependencies of their complex impedance upon the frequency of the AC signal. The particular linking of these impedance elements is based on the relationship between the processes represented by these elements. Subsequently occurring steps are represented by a series connection of the elements while steps occurring simultaneously are represented by a connection in parallel. In the case of fuel cells with conversion of chemical energy into electrical energy generally an equivalent consists of the so-called Faradaic impedance Z_F in parallel to a capacitive element C (Figure 3.5).

The simplest electrode/electrolyte interface is the case when the Faradaic impedance can be described by only a charge transfer resistance R_{ct} and the capacitive element is the double-layer capacity C_{dl}. The impedance spectrum of such an equivalent circuit is represented in Figure 3.6 as Bode plot and Figure 3.7 as Nyquist plot.

FIGURE 3.6 Bode plot of the impedance spectra simulated in the frequency range from 10 mHz to 10 kHz with the equivalent circuit from Figure 3.5.

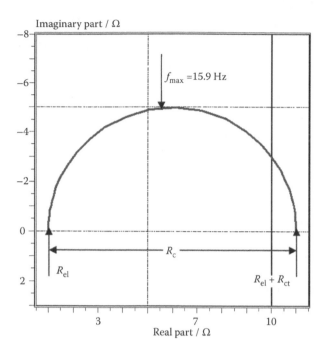

FIGURE 3.7 Nyquist plot of the impedance spectra simulated in the frequency range from 10 mHz to 10 kHz with the equivalent circuit from Figure 3.5.

From Figure 3.6 one can estimate the electrolyte resistance and charge transfer resistance. The value of the double-layer capacity can be calculated with the simple equation given below:

$$2\pi f_{max} = \left(\frac{1}{R_{ct}C_{dl}}\right)\left(\frac{1+R_{ct}}{R_{el}}\right)^{1/2}$$

(3.1)

In a similar way one can estimate from the Nyquist plot the electrolyte resistance, the charge transfer resistance is given by the diameter of the semicircle and from f_{max} the double-layer capacity C_{dl} can be calculated with another simple equation:

$$2\pi f_{max} = \omega_{max} = \frac{1}{C_{dl}R_{ct}}$$

(3.2)

For more complicated processes when the Faradaic impedance consisting of more than only the charge transfer resistance the evaluation of measured EIS is more complicated and is performed via a fit procedure that is usually implemented in the software of commercial available electrochemical workstations. Other processes during fuel cell reactions can be diffusion and adsorption processes. These particular diffusion processes can be described by different impedance elements like Warburg impedance, Nernst impedance, spherical diffusion impedance, and so on. For details see one of the textbooks.

The Faradaic impedance can be described as connections of different impedance elements (Göhr and Schiller, 1986), each of which is associated with a single process. Such an impedance element is the relaxation impedance, describing the surface relaxation of the interface and explains the development of the pseudo-inductive behavior in the low-frequency range (frequency <3 Hz) in the impedance spectra of the fuel cell. This behavior was first found by Müller et al. (1999) during poisoning the anode of a PEMFC

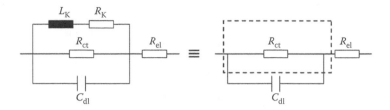

FIGURE 3.8 Equivalent circuit with relaxation impedance (Z_k).

with a mixture of H_2/CO and will be discussed in detail in Section 3.3.2. The surface relaxation impedance represents a Faradaic impedance (Z_F) at nonequilibrium potential with a potential-dependent transfer reaction rate: $k = k(\varepsilon)$ and its time-dependent relaxation according to Equations 3.3 and 3.4:

$$Z_F = \frac{R_{ct}}{1 + R_{ct}/Z_K} \tag{3.3}$$

$$Z_K = \frac{1 + j\omega\tau_K}{I_F \cdot d\ln k/d\varepsilon} \tag{3.4}$$

In Equation 3.3 R_{ct} denotes the charge transfer resistance, Z_K is defined as the relaxation impedance and is schematically shown as series combination of R_k and L_k as well as the box surrounding R_{ct} in Figure 3.8. According to Equation 3.4, where I_F denotes the Faraday current, τ_K the time constant of relaxation and the expression of $d\ln k/d\varepsilon$ is the first derivative of the logarithm of the reciprocal relaxation time constant ($k = 1/\tau_K$) against the potential ε. According to its frequency dependence, Z_K can be split up into the relaxation resistance R_K and the relaxation inductivity X_K, with the pseudo-inductance $L_K = \tau_K \cdot R_K$, which is proportional to the relaxation time constant τ_K.

$$R_K = \frac{1}{I_F \cdot d\ln k/d\varepsilon} \tag{3.5}$$

$$X_K = j\omega\tau_K \cdot R_K = j\omega \cdot L_K \tag{3.6}$$

3.2.2 Modeling and Theory of Porous Electrodes

Porous electrodes are used in electrocatalysis. In catalysis, generally, there is a great advantage in increasing the real surface area, especially the electrochemical active area of electrodes. First, investigations of porous electrodes with EIS were applied by De Levie (1963, 1964, and 1967), presenting the transmission line model, a model describing the pores of a porous electrode as essentially circular cylindrical channels of uniform diameter of semi-infinite length (Figure 3.9).

On the basis of AC impedance measurements, Göhr et al. (1983) and Göhr (1997) investigated porous lead electrodes in sulfuric acid, and proposed another model of the porous electrode represented

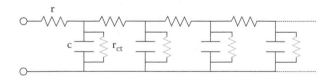

FIGURE 3.9 Transmission line model proposed by de Levie.

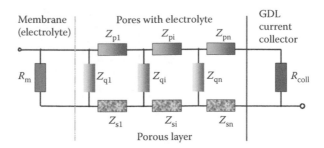

FIGURE 3.10 Cylindrical homogeneous porous electrode model. (Adapted from Göhr, H. 1997. *Electrochemical Applications* 1: 2–3. Available from http://www.zahner.de)

schematically in Figure 3.10. In this model the cylindrical pore was considered as a transmission line made of a large number of infinitesimally thin sections, with impedance elements of pore's ground surface (Z_n), pores electrolyte (Z_p), pore's wall surface (Z_q), porous layer (Z_s), and surface layer (Z_o).

$$R_{electrode} = \frac{(R_{por} \cdot R_{ct})^{1/2}}{\tanh\left\{\left(R_{por}/R_{ct}\right)^{1/2}\right\}} \tag{3.7}$$

In the case of Göhrs's model (Figure 3.10) a simple equation (Equation 3.7) for the resistance $R_{electrode}$ of the porous electrode can be derived if only the pores electrolyte resistance ($Z_p = R_{por}$) and from the pore's wall surface (Z_q) the charge transfer resistance (R_{ct}) are considered.

A review, describing gas diffusion electrodes and flooded electrodes is given in Szpak (1991). The great number of very interesting publications, regarding AC impedance measurements on porous electrodes and model electrodes known in the literature, shows the importance and wide applications of the method, from porous layers to fuel cell electrodes. The existing models can be classified into different groups:

- Simple cylindrical pore model (Levie, 1963, 1964, 1967; Göhr et al., 1983; Göhr, 1997)
- Thin-film model (Will, 1963)
- Surface migration model (Winsel, 1962; Mund, 1975)
- Biporosity model (Grens, 1970; Giordano et al., 1991)
- Flooded agglomerates (Giner and Hunter, 1969; Jaouen et al., 2002)
- Thin film agglomerate model (Raistrick, 1990)
- Cantor block model (Sarangapani et al., 1996; Kötz and Carlen, 2000; Eikerling et al., 2005)
- Triple pore structure (Itagaki et al., 2010)

3.3 Applications of EIS

3.3.1 Steady-State Applications

The most used method to characterize the performance of a fuel cell consists of measuring a steady-state current/potential curve. From such a graph, one obtains information about the entire fuel cell comprising the sum of the electrochemical behavior of the electrode/electrolyte interfaces, conductivity of the electrolyte, the influences of the gas supply and the electrical contacts between the individual components.

In order to determine current/potential curves, one either scans the whole potential range continuously, beginning from the open-circuit potential (OCP) to full load, with a constant voltage scan rate

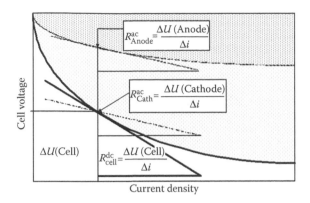

FIGURE 3.11 Schematic representation of the correlation between DC resistances (anode, cathode, and cell) of fuel cell and measured current/voltage curve.

lying between 0.1 and 10 mV s^{-1} and records the resulting current, or one gradually alters the potential and the steady-state current is recorded after reaching a steady state or waiting for a definite time period. Another method to characterize the performance of fuel cells is EIS. In this chapter different applications of the EIS for the characterization of fuel cells will be given.

Conversion of Resistance to Performance Losses of the Fuel Cell during Operation

With measurements over the entire performance range of operation one tries to separate the single contributions to the performance loss of a fuel cell during load or even at OCP.

The correlation between impedance measurements and current/potential curve was first described by Wagner (2002) and is given schematically in Figure 3.11.

The DC resistance of the cell measured at U_n corresponds to the tangent to the current/potential curve at that potential. The DC resistance of the cell (R_{Cell}) is the impedance at frequencies near 0 Hz where only ohmic parts attract attention. To obtain the DC resistance of the cell, one have to extrapolate the measured data or the simulated impedance (model) at very low frequency (e.g., 1 nHz) or summing up the individual resistances, obtained after fitting the measured spectra with an equivalent circuit.

Assuming that the current/potential curve can be expressed by an Equation 3.8 of second order and the resistance is defined by Equation 3.9, then, the parameters a_n, b_n, and c_n from Equation 3.4 are given by Equations 3.10 through 3.12.

$$U_n = a_n I_n^2 + b_n I_n + c_n \tag{3.8}$$

$$R_n = \frac{\partial U}{\partial I}\Big|_n \tag{3.9}$$

with

$$a_n = \frac{R_{n+1} - R_n}{2(I_{n+1} - I_n)} \tag{3.10}$$

$$b_n = R_{n+1} - 2a_n I_{n+1} \tag{3.11}$$

$$c_n = U_{n-1} - a_n I_{n-1}^2 - b_n I_{n-1} \tag{3.12}$$

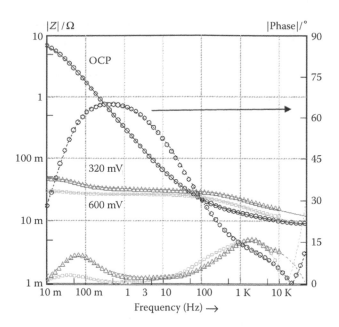

FIGURE 3.12 Bode diagram of the measured impedance spectra, PEMFC at 80°C, at different cell voltages: (○) OCV (1024 mV), (□) 600 mV, and (Δ) 320 mV.

Over the last 10–15 years a lot of key barriers for the development of PEMFC for terrestrial applications have been successfully overcome. However, there is still need for further research and development effort, an important goal is to increase the efficiency and to produce efficient and cheap electrodes, and membrane-electrode assemblies (MEA) with low noble metal catalyst loading, high CO tolerance, and long lifetime. Actually, the R&D efforts are focused on the development of nanodispersed catalysts, nonnoble metal catalyst, and production techniques, which allow production of MEAs with low noble metal catalyst loading and are suitable for mass production, development of high-temperature membranes to enhance the electrode kinetics and to lower the CO influence (up to 160°C), and integration of the MEA into a fuel cell system with simplified water management, gas conditioner, bipolar plate, fuel storage systems, and so on.

To investigate the physical and electrochemical origins of the performance loss in PEMFCs, operated at different conditions like high-current densities, fuel composition (neat H_2, H_2 + 100 ppm CO, H_2O), flow rates, temperatures, air or pure oxygen, and so on, electrochemical impedance studies on different PEMFC systems with different electrodes and membranes were performed (Wagner, 2005).

Applying a classical three electrode cell with one reference electrode is extremely difficult for the investigation of electrochemical systems with solid electrolytes. However, since the anode and cathode transfer functions at OCP can be determined independently without a reference electrode using a symmetrical gas supply of hydrogen or oxygen at the two electrodes of the cell, cathode and anode impedance at OCP can be determined directly with two independent experiments (Wagner, 2002).

By varying some experimental conditions such as current load (Figures 3.12 and 3.13), temperature, gas composition, hydrogen humidification, and membrane thickness (Andreaus et al., 2002), electrode composition (Wagner et al., 2008) measured cell impedance can be split up into anode impedance, cathode impedance, and electrolyte resistance without using reference electrodes. These results were used to derive appropriate equivalent circuits for the analysis of impedance spectra measured on fuel cells operated with H_2/O_2, H_2/air, and H_2 + 100 ppm CO/O_2. The variation of the experimental conditions is also a useful method to confirm the accuracy of the equivalent circuit.

To identify and separate the different diffusion processes, it is useful to represent the measured impedance spectra in both the Bode and Nyquist diagrams (Figure 3.13). In the Nyquist diagram one

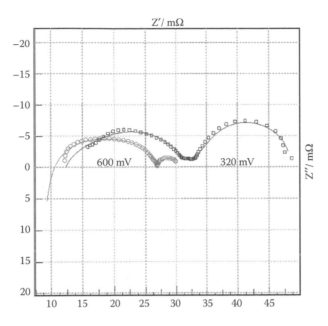

FIGURE 3.13 Bode diagram of the measured impedance spectra, PEMFC at 80°C, at different cell voltages: (○) 600 mV and (□) 320 mV.

can observe the finite diffusion as an additional loop at the lowest part of the frequency range and the infinite diffusion as a straight line with a slope of 1 (real part = imaginary part). In the Bode diagram (Figure 3.12) the difference between the two kinds of diffusion cannot be seen so clearly due to the logarithmic scale. In general, the Bode plot provides a clearer description of the electrochemical system's frequency-dependent behavior than the Nyquist plot, in which frequency values are implicit.

Starting with EIS measured with symmetrical gas supply, an equivalent circuit for the complete PEMFC can be applied for the simulation of the measured impedance spectra of the PEMFC. Each electrode/electrolyte interface can be represented by an equivalent circuit presented in Figure 3.5. In addition to the charge transfer reaction at higher current densities a diffusion process can be seen. For

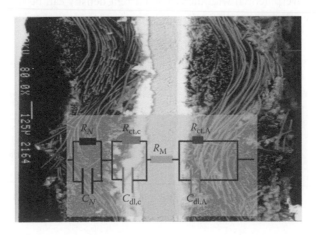

FIGURE 3.14 Equivalent circuit (EC) used for the simulation of PEMFC inserted into SEM picture of a PEMFC cross section.

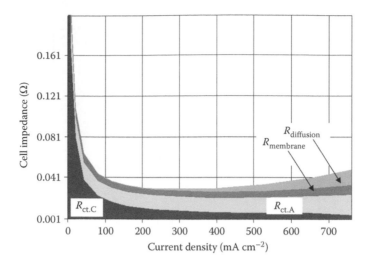

FIGURE 3.15 Current density dependency of the resistances after evaluation of the impedance spectra evaluated with EC from Figure 3.14.

the single fuel cell an equivalent circuit shown in Figure 3.14 was used. Besides a series of resistance (electrolyte or membrane resistance R_{el}), the equivalent circuit contains three time constants of parallel R/C. In the simulation the capacitance (C) was replaced by constant phase element (CPE) due to the porous structure of the electrodes. The cathode can be described using a time constant for the charge transfer through the double layer ($R_{ct(C)}$/CPE$_{dl(C)}$, the exponent of the CPE is around 0.85, for an exponent of 1 the CPE is equal with the capacitance), a time constant for the finite diffusion of water with a Nernst-impedance-like behavior ($R_{(N)}$/CPE$_{(N)}$, the exponent of the CPE is around 0.95), and finally the time constant of the anode ($R_{ct(A)}$/CPE$_{dl(A)}$, the exponent of the CPE is around 0.80).

Applying Equations 3.8 through 3.12 and using the current density dependency of the resistances (R_N, $R_{ct,A}$, $R_{(el)}$, and $R_{ct,C}$) represented in Figure 3.15, gained from the simulation with the equivalent circuit from Figure 3.14 the individual performance losses (overpotentials) can be determined (Figure 3.16).

At low-current densities the cell overpotential is given mainly by the cathodic overpotential. At higher current densities ($i > 400$ mA cm^{-2}) an additional diffusion overpotential becomes noticeable. The increase in the anodic overpotential with increasing current density can be explained by assuming a partially dry out of the interface membrane/anode with simultaneous increasing of the water content

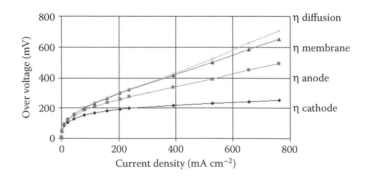

FIGURE 3.16 Individual performance losses of the PEMFC at 80°C in function of current density, calculated by integration of the individual resistances at different current densities.

depending electrolyte resistance inside the pore ($Z_p \equiv R_{pore}$)—applicable only in the case of impregnated electrodes or electrodes with electrolyte powder inside (see Section 3.3.1.1) the electrode and can be minimized by appropriate water management. Therefore, a further refinement of the equivalent circuit is useful and can be made if one uses the porous electrode model.

3.3.1.1 Influence of MEA Composition

In order to optimize the electrode composition and performance of PEMFC and to reduce the production cost of MEAs, different MEAs using different catalyst powders, carbon-supported and -unsupported catalysts with different proton conducting solid electrolyte powder (Nafion) contents were produced by using a dry powder spraying technique (Gülzow et al., 1999, 2000). The electrochemical characterization was performed by recording current–voltage curves and electrochemical impedance spectra in a galvanostatic mode of operation at 500 mA cm^{-2}. The evaluation of the measured impedance spectra (Figure 3.17) using the porous electrode model (Figure 3.10) for each electrode shows that the cathode of the fuel cell is very sensitive to the electrode composition whereas the contribution of the anode is very small and invariant to the electrode composition. Furthermore, it could be shown that using electrolyte powder in the electrodes the charge transfer of the cathode decreasing monotonically with increasing electrolyte content in the cathode. These findings suggest that with increasing electrolyte content in the electrodes, in particular in the cathode, the utilization degree of the catalyst increases linearly with increasing electrolyte content in the electrode (Wagner et al., 2008). Taking into account the change of the catalyst loading with electrode composition and normalizing the charge transfer resistance per unit weight catalyst content ($R_{cath,norm.}$), one can find a very distinct correlation between cell performance and electrode composition. This correlation is shown in Figure 3.18. If one takes into account only the results from MEAs with Nafion one can find even a linear correlation of the cathode

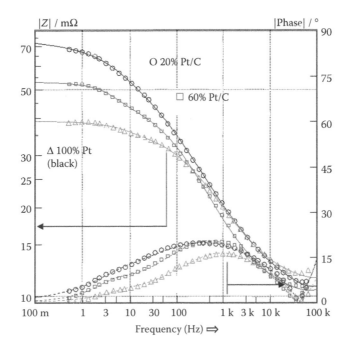

FIGURE 3.17 Impedance spectra (Bode plot) of a MEA produced with different carbon-supported Pt catalyst (20% Pt/C, 60% Pt/C, and Pt black 100%), without electrolyte in the electrode, operated with H$_2$/air at 500 mA cm^{-2}.

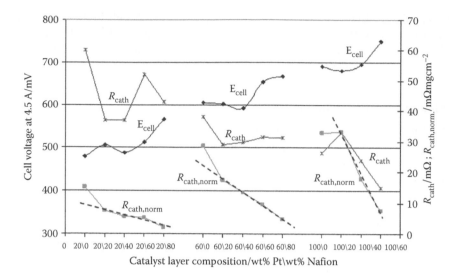

FIGURE 3.18 Dependency of the cathodic charge transfer resistance R_{cath} and normalized cathodic charge transfer resistance $R_{cath,norm.}$ from the catalyst layer composition after evaluation of the EIS measured at different MEAs, operated with H_2/air at 500 mA cm^{-2}.

charge transfer resistance with Nafion content in the electrode (Figure 3.19). These results show for the first time such an explicit behavior and correlation of cell performance and electrode composition.

3.3.1.2 Influence of Cathodic Gas Composition

One of the major sources of performance loss during operation of fuel cells is the slow kinetics of the oxygen reduction reaction (ORR). Therefore, many investigations were performed to understand the

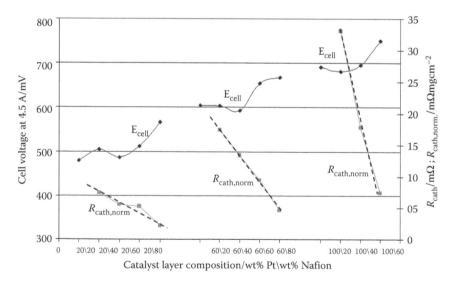

FIGURE 3.19 Cathodic charge transfer resistance R_{cath} and normalized cathodic charge transfer resistance $R_{cath,norm.}$ in function of the catalyst layer composition after evaluation of the EIS measured at different MEAs with Nafion, operated with H_2/air at 500 mA cm^{-2}.

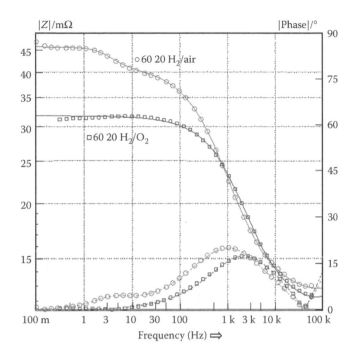

FIGURE 3.20 Comparison of EIS measurements (Bode plot) of the same MEA (60 20) produced with carbon-supported (60 wt% Pt) catalyst and 20 wt% Nafion, operated with H_2/O_2 ($\square$) and with H_2/air ($\bigcirc$) at 500 mA cm^{-2}.

reaction mechanism of the ORR. During operation of the cell with air instead of pure oxygen one can find two sources of performance loss at the cathode:

- Decreasing of charge transfer reaction rate with decreasing partial pressure of oxygen corresponding to an increase in the charge transfer resistance ($R_{ct,C}$)
- Appearance of an additional diffusion impedance term in the low-frequency range of the impedance spectra, for example, at 3 Hz in Figure 3.20

For different fuel cell applications like mobile applications the use of air instead of pure oxygen should be reasonable. For space applications or high-power density applications pure oxygen should be used.

In addition, by changing the cathodic gas composition during galvanostatic mode of fuel cell operation one can distinguish and validate the different contributions to the overall cell impedance.

In Figures 3.21 and 3.22, two EIS spectra measured at the same current are shown. The used MEA was N111 IP from Ion Power Inc., operated with 2.7-fold (λ) oxygen and air, respectively. Given by the high gas flow rate in the case of operation with air, one can recognize in the high-frequency range of the EIS an increase of the membrane resistance determined by an improper humidification of the cathode. Under such operation conditions the cell performance decreases and one would expect a limited durability of the cell.

3.3.2 Time-Resolved Electrochemical Impedance Spectroscopy

A prerequisite for the development and the improvement of fuel cells is the knowledge of the mechanistic processes that take place during operation. The understanding of the kinetic behavior of the fuel cells requires the variation of different experimental parameters. Often, the variation of distinct parameters causes situations where steady-state conditions are no longer fulfilled. In practice, EIS analysis often suffers from the fact that the steady-state condition is violated due to time instability of the examined

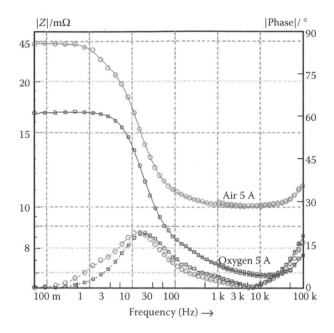

FIGURE 3.21   Bode representation of EIS measured at 80°C, 5 A operated with H_2/O_2 (□) and with H_2/air (○).

systems. While an EIS measurement is running, the examined system should not change its dynamic behavior. Unfortunately, the violation of steady-state conditions complicates the evaluation of experimentally obtained impedance spectra because all relevant physical models for the interpretation of the data are based on steady-state conditions. The time- and frequency-dependent relationship between current and voltage of an electrochemical system is often called "two-pole impedance transfer function" (TTF). All properties, which are influencing the current–voltage dependence, must be stable in this time interval. Otherwise, the TTF will be falsified in a way that is equivalent to a violation of causality.

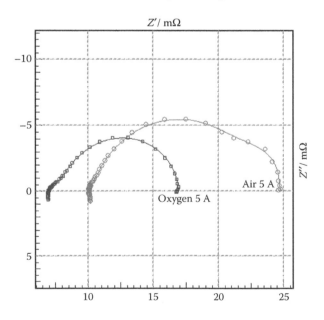

FIGURE 3.22   Nyquist representation of EIS measured at 80°C, 5 A operated with H_2/O_2 (□) and with H_2/air (○).

Nevertheless, it is possible to investigate such "drifting" systems and to obtain relevant data for the development of fuel cells using EIS. Enhanced numerical procedures like real-time-drift compensation, time-course interpolation and causal transfer functions are required to compensate or to eliminate the drift effects of systems with states that change with time.

Real-time-drift compensation: During recording an impedance spectrum a potential drift due to change of the state of the system can occur. The magnitude of drift is frequency dependent and should be compensated by the software during the measurement.

Time-course interpolation: Recording an impedance spectrum one frequency after another requires a finite time, while the measurement at high frequencies requires less time than the measurement at low frequencies. As a consequence, the system at the start is in a different state from at the end if the system changes the state during the impedance measurement.

It is impossible to eliminate the drift that is caused by the finite measuring time, performing only a single spectrum. Especially the data recorded at low frequencies are affected and a mathematical procedure has to be applied to check the data with respect to causality and linearity (Z-HIT) to avoid erroneous interpretations resulting from a fit of the drift-affected data.

According to the idea of Savova-Stoynov (Savova-Stoynov and Stoynov, 1985, 1992; Stoynov, 1990), recording a series of impedance measurements at distinct time intervals offers the possibility to eliminate the drift and therefore to reconstruct an impedance spectrum which is acquired in an "infinite" short time. Additionally, the elapsed time of the experiment is involved as a third parameter. As mentioned above, the measurement at lower frequencies requires a longer time for the registration and therefore, the measured curve is shifted to the back along the time axes. It should be noted, that the absolute value of the shift is frequency dependent because the acquisition time for each measured frequency is different. For the lowest frequency, a single impedance spectrum can be reconstructed by interpolating the impedance value from the time course of the series using an appropriate smoothing function at the time of the start of this distinct measurement. The interpolation procedure is repeated for each measured frequency resulting in a data set where the effect of finite measurement time is significantly reduced or even eliminated for each recorded spectrum of the series.

Causal transfer functions: Causality in the meaning of system theory forces couplings between the real and imaginary part, which are known as Kramers–Kronig relations (KKT) or Hilbert relations (HT).

These relations offer the possibility to examine measured transfer functions (impedance spectra) on errors caused by time instability or time drift. KK-checking techniques have fundamental problems in the application in practical measurements. Therefore, many attempts have been made to overcome these limitations by means of different interpolation procedures. An attempt is the Z-HIT (Agarwal et al., 1995; Ehm et al., 2000; Schiller et al., 2001) approximation, an approximation formula for the calculation of the impedance modulus course from the phase angle by integration.

During operation of fuel cells different time-dependent effects can occur, depending on the type of fuel cell and experimental conditions such as gas flow rates (flooding of the electrode, changing of the gas composition), temperature (corrosion or sintering), impurities in the hydrogen (CO) leading to poisoning of the anode in the case of PEMFCs, and degradation of the electrodes during long time of operation, and so on.

One has to assume that the system changes its state not only between two measurements but also during the recording of a single spectrum. The latter fact causes problems for the evaluation of the spectrum, because the recording of an impedance spectrum one frequency after another requires a finite time, while the measurement at high frequencies requires less time than the measurement at low frequencies. Due to the fact that the recording of a single spectrum in the frequency range, for example, from 10 kHz to 50 mHz requires about 20 min, the influence of the changed state on the measured spectrum is not negligible. For this reason, one of the fundamental prerequisites for the evaluation of impedance measurements is violated. Nevertheless, it is possible, to reconstruct "quasi steady state" (and therefore "quasi causal") spectra from drift affected impedance data using *improved evaluation techniques* which are denoted as the real-time-drift compensation, the time-course interpolation, and the

Z-HIT refinement. These techniques were applied successfully to the interpretation of time-dependent impedance spectra of a fuel cell which exhibits nonsteady-state behavior.

Applying all three techniques to time-drifting systems, quasi steady-state impedance spectra, TREIS, at defined times can be obtained.

3.3.2.1 EIS during CO Poisoning of PEMFC Pt and Pt/Ru Anodes

Owing to the high-energy conversion rate and the harmless emission products, PEMFCs receive more and more attention especially in the case for powering electric vehicles. The highest performance is achieved with hydrogen (H_2) which is the preferred fuel for low-temperature fuel cells and pure oxygen. However, H_2 has several limitations. The storage systems for liquid or compressed H_2 are heavy and bulky. Furthermore, H_2 refuelling is costly and takes time. An additional obstacle is actually the lack of an infrastructure to distribute H_2 to the consumer (Cleghorn et al., 1997).

An alternative to the use of H_2 as fuel are methanol or hydrocarbons (e.g., natural gas, biogas) that can be transformed to hydrogen on board of the electric vehicle by a reformation reaction (Figure 3.23). This allows using the H_2-PEMFC cell which has a higher level of development. The reformate feed gas may contain up to 2.5% carbon monoxide (CO) by volume, which can be reduced to about 50 ppm CO using a selective oxidizer (Wilkinson and Thompsett, 1997).

The performance of platinum which is known as one of the most effective catalysts for the hydrogen oxidation in PEMFCs is influenced even by traces of CO. Compared with the use of pure hydrogen, the maximum power density is more than halved in the presence of only 5 ppm CO. One possible explanation for the decrease of the fuel cell performance is that the CO blocks or limits the active sites of the platinum catalyst due to adsorption which leads to an inhibition of the hydrogen oxidation reaction. In the last two decades, intensive work has been devoted to find electrocatalysts that are tolerant to CO in

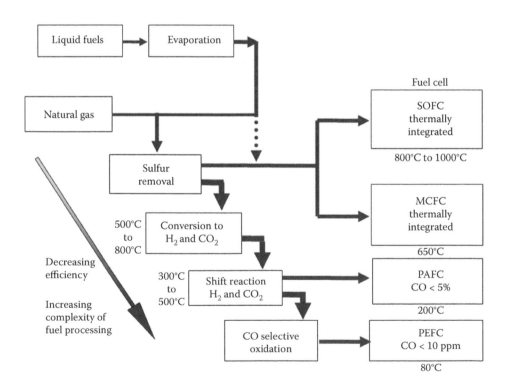

FIGURE 3.23 Schematic representation of fuel processing.

hydrogen at operating temperatures below 100°C. This system has been the object to numerous studies using electrochemical methods like potentiostatic measurement, potentiodynamic measurement, and stripping voltammetry. Detailed kinetic studies were given in Vogel et al. (1975), Schmidt et al. (1998), and Koper et al. (2001).

First *in situ* impedance measurements were reported by Müller et al. (1999), who demonstrated the time dependence of impedance spectra in galvanostatic conditions for CO poisoning Pt anodes. Other publications, with detailed kinetic data were published later (Ciureanu and Wang, 1999, 2000; Kim et al., 2001; Wang et al., 2001; Leng et al., 2002; Wagner and Gülzow, 2004).

For the development of improved electrocatalysts that are less sensitive with respect to the presence of CO, a mechanistically understanding of the poisoning process of the anode is desirable. The progressive poisoning with CO of a fuel cell can be monitored using TREIS.

As shown in detail (Wagner and Schulze, 2003), the CO poisoning of the Pt- and PtRu-anode during galvanostatic mode of operation with H_2 + 100 ppm CO causes a change in the state of the fuel cell that is reflected by the decrease in the cell voltage during operation (Figure 3.24) and by time dependency on the recorded impedance spectra. Besides, an increase of the total impedance of the fuel cell, in the case of galvanostatic mode of operation, the occurrence and the increase of a pseudo-inductive behavior at frequencies lower than 3 Hz is observed. It is useful to operate the fuel cell in the galvanostatic mode of operation because at a constant current density, the impedance of the cathode and the membrane resistance can be assumed to be constant during the course of impedance measurements and the changes in the impedance spectra during poisoning the anode with CO can be attributed exclusively to the impedance of the anode.

Representative time-dependent impedance spectra of the series, that is, time-resolved impedance spectra are depicted in Figure 3.25 for the fuel cell with Pt anode as Nyquist plots whereby the experimental data are represented by dots and the solid lines in the Figure represents the modeled curves after fitting the experimental data with the equivalent circuit from Figure 3.26. For the PtRu, similar time-dependent impedance spectra were measured.

In series to both half cells, the resistance of the membrane itself—denoted as the electrolyte resistance (R_{el})—as well as a parasitic wiring inductance (L_w) due to the mutual induction effect has to be taken into account. The impedance of the cathodic half cell (oxygen reduction) is approximated using a charge transfer resistance ($R_{ct,C}$) in parallel to a constant phase element (CPE$_C$). This simple equivalent circuit describes the partial impedance of the cathodic half-cell with sufficient accuracy. In contrast, the impedance of the anode (hydrogen oxidation) is more complicated due to the CO poisoning and appearance

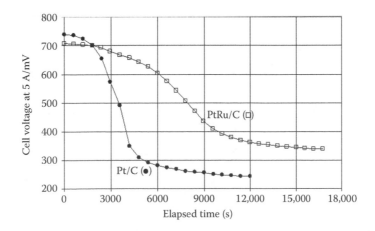

FIGURE 3.24 Evolution of cell voltage during CO poisoning of Pt/C (●) and PtRu/C (□) anodes in galvanostatic mode of fuel cell operation at 217 mA cm^{-2} (5 A), 80°C.

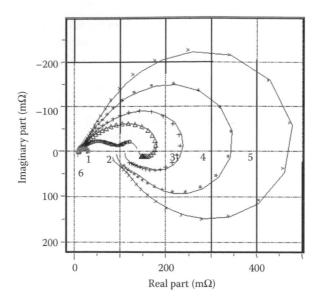

FIGURE 3.25 Nyquist plot of EIS measured at: (1) 0 s, (2) 3601 s, (3) 5402 s, (4) 7204 s, (5) 9605 s, and (6) 11,404 s during CO poisoning of Pt/C anode during galvanostatic mode of operation with H_2 + 100 ppm CO (anode), at 5 A (cell surface = 23 cm²), 80°C, and oxygen (cathode).

of an inductive loop in the low-frequency range of the impedance spectra. The impedance of the anode is modeled using a porous electrode (PE) in series to a double-layer capacity ($C_{dl,A}$) which is in parallel to the Faradaic impedance (Z_F) that contains a surface relaxation impedance Z_K parallel to the charge transfer resistance ($R_{ct,A}$) in series with a finite diffusion impedance element (Nernst-impedance, Z_N) so that the expression for the Faradaic impedance from Equation 3.3 was extended and is given by

$$Z_F = \frac{R_{ct} + Z_N}{1 + R_{ct}/Z_K} \tag{3.13}$$

On the basis of this model and equivalent circuit shown in Figure 3.26, the changes and differences, depending on the anode used in the fuel cell (Pt/C or PtRu/C) in the impedance spectra during the

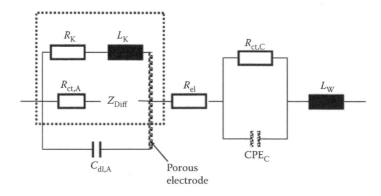

FIGURE 3.26 Equivalent circuit used for evaluation of interpolated (time-dependent) EIS measured during hydrogen oxidation and CO poisoning of the anode and oxygen reduction at the fuel cell cathode in galvanostatic mode of cell operation.

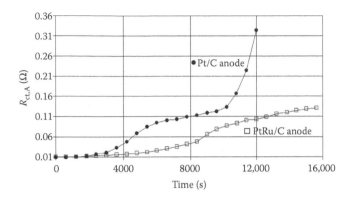

FIGURE 3.27 Time elapsed of the charge transfer resistance of the (●) Pt/C and (□) PtRu/C anodes after evaluation of the time-dependent impedance spectra with the equivalent from Figure 3.26.

experiment are dominated by the changes of the charge transfer resistance of the anode ($R_{ct,A}$), the surface relaxation impedance (R_K, τ_K) and the finite diffusion impedance Z_N.

The time evolution of the charge transfer resistance of the anodes is shown in Figure 3.27 (Pt/C anode (●) and PtRu/C anode (□) as a function of the elapsed time). The Warburg parameter from the Nernst impedance (Z_N) represented in Figure 3.28 shows the greatest difference between the two anodes. The Nernst impedance Z_N contains two parameters: the Warburg parameter W and a diffusion time constant (k_N), determined by the constant of diffusion (D_k) and diffusion layer thickness (d_N). The Nernst impedance is calculated by Equations 3.14 and 3.15 and with Equation 3.16, one can calculate the Nernst resistance R_N:

$$Z_N = \frac{W}{\sqrt{j\omega}}\tanh\sqrt{\frac{j\omega}{k_N}} \tag{3.14}$$

with

$$R_N = \frac{w}{\sqrt{K_N}} \tag{3.15}$$

$$k_N = D_k/d_N^2 \tag{3.16}$$

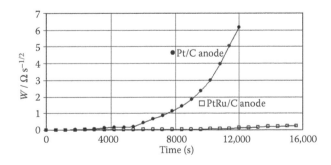

FIGURE 3.28 Time elapsed of the Warburg parameter W from the Nernst impedance for the (●) Pt/C and (□) PtRu/C anodes after evaluation of the time-dependent impedance spectra with the equivalent from Figure 3.26.

3.3.2.2 EIS during the PEMFC Pt-Cathode Flooding

One of the major issues during PEMFC operation is the water management. Despite a proper humidification of the cell, often a drying out or flooding of the cell can occur. A drying out of the cell induces not only an increase in the membrane resistance because the proton conductivity strongly depends on the water content in the membrane but also has a major impact on the degree of reaction zone (triple phase boundary) especially in the case of electrodes impregnated with electrolyte suspension or electrolyte in form of electrolyte powder to increase the triple phase boundary. The other extreme condition during fuel cell operation is the case of flooding of the pores of the electrode with water. This water can result from improper humidification of the electrodes if the relative humidification is more than 100% or if the electrode surface is too hydrophilic and water is accumulated in the pores flooding the electrode. To examine the flooding of the cathode, a PEMFC was operated for 8 h with the cathode in the "dead end" mode of operation at 2 A, 80°C, oxygen and hydrogen at 2 bar. In the "dead end" mode of operation the outlet valve of the cathode was closed so that the formed water could not flow out of the cell. During the time-dependent flooding of the electrode also time-dependent impedance spectra were recorded (Figures 3.29 and 3.30). After time interpolation of the measured impedance spectra they were evaluated with the equivalent circuit described previously. In Figure 3.31, one can observe an increase in the charge transfer resistance of the cathode caused by the flooding of the electrode and a decrease of the active electrode surface. Also from Figure 3.32 an increase of the diffusion hindrance with operational time during the cathode flooding is observed.

During flooding of the cathode, a part of the catalyst can be dissolved and under certain conditions the catalyst can be deposited at other parts of the electrode, for example, in the gas diffusion layer. This loss of catalyst has an important impact on the cell durability and should be avoided during fuel cell operation.

3.3.2.3 EIS during Long-Term Operation (Degradation) of PEMFCs

One of the major key problems to be solved for the application of fuel cells is the long-term behavior and durability. To improve the durability one has to know and understand first the degradation mechanism.

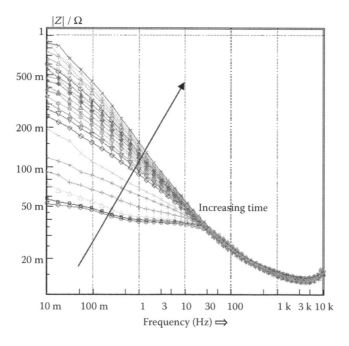

FIGURE 3.29 Bode representation of EIS (impedance vs. frequency) during flooding the cathode 400 min with water at 2 A, dead end.

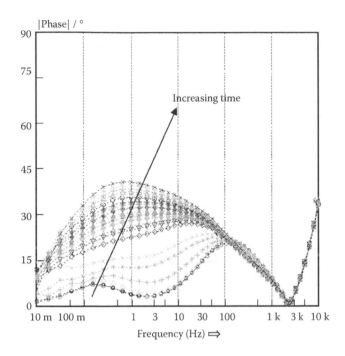

FIGURE 3.30 Bode representation of EIS (phase shift vs. frequency) during flooding the cathode 400 min with water at 2 A, dead end.

To investigate in detail the degradation mechanism of the electrodes in a PEM single cell, a MEA was operated over 1000 h galvanostatically at a current density of 500 mA cm^{-2}. At different times during operation several EIS measurements were performed without interrupting the fuel cell operation. The corresponding measurement time is indicated by arrows in Figure 3.33. As shown in Figure 3.33, after an operation time of 1000 h an additional nearly linear time-dependent voltage loss of about 140 mV

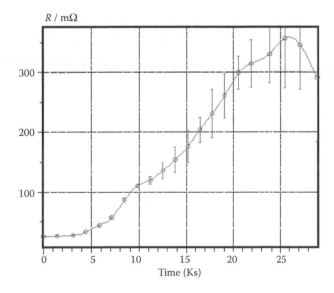

FIGURE 3.31 Time dependency of the charge transfer resistance of the cathode during flooding with water at 2 A, dead end.

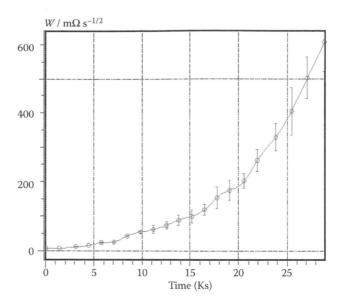

FIGURE 3.32 Time dependency of diffusion impedance element during flooding with water at 2 A, dead end.

could be observed, corresponding to a voltage loss (degradation) rate of 140 μV h^{-1} equivalent to a power density loss rate of 70 μW h^{-1} cm^{-2}. The measured data represented in the Bode diagram (Figure 3.34) as symbols and the modeled curves as lines also show a strong time dependency. A linear increase in the impedance with operating time, especially in the higher frequency range (1–20 kHz) of the impedance spectra could be observed.

After a complete shut down of the cell after an operating time of 1000 h and restarting the cell after 24 h at OCV, the cell showed nearly the same performance as at the beginning of the longtime test and the impedance spectra recorded directly after the restart shows similar values to the impedance spectra recorded at the beginning of the lifetime test.

For the evaluation of the measured spectra the same equivalent circuit as shown in Figure 3.14 was used. Besides a series resistance (membrane resistance R_M), the equivalent circuit contains three time

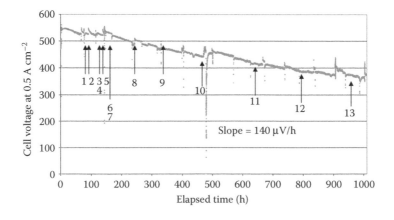

FIGURE 3.33 Time dependency of cell voltage during long-term operation at 80°C and 0.5 A cm^{-2}, arrows indicating the time when EIS is measured.

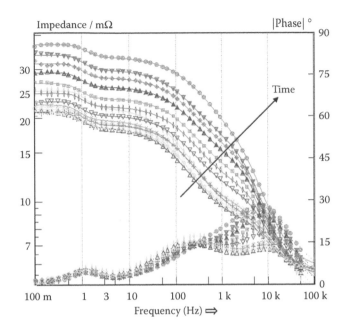

FIGURE 3.34 Bode diagram of EIS measured at different times during 1000 h at 80°C and 0.5 A cm⁻².

constants of parallel R/C. In the equivalent circuit $R_{ct(A)}$ and $R_{ct(C)}$ are related to charge transfer resistance on the anode and on the cathode, the capacitances $C_{dl(A)}$ and $C_{dl(C)}$ are related to the double-layer capacity of both electrodes. The diffusion processes will be simulated also by an RC-element (R_N and C_N) and the membrane resistance by R_M. In the simulation, the capacitance (C) was replaced by CPE due to the porous structure of the electrodes. The cathode can be described using two time constants, one for the charge transfer through the double layer ($R_{ct(C)}$/CPE$_{dl(C)}$, the exponent of the CPE is around 0.85, for an exponent of 1 the CPE is equal with the capacitance) and one for the finite diffusion of water with a Nernst-impedance-like behavior ($R_{(N)}$/CPE$_{(N)}$, the exponent of the CPE is around 0.95). The time constant of the anode ($R_{ct(A)}$/CPE$_{dl(A)}$, the exponent of the CPE is around 0.80) is given by the charge transfer through the anode double layer.

The resistance of the membrane $R_M = 5.65 \pm 0.10$ mΩ and the resistance related to the diffusion $R_N = 2.50 \pm 0.5$ mΩ were both nearly constant during the whole experiment.

The most time-sensitive impedance elements are the charge transfer resistance of the anode $R_{ct(A)}$ and the charge transfer resistance of the cathode $R_{ct(C)}$. The time dependence of $R_{ct(A)}$ is shown in Figure 3.35. The contribution of $R_{ct(A)}$ to the overall cell impedance increased at the beginning of the experiment from 2.5 to nearly 9 mΩ after 1000 h of operation. After a new start-up of the cell, the resistance reached nearly at the same value (3.0 mΩ) as at the beginning of the experiment.

The time dependence of $R_{ct(C)}$ is shown in Figure 3.36. The contribution of $R_{ct(C)}$ to the overall cell impedance increased at the beginning of the experiment from 10.5 to nearly 16 mΩ after 1000 h of operation. After the new start-up of the cell, the charge transfer resistance of the cathode decreased at a value of 14 mΩ, thus the reversible part, defined as the difference between the value after 1000 h of operation (16 mΩ) and the value after start-up (14 mΩ) is 2 mΩ. The irreversible part is defined as the difference between the value after start-up and the value at the beginning of the long-term operation (10.5 mΩ).

Taking into account that the surface area of the cell is 23 cm² and the current density is 0.5 A cm⁻², we obtain the total current of 11.5 A. Using Ohm's law and the exact resistance values from Figures 3.35 and 3.36, we are able to calculate and to separate the voltage losses of the cell during long-term operation

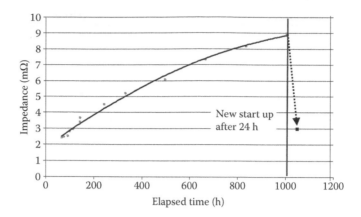

FIGURE 3.35 Time dependence of charge transfer resistance of the anode $R_{ct(A)}$.

into reversible and irreversible voltage losses and furthermore into contributions of the anode and cathode. The result of the separation into different voltage losses is shown in Figure 3.37.

From this we can calculate that total voltage loss of the cell during 1000 h of operation is 137.4 mV. The greatest part of the voltage loss is a reversible loss (88.8 mV) and 48.6 mV is an irreversible voltage loss. The voltage loss related to the anode can be revoked and the irreversible part of voltage loss is small at the anode (5.4 mV) compared to the irreversible voltage loss at the cathode (43.2 mV).

In addition to EIS surface science investigations were performed that allow one to identify the degradation processes. Two different degradation processes were identified: the agglomeration of the platinum catalyst mainly in the cathode and the disintegration of the PTFE and the correlated decrease of the hydrophobic degree. The loss of the hydrophobicity is more significant on the anode than on the cathode (Schulze et al., 2007).

The irreversible degradation is probably attributable to a catalyst structure change that is detected by surface analytical methods. The platinum catalyst agglomerates during fuel cell operation. As a consequence the active surface of the catalyst in the cathode decreases and the electrochemical performance decreases concurrently. In contrast to the decrease of the hydrophobicity the loss of active surface area cannot be compensated by modification of the operation conditions. This degradation is therefore an irreversible process. Under extreme operation conditions when the electrodes are flooded with water the platinum can move across the gas diffusion layer. This indicates that the mobility of the platinum is related to a liquid water phase.

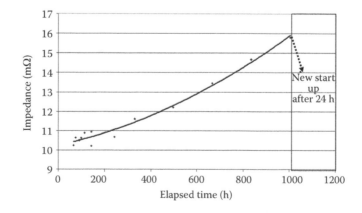

FIGURE 3.36 Time dependence of charge transfer resistance of the cathode $R_{ct(C)}$.

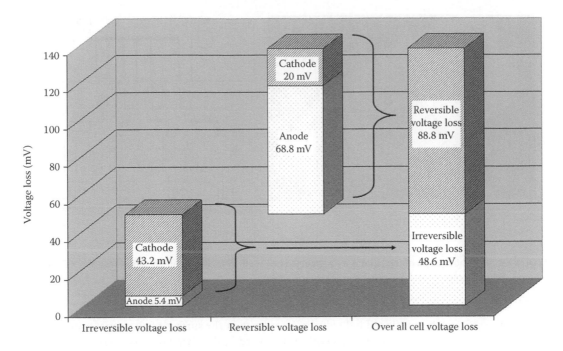

FIGURE 3.37 Breakdown of voltage (performance) losses resulting from long-term operation into reversible and irreversible degradation after evaluation of EIS.

The reversible degradation is related to the decomposition of the PTFE and, respectively, the decrease in the hydrophobicity and the correlated alteration of the water balance. This degradation process will mainly affect the performance of the anodes. So the electrochemical performance increases after an interruption of operation and drying of the fuel cell. The changed hydrophobicity or the related alteration of the PTFE yields a decrease of the electrochemical performance that can be compensated by modification of the water balance; for example, the water balance can be modified by adding purging intervals and adapting the periods between purging and the length of the purging intervals.

The quantification of the effects of both degradation processes has shown that the reversible degradation processes are more significant than the irreversible processes. Therefore, it is very important for long-term experiments and for the long-term stability to take the experimental conditions into account that are used to determine degradation values, especially if the experimental setup eliminates the effect of the reversible degradation process.

3.3.3 Locally Resolved Applications: EIS Using Segmented Cells

To achieve optimum performance and long lifetime of fuel cells for commercial use, a homogeneous electrochemical activity over the electrode area is obligatory. Inhomogeneous current distribution causes low reactant and catalyst utilization as well as reduced efficiency, fast degradation, and low durability. Therefore, the knowledge of the current density distribution is essential for the development of MEAs and stack design as well as for the adaptation of the operation conditions in fuel cells.

In the following, two tools for locally resolved electrochemical measurements is described. The first technique is based on individual segments that are externally connected with resistances for the current measurements in each segment. A photograph and a schematic diagram of this approach are shown in Figure 3.38, respectively. The cell shown in Figure 3.38 has an active are of 5×5 cm² and 16 segments. All segments are isolated with PTFE among each other and the frame. Each segment is connected by a resistance to the current collector.

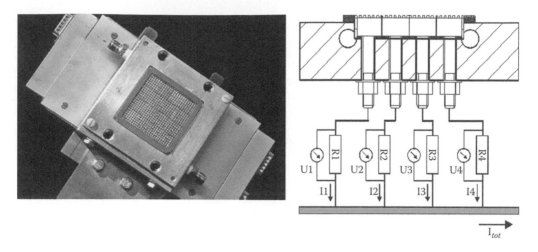

FIGURE 3.38 Photograph of the laboratory segmented cell with 16 segments on an active area of 5×5 cm^2 and scheme of the laboratory cell for current density measurements.

All segments must be positioned individually. In order to fit the surface of all segments in the same plane on the backside of the segments an elastic plate is used. If the elastic plate is too soft it is not possible to apply a high contact pressure from the flow field onto the GDL, which can result in high electrical contact resistance. As a result, the variation of the contact resistances for the individual segments with the GDL can be high. If the elastic plate is too hard, the segments will not be assembled in the same plane, and consequently the contact pressures and contact resistances for the individual segments with the GDL can vary as well. Therefore, it is necessary to choose the elastic material according to the specific MEA and GDL. In addition, the segments have to be manufactured very precisely in order to minimize the in plane positioning problem. Therefore, this approach to locally resolved measurements is time intensive and cost intensive.

An additional problem of the individual positioning of the segments is that a complex flow field increases the experimental problems. Therefore, a chocolate wafer structure with perpendicular channels of 1 mm depth and 1 mm width is used. Typically, the segmented side is used as anode because in a hydrogen-supplied fuel cell the effect of the transport processes on the anode side has a lower influence on the cell performance compared to the cathode. Alternately, a segmented cathode side is used with pure oxygen in order to investigate the anode flow fields. Caused by the external connection of the segments and the thickness of this measurement device, the technique is only suitable for single cells or at outer cells in short stacks.

An advantage of this approach is that each segment can be operated as an individual cell, whereas all cells are parallely connected by the in-plane conductivity of the GDL. This tool can not only be used to measure current density distribution but also to impress a current density onto the cell by using individual loads. The current generated in a certain region of the MEA flows through the corresponding segment and a small series resistance. Its voltage drop along the 16 resistances is scanned and the resulting current distribution density is registered. Exemplary, the change in current distribution as a result of cathode flooding in a PEMFC is shown in Figure 3.39. This routinely used tool helps to detect unfavorable operating conditions, to avoid damage of the PEMFC due to hot spots in the MEA and finally to improve gas distributor structures.

Using the segmented cell from Figure 3.38 locally resolved impedance measurements were performed. Depending on the problem to be solved different cells should be operated under the relevant conditions. In Figures 3.40 and 3.41, the locally resolved impedance spectra measured at open cell voltage are shown. Similar measurements with segmented cells are reported by Schneider et al. (2009).

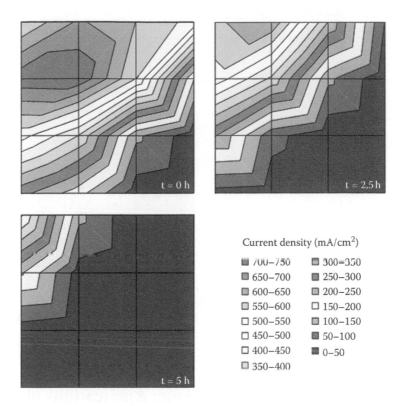

FIGURE 3.39   Change in distribution of current density while flooding the cathode of a PEMFC operating in dead-end mode.

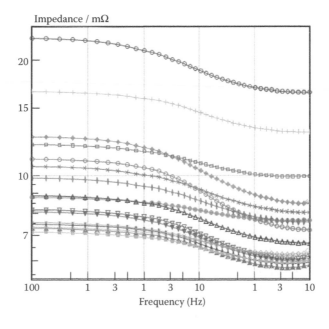

FIGURE 3.40   Bode plot (impedance vs. frequency) of EIS measured with the segmented cell.

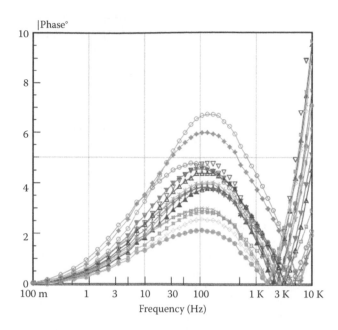

FIGURE 3.41   Bode plot (phase-shift vs. frequency) of EIS measured with the segmented cell.

3.4 Recent Advances of EIS Technique: Applications of EIS for PEMFC Stack Characterization

There is a great interest in characterizing and monitoring fuel cell stacks during operation. Fuel cells stacks have a higher power output than single cells and therefore the hardware used for electrochemical characterization should be able to handle high currents and high voltages. At the moment only a few labs worldwide are able to measure impedance spectra at a stack level.

Given by the serial connection of single cells in the case of fuel cell stack it is very important to monitor the individual cell voltage. The failure of one cell can damage the complete stack. Therefore, we build a test rig (Figure 3.42) with four electronic loads: three work as slaw consuming only power and the fourth as the measuring unit.

To reduce the measuring time of the stack a new electrochemical workstation is used with 16 parallel impedance channels so that 16 cells of the stack can be measured simultaneously. These measurements are performed in a galvanostatic operation mode of the stack. The AC current excitation signal is applied over the whole fuel cell stack and the AC response voltage is collected at the corresponding single cell of the stack. This technique is a great improvement compared to the sequential multiplexer technique.

3.5 Concluding Remarks

EIS is a very powerful method for the characterization of fuel cells and stacks. It is the most used, *in situ* nondestructive steady-state measurement technique that provides detailed diagnostic information over a wide range of electrochemical phenomena occurring during PEMFC operation. EIS can be applied for *in situ* failure analysis using a more or less complex equivalent circuit, as shown in detail in this chapter. There are a great number of advantages using EIS but also some limitations of the method. Given by the high currents flowing through a large cell and low impedance in the high-frequency range of the spectra some artifacts restrict the application to fuel cells with large surface. This restriction can be overcome

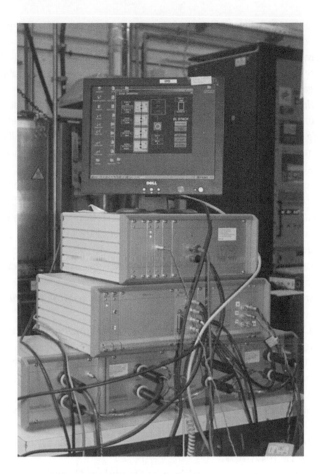

FIGURE 3.42 Test rig with four electronic loads, power potentiostat, and electrochemical workstation for 16 simultaneous impedance measurements on fuel cell stacks.

by twisting the cell connectors or by using in addition the current interrupt method to extend the high-frequency range of EIS. EIS is well established for single fuel cell and cell components characterization. In the near future, EIS is expected to be used for monitoring PEMFC stacks and individual cells of the stack during operation.

References

Agarwal, P., Orazem, M. E., and Garcia-Rubio, L. H. 1995. Application of measurement models to impedance spectroscopy. III. Evaluation of consistency with the Kramers-Kronig relations. *Journal of the Electrochemical Society* 142: 4159–4168.

Andreaus, B., McEvoy, A. J., and Scherer, G. G. 2002. Analysis of performance losses in polymer electrolyte fuel cells at high current densities by impedance spectroscopy. *Electrochimica Acta* 47: 2223–2229.

Bard, A. J. and Faulkner, L. R. 2001. *Electrochemical Methods, Fundamentals and Applications*, 2nd Edition. Hoboken: John Wiley & Sons, pp. 368–416.

Ciureanu, M. and Wang, H. 1999. Electrochemical impedance study of electrode-membrane assemblies in PEM fuel cells I. Electro-oxidation of H_2 and H_2/CO mixtures on Pt-based gas-diffusion electrodes. *Journal of the Electrochemical Society* 146: 4031–4040.

Ciureanu, M. and Wang, H. 2000. Electrochemical impedance study of anode CO-poisoning in PEM fuel cells. *Journal of New Materials for Electrochemical Systems* 3: 107–119.

Cleghorn, S. J. C., Springer, T. E., Wilson, M. S., Zawodzinski, C., Zawodzinski, T. A., and Gottesfeld, S. 1997. PEM fuel cells for transportation and stationary power generation applications. *International Journal of Hydrogen Energy* 22: 1137–1144.

Delahay, P. 1965. *Double Layer and Electrode Kinetics.* New York, NY: Wiley-Interscience.

Ehm, W., Göhr, H., Kaus, R., Röseler, B., and Schiller, C. A. 2000. The evaluation of electrochemical impedance spectra using a modified logarithmic Hilbert transform. *ACH-Models in Chemistry* 137: 145–157.

Eikerling, M., Kornyshev, A. A., and Lust, E. 2005. Optimized structure of nanoporous carbon-based double-layer capacitors. *Journal of the Electrochemical Society* 152: E24–E33.

Fletcher, S. 1992. An electrical model circuit that reproduces the behaviour of conducting polymer electrodes in electrolyte solutions. *Journal of Electroanalytical Chemistry* 337: 127–145.

Giner, J. and Hunter, C. 1969. The mechanism of operation of Teflon-bonded gas diffusion electrode: A mathematical model. *Journal of the Electrochemical Society* 116: 1124–1130.

Giordano, B., Passalacqua, E., Alderucci, V., Staiti, P., Pino, L., Mirzaian, H., Taylor, E. T., and Wilemski, G. 1991. Morphological characteristics of PTFE bonded gas diffusion electrodes. *Electrochimica Acta* 36: 1049–1055.

Göhr, H. 1997. Impedance modelling of porous electrodes. *Electrochemical Applications* 1: 2–3. Available from http://www.zahner.de

Göhr, H. and Schiller, C. A. 1986. Faraday-impedanz als verknüpfung von impedanzelementen. *Z. Phys. Chemie, Neue Folge 148:* 105–124.

Göhr, H., Söllner, J., and Weinzierl, H. 1983. Kinetic properties of smooth and porous lead /lead sulphates electrodes in sulphuric acid during charging and discharging. *Proc. 34th ISE Meeting*, Erlangen, Germany, Poster 0715.

Graham, D. C. 1947. The electrical double layer and the theory of electrocapillarity. *Chemical Reviews* 41: 441–501.

Grens, E. A. II. 1970. On the assumptions underlaying theoretical models for flooded porous electrodes. *Electrochimica Acta* 15: 1047–1057.

Gülzow, E., Schulze, M., Wagner, N., Kaz, T., Reißner, R., Steinhilber, G., and Schneider, A. 2000. Dry layer preparation and characterization of polymer electrolyte fuel cell components. *Journal of Power Sources* 86: 352–362.

Gülzow, E., Schulze, M., Wagner, N., Kaz, T., Schneider, A., and Reissner, R. 1999. New dry preparation technique for membrane electrode assemblies for PEM fuel cells. *Fuel Cells Bulletin* 15: 8–12.

Itagaki, M., Hatada, Y., Shitanda, I., and Watanabe, K. 2010. Complex impedance spectra of porous electrode with fractal structure. *Electrochimica Acta* 55: 6255–6262.

Jaouen, F., Lindbergh, G., and Sundholm, G. 2002. Investigation of mass-transport limitations in the solid polymer fuel cell cathode. *Journal of the Electrochemical Society* 149: A437–A447.

Kim, J.-D., Park, Y.-I., Kobayashi, K., Nagai, M., and Kunimatsu, M. 2001. Characterization of CO tolerance of PEMFC by AC impedance spectroscopy. *Solid State Ionics* 140: 313–325.

Koper, M. T. M., Schmidt, T. J., Markovic, N. M., and Ross, P. N. 2001. Potential oscillations and S-shaped polarization curve in the continuous electro-oxidation of CO on Platinum single-crystal electrodes. *The Journal of Physical Chemistry B* 105: 8381–8386.

Kötz, R. and Carlen, M. 2000. Principles and applications of electrochemical capacitors. *Electrochimica Acta* 45: 2483–2498.

Lasia, A. 1999. Electrochemical impedance spectroscopy and it's applications. In *Modern Aspects of Electrochemistry*, eds. B. E. Conway, J. Bockris, and R. E. White, Vol. 32. New York, NY: Kluwer Academic/Plenum Publishers, pp. 143–248.

Leng, Y.-J. Wang, X., and Hsing, I.-M. 2002. Assessment of CO-tolerance for different Pt-alloy anode catalysts in a polymer electrolyte fuel cell using AC impedance spectroscopy. *Journal of Electroanalytical Chemistry* 528: 145–152.

Levie, R. de. 1963. On porous electrodes in electrolyte solutions. *Electrochimica Acta* 8: 751–780.

Levie, R. de. 1964. On porous electrodes in electrolyte solutions IV. *Electrochimica Acta* 9: 1231–1245.

Levie, R. de. 1967. Electrochemical response of porous and rough electrodes. In *Advances in Electrochemistry and Electrochemical Engineering*, eds. P. Delahay and C. W. Tobias, Vol. 6. New York, NY: Interscience, pp. 329–397.

Müller, B., Wagner, N., and Schnurnberger, W. 1999. Change of electrochemical impedance spectra (EIS) with time during CO-poisoning of the Pt-anode in a membrane fuel cell. In *Proton Conducting Membrane Fuel Cells II*, eds. S. Gottesfeld and T. F. Fuller, *Electrochem. Soc. Proc.*, Vol. 98-27, pp. 187–199.

Mund, K. 1975. Impedanzmessungen an porösen gestützten Raney-Nickel-Elektroden für alkalische Brennstoffzellen. *In Siemens Forsch.- und Entwickl.- Ber.*, Vol. 4, pp. 68–74. Erlangen: Siemens AG.

Orazem, M. E. and Tribollet, B. 2008. *Electrochemical Impedance Spectroscopy*. Hoboken: John Wiley & Sons.

Poltarzewski, Z., Staiti, P., Alderucci, V., Wieczorek, W., and Giordano, N. 1992. Nafion distribution in gas diffusion electrodes for solid–polymer–electrolyte–fuel-cell applications. *Journal of the Electrochemical Society* 139: 761–765.

Raistrick, I. D. 1990. Impedance studies of porous electrodes. *Electrochimica Acta* 35: 1579–1586.

Sarangapani, S., Tilak, B. V., and Chen, C. P. 1996. Materials for electrochemical capacitors. *Journal of the Electrochemical Society* 143: 3791–3799.

Savova-Stoynov, B. and Stoynov, Z. 1985. Computer analysis of non-stationary impedance data. *Proc. Symp., Computer Aided Acquisition and Analysis of Corrosion Data*; eds. M. W. Kendig, U. Bertocci, and J. E. Strutt, Vol. 85-3. Pennington: Electrochem. Soc. Inc., pp. 152–158.

Savova-Stoynov, B. and Stoynov, Z. 1992. Four-dimensional estimation of the instantaneous impedance. *Electrochimica Acta* 37: 2353–2355.

Schiller, C. A., Richter, F., Gülzow, E., and Wagner, N. 2001. Validation and evaluation of electrochemical impedance spectra of systems with states that change with time. *J. Phys. Chem. Chem. Phys.* 3: 374–378.

Schmidt, H. and Stackelberg, M. von, 1963. *Modern Polarographic Methods*. New York, NY: Academic Press Inc.

Schmidt, T. J., Gasteiger, H. A., Stäb, G. D., Urban, P. M., Kolb, D. M., and Behm, R. J. 1998. Characterization of high-surface-area electrocatalysts using a rotating disk electrode configuration. *Journal of the Electrochemical Society* 145: 2354–2358.

Schneider, I. A., Bayer, M. H., Wokaun, A., and Scherer, G. G. 2009. Negative resistance values in locally resolved impedance spectra of polymer electrolyte fuel cells (PEFCs). *ECS Transactions* 25: 937–948.

Schulze, M., Wagner, N., Kaz, T., and Friedrich, K. A. 2007. Combined electrochemical and surface analysis investigation of degradation processes in polymer electrolyte membrane fuel cells. *Electrochimica Acta* 52: 2328–2336.

Srinivasan, S., Ticianelli, E. A., Derouin, C. R., and Redondo, A. 1988. Advances in solid polymer electrolyte fuel cell technology with low platinum loading electrodes. *Journal of Power Sources* 22: 359–375.

Stoynov, Z. 1990. Impedance modelling and data processing: Structural and parametrical estimation. *Electrochimica Acta* 35: 1493–1499.

Szpak, S. 1991. Characterization of Electrodes and Electrochemical Processes. eds. R. Varma and J. R. Selman, pp. 677–716. New York, NY: Wiley-Interscience.

Vogel, W., Lundquist, J., Ross, P., and Stonehart, P. 1975. Reaction pathways and poisons-II. The rate controlling step for electrochemical oxidation of hydrogen on Pt in acid and poisoning of the reaction by CO. *Electrochimica Acta* 20: 79–93.

Wagner, N. 2002. Characterization of membrane electrode assemblies in polymer electrolyte fuel cells using a.c. impedance spectroscopy. *Journal of Applied Electrochemistry* 32: 859–863.

Wagner, N. 2005. Electrochemical power sources—Fuel cells. In *Impedance Spectroscopy: Theory, Experiment, and Applications*, eds. E. Barsoukov and J. R. Macdonald, 2nd Edition. Hoboken: John Wiley & Sons Inc., pp. 497–537.

Wagner, N. and Friedrich, K. A. 2009. Dynamic response of polymer electrolyte fuel cells. In *Encyclopedia of Electrochemical Power Sources*, Vol. 2, eds. Garche J. et al. Amsterdam: Elsevier, pp. 912–930.

Wagner, N. and Gülzow, E. 2004. Change of electrochemical impedance spectra (EIS) with time during CO-poisoning of the Pt-anode in a membrane fuel cell. *Journal of Power Sources* 127: 341–347.

Wagner, N. and Schulze, M. 2003. Change of electrochemical impedance spectra (EIS) during CO-poisoning of the Pt and Pt-Ru -anodes in a membrane fuel cell (PEFC). *Electrochimica Acta* 48: 3899–3907.

Wagner, N., Kaz, T., and Friedrich, K. A. 2008. Investigation of electrode composition of polymer fuel cells by electrochemical impedance spectroscopy. *Electrochimica Acta* 53: 7475–7482.

Wang, X., Hsing, I.-M., Leng, Y.-J., and Yue, P.-L. 2001. Model interpretation of electrochemical impedance spectroscopy and polarization behavior of H_2/CO mixture oxidation in polymer electrolyte fuel cells. *Electrochimica Acta* 46: 4397–4405.

Wilkinson, D. P. and Thompsett, D. 1997. *Proceedings of the Second National Symposium on New Materials for Fuel Cell and Modern Battery Systems*, eds. O. Savadogo and P. R. Roberge, pp. 268, Montreal (Quebec), Canada.

Will, F. G. 1963. Electrochemical oxidation of hydrogen on partially immersed platinum electrodes. *Journal of the Electrochemical Society* 110: 145–152.

Wilson, M. S., Garzon, F. H., Sickafus, K. E., and Gottesfeld, S. 1993. Surface area loss of supported platinum in polymer electrolyte fuel cells. *Journal of the Electrochemical Society* 140: 2872–2877.

Winsel, A. 1962. Beiträge zur Kenntnis der Stromverteilung in porösen Elektroden. *Z. Elektrochemie* 66: 287–304.

Yuan, X. Z., Song, C., Wang, H., and Zhang, J. 2010. *Electrochemical Impedance Spectroscopy in PEM Fuel Cells*. London: Springer-Verlag.

4

Cyclic Voltammetry

Jinfeng Wu
Institute for Fuel Cell Innovation

Xiao-Zi Yuan
Institute for Fuel Cell Innovation

Haijiang Wang
Institute for Fuel Cell Innovation

4.1 Introduction

Cyclic voltammetry is an important and very frequently utilized electrochemical technique because it offers a wealth of experimental information and insights into the kinetic and thermodynamic details of many chemical systems. It was first reported in 1938, and then described theoretically by Randles in 1948. Cyclic voltammetry measurement is accomplished with a two- or three-electrode arrangement, whereby the potential relative to some reference electrode is scanned at a working electrode while the resulting current flowing through a counter electrode is monitored. Cyclic voltammetry is often the first experiment performed in an electroanalytical study. It is rarely used for quantitative determination, but it is ideally suited for a quick search of redox couples, for understanding reaction intermediates, and for obtaining stability in reaction products.

4.2 Principle

4.2.1 Typical Cyclic Voltammogram Analysis

The curve resulting from a particular set of cyclic voltammetry measurements is known as a cyclic voltammogram; a typical cyclic voltammogram recorded for the reversible redox system is shown in Figure 4.1. The characteristics of the peaks in a cyclic voltammogram can be used to acquire qualitative information about the relative rates of reaction and reactant diffusion in a given electrochemical system. In Figure 4.1, it can be observed that when the potential of the working electrode is more positive than that of a redox couple, the corresponding reactants may be oxidized and produce an anodic current. The peak current occurs when the potential reaches a value at which all the electro-chemically active reactants at the electrode surface are completely consumed. When the potential is controlled at this value, the mass transport rate from the bulk to the electrode surface reaches a maximum, driven by the largest concentration gradient between them. After this peak, the current will decline because the double-layer thickness increases, resulting in a less steep concentration gradient for the active reactant. On the reverse voltage scan, as the working electrode potential becomes more negative than the reduction potential of a redox couple, reduction may occur and cause a cathodic current. A cyclic voltammogram can have several cathodic and anodic peaks due to intrinsic reaction mechanisms.

The important parameters in a cyclic voltammogram are the peak potential and peak currents. If the electron-transfer process is fast compared to other processes (such as diffusion), the reaction is regarded as electrochemically reversible, and the peak separation is

$$\left| E_p^{ox} - E_p^{red} \right| = 2.303 \frac{RT}{nF} \tag{4.1}$$

where E_p^{ox} and E_p^{red} are the potential at which the oxidation and reduction processes occur, R is the universal gas constant, T is the absolute temperature (in K), n is the number of exchanged electrons, and F is Faraday's constant.

For a reversible reaction, concentration is related to peak current by the Randles–Sevcik expression (at 25°C):

$$i_p = 2.686 \times 10^5 n^{3/2} A D^{1/2} v^{1/2} C^0 \tag{4.2}$$

where i_p is the peak current (in amperes), A is the electrode area (in cm²), D is the diffusion coefficient (in cm² s⁻¹), v is the potential scan rate (in V s⁻¹), and C^0 is the concentration of the bulk solution (in mol cm⁻³). Thus, if the electrochemical reactions are reversible, the positions of the oxidation and reduction peaks do not change with the potential scan rate, as is shown in Figure 4.2a. The peak height is proportional to the square root of the potential scan rate, according to Equations 4.1 and 4.2.

If the rate constant of the electron-transfer process is slow compared to other processes (such as diffusion), the electrochemical reactions will be quasi-irreversible or irreversible rather than completely

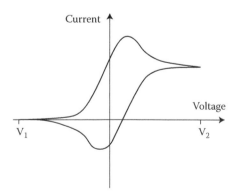

FIGURE 4.1 Typical plot of a cyclic voltammogram.

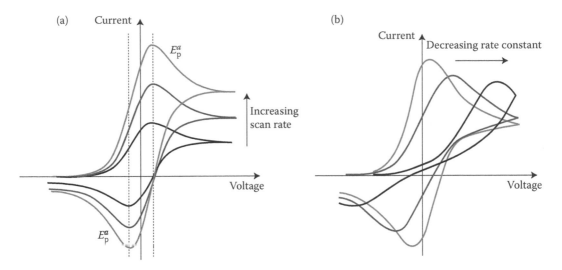

FIGURE 4.2 Dependence of (a) scan rate and (b) rate constant of electron-transfer process in a cyclic voltammogram.

reversible. Therefore, the reversibility of an electrochemical reaction is always a relative term, related to the potential scan rate. A reaction that is reversible at low-scan rates may become quasireversible or even irreversible at high-scan rates. In these cases, the anodic peak potential becomes more positive and the cathodic peak potential becomes more negative. This occurs because the current takes more time to respond to the applied voltage than in the reversible case. The separation of the two peaks also becomes larger than in the reversible case, as shown in Figure 4.2b.

Therefore, cyclic voltammetry can be used to elucidate the kinetics of electrochemical reactions taking place at electrode surfaces. From the sweep-rate-dependent peak amplitudes, and the widths and potentials of the peaks in the voltammogram, information can be obtained about adsorption, desorption, diffusion, and coupled homogeneous electrochemical reaction mechanisms.

4.2.2 Hydrogen Oxidation

The hydrogen oxidation reaction (HOR) is a very fast reaction at the Pt surface. The accepted mechanism of the HOR is the formation of a primary chemical adsorption step:

$$H_2 + 2M \leftrightarrow MH_{ads} \left(\text{Tafel reaction}\right) \tag{4.3}$$

and/or an electrochemical adsorption step:

$$H_2 + M \leftrightarrow MH_{ads} + H^+ + e^- \left(\text{Heyrovsky reaction}\right) \tag{4.4}$$

followed by a discharge step of adsorbed hydrogen, given by

$$MH_{ads} \leftrightarrow M + H^+ + e^- \left(\text{Volmer reaction}\right) \tag{4.5}$$

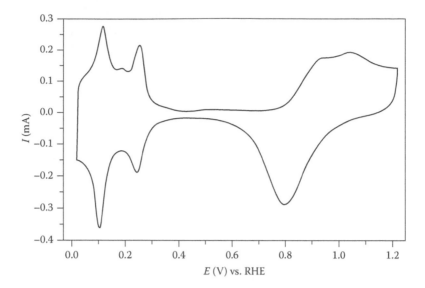

FIGURE 4.3 Cyclic voltammogram of Pt-black/Nafion electrode (Pt-black loading: 25 µg; Nafion film: 0.2 µm); scan rate 25 mV s⁻¹. (With kind permission from Springer Science+Business Media: *Journal of Solid State Electrochemistry*, Kinetic analysis of the hydrogen oxidation reaction on Pt-black/Nafion electrode, 10, 2006, 243–249, Lin, R. B. and Shih, S. M.)

For a given electrode material and electrolyte, the HOR usually takes place through the Tafel–Volmer or the Heyrovsky–Volmer mechanism. A less probable mechanism is

$$H_2 \leftrightarrow 2H^+ + 2e^- \left(\text{direct discharge}\right) \tag{4.6}$$

Direct discharge can be irreversible or reversible, depending on the results of treatment with the Nernst equation. Figure 4.3 shows a cyclic voltammogram obtained from a Pt-black/Nafion electrode. Two well-resolved H adsorption/desorption peaks can be observed, one located at about 125 mV and the other at about 275 mV. These two peaks are associated with the desorption of weakly and strongly adsorbed hydrogen atoms on Pt(110) and Pt(100) crystal surfaces, respectively.

4.2.3 Carbon Monoxide Oxidation

A PEM fuel cell uses hydrogen as its fuel and oxygen (usually from air) as its oxidant. However, hydrogen is not a naturally available fuel, and is often generated through reforming hydrocarbon fuels such as natural gas, methanol, and ethanol. Trace CO contaminants in the resulting reformate are unavoidable and can strongly adsorb on the Pt surface, causing significant overpotential and a decrease in the catalytic activity for the HOR. One of the well-known ways for developing new CO-tolerant catalysts is to modify Pt with metals such as Ru, Mo, Sn, and so on. Figure 4.4 shows the cyclic voltammograms of different catalysts (Pt/C, PtRu/C, and PtSn/C) during the oxidation of H_2 in the presence of 2% CO (Leng et al., 2002). The onset CO oxidation potentials were about 0.3, 0.5, and 0.8 V for PtSn/C, PtRu/C, and Pt/C, respectively. Therefore, the electrocatalytic activity of these three catalysts for CO oxidation increased in the order of Pt/C < PtRu/C < PtSn/C. The higher CO tolerance of PtSn is generally attributed to the weaker adsorption of CO on the alloy catalyst surface

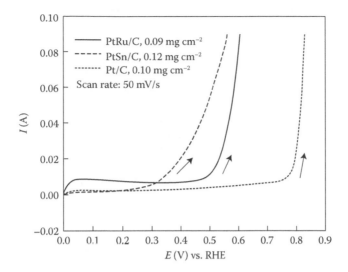

FIGURE 4.4 Cyclic voltammograms of H_2 oxidation in the presence of 2% CO on Pt/C (0.10 mg cm^{-2}), $Pt_{50}Ru_{50}$/C (0.09 mg cm^{-2}), and Pt_3Sn/C (0.12 mg cm^{-2}); $T = 50°C$; scan rate = 50 mV s^{-1}. (Reprinted from *Journal of Electroanalytical Chemistry*, 158, Leng, Y. J., Wang, X., and Hsing, I. M., Assessment of CO tolerance for different Pt-alloy anode catalysts in polymer electrolyte fuel cell using AC impedance spectroscopy, 145–152, Copyright (2002), with permission from Elsevier.)

(Crabb et al., 2000). The action of Ru is ascribed to its ability to weaken CO bonding on Pt due to d-electron deficiency (Yano et al., 2005), or to be the source of oxygen-containing species (Gasteiger et al., 1994), as shown in

$$Ru + H_2O \leftrightarrow (OH)_{ads} + H^+ + e^- \tag{4.7}$$

$$(CO)_{ads} + (OH)_{ads} \xrightarrow{rds} CO_2 + H^+ + e^- \tag{4.8}$$

4.2.4 Oxygen Reduction

In PEM fuel cells, including direct methanol fuel cells, the oxygen reduction reaction (ORR) occurs at the cathode. However, oxygen reduction is a sluggish process even with active noble catalysts, such Pt and some of its alloys. The reaction is further slowed down if some electrolyte species adsorb onto the catalyst surface, or the catalyst is contaminated by strongly adsorbing impurities. Figure 4.5 shows the voltammograms for oxygen reduction on Pt in 0.1 M H_3PO_4, $(FSO_2)_2NH$, and $(CF_3SO_2)_2NH$ solutions saturated with oxygen. The peak current density for oxygen reduction was highest in $(CF_3SO_2)_2NH$, but it was lower in $(FSO_2)_2NH$ compared to the current density observed in H_3PO_4. The half-wave potential (the potential at 85% of the peak current), $E_{1/2}$, in $(CF_3SO_2)_2NH$ was 30 mV more positive compared to the $E_{1/2}$ observed in H_3PO_4, whereas in $(FSO_2)_2NH$ it was 160 mV more negative. The lower current density and high overpotential for oxygen reduction in $(FSO_2)_2NH$ was ascribed to the occupation of the active sites on the electrode by the impurities and/or by the acid itself, resulting in the decreased oxygen reduction kinetics. Conversely, the high-current density and lower overpotential in $(CF_3SO_2)_2NH$ indicated that the electrolyte was relatively free of residual impurities and did not adsorb strongly on the Pt electrode.

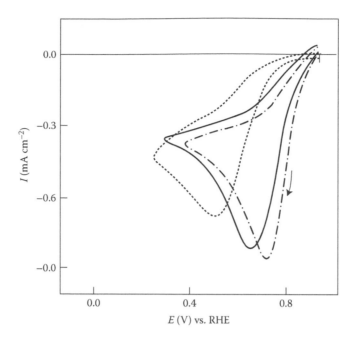

FIGURE 4.5 Cyclic voltammograms of the ORR on Pt in 0.1 M H_3PO_4 (—), $(FSO_2)_2NH$ (· · · ·), and $(CF_3SO_2)_2NH$ (—· —· —·) solutions saturated with oxygen; $T = 20°C$; scan rate = 10 mV s^{-1}. (With kind permission from Springer Science+Business Media: *Journal of Applied Electrochemistry*, Oxygen electroreduction in perfluorinated sulphonylimides, 17, 1987, 1057–1064, Razaq, M., Razaq, A., and Yeager, E.)

4.2.5 Oxidation of Other Anodic Fuels: Methanol, Ethanol, and Formic Acid

In recent years, liquid short-chain alcohols and acids, including methanol (Sivakumar et al., 2006), ethanol (Zhang et al., 2006), and formic acid (Liu et al., 2006) have been actively investigated in fuel cells due to several potential advantages over hydrogen, such as higher specific volume energy density as well as ease of transport and storage. The search for a better catalyst for the oxidation of liquid fuels has been ongoing for decades. As a powerful electroanalytical technique, cyclic voltammetry has played an important role in the development of novel catalysts. Cyclic voltammograms of formic acid oxidation on Pt/C and Pd/C catalysts are given in Figure 4.6. For the Pt/C catalyst, the reaction commenced in the hydrogen region and proceeded slowly during the forward scan direction to reach a plateau at 0.3 V. This corresponded to formic acid oxidation through the dehydrogenation path, but coverage by CO_{ads} simultaneously continued to grow, causing the current to be relatively small. At potentials more positive than 0.65 V, the reaction was significantly accelerated and an anodic peak emerged at 0.75 V. The latter was attributed to the oxidative removal of CO_{ads} together with the oxidation of formic acid on sites previously blocked by CO_{ads}. At higher potentials, formic acid oxidation was deactivated as result of Pt surface oxidation. On the reverse potential scan, the surface remained inactive until the partial reduction of the irreversibly formed surface oxides. The cathodic peak near 0.5 V was due to the oxidation of formic acid after the reduction of Pt oxides. As shown in Figure 4.6, compared to Pt/C the oxidation peak of the Pd/C catalyst demonstrated a negative potential shift and a higher anodic current density, which indicated the better electrocatalytic activity of Pd/C over Pt/C for formic acid oxidation.

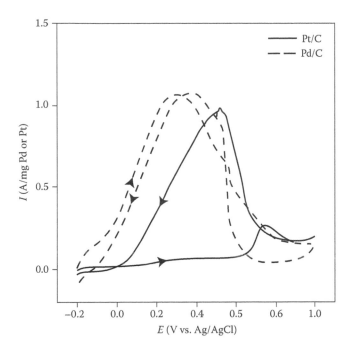

FIGURE 4.6 Cyclic voltammograms of microwave-synthesized Pt/C and Pd/C catalysts in 3 M HCOOH, 1 M H_2SO_4; $T = 25°C$; scan rate = 10 mV s^{-1}. (Reprinted from *Journal of Power Sources*, 161, Liu, Z. et al., Nanostructured Pt/C and Pd/C catalysts for direction formic acid fuel cells, 831–835, Copyright (2006), with permission from Elsevier.)

4.3 Instruments and Measurements

4.3.1 Voltammetry Instrumentation

The instrument used to perform cyclic voltammetry and potential step experiments is a potentiostat/galvanostat. In a three-electrode configuration the potentiostat allows the potential difference between the working and the reference electrodes to be controlled at a desired value by varying the resistance of the resistor (R_m), as shown in Figure 4.7. An electrometer measures the voltage difference between the reference and the working electrodes. The electrometer has extremely high-input impedance so that the current flowing through the reference electrode can be minimized, which avoids polarization of the reference electrode and hence enables it to maintain a constant potential. The voltage sweep is normally linear with time, as shown in Figure 4.8. The resultant current passing through the working and the counter electrodes is determined and recorded in response to the potential of the working electrode. In this technique, the potential applied to the working electrode is swept back and forth between two set voltage limits and the sweep rate can vary from a few millivolts per second to hundred volts per second.

4.3.2 *In Situ* and *ex Situ* Cyclic Voltammetry Measurements

In PEM fuel cell electrocatalyst and electrode research, either *ex situ* or *in situ* voltammetry can be conducted, depending on the purposes of the experiments. In the case of *ex situ* experiments, also referred to as half-cell experiments, the properties of the catalyst are evaluated using a standard

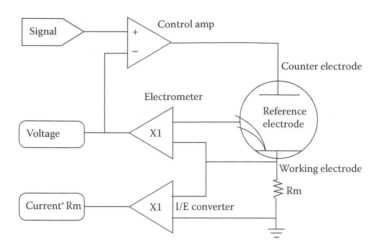

FIGURE 4.7 Schematic of the basic circuitry of a potentiostat.

three-electrode configuration with a catalyst-coated glass carbon disk electrode as the working electrode. The counter electrode, also known as the auxiliary or second electrode, is often made of highly conductive materials with good resistance to the redox solution, such as Pt wire, graphite, or Au. The calomel electrode (Ag/AgCl or Hg/Hg_2Cl_2) has been most commonly used as the reference electrode, as it provides a reversible half-reaction with Nernstian behavior, has a constant potential over a period of time, and is easy to assemble and maintain. An aqueous solution (e.g., perchloric acid or sulfuric acid) simulates the proton-conducting electrolyte in a PEM fuel cell. The three-electrode configuration can be employed to evaluate either the relative activities of a series of electrocatalysts toward the same reaction, or the activity of one electrocatalyst toward several different reactions, as will be described in Section 4.4.1.2. The *ex situ* cyclic voltammetry experiment is a convenient and relatively fast method of screening electrocatalysts but is not suitable for evaluation of fuel cell electrodes under operating conditions.

In situ experiments use a two-electrode configuration in which the electrode of interest is chosen to be the working electrode and the other in the PEM fuel cell serves as both a counter electrode and a pseudo-reference electrode. The electrochemical activity of the fuel cell cathode is typically of most interest because of the sluggish kinetics of the ORR. The fuel cell anode is used as the counter/reference electrode, with the inherent assumption that polarization of this electrode is small relative

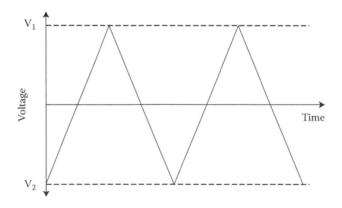

FIGURE 4.8 Schematic of a typical voltage sweep back and forth between two voltage limits.

to the polarization imposed on the fuel cell cathode, that is, the working electrode. *In situ* cyclic voltammetry has been used widely for determining the ECSA and catalyst utilization of PEM fuel cell electrodes.

4.3.3 Hydrogen Adsorption/Desorption Method and CO Stripping Method

The common reactant used for characterizing PEM fuel cell electrodes is hydrogen, and the hydrogen adsorption/desorption process follows the reaction in Equation 4.5. During testing, H_2 is fed to one side of the fuel cell, acting as both the counter electrode and the reference electrode, to function as a dynamic hydrogen electrode (DHE). The other side is flushed with inert gas (N_2 or Ar) and connected to the working electrode. In the typical PEM fuel cell cyclic voltammogram shown in Figure 4.3, two well-defined peaks represent the hydrogen adsorption/desorption on different Pt catalyst surfaces. A lower sweep rate is often used, for example, 10 mV s^{-1}, to minimize the impedance losses in the porous electrodes (Wu et al., 2008). The disadvantage of this technique for assessing supported electrocatalysts is that the carbon features mask the H_2 adsorption and desorption characteristics, for example, double-layer charging and the redox behavior of surface-active groups on carbon. To avoid carbon oxidation, the anodic limit is always set below 1.0 V (vs. DHE).

It is well known that CO is a potent poison in fuel cells due to its ability to strongly adsorb on Pt and other Pt-group metals. This preference for occupying catalyst active sites can be used to determine the ECSA of a PEM fuel cell electrode through the oxidation of adsorbed CO at room temperature, operating under the same principle as the hydrogen adsorption/desorption method. One side of the fuel cell is supplied with CO plus inert gas, or humidified high-purity inert gas (Ar or N_2), and connected to the working electrode while humidified H_2 is fed to the other side, serving as the DHE. During the CO adsorption process, CO plus inert gas is supplied to the anode at a certain flow rate, while the electrode potential is kept at about 0.1 V (vs. DHE). Then, the gas is switched to high-purity Ar for a long time to remove any CO from the gas phase. To record the CO stripping voltammogram, the potential is scanned from the adsorption value to near 0.9 V at a scan rate of 5 or 10 mV s^{-1} (Wu et al., 2008). The ESCA determined by CO stripping voltammetry is normally comparable to that measured by the hydrogen adsorption/desorption method, with a relative error of less than 10% (Pozio et al., 2002; Chaparro et al., 2009).

However, one limitation of this method is the uncertainty regarding the types of CO bonding taking place at the catalyst surface. Information about CO bonding is valuable in determining the number of CO atoms per metal atom. For example, if CO is on-top bonded, two electrons will be exchanged per surface site during CO oxidation, whereas for bridge-bonded CO, only one electron will be exchanged per surface site. Some methods, such as Fourier transform infrared (FTIR) spectroscopy (Cuesta et al., 2006) and differential electrochemical mass spectroscopy (Jusys et al., 2002) have been utilized to obtain information about CO bonding on the electrode surface. Another uncertainty is CO charge correction with respect to other contributions like double-layer charging and charging due to metal oxide formation. To calculate the CO stripping charge, the area under the CO oxidation peak has to be integrated and the charge due to double-layer charging and oxidate formation has to be subtracted. The practical approach is to consider double-layer charging and oxide formation in the same way as in the absence of CO, that is, to use the second cycle in the same experiment for baseline subtraction and to contribute the difference between the first and the second cycle only to CO oxidation, as shown in Figure 4.9. For a pure Pt electrode the deviation is not so significant since CO oxidation on Pt does not overlap to a great extent with oxide formation on Pt, whereas in the case of less noble metal like Ru or Ru alloys, the overlap is significant because oxide formation commences early. Recently, Vidaković et al. (2007) suggested a new constructed base line for accurately determining the surface-active area of PtRu catalysts, in which the corrected CO stripping line is integrated up to the half-peak potential and multiplied by the factor determined in a similar experiment with unsupported Pt.

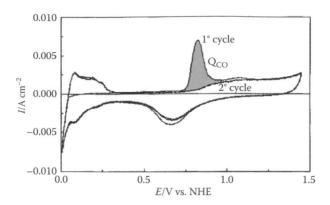

FIGURE 4.9 Cyclic voltammograms (10 mV s^{-1}) at 25°C in the potential range 0–1400 mV versus normal hydrogen electrode (NHE) on Pt/C (E-TEK) with (first cycle) and without (second cycle) a CO adsorbed ad-layer. The filled area represents the charge related to the CO oxidation reaction. (Reprinted from *Journal of Power Sources*, 105, Pozio, A. et al., Comparison of high surface Pt/C catalysts by cyclic voltammetry, 13–19, Copyright (2002), with permission from Elsevier.)

4.4 Applications in PEM Fuel Cells

4.4.1 Characterization of PEM Fuel Cells

4.4.1.1 Estimating the Extent of Pt Oxidation

To date, Pt-based materials are the most efficient, applicable, and successful catalysts for PEM fuel cells at the current technological stage. However, the problem associated with Pt electrode materials is catalyst oxidation at high thermodynamic potential in the ORR (1.23 V vs. NHE at standard conditions). Pt oxidation is of great importance in studying the oxygen reduction kinetics, as oxygen presumably cannot react on Pt surfaces effectively covered by these oxide species, and thus the oxygen reduction rate is decreased. The presence of PtO on the electrode surface not only slows down the oxygen reduction kinetics, but also results in a lower rest potential on the Pt electrode. The steady-state rest potential of a Pt electrode in oxygen saturated 1 M H_2SO_4 is only 1.06 V, because two reactions occur: Pt oxidation and O_2 reduction (Chevillot et al., 1975). Moreover, the value of this rest potential can be changed by altering the extent to which PtO covers the electrode surface. Using cyclic voltammetry, Xu et al. (2007) investigated the surface oxidation of Pt cathodes in PEM fuel cells at various relative humidities, in which the charge under the PtO_x reduction peak area shown in Figure 4.10 was used to estimate the extent of Pt oxidation. Their studies showed that the degree of Pt oxidation increased significantly with an increase in relative humidity from 20% to 72%. Holding the cathode at high potentials and exposing it to air instead of N_2 resulted in the formation of more Pt oxides.

4.4.1.2 Catalyst Activity Analysis

The high cost of currently used catalysts, mainly Pt- or PtRu-based noble metal catalysts, is becoming one of the main obstacles for scaled applications of PEM fuel cells. Over the past several decades, extensive research has focused on reducing the consumption of Pt catalysts or searching for other low-cost, nonnoble catalytic materials as alternatives. Cyclic voltammetry, including the hydrogen adsorption/desorption and CO stripping methods, has been universally adopted to qualitatively study catalyst activity toward the HOR or ORR. The activity of a catalyst can be evaluated by both the onset potential and the peak (or plateau) current. Higher activity will generate a lower onset potential and a higher peak (or limiting) current. Recently, Cui et al. (2008) synthesized mesoporous WO_3 and WO_3/C composites for potential application as PEM fuel cell anode catalysts. The electrochemical catalytic activity of the

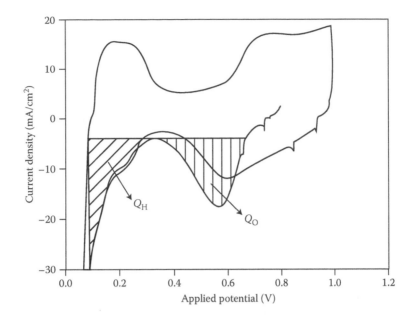

FIGURE 4.10 Cyclic voltammograms to illustrate the charge associated with hydrogen adsorption Q_H and that associated with Pt oxide reduction Q_O. Nafion 112 membrane; Pt/C cathode catalyst; $T = 100°C$; scan rate = 30 mV s⁻¹. (From Xu, H., Kunz, R., and Fenton, J. M. 2007. Investigation of platinum oxidation in PEM fuel cells at various relative humidities. *Electrochemical and Solid-State Letters* 10: B1–B5. Reproduced by permission of The Royal Society of Chemistry.)

prepared materials for hydrogen oxidation was characterized by means of the hydrogen adsorption/desorption method, as shown in Figure 4.11. The WO₃ exhibited stable electrochemical activity for hydrogen desorption/oxidation, and after mixing with the appropriate amount of carbon black the resultant mesostructured WO₃/C composites showed further enhanced electrocatalytic activity and even higher hydrogen oxidation/reduction peak currents than the commercial catalyst 20 wt% Pt/C. Experimental results have also demonstrated that the CO stripping peak potential could provide information on the composition of an unsupported metal alloy surface (Dinh et al., 2000), and that it was useful for exploring the reaction mechanism of a metal alloy with enhanced CO tolerance (Mukerjee et al., 1999).

4.4.1.3 Electrochemical Active Surface Area

The *in situ* cyclic voltammetry technique has proven to be quite valuable for ascertaining the ECSA of gas diffusion electrodes. During the electrode fabrication process, not all of the catalyst in the electrode is able to participate in the electrochemical reaction, due to either insufficient contact with the solid electrolyte or electrical isolation of the catalyst particles. Therefore, the ECSA is one of the most important parameters for evaluating catalyst layers. The ECSA of the electrode is estimated based on the relationship between the total number of reactive surface sites and the charge needed to remove a monolayer of adsorbed hydrogen on the electrode, as determined from the cyclic voltammetry measurement. The charge is practically done by integrating the current in the cathodic or anodic scan of the hydrogen adsorption/desorption region of the cyclic voltammogram. The H₂ adsorption charge on a smooth Pt electrode has been measured to be 210 μC cm⁻² of Pt loading in the catalyst layer. The ECSA of the electrode is then calculated using the following:

$$\text{ECA}\left(\text{cm}^2\text{Pt/gPt}\right) = \frac{\text{Charge}\left(\mu C/cm^2\right)}{210\left(\mu C/cm^2\,Pt\right) \times \text{Catalyst Loading}\left(gPt/cm^2\right)} \tag{4.9}$$

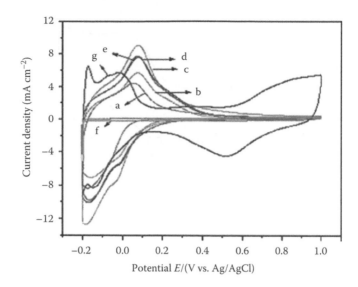

FIGURE 4.11   Cyclic voltammogram curves of various samples in 0.5 M H$_2$SO$_4$ solution with a scan rate of 0.05 V s^{-1} at 298 K: (a) WO$_3$-K, (b) m-WO$_3$/C-0.5, (c) m-WO$_3$/C-1, (d) m-WO$_3$/C-2, (e) m-WO$_3$/C-5, (f) pure carbon black, (g) commercial catalyst 20 wt% Pt/C (E-TEK). (Reprinted from Cui, X. et al. 2008. *Journal of Materials Chemistry* 18: 3575–3580. With permission from the Electrochemical Society.)

As mentioned in Section 4.3.3, one of uncertainties in the CO stripping method is the type of CO bonding on the catalyst surface. On-top or bridge-bonded CO will have an impact on the CO charge that corresponds to oxidation of a monolayer of adsorbed CO on the surface. Recent voltammetric studies (Vidaković et al., 2007) and FTIR results (Cuesta et al., 2006) have shown that at polycrystalline Pt the CO is mainly on-top adsorbed and only a small fraction is bridge bonded. Thus, when assessing ECSA based on CO stripping voltammetry, one can also employ Equation 4.9 by utilizing a value of 424 μC cm^{-2} for polycrystalline Pt (Pozio et al., 2002). Song et al. (2005) used the CO stripping technique to investigate the effect of different electrode fabrication procedures on the structural properties of MEAs. It has also been found that exposing CO to Pt and then subsequently removing that CO by electrochemical stripping is an excellent method of cleaning and activating Pt (Xu et al., 2006). During the CO adsorption and desorption processes, no degradation in the MEA performance or the ECSA of Pt was noted.

4.4.2 Diagnosis of PEM Fuel Cell Degradation

Since the power density specifications of PEM fuel cells for automotive and stationary applications are close to being met, the focus of PEM fuel cell R&D is currently shifting to the other requirements for successful introduction of the PEM fuel cell, that is, cost and durability. With respect to the latter, the requirements for fuel cell lifetime vary significantly for different applications, ranging from 5000 h for cars to 20,000 h for buses and 40,000 h of continuous operation for stationary applications. At present, most PEM fuel cell stacks provided by manufacturers and research institutes cannot achieve these goals. As a powerful electrochemical tool, cyclic voltammetry has been widely employed to determine the ECSA change in an electrode during durability testing, diagnose degradation mechanisms, and explore mitigation strategies.

Cho et al. (2003) investigated the effects that freezing water had on MEA characteristics in PEM fuel cells. Using cyclic voltammetry and other electrochemical diagnostic tools, they found that the change in water volume resulting from repetitive freezing and melting of the PEM fuel cell deformed the catalyst layer structure, thereby decreasing the ECSA as well as the utilization of Pt catalysts. However, purging with a reactant gas of a certain relative humidity (Hou et al., 2006) or an antifreeze solution (Cho et al., 2004) to remove the residual water from the PEM fuel cell significantly reduced the ECSA decay rate, providing effective solutions for fuel cell storage at subzero temperatures.

Inaba (2009) proposed two different accelerated protocols to elucidate Pt catalyst degradation mechanisms due to Pt dissolution–precipitation and carbon corrosion. In Protocol (A), square-wave potential cycling between 0.6 and 1.0 V with a pulse width of 3 s was employed to simulate the frequent load changes in an automotive application. In this case, the repetition of Pt oxide formation and reduction with steep voltage changes effectively accelerated Pt dissolution, resulting in a drastic decrease in the ECSA of Pt/C catalyst, as shown in Figure 4.12a. In Protocol (B), triangle-wave potential cycling between 1.0 and 1.5 V at 0.5 V s^{-1} was used to mimic start-up and shutdown cycles. Figure 4.12b shows the cyclic voltammogram change for Pt/C catalyst during testing using Protocol (B). The double-layer capacity in the potential range of 0.3–0.6 V increased up to 5000 cycles, then decreased together with a sudden drop

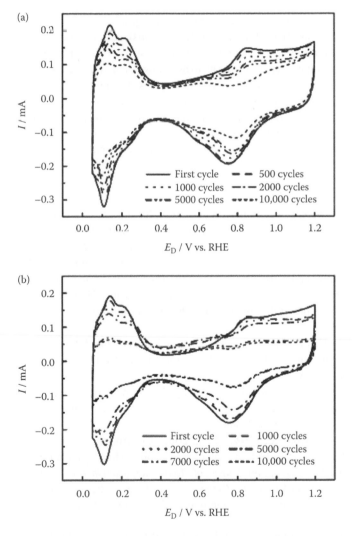

FIGURE 4.12 Variations in cyclic voltammograms of a commercially available Pt/C catalyst during accelerated tests using (a) Protocol (A) and (b) Protocol (B), respectively. Electrolyte: 0.1 M HClO$_4$; temperature: 25°C. (Reprinted from Inaba, M. 2009. *ECS Transactions* 25: 573–581. With permission from the Electrochemical Society.)

in the hydrogen adsorption and desorption peaks. The results clearly indicated that Protocol (B) effectively accelerated corrosion of the carbon support and caused Pt detachment and agglomeration.

Wu et al. (2010) utilized cyclic voltammetry to diagnose the degradation of a six-cell PEM fuel cell stack under idle conditions. The results showed that the decay rate of the average ECSA of the stack for the first 800 h was much lower than after 800 h, which was ascribed to the dominance of different degradation mechanisms. Zhang et al. (2010) studied the effect of open-circuit operation on membrane and catalyst layer degradation in PEM fuel cells. A downward ECSA trend was observed in the first 206 h, which was attributed to degradation processes in the catalyst layer such as dissolution/redeposition of Pt, corrosion of carbon, and adsorption of impurities. Further testing also showed that degradation due to Pt oxidation or catalyst contamination could be partially recovered by potential cycling processes after 206 and 256 h of open-circuit operation, resulting in a rise in the ECSA.

4.5 Concluding Remarks

Cyclic voltammetry is the most widely used technique for acquiring versatile information on the thermodynamics of redox processes and the kinetics of heterogeneous electron-transfer reactions, and on coupled chemical reactions or adsorption processes. It is often the first experiment performed in an electroanalytical study. *Ex situ* cyclic voltammetry has proven to be a convenient and relatively fast method of screening electrocatalysts, while *in situ* cyclic voltammetry has been widely used for evaluating electrode preparation procedures and characterizing electrode ECSA changes during fuel cell durability testing. However, the limitation of cyclic voltammetry is that it mainly provides qualitative information rather than quantitative results, and so to thoroughly investigate the mechanisms of electrochemical reactions and performance degradation it is always associated with other electroanalytical methods.

References

Chaparro, A. M., Martín, A. J., Folgado, M. A., Gallardo, B., and Daza, L. 2009. Comparative analysis of the electroactive area of Pt/C PEMFC electrodes in liquid and solid polymer contact by underpotential hydrogen adsorption/desorption. *International Journal of Hydrogen Energy* 34: 4838–4846.

Chevillot, J. P., Farcy, J., Hinnen, C., and Rousseau, A. 1975. Electrochemical study of hydrogen interaction with palladium and platinum. *Journal of Electroanalytical Chemistry* 64: 39–62.

Cho, E. A., Ko, J. J., Ha, H. Y., Hong, S. A., Lee, K. Y., Lim, T. W., and Oh, I. H. 2003. Characteristics of the PEMFC repetitively brought to temperatures below 0°C. *Journal of the Electrochemical Society* 150: A1667–A1670.

Cho, E. A., Ko, J. J., Ha, H. Y., Hong, S. A., Lee, K. Y., Lim, T. W., and Oh, I. H. 2004. Effects of water removal on the performance degradation of PEMFCs repetitively brought to <0°C. *Journal of the Electrochemical Society* 151: A661–A665.

Crabb, E. M., Marshall, R., and Thompsett, D. 2000. Carbon monoxide electro-oxidation properties of carbon-supported PtSn catalysts prepared using surface organometallic chemistry. *Journal of the Electrochemical Society* 147: 4440–4447.

Cuesta, A., Couto, A., Rincón, A., Pérez, M. C., López-Cudero, A., and Gutiérrez, C. 2006. Potential dependence of the saturation CO coverage of Pt electrodes: The origin of the pre-peak in CO-stripping voltammograms. Part 3: Pt(poly). *Journal of Electroanalytical Chemistry* 586: 184–195.

Cui, X., Zhang, H., Dong, X., Chen, H., Zhang, L., Guo, L., and Shi, J. 2008. Electrochemical catalytic activity for the hydrogen oxidation of mesoporous WO_3 and WO_3/C composites. *Journal of Materials Chemistry* 18: 3575–3580.

Dinh, H. N., Ren, X. M., Garzon, F. H., Zelenay, P., and Gottesfeld, S. 2000. Electrocatalysis in direct methanol fuel cells: *In-situ* probing of PtRu anode catalyst surfaces. *Journal of Electroanalytical Chemistry* 491: 222–233.

Gasteiger, H. A., Marković, N., Ross, P. N. Jr., and Cairns, E. J. 1994. Carbon monoxide electrooxidation on well-characterized platinum-ruthenium alloys. *Journal of Physical Chemistry* 98: 617–625.

Hou, J., Yu, H., Zhang, S., Sun, S., Wang, H., Yi, B., and Ming, P. 2006. Analysis of PEMFC freeze degradation at −20°C after gas purging. *Journal of Power Sources* 162: 513–520.

Inaba, M. 2009. Durability of electrocatalysts in polymer electrolyte fuel cells. *ECS Transactions* 25: 573–581.

Jusys, Z., Schmidt, T., Dubau, L., Lasch, K., Jörissen, L., Garche, J., and Behm, R. 2002. Activity of PtRuMeO$_x$ (Me = W, Mo or V) catalysts towards methanol oxidation and their characterization. *Journal of Power Sources* 105: 297–304.

Leng, Y. J., Wang, X., and Hsing, I. M. 2002. Assessment of CO tolerance for different Pt-alloy anode catalysts in polymer electrolyte fuel cell using AC impedance spectroscopy. *Journal of Electroanalytical Chemistry* 158: 145–152.

Lin, R. B. and Shih, S. M. 2006. Kinetic analysis of the hydrogen oxidation reaction on Pt-black/Nafion electrode. *Journal of Solid State Electrochemistry* 10: 243–249.

Liu, Z., Hong, L., Tham, M. P., Lim, T. H., and Jiang, H. 2006. Nanostructured Pt/C and Pd/C catalysts for direction formic acid fuel cells. *Journal of Power Sources* 161: 831–835.

Mukerjee, S., Lee, S. J., Ticianelli, E. A., McBreen, J., Grgur, B. N., Markovic, N. M., Ross, P. N., Giallombardo, J. R., and Castro, E. S. De. 1999. Investigation of enhanced CO tolerance in proton exchange membrane fuel cells by carbon supported PtMo alloy catalyst. *Electrochemical and Solid-State Letters* 2: 12–15.

Pozio, A., Francesco, M. De., Cemmi, A., Cardellini, F., and Giorgi, L. 2002. Comparison of high surface Pt/C catalysts by cyclic voltammetry. *Journal of Power Sources* 105: 13–19.

Randles, J. E. B. 1948. A cathode ray polarograph. Part II The current–voltage curves. *Transactions of the Faraday Society* 44: 327–338.

Razaq, M., Razaq, A., and Yeager, E. 1987. Oxygen electroreduction in perfluorinated sulphonylimides. *Journal of Applied Electrochemistry* 17: 1057–1064.

Sivakumar, P. and Tricoli, V. 2006. Pt–Ru–Ir nanoparticles prepared by vapor deposition as a very efficient anode catalyst for methanol fuel cells. *Electrochemical and Solid-State Letters* 9: A167–A170.

Song, S. Q., Liang, Z. X., Zhou, W. J., Sun, G. Q., Xin, Q., Stergiopoulos, V., and Tsiakaras, P. 2005. Direct methanol fuel cells: The effect of electrode fabrication procedure on MEAs structural properties and cell performance. *Journal of Power Sources* 145: 495–501.

Vidaković, T., Christov, M., and Sundmacher, K. 2007. The use of CO stripping for *in situ* fuel cell catalyst characterization. *Electrochimica Acta* 52: 5606–5613.

Wu, J. F., Yuan, X. Z., Wang, H. J., Blanco, M., Martin, J. J., and Zhang, J. J. 2008. Diagnostic tools in PEM fuel cell research: Part I Electrochemical techniques. *International Journal of Hydrogen Energy* 33: 1735–1746.

Wu, J. F., Yuan, X. Z., Martin, J. J., Wang, H. J., Yang, D. J., Qiao, J. L., and Ma, J. X. 2010. Proton exchange membrane fuel cell degradation under close to open-circuit conditions: Part I: *In situ* diagnosis. *Journal of Power Sources* 195: 1171–1176.

Xu, H., Kunz, R., and Fenton, J. M. 2007. Investigation of platinum oxidation in PEM fuel cells at various relative humidities. *Electrochemical and Solid-State Letters* 10: B1–B5.

Xu, Z., Qi, Z., and Kaufman, A. 2006. Activation of proton-exchange membrane fuel cell via CO oxidation stripping. *Journal of Power Sources* 156: 281–283.

Yano, H., Ono, C., Shiroishi, H., and Okada, T. 2005. New CO tolerant electro-catalysts exceeding Pt–Ru for the anode of fuel cells. *Chemical Communication* 2005: 1212–1214.

Zhang, D. Y., Ma, Z. F., Wang, G., Konstantinov, K., Yuan, X., and Liu, H. K. 2006. Electro-oxidation of ethanol on Pt-WO$_3$/C electrocatalyst. *Electrochemical and Solid-State Letters* 9: A423–A426.

Zhang, S. S., Yuan, X. Z., Hin, J. N. C., Wang, H. J., Wu, J. F., Friedrich, K. A., and Schulze, M. 2010. Effects of open-circuit operation on membrane and catalyst layer degradation in proton exchange membrane fuel cells. *Journal of Power Sources* 195: 1142–1148.

5

Linear Sweep Voltammetry

Shengsheng Zhang
Institute for Fuel Cell Innovation

Xiao-Zi Yuan
Institute for Fuel Cell Innovation

Haijiang Wang
Institute for Fuel Cell Innovation

5.1 Introduction

The membrane is a key component in a PEM fuel cell, functioning as an electrolyte for transferring protons from anode to cathode, as well as a barrier to prevent gas permeation and electron transfer between anode and cathode. In a traditional membrane electrode assembly (MEA) it is always sandwiched by two catalyst layers (CLs) and gas diffusion layers (GDLs). Due to their high hydrolytic and oxidative stability and excellent proton conductivity, perfluorosulfonic acid (PFSA) membranes such as Nafion® have been widely used in PEM fuel cells. During the last few decades of development in fuel cell technology, the properties of the Nafion membrane and other components have been greatly improved. For example, the ordinary thickness of Nafion membrane has been lowered to just ~25 μm, which significantly enhances cell performance due to reduced membrane resistance. Several alternative membranes are also in the process of development, including reinforced or modified PFSA membranes (Ralph et al., 2003; Zhu et al., 2006; Wang et al., 2007), and alternative ionomer polymers and their composite membranes, such as sulfonated poly-sulfone (SPSF) (Genova-Dimitrova et al., 2001), sulfonated poly-ether ether ketone (SPEEK) (Park et al., 2008), modified poly-benzimidazole (PBI) (Zhai et al., 2007), and modified poly-vinylidene fluoride (PVDF) (Gode et al., 2003). However, none of these materials are perfect for achieving high performance and reliable longevity under fuel cell operating conditions.

The greatest concern for membranes is gas crossover. Although the membrane should act as a nonpermeable component in PEM fuel cells, at present all the commonly used membranes encounter the problem of gas permeation, especially when a thinner membrane is pursued for its significant electrical efficiency benefits. Both oxidant and fuel can cross through the membrane to the opposite side during cell operation. Gas crossover impacts several factors in PEM fuel cell development. In terms of reactant usage, reactant that crosses over through the membrane is consumed without generating useful power.

This inevitably leads to fuel inefficiency. For instance, fuel crossover of 1 mA cm^{-2} equates to a loss in current efficiency of 0.25% at an operating current density of 400 mA cm^{-2}. As gas crossover increases, cell performance is expected to decrease. The mixing of oxygen and hydrogen potentially leads to the formation of peroxide and hydroperoxide radicals, which harm the membrane by reducing the lifespan of PEM fuel cell stacks (LaConti et al., 2003). Furthermore, mixed potential at the electrodes, which may decrease the open-circuit potential, is another negative effect on cell performance. Weber (2008) has also addressed some other impacts of increased gas crossover, including substantial hydrogen dilution by nitrogen, which may lead to fuel starvation at the end of the cell and subsequent carbon corrosion. Therefore, gas crossover in a PEM and its impact on fuel cell operation is an important parameter that needs to be characterized in the development and advancement of PEM fuel cells. Accordingly, we dedicate an independent chapter to introduce a characterization method that is highly important and widely used.

Although crossover of both fuel (e.g., hydrogen) and oxidant (oxygen) takes place, the latter generally occurs at a lower rate. For instance, it has been found that under the same conditions, the crossover rate of oxygen is about half the rate of hydrogen through Nafion 125 membrane (with a thickness of 125 µm) at 25°C (Ogumi et al., 1984a). In addition, based on Weber's simulation (Weber, 2008), it has also been demonstrated that hydrogen crossover is more detrimental to cell operation than oxygen crossover, due to the facile kinetics of hydrogen oxidation and the fact that most fuel gases are pure hydrogen. Therefore, fuel crossover is most often the main issue of interest, and has been investigated by various research groups (Ogumi et al., 1984a,b; Sakai et al., 1985; Broka et al., 1997; Inaba et al., 2006).

Several methods have been developed to characterize hydrogen crossover through the membrane in PEM fuel cells, such as the volumetric, gas chromatography, and morphology methods. For the volumetric method, higher pressure is applied at one side of the membrane and the gas permeation rates are measured at the other side (Sakai et al., 1985). During testing, humidified penetrating gases such as H$_2$ and O$_2$ are introduced into the higher-pressure side. The gas crossover rate is measured by a precise gas flow meter in a steady-state condition at the lower-pressure side. To prevent the wet membrane from drying out during the experiment, filter paper soaked in distilled water is placed on both sides of the membrane.

Unlike the pressure gradient used in the volumetric method, the concentration difference is the driving force for the gas chromatography method. This method measures the change in gas concentration for one side of the membrane when different gas concentrations are applied to both sides (Broka and Ekdunge, 1997). The amount of oxygen or hydrogen diffusing through the membrane is determined by using a gas chromatograph to measure the concentration in the gas outlet from the right chamber.

With newly developed morphology methods such as IR imaging (see Chapter 16 for details), it is possible to detect pinholes in the membrane that can result in huge crossover within the fuel cell. However, rather than yielding a quantitative result, morphology methods can only tell the position and relative degree of the crossover. No accurate gas crossover data are available.

All these aforementioned methods can provide useful information on a membrane's gas permeability and gas crossover. However, they require additional equipment or special designs and they are difficult to carry out under conditions close to those in real fuel cells. In comparison, linear sweep voltammetry (LSV), an electrochemical method, is the most frequently used technique to estimate gas crossover through membranes in PEM fuel cells, as it can be applied directly under different, realistic fuel cell operating conditions (Kocha et al., 2006). During testing, a linear potential scan is applied to the fuel cell electrode to obtain a limiting current, which is useful for calculating the crossover rate of hydrogen. The test can be carried out online without disassembling the cell system. Thus, the gas crossover rate provided by LSV expresses realistic membrane situations under actual working conditions. Due to its convenience and practicability, LSV has become a basic testing method to determine the applicability of a material when assessing new membranes. During degradation tests under harsh conditions or long-term fuel cell operation, it is also easy to conduct LSV before and after operation, or during any operation period, to assess gas crossover evolution. Such comparison provides information about membrane degradation over a period of time, which is helpful in monitoring failure modes and investigating fuel cell degradation mechanisms.

5.2 Principles and Measurements

5.2.1 Principles

The primary advantage of LSV is its practicability, particularly as measurements can be obtained under various realistic fuel cell operating conditions. With regard to gas crossover measurements in PEM fuel cells, normally the testing setup for LSV is similar to that for cyclic voltammetry (CV), with the supplement of humidified H_2 and N_2 (or other inert gases) to the fuel cell's anode and cathode, respectively. In this case, the anode serves as the reference electrode (RE) as well as the counter electrode (CE), and the cathode serves as the working electrode (WE). Fully or supersaturated humidified conditions are generally used for both the fuel and the inert gas to guarantee sufficient water is present for full membrane hydration. Using a potentiostat or other controllable power supply, which can generate slow-scan linear sweep voltammograms, the working electrode is swept by a linear potential scan from the initial potential to the final potential (e.g., 0–0.5 V) at a low scan rate (e.g., 4 mV s^{-1}). Typically, potentials higher than 0.8 V are avoided to prevent irreversible Pt or carbon oxidation (Ramani et al., 2004).

When controlling the applied potential of the fuel cell cathode, electrochemical activity occurring in the form of a current is monitored through the two electrodes at the same time. Under the chosen high potential, hydrogen that has crossed through the membrane from anode to cathode should be totally oxidized under mass transfer-limited conditions, which can instantaneously generate a limiting hydrogen oxidation current density. Since N_2 gas is the only substance introduced into the cathode side, any current generated in the given potential range is solely attributable to the electrochemical oxidation of H_2 gas that crosses over from the anode side through the membrane. Thus, the LSV can be used to estimate the hydrogen crossover from anode side to cathode side. A schematic diagram of LSV measurement is shown in Figure 5.1. According to the reaction shown in Figure 5.1, the hydrogen crossing to the cathode side is electrochemically oxidized into protons. On the other side, hydrogen is produced through the reaction of protons and electrons at the anode. The overall fuel cell reaction during the test can be expressed as

$$H_2\left(\text{Cathode}\right) \rightarrow H_2\left(\text{Anode}\right) \tag{5.1}$$

5.2.2 Measurements

Like CV measurements, in LSV measurements the current response is plotted as a function of voltage rather than time. As an example, Figure 5.2 shows a typical LSV curve (hydrogen crossover current density vs. potential). One can find from the current density curve that as the applied voltage increases from a low potential and the oxidation of hydrogen at the cathode electrode, the current density increases rapidly and then reaches a limiting value (i_{lim}) when the potential grows to around 0.3–0.35 V, a potential range that is free from the effects of hydrogen adsorption/desorption on platinum (Inaba et al., 2006). The hydrogen crossover rates can be expressed by the limiting currents, which are in the order of a few milliamps to tens of milliamps per square centimeter. For example, the hydrogen crossover current of a Nafion 112 membrane of ~50 μm thickness is reported to be 1 mA cm^{-2} at the beginning of the membrane's life under atmospheric pressure (Cleghorn et al., 2006).

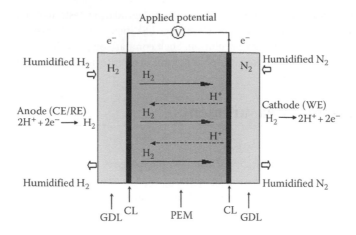

FIGURE 5.1 Schematic of the principles of LSV measurement in PEM fuel cells.

On the basis of the limiting current density, the hydrogen crossover flux $J_{x\,over}$ (mol cm^{-2} s^{-1}) can be determined by Faraday's law,

$$J_{xover} = \frac{i_{lim}}{nF} \tag{5.2}$$

where i_{lim} is the transport limiting current density (A cm^{-2}), n is the number of electrons taking part in the reaction ($n = 2$ for the hydrogen oxidation reaction), and F is Faraday's constant (96,485 C mol^{-1}). For example in Figure 5.2, the calculated flux $J_{x\,over}$ is 2.59×10^{-8} mol cm^{-2} s^{-1}.

5.3 Applications

5.3.1 Hydrogen Crossover Determination

A considerable number of LSV experiments have been carried out to investigate various types of membranes under different conditions. These investigations have proven that the crossover rate is determined both by membrane properties (such as composition and thickness) and operational environment factors (such as gas pressure, temperature, and humidification level).

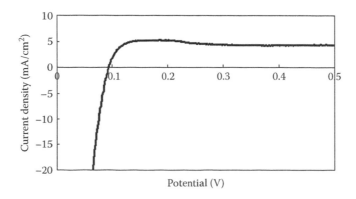

FIGURE 5.2 A typical LSV curve measured at 60°C for a commercial Nafion membrane under 100% humidification.

5.3.1.1 Effect of Pressure and Temperature

Figure 5.3 shows the temperature and pressure effects on hydrogen crossover currents for Nafion 112 membrane in a PEM fuel cell supplied with saturated gases (Kocha et al., 2006). From the curves, linearly increasing crossover current is observable with increases in hydrogen partial pressure and temperature. The temperature effect might be explained by mechanical stress of the membrane, which would be greater at higher temperatures and would allow hydrogen molecules to permeate more easily through the membrane (Solasi et al., 2007). Inaba et al. (2006) also reported that the hydrogen crossover current density increased dramatically with increasing hydrogen pressure, under both iso-pressure and differential pressure conditions. In addition, the pressure difference across the membrane further enhanced hydrogen crossover. As shown in Figure 5.4, when the hydrogen pressure was 0.2 MPa higher than that of the inert gas at the cathode, the hydrogen crossover was about 10 times higher than at atmospheric pressure.

5.3.1.2 Effect of Membrane Thickness

A recent durability test under idle conditions has reported the gas crossover rates for membranes of different thickness, as shown in Figure 5.5 (Yuan et al., 2010). The corresponding properties of these membranes are listed in Table 5.1. It is worth noting that LSV cannot measure the crossover of the entire cell stack. Instead, each crossover datum listed was obtained separately in a four-cell stack. It is evident that the thinner the membrane, the more hydrogen crosses over when other conditions are kept the same.

5.3.1.3 Effect of Humidity and Flow Rate

Gas permeability through the membrane is also a function of the relative humidity (RH) of the gas, which determines the water content in the membrane. According to the LSV measurements by Inaba et al. (2006), hydrogen crossover through the membrane clearly increases with greater humidity and

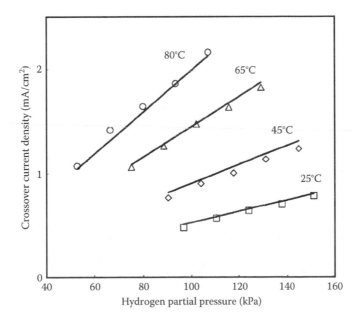

FIGURE 5.3 Hydrogen crossover current versus hydrogen partial pressure through Nafion 112 membrane at different temperatures. (From Kocha, S. S., Yang, J. D., and Yi, J. S. Characterization of gas crossover and its implications in PEM fuel cells. *AIChE J.* 2006. 52: 1916–1924. Copyright Wiley-VCH Verlag GmbH & Co. KGaA. Reproduced with permission.)

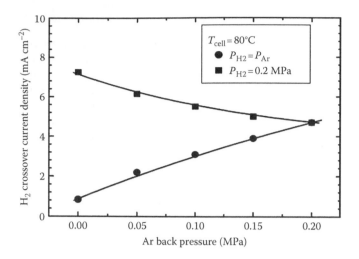

FIGURE 5.4 Effects of gas pressure on H_2 crossover at 80°C. H_2 and Ar were fed at iso-pressure ($P_{H2} = P_{Air}$) or at different pressures ($P_{H2} = 0.2$ MPa, $P_{H2} \geq P_{Air}$). (Reproduced from *Electrochim. Acta*, 51, Inaba, M. et al. Gas crossover and membrane degradation in polymer electrolyte fuel cells, 5746–5753, Copyright (2006), with permission from Elsevier.)

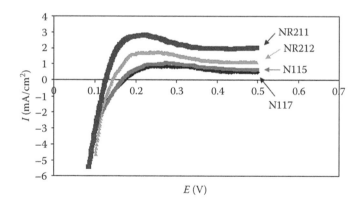

FIGURE 5.5 Hydrogen crossover comparison for Nafion membranes with different thicknesses at 70°C and 100%RH. (Reproduced from *J. Power Sources*, 195, Yuan, X.-Z. et al. Degradation of a polymer exchange membrane fuel cell stack with Nafion® membranes of different thicknesses: Part I. *In situ* diagnosis, 7594–7599, Copyright (2010), with permission from Elsevier.)

TABLE 5.1 Comparison of Properties and Hydrogen Crossover of Nafion PFSA membranes

Membrane Type	Typical Thickness (μm)	Basis Weight (g m⁻²)	Limiting Current Density (mA cm⁻²)	Hydrogen Crossover Flux (mol cm⁻² s⁻¹)
NR211	25.4	50	2.0	1.04×10^{-8}
NR212	50.8	100	1.1	5.7×10^{-9}
N115	127	250	0.72	3.73×10^{-9}
N117	183	360	0.57	2.95×10^{-9}

Source: Reproduced from *J. Power Sources*, 195, Yuan, X.-Z. et al. Degradation of a polymer exchange membrane fuel cell stack with Nafion® membranes of different thicknesses: Part I. *In situ* diagnosis, 7594–7599, Copyright (2010), with permission from Elsevier.

temperature, as shown in Figure 5.6. The authors explained this dependence by the increase in membrane flexibility at higher temperatures and humidities. Under these conditions, the hydration level of the membrane is enhanced by the higher water content, which might increase the expansion and even the pore sizes of the membrane, thereby facilitating hydrogen crossover. More recently, Wasterlain et al. (2010) also examined the impacts of flow rates, operating temperatures, and RHs on hydrogen crossover in Nafion membranes. Similar results were found: a higher hydrogen crossover rate was detected under conditions of elevated temperature and RH, while the reactant flow rate showed no significant effect on gas crossover.

5.3.2 Methanol Crossover Determination

Methanol crossover can also be determined using this basic voltammetric method by replacing hydrogen with liquid methanol (Ren et al., 2000).

Methanol permeation through the membrane is an important factor affecting direct methanol fuel cell (DMFC) performance due to the reactant consumption and catalyst poisoning. Liquid reactant crossover is more serious than gas crossover. Therefore, characterization of and mitigation methods for methanol crossover are urgently needed to enhance DMFC development. In practice, methanol that has diffused to the cathode side can be determined by gas chromatography (GC) at the cathode exhaust. For instance, Gurau and Somtkin (2002) used this method and concluded that methanol crossover depends on operating parameters, including MEA temperature, methanol solution concentration, and flow rate.

However, a number of different investigations have proven that LSV can be employed as a reliable, convenient method for measuring the methanol permeation rate through a membrane under operating conditions using fuel cell hardware directly (Jia et al., 2000; Ren et al., 2000; Seo and Lee, 2010). Similar to measurements taken under hydrogen and nitrogen conditions, the flux rate of methanol crossover is determined by measuring the steady-state limiting current density resulting from complete electrooxidation at the Nafion membrane/Pt catalyst interface under applied potentials in an inert atmosphere, such as well-humidified nitrogen or pure water. Figure 5.7 shows a schematic diagram of LSV to measure methanol crossover in DMFCs. This method has been applied to membrane modification analysis and investigations on the factors influencing methanol crossover. Figure 5.8 is an example of the application of LSV in methanol crossover detection. The limiting current at higher potentials is proportional to the

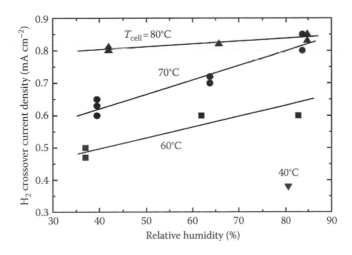

FIGURE 5.6 Effect of cell temperature and humidification on H_2 crossover current density at atmospheric pressure. (Reproduced from *Electrochim. Acta*, 51, Inaba, M. et al. Gas crossover and membrane degradation in polymer electrolyte fuel cells, 5746–5753, Copyright (2006), with permission from Elsevier.)

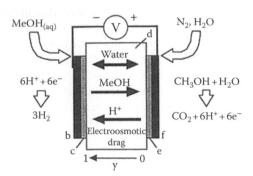

FIGURE 5.7 Schematic diagrams showing methanol permeation measurement in a DMFC configuration. (Reproduced from Ren, X. et al. 2000. *J. Electrochem. Soc* 147: 466–474. With permission from Electrochemical Society.)

rate of methanol crossover (i.e., the flux of methanol across the membrane). Therefore, the result indicates that methanol crossover increases with operating temperature. Ren et al. (2000) obtained similar results for a preswollen membrane and proposed that the measured methanol flux rates must be corrected for the effect of electro-osmotic drag from the membrane protonic current, in order for them to represent the methanol crossover rate in a DMFC. Conversely, for a predried membrane they observed that the methanol flux rate decreased above 90°C, possibly due to changes in the membrane structure under this condition.

5.3.3 Membrane Evaluation and Degradation

From the viewpoint of component performance, gas crossover reflects the working condition of a membrane, which greatly influences a PEM fuel cell's performance, efficiency, and durability (Cooper, 2008). As hydrogen across the membrane from the anode can be oxidized chemically or electrochemically at the cathode in operating fuel cells, LSV data are helpful for understanding a fuel cell's performance loss

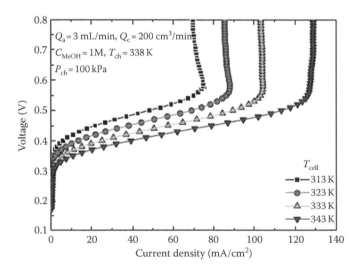

FIGURE 5.8 Effects of cell temperature on methanol crossover and efficiency of a DMFC. (Reproduced from *Appl. Energy*, 87, Seo, S. H. and Lee, C. S., A study on the overall efficiency of direct methanol fuel cell by methanol crossover current, 2597–2604, Copyright (2010), with permission from Elsevier.)

and degradation. For instance, using this diagnostic method, Zhang et al. (2006) determined the hydrogen crossover rate through Nafion 112 membrane in the temperature range of 23–120°C. Based on their measurements, the authors proposed that hydrogen crossing from anode to cathode could react with oxygen and reduce the oxygen surface concentration, resulting in an open-circuit voltage (OCV) drop. This potential drop is pointed to as one important source of the difference between theoretical and measured OCV. Another part of the potential drop is the mixed potential arising from the reaction between the Pt surface and oxygen in the cathode.

As a membrane thins, the gas crossover effect becomes a critical factor in the lifetime or durability of a fuel cell. From this point of view, LSV also plays an important role in failure diagnosis by measuring crossover currents before and after durability testing; it is sufficiently sensitive to diagnose membrane integrity under diverse conditions. Zhang et al. (2010) and Wu et al. (2010) measured hydrogen crossover changes under both OCV and idle conditions using LSV in a single cell and a six-cell stack, respectively. Figure 5.9 compares the hydrogen crossover current density measured by LSV when the cell suffered different periods of open circuit (OC) degradation during 256 h of OCV testing. The two recovery processes of potential cycling, consisting of three voltage steps at 0.6 V, 0.3 V, and OCV, were conducted on the cell after 206 h of OCV degradation (first recovery) and after 256 h of OCV degradation (second recovery). It can be observed that the limiting current densities are approximately stable for 206 h during degradation. The calculated hydrogen crossover flux according to the limiting current density (<6 mA cm^{-2}) is about 3×10^{-8} mol cm^{-2} s^{-1}. After 256 h of OC operation, the current density curve exhibits a significant jump in the limiting current density, which indicates a high gas crossover rate through the membrane. Based on the polarization curves measured during the same experiments, cell performance decreases dramatically at the same time. These findings therefore demonstrate that severe membrane structural damage (membrane thinning or pinhole formation) occurred during 206–256 h, which led to fatal failure in the fuel cell.

Similar results can be found elsewhere in the literature. Yuan et al. (2010) studied the hydrogen crossover evolution of fuel cells in a four-cell stack with Nafion membranes of different thicknesses. Their results showed that for thicker membranes (N117 and N115) there was no significant change in hydrogen crossover before versus after degradation. However, the crossover rate jumped significantly for NR212 after 1000 h of operation, whereas for NR211 hydrogen crossover started to increase after 600 h of operation and kept increasing with degradation until it exceeded the equipment limit (2 A) after 1000 h of degradation.

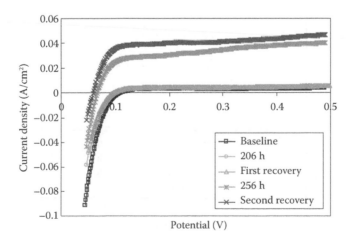

FIGURE 5.9 Hydrogen crossover current density evolution during 256 h of OCV testing. (Reproduced from *J. Power Sources*, 195, Zhang, S. et al. Effects of open-circuit operation on membrane and catalyst layer degradation in proton exchange membrane fuel cells, 1142–1148, Copyright (2010), with permission from Elsevier.)

To develop and explore new materials for the PEMs used in fuel cells, LSV can also help by detecting the gas crossover rate, since gas permeability is as important as proton conductivity for PEM fuel cell membranes. For example, Zhai et al. (2007) employed LSV at 150°C and 0.1 MPa pressure in characterizing the hydrogen permeability of the novel composite membrane H₃PO₄/Nafion-PBI, and thereby indicated its enhanced durability. In their 720 h life test, hydrogen crossover was tested by LSV analysis at different times, as shown in Figure 5.10. During the first test at about 360 h, the hydrogen crossover current was very small. It then increased markedly as degradation time continued. In the last life test, at about 760 h the hydrogen crossover current rose rapidly with the increase in working electrode potentials, and no limiting current was found with increasing overpotential. Similarly, Ramani et al. (2004) also used LSV to investigate the gas crossover performance of Nafion/HPA composite membranes at room temperature. The results revealed that the hydrogen crossover current of different kinds of Nafion/HPA composite membranes was very small, indicating that these membranes were all sufficiently robust. No evidence of internal shorting was found in any of the MEAs. Further study demonstrated the ability of these membranes to perform satisfactorily in high-temperature and low-RH environments.

5.3.4 Short-Circuit Detection

Apart from enabling the detection of hydrogen crossover, LSV is also a powerful diagnostic method to provide useful information about any internal short circuit between the anode and the cathode. Although Nafion electrolyte is designed to be an ionic conductor but not an electronic conductor, finite electrical shorts between anode and cathode may occur as a result of electrolyte thinning or pinhole formation. Fuel crossover and internal short circuits are essentially equivalent in terms of their impact on efficiency. According to the literature (Vilekar and Datta, 2010), under normal conditions, the electrical short-circuit current is an order of magnitude smaller than the hydrogen crossover current. Direct conduction of electrons between the electrodes through the electrolyte is a source of loss within a fuel cell. Figure 5.11 shows the simulated current responses for various magnitudes of electrical shorts. The significant difference in these curves from the ordinary hydrogen crossover curve is the linearly increasing current superimposed in the limiting current region. Under these conditions, both the crossover current and the current flow through the electrical short and contribute to the measured current.

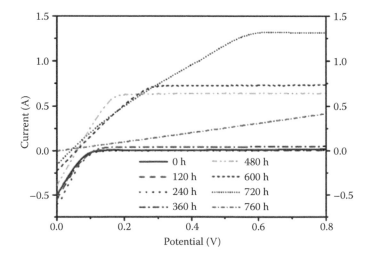

FIGURE 5.10 LSV curves of an MEA with H₃PO₄/Nafion-PBI composite membrane. (Reproduced from *J. Power Sources*, 169, Zhai, Y. et al. A novel H₃PO₄/Nafion-PBI composite membrane for enhanced durability of high temperature PEM fuel cells, 259–264, Copyright (2007), with permission from Elsevier.)

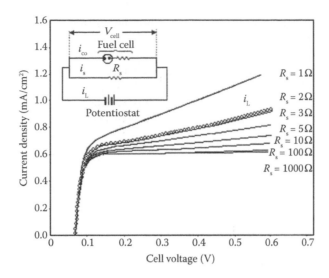

FIGURE 5.11 Simulated current responses for various magnitudes of electrical shorts. (From Kocha, S. S., Yang, J. D., and Yi, J. S. Characterization of gas crossover and its implications in PEM fuel cells. *AIChE J.* 2006. 52: 1916–1924, Copyright Wiley-VCH Verlag GmbH & Co. KGaA. Reproduced with permission.)

The electrical short current linearly increases with the applied voltage and is inversely proportional to the resistance of the short. In calculations, the magnitude of the electrical short is equal to the inverse of the slope (*V/I*). In practice, the short-circuit resistance is usually >>1000 Ω cm^2, and so the current density measurement errors are <<1 mA cm^{-2} and can be disregarded (Gasteiger et al., 2005).

Wasterlain et al. (2010) observed internal short-circuit resistance in their study of temperature, air dew point temperature, and reactant flow effects on PEM fuel cell performance. According to their report, the short-circuit resistances calculated from the inverse of the slope in the linear region (0.4–0.6 V) under different conditions proved that the short-circuit resistance does not show any strong dependence on temperature, air dew point temperature, or reactant flow. In general, a drop in short-circuit resistance frequently occurs during durability testing for membrane degradation. In the OCV degradation test described earlier and illustrated in Figure 5.9, after 256 h of degradation the current curve showed a slight linear increase with increasing potential, which is attributable to an electrical short circuit due to membrane thinning or pinholes.

5.4 Chronocoulometry

Another electrochemical method to measure hydrogen crossover through the membrane in PEM fuel cells is chronocoulometry (CC), which is quite similar to LSV (Cleghorn et al., 2003; Wu et al., 2008). As one of the classic electrochemical techniques frequently used in electroanalytical chemistry, CC measures the charge (in Coulombs) as a function of time. In CC, hydrogen and nitrogen are supplied to the anode and cathode of the fuel cell, respectively. In addition, the anode serves as RE and CE while the cathode serves as WE. The only difference between CC and LSV is that instead of a potential scan, a certain potential, for example, 0.2 V, is applied to the cathode side of the cell. Under this potential, the hydrogen crossing over the membrane from the anode side can be completely oxidized. It is possible to record the electrical charge, which arises from oxidation, passing through the electrode. By the coulombs measured, the hydrogen crossover rate can be calculated. Although it is a simple electroanalytical technique, CC is not as widely used as LSV in the investigation of gas crossover in fuel cells.

In terms of application, CC can also be used to evaluate the crossover rate of different membranes and the chronological degradation of membranes during lifetime tests. Several examples can be found in the literature. Using this method, Yu et al. (2005) determined the gas crossover rate of a membrane during lifetime testing under low-humidification conditions. The authors observed that hydrogen crossover noticeably increased with lifetime, which proved that the membrane degradation rate accelerated during long-term operation. Further analysis of how increased gas crossover and pinhole production can dramatically affect membrane decay, and thereby cell performance, was also illustrated by their experimental results. A series of novel membrane investigations have also employed this method to measure hydrogen crossover under diverse conditions (Liu et al., 2006, 2007; Wang et al., 2006).

5.5 Summary

To sum up, LSV is a powerful analytical tool to characterize the fuel crossover level of membranes in PEM fuel cells. Hydrogen and methanol crossover measurements can be effectively carried out under real fuel cell operating conditions (i.e., temperature, flow rate, RH, back pressure, etc.) based on the configurations of both single cells and cell stacks. This is important for the convenience and accuracy of membrane evaluation. In addition, LSV testing can yield quantitative results for the membrane that encompass the whole active surface area without incurring any damage to the membrane itself. LSV testing also permits the measurement of short-circuit resistance. Currently, it is the most frequently used method to investigate membrane properties and the related performance of PEM fuel cells.

LSV is an online-served method, which means measurement is conducted based on the entire cell configuration. This makes LSV the most convenient characterization tool for hydrogen crossover evaluation. However, this aspect is also the source of its limitation, because other components such as CLs and GDLs are also in the testing system together with the membrane during LSV. Fortunately, the influence of these components is very limited and can be neglected under most conditions. Cheng et al. (2007) has constructed a hydrogen crossover model based on an MEA consisting of five layers: anode GDL, anode CL, PEM (Nafion 112 or Nafion 117), cathode CL, and cathode GDL. Their model analysis suggests that the dominant factor in overall hydrogen crossover is the step when hydrogen diffuses through the PEM. But these factors must be the subject of future investigations with the capacity for extremely high accuracy.

In general, LSV is an excellent diagnostic method for membrane evaluation and failure diagnosis in PEM fuel cells. Although it only reflects gas crossover at an average level, the sensitivity and accuracy are unassailable for membrane diagnosis, which can be conducted whenever a new cell is assembled, and periodically during testing. As a simple, nondestructive technique, LSV is also useful in mechanism analysis and degradation prediction, by yielding information prior to postmortem analysis.

References

Broka, K. and Ekdunge, P. 1997. Oxygen and hydrogen permeation properties and water uptake on Nafion® 117 membrane and recast film for PEM fuel cell. *J. Appl. Electrochem.* 27: 117–123.

Cheng, X., Zhang, J., Tang, Y. et al. 2007. Hydrogen crossover in high-temperature PEM fuel cells. *J. Power Sources* 167: 25–31.

Cleghorn, S., Kolde, J., and Liu, W. 2003. Catalyst-coated composite membranes. In *Handbook of Fuel Cells: Fundamentals, Technology and Applications*, eds. W. Vielstich, H. A. Gasteiger, and A. Lamm; pp. 566–575. New York, NY: Wiley.

Cleghorn, S. J. C., Mayfield, D. K., Moore, D. A., and Moore, J. C. 2006. A polymer electrolyte fuel cell life time test: 3 years of continuous operation. *J. Power Sources* 158: 446–454.

Cooper, K. R. 2008. *In situ* PEM fuel cell: Fuel crossover and electrical short circuit measurement. *Fuel Cell* 8: 34–35.

Gasteiger, H. A., Kocha, S. S., Sompalli, B., and Wagner, F. T. 2005. Activity benchmarks and requirements for Pt, Pt-alloy, and non-Pt oxygen reduction catalysts for PEMFCs. *Appl. Catal. B* 56: 9–35.

Genova-Dimitrova, P., Baradie, B., Foscallo, D., Poinsignon, C., and Sanchez, J. Y. 2001. Ionomeric membranes for proton exchange membrane fuel cell (PEMFC): Sulfonated polysulfone associated with phosphatoantimonic acid. *J. Membr. Sci.* 185: 59–71.

Gode, P., Ihonen, J., Strandroth, A. et al. 2003. Membrane durability in a PEM fuel cell studied using PVDF based radiation grafted membranes. *Fuel Cells* 3: 21–27.

Gurau, B. and Smotkin, E. S. 2002. Methanol crossover in direct methanol fuel cells: A link between power and energy density. *J. Power Sources* 112: 339–352.

Inaba, M., Kinumoto, T., Kiriake, M., Umebayashi, R., Tasaka, A., and Ogumi, Z. 2006. Gas crossover and membrane degradation in polymer electrolyte fuel cells. *Electrochim. Acta* 51: 5746–5753.

Jia, N., Lefebvre, M. C., Halfyard, J., Qi, Z., and Pickup, P. G. 2000. Modification of Nafion proton exchange membranes to reduce methanol crossover in PEM fuel cells. *Electrochem. Solid-State Lett.* 3: 529–531.

Kocha, S. S., Yang, J. D., and Yi, J. S. 2006. Characterization of gas crossover and its implications in PEM fuel cells. *AIChE J.* 52: 1916–1924.

LaConti, A. B., Hamdan, M., and McDonald, R. C. 2003. Mechanisms of membrane degradation. In *Handbook of Fuel Cells-Fundamentals, Technology and Applications*, eds. W. Vielstich, H. A. Gasteiger, A. Lamm, Vol. 3. Chichester, UK: Wiley.

Liu, G., Zhang, H., Hu, J., Zhai, Y., Xu, D., and Shao, Z. 2006. Studies of performance degradation of a high temperature PEMFC based on H_3PO_4-doped PBI. *J. Power Sources* 162: 547–552.

Liu, Y.-H., Yi, B., Shao, Z., Wang, L., Xing, D., and Zhang, H. 2007. Pt/CNTs-Nafion reinforced and self-humidifying composite membrane for PEMFC applications. *J. Power Sources* 163: 807–813.

Ogumi, Z., Kuroe, T., and Takehara, Z. -I. 1984a. Gas permeation in SPE method II. Oxygen and hydrogen permeation through Nafion. *J. Electrochem. Soc.* 132: 2601–2605.

Ogumi, Z., Takehara, Z., and Yoshizawa, S. 1984b. Gas permeation in SPE method I. Oxygen permeation through Nafion and NEOSEPTA. *J. Electrochem. Soc.* 131: 769–773.

Park, J.-S., Krishnman, P., Park, S.-H. et al. 2008. A study on fabrication of sulfonated poly(ether ether ketone)-based membrane-electrode assemblies for polymer electrolyte membrane fuel cells. *J. Power Sources* 178: 642–650.

Ralph, T. R., Barnwell, D. E., Bouwman, P. J., Hodgkinson, A. J., Petch, M. I., and Pollington, M. 2003. Reinforced membrane durability in proton exchange membrane fuel cell stacks for automotive applications. *J. Electrochem. Soc* 155: B411–B422.

Ramani, V., Kunz, H. R., and Fenton, J. M. 2004. Investigation of Nafion®/HPA composite membranes for high temperature/low relative humidity PEMFC operation. *J. Membr. Sci* 232: 31–44.

Ren, X., Springer, T. E., Zawodzinski, T. A., and Gottesfeld, S. 2000. Methanol transport through Nafion membranes—Electro-osmotic drag effects on potential step measurements. *J. Electrochem. Soc* 147: 466–474.

Sakai, T., Takenaka, H., Wakabayashi, N., Kawami, Y., and Torikai, E. 1985. Gas permeation properties of solid polymer electrolyte (SPE) membranes. *J. Electrochem. Soc* 132: 1328–1382.

Seo, S. H. and Lee, C. S. 2010. A study on the overall efficiency of direct methanol fuel cell by methanol crossover current. *Appl. Energy* 87: 2597–2604.

Solasi, R., Zou, Y., Huang, X., Reifsnider, K., and Condit, D. 2007. On mechanical behaviour and in-plane modeling of constrained PEM fuel cell membranes subjected to hydration and temperature cycles. *J. Power Sources* 167: 366–377.

Vilekar, S. A. and Datta, R. 2010. The effect of hydrogen crossover on open-circuit voltage in polymer electrolyte membrane fuel cells. *J. Power Sources* 195: 2241–2247.

Wang, L., Xing, D. M., Liu, Y. H. et al. 2006. Pt/SiO_2 catalyst as an addition to Nafion/PTFE self-humidifying composite membrane. *J. Power Sources* 161: 61–67.

Wang, L., Yi, B. L., Zhang, H. M. et al. 2007. Sulfonated polyimide/ PTFE reinforced membrane for PEMFCs. *J. Power Sources* 167: 47–52.

Wasterlain, S., Candusso, D., Hissel, D. et al. 2010. Study of temperature, air dew point temperature and reactant flow effects on proton exchange membrane fuel cell performances using electrochemical spectroscopy and voltammetry techniques. *J. Power Sources* 195: 984–993.

Weber, A. Z. 2008. Gas-crossover and membrane-pinhole effects in polymer-electrolyte fuel cells. *J. Electrochem. Soc* 155: B521–B531.

Wu, J., Yuan, X.-Z., Wang, H., Blanco, M., Martin, J., and Zhang, J. 2008. Diagnostic tools in PEM fuel cell research: Part I Electrochemical techniques. *Int. J. Hydrogen Energy* 33: 1735–1746.

Wu, J., Yuan, X.-Z., Martin, J. J. et al. 2010. Proton exchange membrane fuel cell degradation under close to open-circuit conditions, Part I. *In situ* diagnosis. *J. Power Sources* 195: 1171–1176.

Yu, J., Matsuura, T., Yoshikawa, Y., Islam, M. N., and Hori, M. 2005. Lifetime behaviour of a PEM fuel cell with low humidification of feed stream. *Phys. Chem. Chem. Phys.* 7: 373–378.

Yuan, X.-Z., Zhang, S., Wang, H. et al. 2010. Degradation of a polymer exchange membrane fuel cell stack with Nafion® membranes of different thicknesses: Part I. *In situ* diagnosis. *J. Power Sources* 195: 7594–7599.

Zhai, Y., Zhang, H., Zhang, Y., and Xing, D. 2007. A novel H_3PO_4/Nafion-PBI composite membrane for enhanced durability of high temperature PEM fuel cells. *J. Power Sources* 169: 259–264.

Zhang, J., Tang, Y., Song, C., Zhang, J., and Wang, H. 2006. PEM fuel cell open circuit voltage (OCV) in the temperature range of 23°C to 120°C. *J. Power Sources* 163: 532–537.

Zhang, S., Yuan X.-Z., Hin, J. N. C. et al. 2010. Effects of open-circuit operation on membrane and catalyst layer degradation in proton exchange membrane fuel cells. *J. Power Sources* 195: 1142–1148.

Zhu, X. B., Zhang, H. M., Liang, Y. M., Zhang, Y., and Yi, B. L. 2006. A novel PTFE-reinforced multilayer self-humidifying composite membrane for PEM fuel cells. *Electrochem. Solid-State Lett* 9: A49–A52.

6

Current Interruption

Miguel A. Rubio
Universidad Nacional de Educacian a Distancia

Alfonso Urquia
Universidad Nacional de Educacian a Distancia

Sebastian Dormido
Universidad Nacional de Educacian a Distancia

6.1 Introduction

The fuel cell diagnosis is performed in CI methods by fitting experimental data to a dynamic linear model of the PEMFC impedance and analyzing whether the obtained model indicates the existence of performance degradation phenomena. The diagnosis is based on the fitted value of the impedance transfer function coefficients or on the value of certain PEMFC electrochemical parameters, which are calculated from the fitted coefficients of the impedance transfer function.

The dynamic linear model of the PEMFC impedance describes the relationship between the fuel cell current and voltage, over a certain frequency range. The impedance model is commonly represented as an electric circuit composed of linear resistors, capacitors, inductances, and other electrochemical impedance elements, or equivalently, as a transfer function relating the PEMFC current and voltage: $Z(s) = V(s)/I(s)$.

System identification techniques are applied to estimate from experimental data the coefficients of the impedance transfer function, $Z(s)$. The current, $I(s)$, is the input signal and the output signal is the voltage, $V(s)$. The experimental data used to perform the estimation are the transient I–V (current–voltage) response of the fuel cell to certain electrical stimuli, such as CI, current pulse and pseudo random binary sequence (PRBS) signal, and so on. These stimuli are usually generated, thereby forcing changes in the electrical load applied to the fuel cell. Different experimental procedures have been devised. Some of them affect the fuel cell operation (e.g., sequences of CIs), while others require performing small changes in the electric load and consequently do not interfere appreciably with the normal operation of the fuel cell.

Other diagnosis methods based on the estimation of the PEMFC impedance are electrochemical impedance spectroscopy (EIS) and polarization curve. The polarization curve is the steady-state I–V characteristic of the PEMFC. It is obtained by measuring the steady-state value of the PEMFC voltage and current for different values of the electric load. In EIS, voltage or current sinusoidal signals are applied to the PEMFC and the fuel cell impedance is estimated using frequency-domain analysis techniques. On the contrary, CI methods employ input signals with rich frequency content (e.g., CIs and PRBS signal, among other signals) and the impedance is estimated using time-domain analysis techniques.

Cyclic voltammetry and linear sweep voltammetry are also used for diagnosis. They are conducted introducing hydrogen and nitrogen through the cell anode and cathode, respectively. For this reason, these two methods are not suited for the diagnosis of operating fuel cells.

The degradation phenomena in the PEMFC have different time constants. Therefore, the diagnosis methods based on the analysis of the PEMFC dynamic response, such as EIS and CI, need to provide a procedure to select the frequency range under analysis. This frequency range is set in EIS by selecting the frequency of the voltage or current sinusoidal signal used to excite the fuel cell. One single frequency is stimulated in each measurement period. On the contrary, the input signals used in CI methods can be designed in a way that they cover the complete frequency range of interest.

The application of the CI methods is not restricted to the diagnosis of PEMFCs. On the contrary, applications have been found in the diagnosis for different electrochemical devices. Among the pioneer works, Kordesch and Marko (1960) applied CI to batteries on the characterization of the polarization and ohmic resistances, where current pulses and half-wave rectifiers at 60 Hz are used. Galvanostatic and potentiostatic control with small switching times was employed to analyze the device dynamic response (Warner and Schuldin, 1967). The parameters of a simple equivalent-circuit model of a rotating disk electrode (RDE) were estimated using CI (Mcintyre and Peck, 1970; Newman, 1970), and the effect of the frequency and length of the interruption periods on the value of the equivalent-circuit model parameters was analyzed (Wruck et al., 1987). One of the first works on the diagnosis of PEMFCs using current pulses was done by Buchi et al. (1995), where the estimated value of the membrane resistance is used as an indicator of the chemical stability of the fuel cell.

6.2 Principle

The dynamic behavior of a PEMFC over certain frequency range can be described by a linear time-invariant model, which relates the current and voltage of the PEMFC. This model is described in the time domain by the differential equation shown in

$$\frac{d^n y}{dt^n} + a_1 \frac{d^{n-1} y}{dt^{n-1}} + \cdots + a_n y = b_1 \frac{d^{n-1} u}{dt^{n-1}} + b_2 \frac{d^{n-2} u}{dt^{n-2}} + \cdots + b_n u \tag{6.1}$$

Equivalently, Equation 6.1 can be formulated in the Laplace domain as shown in Equation 6.2, where $H(s)$ is a single-input single-output (SISO) transfer function, and $U(s)$ and $Y(s)$ are the Laplace-domain description of the input and output signals, respectively.

$$H(s) = \frac{Y(s)}{U(s)} = \frac{b_1 s^{n-1} + b_2 s^{n-2} + \cdots + b_{n-1} s + b_n}{s^n + a_1 s^{n-1} + a_2 s^{n-2} + \cdots + a_{n-1} s + a_n} \tag{6.2}$$

The goal of the identification process is to estimate the transfer function parameters (i.e., $a_1, \ldots, a_n$, $b_1, \ldots, b_n$) by fitting the model to the experimental input and output data. The input $U(s)$ is usually the PEMFC current and the output $Y(s)$ is the PEMFC voltage.

The experimental data set is obtained by measuring the time evolution of the PEMFC voltage and current in response to changes in the electric load. The selection of the input signals is discussed in Section 6.2.1. The voltage response for certain impedance models and input signals (i.e., unit impulse, step, and CI) are described in Section 6.2.2. Fundamentals of system identification are discussed in Box and Jenkins (1976), Bohlin (1991). The selection of input signals is addressed in Godfrey (1993).

6.2.1 Input Signal

The selection of the input signal for the identification process is made according to the following observation: the system identification process cannot model behavior that it is not contained in the measured data. As the goal is to obtain a transfer function that represents the PEMFC dynamic behavior over a certain frequency range, the selection of the input signal for the identification process needs to be done to ensure that the signal excites this frequency range of interest.

6.2.1.1 Impulse and Step Signals

Impulse and step signals are frequently used in system identification because the response analysis is relatively simple to conduct and these signals are rich in frequency content. To illustrate this fact, the waveform and spectrum of two signals are shown in Figure 6.1. The impulse signal has an impulse time smaller than 0.2 s and the step signal has a transition time smaller than 0.1 ms. The sampling period is 0.1 ms in both cases. The high-frequency domain of the signals is broadened as the interruption time decreases.

The waveform of the unit impulse signal at $t = 0$ is shown in Equation 6.3, where the constraint shown in Equation 6.4 has to be satisfied. This signal can be considered as a practical implementation of Dirac's delta.

$$I_{\text{impulse}}(t) = \frac{1}{\sigma\sqrt{2\pi}} e^{-t^2 2\sigma^2} \tag{6.3}$$

$$\int_{-\infty}^{\infty} \frac{1}{\sigma\sqrt{2\pi}} e^{-t^2 2\sigma^2} \, dt = 1 \tag{6.4}$$

The unit step signal is defined in Equation 6.5. The CI signal is the complementary step signal, which is defined in Equation 6.6. Both signals (i.e., unit step and complementary unit step) have the same spectrum.

$$I_{\text{step}}(t) = \begin{cases} 0 & \text{if } t < 0 \\ 1 & \text{otherwise} \end{cases} \tag{6.5}$$

$$I_{\text{CI}}(t) = \begin{cases} 1 & \text{if } t < 0 \\ 0 & \text{otherwise} \end{cases} \tag{6.6}$$

6.2.1.2 PRBS Signal

CI and current impulse signals have good frequency properties as inputs for the identification process, that is, their spectrum is broad enough to excite the frequency range of interest. However, current and voltage measurements are affected by noise, which is a source of error in the estimation. A method to reduce the noise-induced error is to perform several runs of the experiment and statistical analysis, assuming that the PEMFC state does not change significantly during the time required to perform all the experiments.

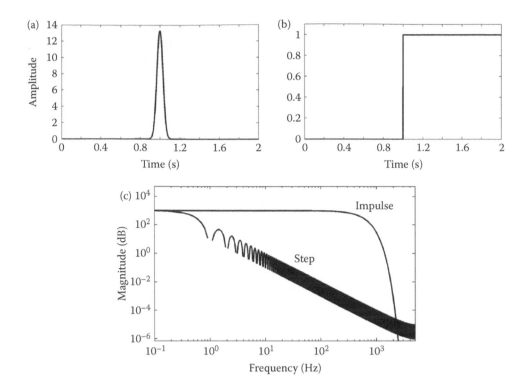

FIGURE 6.1 (a) Impulse signal with pulse time smaller than 0.2 s; (b) pulse signal with transition time smaller than 0.1 ms; and (c) spectrum of the signals (the sampling period is 0.1 ms).

Other approaches are to employ input signals of longer duration, such as white-noise and PRBS signals. The use of a long-duration signal makes the estimation method less sensitive to measurement errors, allowing consequently the signal amplitude to be smaller. The smaller the input signal amplitude, the smaller the perturbation introduced in the fuel cell operation, which is advantageous for the diagnosis of an operating PEMFC.

The PRBS signal consists in a controlled sequence of CI. It has similar properties to the white-noise signal. Two properties make the PRBS signal well suited as input of the identification process:

1. The PRBS signal can be generated using simple electronic circuits. An example of digital circuit that generates a PRBS signal is shown in Figure 6.2 (Braun et al., 1999). It is composed of a shift register and a logic OR gate.
2. The PRBS signal can be designed to have the desired frequency range, $[\omega_{\mathrm{inf}}, \omega_{\mathrm{sup}}]$. The spectral power of a PRBS signal generated using a shift register is described by Equation 6.7 (Braun et al., 1999). The signal frequency range, $[\omega_{\mathrm{inf}}, \omega_{\mathrm{sup}}]$, is shown in Equation 6.8, where N is the cycle length, $T_{s\omega}$ is the inverse of the signal rate, and α is the signal amplitude.

$$\Phi(\omega) = \frac{\alpha^2(N+1)T_{s\omega}}{N}\left[\frac{\sin(\omega T_{s\omega})}{\omega T_{s\omega}/2}\right]^2 \tag{6.7}$$

$$\omega_{\mathrm{inf}} = \frac{2\pi}{T_{s\omega}N} \leq \omega \leq \frac{2.8}{T_{s\omega}} = \omega_{\mathrm{sup}} \tag{6.8}$$

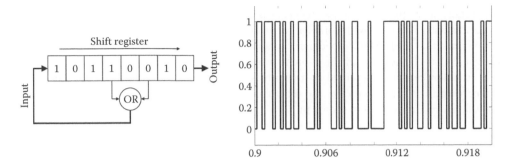

FIGURE 6.2 Circuit that generates a PRBS signal (left) and generated signal (right).

Once the desired frequency range is selected (i.e., the values of ω_{inf} and ω_{sup} have been set), the corresponding values of $T_{s\omega}$ and N can be calculated from Equation 6.8. For instance, if the frequency range for the relevant cell phenomena is considered to be 1 Hz–2 kHz, then the frequency range of the PRBS signal should be $[\omega_{inf} = 2\pi, \omega_{sup} = 4\pi 10^3]$ rad s^{-1}.

6.2.1.3 Rectangular Pulse Train

Rectangular pulse trains can be used to characterize specific frequencies. In particular, the frequency T^{-1} can be stimulated using a rectangular pulse train with a constant period T. The use of CI sequences to estimate the PEMFC membrane resistance will be discussed in Section 6.4.

6.2.2 PEMFC Voltage Response

The impulse response of an n-order system can be written as the partial fraction expansion shown in Equation 6.9. This response can be expressed in the time domain as shown in Equation 6.10, where both real and complex conjugate poles are considered, and $q + 2p = n$ is satisfied. This dynamic response may exhibit oscillations and overshoot.

$$Y(s) = \sum_{i=1}^{q} \frac{B_j}{s + A_j} + \sum_{k=1}^{p} \frac{C_k(s + \zeta_k\omega_k) + D_k\omega_k\sqrt{1 - \zeta_k^2}}{s^2 + 2\zeta_k\omega_k s + \omega_k^2} \tag{6.9}$$

$$Y(t) = \sum_{j=1}^{q} B_j e^{-A_j t} + \sum_{k=1}^{p} C_k e^{-\zeta_k\omega_k t} \cos\left(\omega_k\sqrt{1 - \zeta_k^2}t\right)$$
$$+ \sum_{k=1}^{p} D_k e^{-\zeta_k\omega_k t} \sin\left(\omega_k\sqrt{1 - \zeta_k^2}t\right) \tag{6.10}$$

6.2.2.1 Impedance with Only Real Poles

Most equivalent-circuit models of the PEMFC impedance have only real poles. In particular, performance degradation processes such as cathode flooding, membrane drying, and high-frequency effect of CO poisoning can be represented by impedances with only real poles. The dynamics represented by complex conjugate poles arise when there are inductances in the measurement or polarization equipment, or when the effect of CO poisoning at low frequency is modeled. In addition, the time-domain response of high-order transfer functions with complex conjugate poles is difficult to analyze. For these reasons, it will be assumed that the PEMFC impedance contains only real poles.

Supposing that all the poles of the n-order transfer function shown in Equation 6.2 are real, the transfer function can be represented as a sum of simpler fractions. This partial fraction expansion allows representing the impedance as the series connection of n RC parallel circuits (see Figure 6.3a), or equivalently, as the expression shown in

$$Z(s) = \sum_{i=1}^{n} \frac{R_i}{1 + R_i C_i s} \tag{6.11}$$

The voltage drop across the fuel cell (V) can be written as a function of the steady-state open-circuit voltage of the cell (V_{oc}), and the overvoltages of the anode (η_a), the cathode (η_c), and the membrane (η_m). This relationship is shown in

$$V(t) = V_{oc} - \eta_a(t) - \eta_a(t) - \eta_m(t) - \eta_a(t) \tag{6.12}$$

The property of linearity shown in Equation 6.12 is particularly useful, because it allows calculating the total dynamic behavior as the sum of the individual contributions. The overvoltage can be interpreted as the voltage drop due to the impedance of each PEMFC layers (i.e., the anode, the cathode, and the membrane). Also, each layer overvoltage can be written as the sum of the overvoltages associated with different electrochemical phenomena that take place in the layer, for example, activation, ohmic, and mass transfer overvoltages.

Equation 6.12 can be written as Equation 6.13, where $Z(t)$ is the PEMFC impedance and $I(t)$ the PEMFC current.

$$V(t) = V_{oc} - I(t)Z(t) = V_{oc} - V_{cell}(t) \tag{6.13}$$

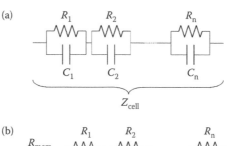

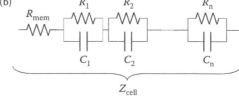

FIGURE 6.3   (a) Partial fraction expansion of impedance with n real poles and (b) membrane resistance (R_{mem}) in series with anode and cathode impedances.

The voltage in response to an input impulse, $I(t) = T_0\delta(t)$, is shown in Equation 6.14. It is expressed in the time domain in Equation 6.15.

$$V_{\text{cell}}(s) = I(s)Z(s) = T_0 Z(s) = T_0 \sum_{i=1}^{n} \frac{1}{C_i} \frac{1}{s + 1/R_i C_i} \tag{6.14}$$

$$V_{\text{cell}}(t) = T_0 \sum_{i=1}^{n} \frac{1}{C_i} e^{-t/R_i C_i} \tag{6.15}$$

The response to an input step with amplitude I_0 is shown in Equations 6.16 and 6.17.

$$V_{\text{cell}}(s) = \frac{I_0}{s} Z(s) = \frac{I_0}{s} \sum_{i=1}^{n} \frac{1}{C_i} \frac{1}{s + 1/R_i C_i} \tag{6.16}$$

$$V_{\text{cell}}(t) = I_0 \sum_{i=1}^{n} R_i (1 - e^{-t/R_i C_i}) \tag{6.17}$$

The response to the CI signal defined by Equation 6.6 can be calculated considering its relationship with the step signal. This relationship is shown in Equation 6.18, where the CI and step signals are represented as $\underset{\text{CI}}{I}(s)$ and $\underset{\text{STEP}}{I}(s)$, respectively.

$$\underset{\text{CI}}{I}(s) = I_0 - \underset{\text{STEP}}{I}(s) = I_0\left(1 - \frac{1}{s}\right) \tag{6.18}$$

The response to the CI signal can be calculated as shown in Equation 6.19, where $\underset{\text{STEP}}{V_{\text{cell}}}(t)$ was replaced by its expression in Equation 6.17. V_{0^-} represents the steady-state voltage drop in the PEMFC impedance prior to the CI.

$$\left.\begin{array}{l} \underset{\text{CI}}{V_{\text{cell}}}(t) = V_{0^-} - \underset{\text{STEP}}{V_{\text{cell}}}(t) \\[2mm] V_{0^-} = I_0 \sum_{i=1}^{n} R_i \end{array}\right\} \rightarrow \underset{\text{CI}}{V_{\text{cell}}}(t) = I_0 \sum_{i=1}^{n} R_i e^{-t/R_i C_i} \tag{6.19}$$

The voltage drops considering the steady-state open-circuit voltage (V_{oc}) are shown in Table 6.1.

6.2.2.2 Membrane Resistance

The membrane resistance can be modeled as an ohmic resistance connected in series with the anode and cathode impedances (see Figure 6.3b).

$$Z_{\text{cell}}(s) = R_{\text{mem}} + \sum_{i=1}^{n} \frac{R_i}{1 + R_i C_i s} \tag{6.20}$$

TABLE 6.1 PEMFC Voltage for Different Input Currents, Assuming That the Fuel Cell Impedance Is Given by Equation 6.11

Current		Voltage, $V(t) = V_{oc} - V_{cell}(t)$
Impulse	$I(t) = T_0\delta(t)$	$V(t) = V_{oc} - T_0 \sum_{i=1}^{n} \frac{1}{C_i} e^{-t/R_i C_i}$
Step	$I(t) = I_0 I_{step}(t)$	$V(t) = V_{oc} - I_0 \sum_{i=1}^{n} R_i + I_0 \sum_{i=1}^{n} R_i e^{-t/R_i C_i}$
CI	$I(t) = I_0(1 - I_{step}(t))$	$V(t) = V_{oc} - I_0 \sum_{i=1}^{n} R_i e^{-t/R_i C_i}$

The PEMFC voltage response to an impulse, step and CI signal can be calculated, assuming that the impedance is described by Equation 6.20. The response to a unit impulse signal is shown in Table 6.2.

The response to a step input can be expressed in the frequency and time domain as shown in Equations 6.21 and 6.22, respectively. The steady-state cell impedance (R_T) is calculated in Equation 6.23.

$$V_{cell}(s) = \frac{I_0}{s} Z(s) = \frac{I_0}{s} \left[R_{mem} + \sum_{i=1}^{n} \frac{1}{C_i} \frac{1}{s + 1/R_i C_i} \right] \tag{6.21}$$

$$V_{cell}(t) = I_0 \left[R_{mem} + \sum_{i=1}^{n} R_i(1 - e^{-t/R_i C_i}) \right] = I_0 \left[R_T - \sum_{i=1}^{n} R_i e^{-t/R_i C_i} \right] \tag{6.22}$$

$$R_T = R_{mem} + \sum_{i=1}^{n} R_i \tag{6.23}$$

TABLE 6.2 PEMFC Voltage for Different Input Currents, Assuming That the Fuel Cell Impedance Is Given by Equation 6.20

Current		Voltage, $V(t) = V_{oc} - V_{cell}(t)$
Impulse	$I(t) = T_0\delta(t)$	$V(t) = V_{oc} - T_0 \sum_{i=1}^{n} \frac{1}{C_i} e^{-t/R_i C_i}$
Step	$I(t) = I_0 I_{step}(t)$	$V(t) = V_{oc} - I_0 R_T + I_0 \sum_{i=1}^{n} R_i e^{-t/R_i C_i}$
CI	$I(t) = I_0(1 - I_{step}(t))$	$V(t) = V_{oc} - I_0 \sum_{i=1}^{n} R_i e^{-t/R_i C_i}$

Finally, the response to a CI input is calculated in Equation 6.24, where $V_{\text{cell}}^{\text{STEP}}(t)$ has been replaced by its expression in Equation 6.22.

The voltage drops considering the steady-state open-circuit voltage (V_{oc}) are shown in Table 6.2.

$$\left.\begin{array}{l} V_{\text{cell}}^{\text{CI}}(t) = V_{0^-} - V_{\text{cell}}^{\text{STEP}}(t) \\ V_{0^-} = I_0 R_T \end{array}\right\} \rightarrow V_{\text{cell}}^{\text{CI}}(t) = I_0 R_T - \left(I_0 R_T - I_0 \sum_{i=1}^{n} R_i e^{-t/R_i C_i}\right) = I_0 \sum_{i=1}^{n} R_i e^{-t/R_i C_i} \tag{6.24}$$

6.2.3 Estimation of the Membrane Resistance

The membrane resistance can be estimated using a CI input signal. However, as the voltage step due to the membrane resistance is instantaneous (see Figure 6.4), the measurement result is strongly affected by the difference between the time when the current is interrupted and the time when the voltage is measured. This difference, which should be reduced as much as possible, may be a function, for instance, of the oscilloscope sampling time.

A method was proposed by Buchi et al. (1995) and Mennola et al. (2002) to address this and other difficulties, for instance, the oscillations generated by inductances in the measurement circuitry. The voltage step height is estimated by fitting the data before and after the CI to two different linear functions. The method is illustrated in Figure 6.5.

The relationship between the membrane resistance and the voltage step height is shown in

$$\Delta V_{\text{CI}}(t) \Big|_{t \to 0} = \left(V_{oc} - I_0 \sum_{i=1}^{n} R_i\right) - \left(V_{oc} - I_0 R_T\right) = I_0 R_{\text{mem}} \tag{6.25}$$

Other methods to estimate the membrane resistance are based on the use of a short-period square signal as input. The estimation error is reduced by using the average values of the current and voltage. An example of application of this technique is shown in Section 6.4.3.

6.2.4 Estimation of the Impedance Parameters

Two different types of identification methods are briefly discussed in this section: graphical and numerical methods. Graphical methods are well suited for low-order transfer functions and simple input signals. Numerical methods allow dealing with high-order transfer functions and more elaborate input signals.

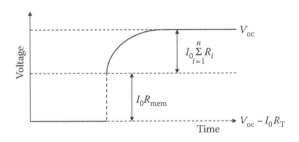

FIGURE 6.4 Estimation of the PEMFC membrane resistance (R_{mem}).

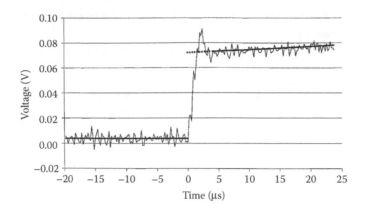

FIGURE 6.5 Estimation of the PEMFC membrane resistance based on linear fitness functions. The current was interrupted at $t = 0$. (Reprinted from *J. Power Sources*, 112(1), Mennola, T. et al. Measurement of ohmic voltage losses in individual cells of a PEMFC stack, 261–272, Copyright (2002), with permission from Elsevier.)

6.2.4.1 Graphical Methods

The goal is to identify the impedance model shown in Figure 6.6a (i.e., a first-order transfer function) using a CI input signal. The impedance is modeled by the series connection of an ohmic resistor, which represents the membrane resistance, and an RC parallel circuit.

The first step is to estimate the value of the membrane resistance. This allows calculating the contribution of the membrane resistance to the system response and to subtract its effect of the system response to the CI input signal.

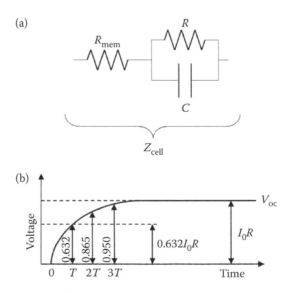

FIGURE 6.6 (a) Equivalent-circuit model of a PEMFC impedance; and (b) estimation of the response time constant, T, using a graphical method.

A graphical procedure to estimate the time constant of the response is shown in Figure 6.6b. The relationship between the impedance model parameters (R, C) and the time constant (T) is shown in

$$T = RC \tag{6.26}$$

This method can also be applied to higher-order impedances, if one of the poles is much slower than the others. In this case, the fast-pole dynamics evolve fast and the graphical method can be used to identify the dynamics of the slow pole, in a fashion similar to the identification of first-order impedance.

6.2.4.2 Numerical Methods

Numerical methods can be used to estimate the parameters of low-order transfer functions using only a single CI input signal. However, the potential of these methods is demonstrated when applied to the estimation of higher-order transfer functions using elaborate input signals, such as PRBS signals.

Numerical estimation methods can be classified into offline and online methods. Online methods, also called recursive identification methods, are of special interest because they allow reestimating the PEMFC impedance in real time. Fitting algorithms and recursive identification are explained in Norton (1986), Soderstrom and Stoica (1989), Unbehauen and Rao (1987), and Ljung and Soderstrom (1987).

Several software tools support parametric identification algorithms. For instance, a MATLAB toolbox for offline identification was employed in the application described in Section 6.4.

6.2.5 Stack Impedance

Frequently, PEMFC are connected to form a stack. From the electrical standpoint, the stack can be considered as a set of single fuel cells connected in series, that is, the stack impedance is equal to the sum of the single-cell impedances. This is represented in the frequency and time domain by Equations 6.27 and 6.28, where Z_T represents the impedance of an m-cell stack. The order of the stack transfer function is equal to the sum of the orders of all the single-cell transfer functions. Some experimental setups for PEMFC stacks are discussed in Section 6.3.2.

$$Z_T(s) = Z_1(s) + Z_2(s) + \cdots + Z_m(s) \tag{6.27}$$

$$Z_T(t) = Z_1(t) + Z_2(t) + \cdots + Z_m(t) \tag{6.28}$$

6.3 Instruments and Measurements

6.3.1 Single PEMFCs

Two experimental setups are schematically represented in Figure 6.7. In the first configuration, shown in Figure 6.7a, the CI circuit is connected in series with the polarization circuit. The CI signal $L(t)$ is directly related to the current signal $I(t)$. $L(t)$ equals one while the switch is closed and equals zero while it is open. This configuration exhibits a disadvantage. The current through the load is zero during the CI. Consequently, the data acquisition procedure disturbs the system normal operation.

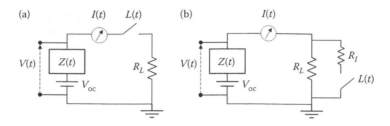

FIGURE 6.7 Experimental setups for single PEMFC. The CI circuit is connected: (a) in series and (b) in parallel.

The configuration shown in Figure 6.7b does not have this disadvantage. In this case, the CI circuit is connected in parallel with the polarization circuit. The value of the resistance R_I is selected satisfying $R_I \gg R_L$, so that the perturbation in the current that flows through the load is negligible. This allows performing the data acquisition without interrupting its operation. Also, small amplitude changes in the current produce small variations around the operating point. The membrane resistance can be assumed to be constant for small variations around any operating point.

6.3.2 PEMFC Stacks

Experimental configurations have been devised to measure the individual response of each fuel cell and the response of the complete stack. Three experimental setups are shown in Figure 6.8.

The experimental configuration shown in Figure 6.8a allows measuring the response of the total stack impedance using a single CI cycle. The circuit is simple. However, if a failure is detected, this configuration does not provide information about which individual cell or cells are failing.

The experimental configuration shown in Figure 6.8b allows measuring each cell separately. While one cell is being measured, the CI circuits of other cells need to remain opened. As only one cell is measured at a time, the measurement process takes longer. Also, the circuit is more complex.

The configuration shown in Figure 6.8c allows measuring the individual response of all the cells simultaneously. However, as the stimulus is divided by n cells, its amplitude needs to be higher in this case than in the configuration shown in Figure 6.8b.

6.3.3 Commercial and Laboratory-Made Equipment

6.3.3.1 Switching Device

Up to this point, the device that interrupts the current flow has been conceptually represented as a switch. However, the right choice of this device is crucial. The selection is done according to specifications on the switching time and on the maximum current that the device is able to handle.

The mechanical relays commonly used in industrial applications satisfy some of these specifications. In particular, they can handle high-enough currents and can be easily controlled. However, their switching time is in the order of 1 ms, which is too long for CI applications.

A better approach consists in using transistors. metal oxide semiconductor field effect transistors (MOSFETs) allow handling high-enough currents, are voltage controlled, and exhibit fast-enough switching frequencies (e.g., in the order of 1 MHz). Bipolar junction transistors (BJTs) allow handling higher currents than MOSFETs. However, their switching frequency is of the order of 10 kHz and they are current controlled. Insulated gate bipolar transistors (IGBTs) allow handling high-enough current and are voltage controlled, but they exhibit a longer switching time than MOSFETs.

As the impedance of a MOSFET operating in the conducting regime can be controlled, the MOSFET can be used to implement the resistance R_I displayed in Figures 6.7b and 6.8.

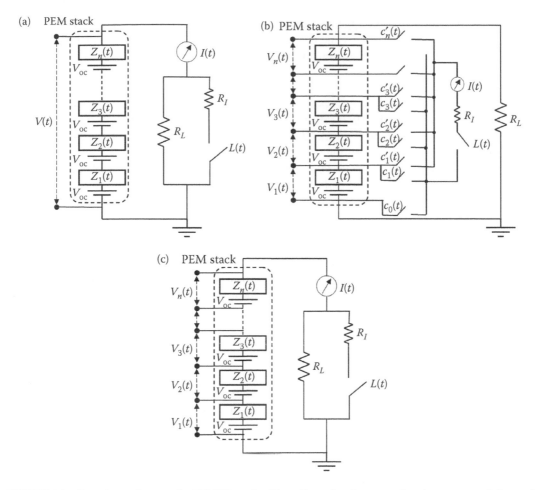

FIGURE 6.8 Experimental setups for PEMFC stacks: (a) configuration for measuring the response of the total stack impedance using a single CI cycle; (b) configuration that allows measuring each cell separately using different CI cycles; and (c) configuration that allows measuring the individual response of all the cells simultaneously using a single CI cycle.

Commercial galvanostats can be used to generate CI signals in the laboratory. The galvanostat needs to comply with the switching time specifications. Also, the additional dynamics associated with the galvanostat circuitry need to introduce negligible errors in the measurements.

Commercial electric loads, which are usually implemented using transistors, can be employed. Also in this case, the electric load needs to be chosen paying attention to the switching time specification.

6.3.3.2 Signal Generator

The signal generator is used to control the switching device, for instance as shown in Figure 6.13b. Commercial signal generators and generators of voltage binary signals (e.g., microcontrollers) can be used for this purpose.

6.3.3.3 Data Acquisition Equipment

The current and voltage measurements obtained during the CI process can be processed online, or stored and processed offline. As PEMFC dynamic phenomena have different timescales, high-speed and high-sensitivity data acquisition equipment needs to be used. Two measurement channels are required per cell (i.e., voltage and current). Some commercial oscilloscopes have the required features. Data acquisition cards and A/D converters can also be used.

6.4 Applications

The application of the CI method to the diagnosis of an operating PEMFC is described in this section. The discussion is based on Rub07b's work (2007 and 2008). The PEMFC impedance is modeled as the sum of two terms: the membrane resistance and a second-order transfer function.

The experimental procedure consists in reading the dynamic response of the cell voltage and current after the occurrence of small changes in the load value. This experimental procedure, intended for application on operating cells, does not significantly interfere with the cell operation. It has been designed supposing that the frequency range of the relevant cell phenomena is 1 Hz–5 k Hz. The equipment required to make the measurements is easily portable and inexpensive.

The parameters of the impedance model are estimated from the experimental data by applying parameter identification techniques. Analytical relationships have been derived in order to calculate the following cell electrochemical parameters from the model parameters: the diffusion resistance, the charge transfer resistance, the diffusion-related time constant, the membrane resistance, and the double-layer capacitance.

6.4.1 Modeling

6.4.1.1 Main Assumptions

The voltage drop across the fuel cell (V) can be written as a function of the steady-state open-circuit voltage of the cell (V_{oc}), and the overvoltages of the anode (η_a), the cathode (η_c), and the membrane (η_m) (Larminie and Dicks, 2000; Mann et al., 2000).

$$V = V_{oc} - \eta_a - \eta_m - \eta_c \tag{6.29}$$

The open-circuit voltage (V_{oc}) is modeled as an ideal voltage source. The external operating conditions and the cell current are considered constant during the experiments. Therefore, the overvoltages (η_a, η_m, η_c) are assumed to be only dependent on the PEMFC internal phenomena. The overvoltages of the different layers can be modeled as electric components that represent the PEMFC dynamic behavior.

The cell voltage in Equation 6.29 can be modeled by the electric circuit shown in Figure 6.9a, which is composed of Randles models connected in series (Macdonald, 1987; Iftikhar et al., 2006). Z_W^a and Z_W^c are the Warburg impedances associated with the gas diffusion in the anode and the cathode, respectively. R_p^a and R_p^c are the charge transfer resistances in the anode and the cathode. C_{dl}^a and C_{dl}^c are the double-layer capacities in the anode and the cathode. Finally, R_m is the membrane resistance.

Two additional hypotheses are made in order to simplify the cell model shown in Figure 6.9a. As a result, the simplified model shown in Figure 6.9b is obtained.

1. The oxygen reduction reaction in the cathode is very slow in comparison with the hydrogen oxidation reaction (Natarajan and Nguyen, 2001; Pisani et al., 2002). Therefore, the anode overvoltage is very small in comparison with the cathode overvoltage. As a consequence, the anode overvoltage contribution to the cell voltage can be neglected in the model.
2. The double-layer capacity, which is usually represented by constant-phase elements, is represented by a pure, single-frequency theoretical capacity (Hombrados et al., 2005; Fouquet et al., 2006; Iftikhar et al., 2006; Sadli et al., 2006). This approximation, which reduces the computational cost of the model, is reasonably accurate at low and medium frequencies (i.e., frequencies below a few hundred Hertz). The error due to this hypothesis is justified for the sake of obtaining a model simple enough to be suited for control applications.

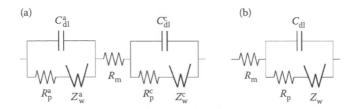

FIGURE 6.9 Equivalent circuit models of the PEMFC: (a) complete model; and (b) simplified model. (Reprinted from *J. Power Sources*, 183, Rubio, M. et al. Electrochemical parameter estimation in operating proton exchange membrane fuel cells, 118–125, Copyright (2008), with permission from Elsevier.)

6.4.1.2 Modeling of the Warburg Impedance

The Warburg impedance (Z_W) can be written in the Laplace domain as a function of the finite-length diffusion (Macdonald, 1987):

$$Z_W(s) = R_d \frac{\tanh\sqrt{s\tau_d}}{\sqrt{s\tau_d}} \tag{6.30}$$

where the diffusion resistance (R_d) and the diffusion time constant (τ_d) can be calculated from the following expressions:

$$R_d \frac{RT\delta}{SC_g Dn^2 F^2} \tag{6.31}$$

$$\tau_d = \frac{\delta^2}{D} \tag{6.32}$$

The approximation of the Warburg impedance shown in Equation 6.33 was proposed by Rubio et al. (2007). This approximation is equivalent to model the cell by using the circuit shown in Figure 6.10. The value of the parameters R_1, R_2, C_1, and C_2 were also calculated by Rubio et al. (2007) and they are shown in Table 6.3.

$$Z_W(s) = R_d \left(\frac{R_1}{1 + R_1 C_1 \tau_d s} + \frac{R_2}{1 + R_2 C_2 \tau_d s} \right) \tag{6.33}$$

The exact value of the impedance, calculated from Equation 6.30, and the approximated value, obtained from Equation 6.33 using the previously calculated values of R_1, R_2, C_1, and C_2, are plotted in Figure 6.11. The absolute error is also shown in Figure 6.11. Frequencies (w) in the range from 1 rad s⁻¹ to 10^4 rad s⁻¹ and diffusion-related time constants (τ_d) in the range from 0.2 s to 1 s have been considered. As a consequence, the range of $j_{\bar{\omega}}\tau_d$ in Figure 6.11 is 0.2–10^4.

6.4.1.3 Modeling of the PEMFC Impedance

The cell impedance (Z_{cell}), calculated from the circuit shown in Figure 6.10, is

$$Z_{cell}(s) = R_m + \frac{1}{sC_{dl} + (1/R_p + Z_W)} \tag{6.34}$$

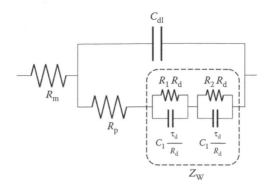

FIGURE 6.10 Equivalent circuit model of the PEMFC with the Warburg impedance calculated from Equation 6.33. (Reprinted from *J. Power Sources*, 183, Rubio, M. et al. Electrochemical parameter estimation in operating proton exchange membrane fuel cells, 118–125, Copyright (2008), with permission from Elsevier.)

It can be written as follows:

$$Z_{cell}(s) = R_m + \frac{as^2 + bs + c}{ds^3 + es^2 + fs + g} \tag{6.35}$$

where the parameters a, b, c, d, e, and f are related to the cell electrochemical parameters as shown below (Rubio et al., 2007).

$$a = 9.76 \times 10^{-4} R_p \tau_d^2 \tag{6.36}$$

$$b = 0.304 \, R_p \tau_d + 3.38 \times 10^{-2} R_d \tau_d \tag{6.37}$$

$$c = R_p + 0.949 \, R_d \tag{6.38}$$

$$d = 9.76 \times 10^{-4} \, C_{dl} R_p \tau_d^2 \tag{6.39}$$

$$e = 3.37 \times 10^{-2} \, C_{dl} R_d \tau_d + 0.3048 \, C_{dl} R_p \tau_d + 9.76 \times 10^{-4} \tau_d^2 \tag{6.40}$$

TABLE 6.3 Fitted Values of the Parameters in Equation 6.33

Parameter	Value	Units
R_1	0.8463	Dimensionless
R_2	0.1033	Dimensionless
C_1	0.3550	Dimensionless
C_2	0.03145	Dimensionless

Source: Reprinted from *J. Power Sources*, 183, Rubio, M. et al. Electrochemical parameter estimation in operating proton exchange membrane fuel cells, 118–125, Copyright (2008), with permission from Elsevier.

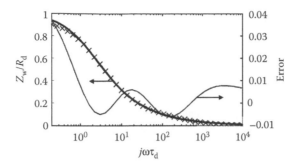

FIGURE 6.11 Warburg impedance calculated from Equations 6.30 (–) and 6.33 (×). Absolute error. (Reprinted from *J. Power Sources*, 183, Rubio, M. et al. Electrochemical parameter estimation in operating proton exchange membrane fuel cells, 118–125, Copyright (2008), with permission from Elsevier.)

$$f = C_{dl}R_p + 0.3048 \, \tau_d + 0.949 \, C_{dl}R_d \tag{6.41}$$

$$g = 1 \tag{6.42}$$

The effect of the electrochemical parameters on the PEMFC dynamic response is shown in Figure 6.12.

A further approximation for the cell impedance is proposed, which reduces the order of the model and, consequently, its computation time. The parameters a and d, which are two orders of magnitude

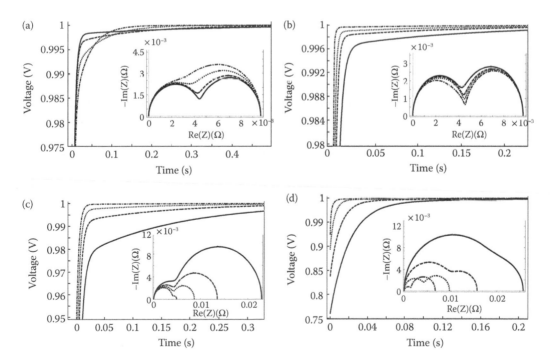

FIGURE 6.12 Simulated cell voltage after CI and impedance spectra for selected parameter values. Baseline: $\tau_d = 0.5$ s, $C_{dl} = 0.1$ F, $R_d = 6$ mΩ and $R_d = 4$ mΩ. (a) $\tau_d = \{1$ s(–), Baseline (—), 0.2 s(...), 0.1 s(–.–)}; (b) $C_{dl} = \{$ Baseline (-), 0.5 F(—), 0.2 F(...), 0.1 F(–.–)}; (c) $R_d = \{20$ mΩ(–), 10 mΩ(—), Baseline (...), 1 mΩ(–.–)}; and (d) $R_p = \{20$ mΩ(–), 10 mΩ(—), Baseline (...), 1 mΩ(–.–)}. (Reprinted from *J. Power Sources*, 183, Rubio, M. et al. Electrochemical parameter estimation in operating proton exchange membrane fuel cells, 118–125, Copyright (2008), with permission from Elsevier.)

smaller than the other parameters (i.e., b, c, e, f, and g), are neglected in Equation 6.35. Making this approximation, the following model for the cell impedance is obtained:

$$Z_{cell}(s) = R_m + \frac{bs + c}{es^2 + fs + 1} \tag{6.43}$$

Once the parameters b, c, e, and f have been estimated from the cell experimental data, the following cell electrochemical parameters can be calculated from Equations 6.37, 6.38, 6.40, and 6.41: the double-layer capacitance (C_{dl}), the diffusion resistance (R_d), the charge transfer resistance (R_p), and the diffusion-related time constant (τ_d). The proposed simplification of the impedance model does not introduce additional error in the estimation of the diffusion resistance (R_d), the charge transfer resistance (R_p), and the diffusion-related time constant (τ_d). The model simplification is made at the cost of increasing the error in the estimation of the double-layer capacitance (C_{dl}), whose value is underestimated.

The error in the estimation of the double-layer capacitance ($\varepsilon_{C_{dl}}$) is approximately described by Equation 6.44. Therefore, the double-layer capacitance is estimated by adding the following two terms: (a) the value calculated from the estimated model parameters (i.e., b, c, e, and f) using Equations 6.37, 6.38, 6.40, and 6.41; and (b) the estimation error calculated from Equation 6.44.

$$\varepsilon_{C_{dl}} = -0.005155\, R_p^{-0.9197}\tau_d - 0.2629\, R_d\tau_d + 0.02855\, \tau_d \tag{6.44}$$

6.4.2 Experimental Setup and Procedure

6.4.2.1 Experimental Setup

The characteristics of the PEMFC used in the study are: 100 cm² electrode surface; GORE 5761 membrane with a membrane thickness of 19 μm; SGL 10 BB gas diffusion layer (GDL); and serpentine flow-field at both the anode and the cathode. The operating conditions are: the cell temperature is not controlled during the experiment; the PEMFC is fed with air and hydrogen at atmospheric pressure; the hydrogen flow is 0.05 mL min⁻¹; and three air flows are used: 0.61 mL min⁻¹, 0.51 mL min⁻¹, and 0.32 mL min⁻¹. The EIS measurements are made using an IM6 (ZAHNER electrik) workstation.

The PEMFC, the load, and the circuit intended to produce small changes in the load value are schematically represented in Figure 6.13a. The PEMFC equivalent circuit is the subcircuit named FC. The load resistance is R_L. It sets the cell operating point. A resistor R_I (with $R_I \gg R_L$) and a switch represent the circuit added to perform the experimental tests. It allows forcing small changes in the load value, which produce small excursions from the cell operating point.

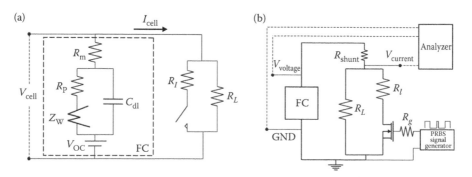

FIGURE 6.13 Experimental setup: (a) schematic representation and (b) implementation using an N-MOSFET device. (Reprinted from *J. Power Sources*, 183, Rubio, M. et al. Electrochemical parameter estimation in operating proton exchange membrane fuel cells, 118–125, Copyright (2008), with permission from Elsevier.)

An implementation of this test bench is shown in Figure 6.13b. The switch is implemented using an N-MOSFET, whose gate-to-source voltage is set by a signal generator. As the N-MOSFET channel resistance can be controlled by manipulating the gate-to-source voltage, in some cases the resistor R_l does not need to be included in the circuit.

A microcontroller with digital output is used to generate the N-MOSFET gate-to-source voltage (see Figure 6.13b). The voltage waveforms used during the identification process are rectangular pulse-trains and PRBS. The PRBS signals are generated by using a shift register (Braun et al., 1999). The reasons behind the selection of this voltage signals will be discussed in Sections 6.4.3 and 6.4.4.

The analyzer device can be designed to perform the parametric identification calculations described in Section 6.4.5. Recursive least-squares estimation, widely used in adaptive control (Ljung and Soderstrom, 1987; Yang et al., 2007), can be used. Nevertheless, the results discussed in this section have been obtained by using offline identification, which requires storing the signals in the memory of the analyzer device.

6.4.2.2 Experimental Procedure

The following two considerations have been taken into account when designing the experimental procedure. The on-state drain current of the N-MOSFET needs to be small enough to justify the linear approximation, while maintaining an acceptable signal-to-noise ratio. The changes in the cell voltage induced by the load changes should be smaller than the thermal voltage (i.e., approximately 26 mV at 300 K) (Macdonald, 1987).

The data acquisition procedure consists of the following three steps:

1. First, a 5 kHz rectangular pulse-train with a 50% duty cycle is applied to the N-MOSFET gate for 80 ms. The cell voltage and current are read with a sampling period of 10 μs. These data are used to estimate the value of the membrane resistance (R_m). The procedure is discussed in Section 6.4.3.
2. Next, the high-frequency PRBS voltage waveform described in Table 6.4 is applied to the N-MOSFET gate for 0.15 s. The cell voltage and current are read with a sampling period of 10 μs.
3. Finally, the low-frequency PRBS voltage waveform described in Table 6.4 is applied to the N-MOSFET gate for 13 s. The cell voltage and current are read with a sampling period of 500 μs.

The data obtained from Steps (2) and (3) is used to estimate the parameters b, c, e, and f in Equation 6.43, by applying the parametric identification techniques described in Section 6.4.5.

6.4.3 Estimation of the Membrane Resistance

As it was described in Section 6.4.2, the experimental data required to estimate the membrane resistance (R_m) is obtained by applying a 5 kHz rectangular pulse-train voltage to the N-MOSFET gate for 80 ms. The cell current and voltage are read with a sampling period of 10 μs.

The R_m term in Equation 6.43 constitutes a proportionality factor between the changes in the cell current and voltage. The effect of the membrane resistance manifests at high frequencies, usually

TABLE 6.4 Parameters Defining the Two PRBS Voltage Waveforms Applied during the Data Acquisition Process

	Low-Freq. PRBS	High-Freq. PRBS	Units
Frequency range	0.77–75	71–2×10^3	Hz
Cycle length (N)	217	63	Bit
Period (T_{so})	5.94×10^{-3}	2.23×10^{-4}	s bit^{-1}
Number of cycles	10	10	

Source: Reprinted from *J. Power Sources*, 183, Rubio, M. et al. Electrochemical parameter estimation in operating proton exchange membrane fuel cells, 118–125, Copyright (2008), with permission from Elsevier.

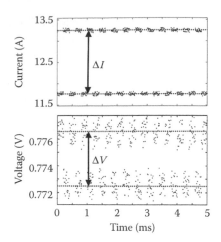

FIGURE 6.14 Simulated change in the cell current (ΔI) and voltage (ΔV) in response to the applied N-MOSFET gate-to-source voltage (i.e., 5 kHz pulse-train with 50% duty cycle). The membrane resistance is $R_m \simeq (|\Delta V|/|\Delta I|)$. (Reprinted from *J. Power Sources*, 183, Rubio, M. et al. Electrochemical parameter estimation in operating proton exchange membrane fuel cells, 118–125, Copyright (2008), with permission from Elsevier.)

overlapping with the high-frequency inductive behavior, which is represented by Z_i in Equation 6.45. The method used to estimate the membrane resistance is described by Equation 6.46. An example is shown in Figure 6.14.

$$\lim_{s \to \infty} Z_{cell}(s) = R_m + Z_i(s) \tag{6.45}$$

$$\mathrm{Re}\left[\lim_{s \to \infty} Z_{cell}(s)\right] = R_m \simeq \frac{|\Delta V|}{|\Delta I|} \tag{6.46}$$

6.4.4 Selection of the PRBS Voltage Signals

In order to estimate the parameters b, c, e, and f in Equation 6.43, two PRBS voltage waveforms are applied to the N-MOSFET gate (see Section 6.4.2). The reason for using two PRBS signals instead of only one PRBS signal is reducing the number of collected experimental samples. In order to justify this assertion, the use of only one PRBS voltage waveform is considered first.

The frequency range for the relevant cell phenomena should be 1 Hz–2 kHz. On the other hand, the spectral power of a PRBS signal generated using a shift register is described by Equation 6.7 (Braun et al., 1999).

Once the desired frequency range has been selected (i.e., the values of ω_{inf} and ω_{sup} have been set), the corresponding values of $T_{s\omega}$ and N can be calculated from Equation 6.8. In order to cover the frequency range of the cell-relevant phenomena (i.e., 1 Hz–2 kHz), the frequency range of the PRBS waveform should be $[\omega_{inf} = 2\pi, \omega_{sup} = 4\pi 10^3]$ rad s^{-1}.

While the PRBS voltage waveforms are applied to the N-MOSFET gate, the cell voltage and current need to be read with a sampling frequency of at least ten times $T_{s\omega}$. As a consequence, the sampling frequency needs to be at least 4.5×10^4 samples s^{-1}. On the other hand, adopting the heuristic criterium of collecting data at least during 10 cycles of the PRBS voltage waveform, the data acquisition process takes at least $10T_{s\omega}N$. For this frequency range selection (i.e., $\omega_{inf} = 2\pi$ rad s$^{-1} \to T_{s\omega}N = 2\pi/\omega_{inf} = 1$ s cycle^{-1}),

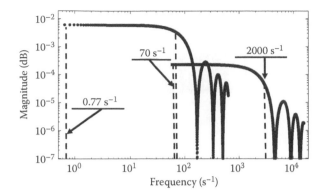

FIGURE 6.15 Frequency spectrum of the two PRBS signals described in Table 6.4. (Reprinted from *J. Power Sources*, 183, Rubio, M. et al. Electrochemical parameter estimation in operating proton exchange membrane fuel cells, 118–125, Copyright (2008), with permission from Elsevier.)

the process should take at least 10 s. Considering that the sampling frequency needs to be at least 4.5×10^4 samples s^{-1}, the total number of samples should be at least 4.5×10^5 samples.

An alternative approach consists of first applying 10 cycles of a low-frequency PRBS signal, whose frequency range is 0.77–75 Hz, and next applying 10 cycles of a high-frequency PRBS signal, whose frequency range is $71-2 \times 10^3$ Hz. The characteristics of these two waveforms are summarized in Table 6.4 and their frequency spectra are shown in Figure 6.15. This approach allows reducing the number of samples in an order of magnitude (i.e., the required number of samples is now approximately 4×10^4 samples). As a consequence, using only one PRBS signal to cover all the frequency range is impractical, because it leads to an unnecessarily large number of samples.

The test circuit shown in Figure 6.13b allows obtaining switching time under 10^{-5} s. As the time constant of the cell response is typically above 10^{-4} s, switching can be considered instantaneous.

The PRBS voltage applied to the N-MOSFET gate induces changes in the cell current whose waveform is not exactly a PRBS (see Figure 6.16a). Nevertheless, the cell current exhibits a frequency response similar to the frequency response of a PRBS (see Figure 6.16b).

The state of the switch that controls the load changes during the simulated experimental run is represented in Figure 16.17. The pulse-train voltage waveform is applied to the N-MOSFET gate in the first place. Next, the two PRBS voltage waveforms are applied.

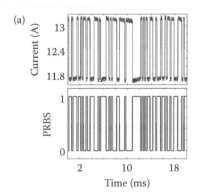
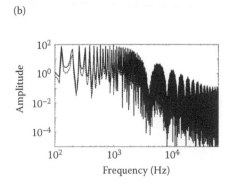

FIGURE 6.16 Response to the high-frequency PRBS voltage waveform applied to the N-MOSFET gate: (a) cell current and applied voltage; and (b) frequency spectrums of the cell current and the applied voltage. (Reprinted from *J. Power Sources*, 183, Rubio, M. et al. Electrochemical parameter estimation in operating proton exchange membrane fuel cells, 118–125, Copyright (2008), with permission from Elsevier.)

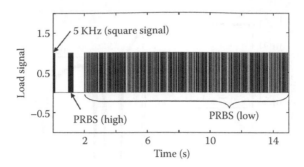

FIGURE 6.17 State of the switch that controls the load changes during the data acquisition process (1: closed; 0: open). (Reprinted from *J. Power Sources*, 183, Rubio, M. et al. Electrochemical parameter estimation in operating proton exchange membrane fuel cells, 118–125, Copyright (2008), with permission from Elsevier.)

6.4.5 Estimation of the Parameters *b, c, e,* and *f*

The cell voltage and current measured during the application of the two PRBS voltage signals are used to estimate the impedance parameters $b, c, e,$ and f. As $Z_{cell}(s) = V_{cell}(s)/I_{cell}(s)$, the parametric identification is performed by considering that the cell current (I_{cell}) is the process input and the cell voltage (V_{cell}) is the process output.

The effect of the membrane resistance is subtracted from the measured cell voltage (V_{cell}) as shown in Equation 6.47. The cell current (I_{cell}) and the calculated cell voltage (V_{cell}^*) are used to estimate the parameters $b, c, e,$ and f on the basis of Equation 6.48. To this end, the frequency-domain system identification toolbox for MATLAB, called *Ident*, has been used.

$$V_{cell}^* = V_{cell} - R_m I_{cell} \tag{6.47}$$

$$V_{cell}^*(s) = \frac{bs + c}{es^2 + fs + 1} I_{cell}(s) \tag{6.48}$$

The *Ident* toolbox for MATLAB allows estimating the coefficients of the transfer function shown in

$$V_{cell}^*(s) = \frac{k(1 + T_z s)}{(1 + T_{p1}s)(1 + T_{p2}s)} I_{cell}(s) \tag{6.49}$$

The parameters $b, c, e,$ and f can be calculated from the coefficients k, T_z, T_{p1}, T_{p2} using Equations 6.50 through 6.53.

$$c = k \tag{6.50}$$

$$b = kT_z \tag{6.51}$$

$$e = T_{p1}T_{p2} \tag{6.52}$$

$$f = T_{p1} + T_{p2} \tag{6.53}$$

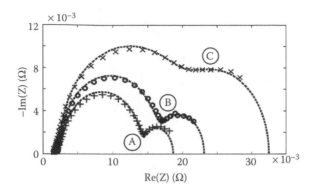

FIGURE 6.18 EIS spectrum corresponding to the following airflows: (A) 0.61 mL min⁻¹, (B) 0.51 mL min⁻¹, and (C) 0.32 mL min⁻¹; and fitted third-order models (- -). (Reprinted from *J. Power Sources*, 183, Rubio, M. et al. Electrochemical parameter estimation in operating proton exchange membrane fuel cells, 118–125, Copyright (2008), with permission from Elsevier.)

6.4.6 Parameter Evolution during the Flooding Process

EIS measurements have been made on the PEMFC under the experimental conditions described in Section 6.4.2. In order to induce different stages of the flooding process, three air flows have been used. The results are shown in Figure 6.18. The third-order impedance model has been fitted to the experimental data corresponding to these flooding process stages, named A, B, and C. The electrochemical parameters of the cell, calculated from the fitted models, are shown in Table 6.5.

First, the data acquisition procedure and the cell voltage have been simulated at these three different stages of the cathode flooding process, using the third-order impedance model. The state of the switch that controls the load changes during the simulated experimental run is represented in Figure 6.17. The pulse-train voltage waveform is applied to the N-MOSFET gate in the first place. Next, the two PRBS voltage waveforms are applied.

Second, uniformly distributed noise has been added to the cell voltage and current values obtained by simulating the experimental procedure. Each simulated value of the cell current has been incremented by a $U(-50, 50)$ mA distributed random variate. Analogously, each simulated cell voltage has been incremented by a $U(-1, 1)$ mV distributed random variate.

For instance, the cell voltage calculated from the third-order impedance model during the application of the low-frequency PRBS voltage, and modified by adding the corresponding $U(-1, 1)$ mV distributed random variate, is represented by dots in Figure 6.19. The impedance model parameters correspond to the A stage of the cathode flooding process (see Table 6.5).

Third, the cell voltage and current values previously calculated have been used to estimate the parameters of the second-order impedance model. Three impedance models are obtained, corresponding to the three flooding process stages. The Nyquist and Bode diagrams for the third-order and the

TABLE 6.5 Cell Parameters Corresponding to Successive Stages in the Cathode Flooding Process

Curve	C_{dl} (F)	R_d (mΩ)	R_p (mΩ)	τ_d (s)	R_m (mΩ)
A	0.253	5.769	10.6	0.751	2.562
B	0.372	7.969	13.2	0.698	2.343
C	0.587	12.69	18.1	0.409	2.304

Source: Reprinted from *J. Power Sources*, 183, Rubio, M. et al. Electrochemical parameter estimation in operating proton exchange membrane fuel cells, 118–125, Copyright (2008), with permission from Elsevier.

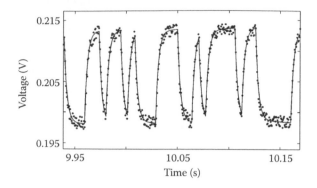

FIGURE 6.19   Cell voltage during the application of the low-frequency PRBS voltage waveform. Simulated experimental data (dots) corresponding to the "A" flooding process stage and values predicted by using the fitted second-order impedance model (continuous line curve). (Reprinted from *J. Power Sources*, 183, Rubio, M. et al., Electrochemical parameter estimation in operating proton exchange membrane fuel cells, 118–125, Copyright (2008), with permission from Elsevier.)

second-order impedance models are shown in Figure 6.20. Continuing with the example shown in Figure 6.19, the corresponding cell voltage predicted by the second-order impedance model is represented by a continuous line curve in Figure 6.19.

Finally, the cell electrochemical parameters predicted by the second-order impedance models are calculated. They are shown in Table 6.6. The values of the double-layer capacitance have been corrected by adding the corresponding estimated errors, which are evaluated from Equation 6.44.

Comparing the results shown in Tables 6.5 and 6.6, the conclusion is that the model simplification and the data-acquisition procedure do not introduce additional error in the estimation of the following electrochemical parameters: the membrane resistance (R_m), the diffusion resistance (R_d), the charge transfer resistance (R_p), and the diffusion-related time constant (τ_d). The impedance model simplification is made at the cost of increasing the error in the estimation of the double-layer capacitance (C_{dl}), whose value is underestimated. A more precise estimation of the double-layer capacitance is obtained by adding the estimated value to the correction term described by Equation 6.44.

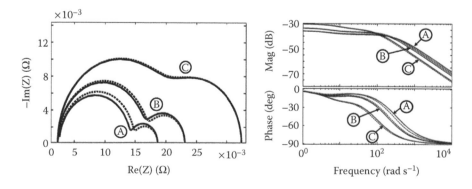

FIGURE 6.20   Nyquist and Bode diagrams for the third-order (—) and the second-order (- -) impedance models, corresponding to three different flooding process stages. (Reprinted from *J. Power Sources*, 183, Rubio, M. et al. Electrochemical parameter estimation in operating proton exchange membrane fuel cells, 118–125, Copyright (2008), with permission from Elsevier.)

TABLE 6.6 Parameters Estimated Using the Diagnosis Method Described in This Section

Curve	C_{dl} (F)	R_d (mΩ)	R_p (mΩ)	τ_d (s)	R_m (mΩ)
A	0.323	5.322	11.02	0.807	2.620
B	0.446	7.866	13.30	0.715	2.402
C	0.678	12.930	17.87	0.453	2.396

Source: Reprinted from *J. Power Sources*, 183, Rubio, M. et al. Electrochemical parameter estimation in operating proton exchange membrane fuel cells, 118–125, Copyright (2008), with permission from Elsevier.

6.5 Literature Review

The CI technique has been applied to the diagnosis of electrochemical devices. The main application to fuel cells has been the estimation of the ohmic losses. The estimation of the membrane resistance was addressed by Abe et al. (2004). They reported that the voltage step due to the CI takes place within 5 µs after the CI.

Current pulses were used by Buchi et al. (1995) and Buchi and Scherer (1996) to estimate the membrane resistance with a parallel circuit. The input signal consisted of 100 pulses, the pulse amplitude and duration being 5 A and 10 µs, and the rise time being below 10 ns. The voltage step was estimated by linear fitness of the voltage measurements recorded within the interval (300, 900) ns after the CI. The voltage before the CI was estimated using the voltage measured before the current pulse within the interval (–100, –20) ns.

The membrane resistance of each PEMFC in a stack was estimated by Mennola et al. (2002). The cell voltage was recorded during 20 ns after the overshoot produced by the parasitic inductances of the measurement circuit. The linear least-squares fit to this data was extrapolated to the overshoot region in order to estimate the voltage step amplitude. The measured total resistance of the stack was checked to be equal to the sum of the estimated resistances of the component fuel cells.

The CI technique has been used to characterize other PEMFC parameters. A dynamic impedance model was proposed by Jaouen and Lindbergh (2003) and a procedure to estimate certain parameters from the CI data was discussed. The mechanistic PEMFC models proposed by Ceraolo et al. (2003) and Rubio et al. (2007) allow reproducing the dynamic behavior after the CI.

An equivalent-circuit model of the PEMFC impedance was estimated using 100 ms current pulses (Sugiura et al., 2006). Also, the relationship between the circuit model parameters and operating parameters was demonstrated. This method was also applied to MCFC (Sugiura et al., 2005).

Other input signals have also been used for PEMFC characterization. Triangular waveform was used to estimate the membrane resistance of a PEMFC stack under flooding and drying operating conditions (Hinaje et al., 2009). PRBS signal was used for the diagnosis of flooding and drying processes in PEMFCs, and equivalent-circuit impedance model and electrochemical parameters were estimated from the experimental data (Rubio et al., 2008).

Different equipment configurations for measurement and control have been proposed. Power circuits to generate pulse sequences and control logic circuits were proposed by Wruck et al. (1987). A review on circuit configurations to perform CI was provided by Gsellmann and Kordesch (1985). Special attention was paid to the switching speed.

EIS and CI have been compared as methods for PEMFC diagnosis. A strong point of CI is that it requires simpler equipment than EIS. In relation to the series resistance estimation, a difficulty associated with CI is that the amplitude of the voltage step depends on the measurement instant. On the other hand, the estimation method based on EIS data, consisting in obtaining the cross point of the spectrum on the horizontal axis at high frequency, has been questioned by Jaouen and Lindbergh (2003). As the iR-drop does not have a linear dependence with respect to the current, it has been suggested that the calculated value of the membrane resistance needs to be corrected.

6.6 Concluding Remarks

CI method can be applied to characterize the dynamic behavior of electrochemical devices, for example, PEM fuel cells. Impedance models of different complexity can be used, according to the required level of detail and the degradation phenomena under analysis. In particular, low-order models, adequate for control applications, have been developed. Experimental procedures that do not significantly interfere with the cell operation have been devised, facilitating the application of the CI method on operating cells. In addition, an advantage of this technique is that it requires inexpensive and easily portable experimental equipment. All these features contribute to make the CI method well suited for being implemented in portable diagnosis systems.

References

Abe, T., Shima, H., Watanabe, K., and Ito, Y. 2004. Study of PEFCs by AC impedance, current interrupt, and dew point measurements. *J. Electrochem. Soc.* 151(1): A101–A105.

Bohlin, T. 1991. *Interactive System Identification: Prospects and Pitfalls.* London: Springer-Verlag.

Box, G. and Jenkins, G. 1976. *Time Series, Forecasting and Control.* Holden-Day.

Braun, M., Rivera, D., Stenman, A., Foslien, W., and Hrenya, C. 1999. Multi-level pseudo-random signal design and "model-on-demand" estimation applied to nonlinear identification of a RTP wafer reactor. *Proceedings of the American Control Conference.* San Diego, California.

Buchi, F. N. and Scherer, G. G. 1996. *In-situ* resistance measurements of Nafion 117 membranes in polymer electrolyte fuel cells. *J. Electroanal. Chem.* 404(1): 37–43.

Buchi, F. N., Marek, A., and Scherer, G. 1995. *In situ* membrane resistance measurements in polymer electrolyte fuel cells by fast auxiliary current pulses. *J. Electrochem. Soc.* 142(6): 1895–1901.

Buchi, F. N., Gupta, B., Haas, O., and Scherer, G. G. 1995. Performance of differently cross-linked, partially fluorinated proton-exchange membranes in polymer electrolyte fuel-cells. *J. Electrochem. Soc.* 142(9): 3044–3048.

Ceraolo, M., Miulli, C., and Pozio, A. 2003. Modelling static and dynamic behaviour of proton exchange membrane fuel cells on the basis of electro-chemical description. *J. Power Sources* 113(1): 131–144.

Fouquet, N., Doulet, C., Nouillant, C., Dauphin-Tanguy, G., and Ould-Bouamama, B. 2006. Model based PEM fuel cell state-of-health monitoring via ac impedance measurements. *J. Power Sources* 159: 905–913.

Godfrey, K. 1993. *Perturbation Signals for System Identification.* New York: Prentice-Hall.

Gsellmann, J. and Kordesch, K. 1985. An improved interrupter circuit for battery testing. *J. Electrochem. Soc.* 132(4): 747–751.

Hinaje, M., Sadli, I., Martin, J.-P., Thounthong, P., Raël, S., and Davat, B. 2009. Online humidification diagnosis of a PEMFC using a static dc–dc converter. *Int. J. Hydrogen Energ.* 34(6): 2718–2723.

Hombrados, A. G., Gonzalez, L., Rubio, M. A., Agila, W., Villanueva, E., Guinea, D., Chimarro, E., Moreno, D., and Jurado, J. R. 2005. Symmetrical electrode mode for PEMFC characterisation using impedance spectroscopy. *J. Power Sources* 151: 25–31.

Iftikhar, M., Riu, D., Druart, F., Rosini, S., Bultel, Y., and Retière, N. 2006. Dynamic modeling of proton exchange membrane fuel cell using non-integer derivatives. *J. Power Sources* 160: 1170–1182.

Jaouen, F. and Lindbergh, G. 2003. Transient techniques for investigating mass-transport limitations in gas diffusion electrodes. *J. Electrochem. Soc.* 150(12): A1699–A1710.

Kordesch, K. and Marko, A. 1960. Sine wave pulse current tester for batteries. *J. Electrochem. Soc.* 107(6): 480–483.

Larminie, J. and Dicks, A. 2000. *Fuel Cell Systems Explained.* West Sussex, England: John Wiley and Sons.

Ljung, L. and Soderstrom, T. 1987. *Theory and Practice of Recursive Identification.* Cambridge, UK: MIT Press.

Macdonald, J. R. 1987. *Impedance Spectroscopy*. New York: John Wiley & Sons.

Mann, R., Amplhett, J., Hooper, M., Jensen, H., Peppley, B., and Roberge, P. 2000. Development and application of a generalised steady-state electrochemical model for a PEM fuel cell. *J. Power Sources* 86: 173–180.

Mcintyre, J. D. and Peck, W. F. 1970. An interrupter technique for measuring uncompensated resistance of electrode reactions under potentiostatic control. *J. Electrochem. Soc.* 117(6): 747–751.

Mennola, T., Mikkola, M., Noponen, M., Hottinen, T., and Lund, P. 2002. Measurement of ohmic voltage losses in individual cells of a PEMFC stack. *J. Power Sources* 112(1): 261–272.

Natarajan, D. and Nguyen, T. V. 2001. A two-dimensional, two-phase, multicomponent, transient model for the cathode of a proton exchange membrane fuel cell using conventional gas distributors. *J. Electrochem. Soc.* 148(12): 1324–1335.

Newman, J. 1970. Ohmic potential measured by interrupter techniques. *J. Electrochem. Soc.* 117(4): 507–508.

Norton, J. P. 1986. *An Introduction to Identification*. London, UK: Academic Press.

Pisani, L., Murgia, G., Valentini, M., and Aguanno, B. D. 2002. A new semi-empirical approach to performance curves of polymer electrolyte fuel cells. *J. Power Sources* 108: 192–203.

Rubio, M., Urquia, A., and Dormido, S. 2007. Diagnosis of PEM fuel cells through current interruption. *J. Power Sources* 171: 670–677.

Rubio, M., Urquia, A., Kuhn, R., and Dormido, S. 2008. Electrochemical parameter estimation in operating proton exchange membrane fuel cells. *J. Power Sources* 183: 118–125.

Sadli, I., Thounthong, P., Martin, J., Raël, S., and Davat, B. 2006. Behaviour of a PEMFC supplying a low voltage static converter. *J. Power Sources* 156: 119–125.

Soderstrom, T. and Stoica, P. 1989. *System Identification*. Hertfordshire, UK: Prentice-Hall.

Sugiura, K., Matsuoka, H., and Tanimoto, K. 2005. MCFC performance diagnosis by using the current-pulse method. *J. Power Sources* 145(2): 515–525.

Sugiura, K., Yamamoto, M., Yoshitani, Y., Tanimoto, K., Daigo, A., and Murakami, T. 2006. Performance diagnostics of PEFC by current-pulse method. *J. Power Sources* 157(2): 695–702.

Unbehauen, H. and Rao, G. P. 1987. *Identification of Continuous Systems*. Amsterdam, Holland: Elsevier.

Warner, T. B. and Schuldiner, S. 1967. Rapid electronic switching between potentiostatic and galvanostatic control. *J. Electrochem. Soc.* 114(4): 359–360.

Wruck, W. J., Machado, R. M., and Chapman, T. W. 1987. Current interruption instrumentation and applications. *J. Electrochem. Soc.* 134(3): 539–546.

Yang, Y., Wang, F., Chang, H., Ma, Y., and Weng, B. 2007. Low power proton exchange membrane fuel cell system identification and adaptive control. *J. Power Sources* 164: 761–771.

7

Cathode Discharge

Herwig Robert Haas
*Automotive Fuel Cell
Cooperation*

7.1 Introduction

A number of *in situ* diagnostic techniques exist today and are adopted widely to characterize fuel cells, membrane electrode assemblies (MEAs), and their subcomponents. One newly emergent field of diagnostic techniques uses cathode discharge measurements to gain information on the properties of electrodes and MEAs as well as on the physical processes occurring during the operation of fuel cell stacks.

During the last decade, cathode discharge methods have been developed and applied at Ballard Power Systems and Automotive Fuel Cell Cooperation for *in situ* diagnostic purposes (Stumper et al., 2005a,b, 2009, 2010). The focus of this chapter is to provide an overview of the state of the art of cathode discharge methods as well as to introduce new variants thereof. The advantages over alternative methods are generally in the areas of ease and speed of test execution, as well as the uniqueness of the captured information. Cathode discharge methods can be applied to different fuel cell types; however, the experiments and examples in this chapter focus on PEM fuel cells.

When the electrochemical potential of a fuel cell electrode is reduced, charge has to be removed from the electrochemical double layers and depending on the experiment, other physical effects may occur in parallel. This process of reducing the cathode electrode potential shall be defined in this chapter as cathode discharge. The experimental procedure of reducing the cathode potential in a cell can be tailored such that specific information about the MEA and cell is obtained. A commonly applied type of electrochemical cathode discharge experiment is cyclic voltammetry (CV). CVs are typically used to measure the properties and characteristics of catalyst surfaces. This chapter describes principles and methods whereby cathode discharge measurements are used to learn about cell performance, as well as gas diffusion processes and reactant distributions within fuel cells.

In recent publications, several different experimental cathode discharge methods have been described. The level of sophistication varies greatly, in experimental execution as well as data analysis and interpretation. Some methods are quick and easily implemented, while others use special test setups, aiming to increase resolution and accuracy.

In this chapter, cathode discharge methods are described with the following goals:

- To increase awareness and understanding of this class of *in situ* techniques
- To enable test engineers and skilled practitioners to derive test protocols and methods to answer specific questions in their systems
- To stimulate ideas in readers to derive new methods for improving fuel cell characterization

7.2 Principles and Essential Theory

Polarization curves under steady-state conditions are widely used to diagnose the performance characteristics and losses of fuel cells. The interpretation of measured polarization curves, particularly at high-current densities (mass transport losses), can be an art. Cell voltage measurements at high load are often insufficient to isolate the effects of gas concentration gradients and diffusion processes present in the cell. The measurement of mass transport-free polarizations is desired to help with the isolation of those effects.

Cathode discharge methods enable the approximate measurement of mass transport-free performance. Experiments with current load steps from open-circuit voltage (OCV) as well as variable ramp rate current sweep rates create favorable boundary conditions such that the mass transport-free performance can be derived.

Load step experiments from OCV can be designed such that the voltage response becomes sensitive to specific occurring processes. At low loads, molar reactant contents in cells can be quantified. At high loads, diffusive properties of electrodes can be determined.

In this chapter, the following variants of cathode discharge methods are discussed in detail:

1. Current step, closed volume (Sections 7.2.1 through 7.2.3)
2. Current sweep, constant flow (fast transient polarization) (Section 7.2.4)

7.2.1 Current Step, Closed-Volume, Mass Transport-Free Polarization

During OCV, most conditions in the cell are known or can be derived with simplifying assumptions. When load current is applied rapidly, multiple processes commence in parallel—reactants are consumed by the oxygen reduction reaction, the electrode oxide coverage changes, heat is generated, gas diffusion through the gas diffusion layer (GDL) initiate, and so on. All these processes have their own characteristic timescales, from milliseconds to seconds. By changing the ramp rate and height of the load step, different processes, such as reaction kinetics and mass transport, can be isolated. The test setup and preconditioning of the cell can greatly impact the voltage response, thereby creating the opportunity for a number of diagnostic methods.

In this initial introduction of the principle, some physical effects are neglected for simplification (e.g., double-layer discharge currents, nonuniformities) but will be discussed later.

Figure 7.1 shows the expected, simplified voltage response as an example of an ideal cathode discharge experiment. Initially, a cell is filled with reactants (e.g., air on cathode, hydrogen on anode), and then the reactant volumes are closed such that no reactants are flowing into or out of the cell for the remainder of the experiment. The cell is initially at OCV. At time $t = 0$, a load is applied and held until the cell is no longer capable of sustaining it. Figure 7.1 shows the expected (a) voltage profile, (b) current profile, and (c) reactant concentration for this experiment.

After the load is applied, the voltage drops rapidly from OCV to a reduced level that depends on MEA and cell attributes as well as the initial gas concentration. This voltage shall be defined as V_{0+} and is primarily determined by ohmic and kinetic losses in the cell. During the time that follows V_{0+}, the voltage drops below V_{0+} due to the drop in oxidant concentration at the catalyst sites, which is caused by reactant consumption. An oxygen concentration gradient builds up between the catalyst layer and the flow channels, as schematically shown in Figure 7.1c. This gradient builds very quickly (i.e., in milliseconds). Once the oxygen concentration drops near 0 at the catalyst sites, the cell voltage drops substantially (Figure 7.1d) and the load cannot be supplied any longer.

From the voltage profile in Figure 7.1a, several attributes can be measured.

- The voltage at time $t = 0+$ is a strong function of ohmic and kinetic losses. The approximate mass transport-free performance can be derived.
- For high-current densities, the slope after $t = 0+$ may be dependent on the *in situ* diffusivity of the cathode GDL and catalyst layer.
- For low-current densities, the duration until the voltage approaches 0 V (assuming constant load) is a measure of the moles of reactant that are initially present in the cell. Indirectly, the volume occupied by other phases and gases can be determined.

One effect that has so far been neglected is the discharge of the cathode double layer. When looking at CV graphs of cells, one observes substantial charge removal during the cathodic scan, marking the transition from oxide-covered to metallic electrode surfaces. During a discharge experiment such as the one described above, charge has to be removed from the electrode surfaces when the voltage is reduced.

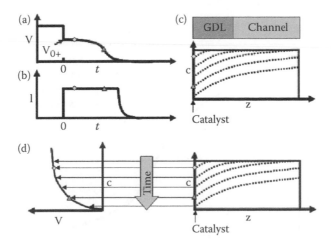

FIGURE 7.1 Cathode discharge experiment: (a) voltage versus time, (b) load versus time, (c) expected reactant concentration versus time, and (d) reactant concentration versus time and expected voltage for a given reactant concentration at the catalyst layer.

This process can take some time, such that the determination of V_{0+} is not straightforward. However, V_{0+} can be derived for practical applications from most step current measurements.

Figure 7.2 shows typical experimental cathode discharge measurements with constant load steps for four different current densities that are obtained on a 20-cell PEM short stack. The stack is preconditioned for 20 min at a steady state of 1 A cm^{-2} in the same way prior to each cathode discharge measurement. For the first few milliseconds after $t = 0+$, deviation from the ideal curve in Figure 7.1 is observed, due to double-layer discharge effects. After that period, the cell response shows the same behavior as the idealized curve.

As shown in Figure 7.2, the double-layer discharge effects do not allow the direct measurement of V_{0+}. In Section 7.3.8, a method is suggested that allows the derivation of V_{0+} from extrapolation of the discharge curve after complete double-layer discharge to $t = 0+$.

Figure 7.3 shows V_{0+} values for different current densities plotted as a polarization curve. Assuming that V_{0+} is mass transport free, then Figure 7.3 can be considered a mass transport-free polarization curve.

7.2.2 Current Step, Closed-Volume, Reactant Content Determination

Reactant content quantification experiments using the cathode discharge method can be designed to quantify the moles of reactants in the cathode volume as well as the anode volume of cells and stacks. When load steps from OCV to low-current density are applied with closed reactant volumes, cell voltages drop to nearly 0 V once either the cathode or the anode is depleted of reactants. It must be ensured that the side of interest is depleted first. For these types of discharge methods, low loads are typically chosen to minimize the effects of gas diffusion. The total charge is proportional to the total number of moles in the sum of all volumes—gas distribution manifolds, channels, pores in gas diffusion media, and catalyst layers. In experiments on stacks, information on moles of reactants within individual cells can be obtained.

In the literature, cathode discharge consumption measurements are described (Stumper et al., 2005a,b). In this chapter, an example of anode consumption measurement is presented.

After the operation of a 20-cell stack, the valves at the anode inlet and outlet are closed. The valves on the cathode side remain open and a small flow of humidified air is continued, to ensure that the anode is concentration limited. The stack is then operated at 0.01 A cm^{-2} and the cell voltages are measured over time. Figure 7.4 shows the experimental results.

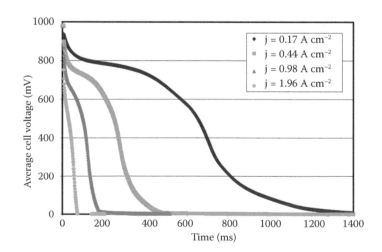

FIGURE 7.2   Measured cathode discharge voltage profiles for different current densities.

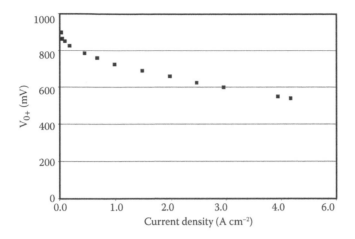

FIGURE 7.3   Mass transport-free polarization from V_{0+} measurements at different loads.

The individual cell voltages within the stack show a very different behavior. A few cells drop in voltage several seconds before other cells, due to insufficient reactants within the cells' anodes, while others remain at a high voltage for nearly twice the time. For this reason, some cells appear to initially have twice the number of moles of reactant present than others.

7.2.3 Current Step, Closed-Volume, Reactant Diffusivity

In addition to the measurement of mass transport-free performance and reactant content, another important characteristic of MEAs can be measured using current step closed-volume measurements—the diffusive characteristics of electrodes.

After the load step is applied and V_{0+} is passed, the reactant content at the cathode catalyst drops due to consumption by the current. At high-current densities, this drop occurs quite rapidly and the slope at which the voltage drops over time should be sensitive to the mass transport of reactants through the

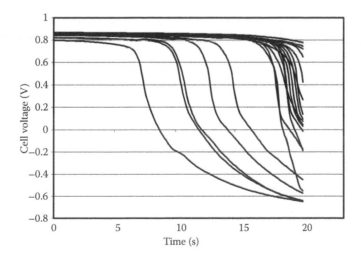

FIGURE 7.4   Voltage versus time in anode consumption measurement on a 20-cell stack with closed anode volume, airflow in cathode, and a current density of 0.01 A cm^{-2}.

catalyst and GDL layers of the MEA. Figure 7.5 shows schematically how the localized reactant concentrations should drop for two electrodes with different effective cathode diffusion coefficients. For a highly diffusive electrode (dotted line), lower concentration gradients are expected than for lower-diffusivity electrodes (solid line). The expected voltage discharge behavior for the two cases is also shown in Figure 7.5.

The principle as shown in Figure 7.5 appears intuitive and theoretically feasible, yet it has not been adopted widely to date. One similar test method has been introduced by Stumper et al., Section 7.4 contains experimental evidence that the diffusivity of electrodes can indeed be measured by this current step method.

7.2.4 Current Sweep, Constant Flow, Fast Transient Polarization

One method to measure the mass transport-free polarization was earlier described—the determination of V_{0+} values for numerous current densities. If local hydration conditions in the MEA of the cell change during V_{0+} measurements, then the cell needs to be reset prior to every V_{0+} measurement to ensure quasiconstant hydration states for all V_{0+} points. This can be cumbersome and time consuming.

A quicker and more convenient method to obtain mass transport-free polarization curves utilizes a variable current sweep, constant flow approach (Lim and Haas, 2006). In this approach, current sweeps are applied to obtain the mass transport-free polarization. They are referred to as fast transient polarizations (FTP) in this chapter. By clever choice of current sweep rates, one can obtain meaningful polarization curves very quickly.

When cathode discharge experiments are carried out with constant current density ramp rates, the obtained polarization curves typically deviate from the mass transport-free performance curve in parts of the polarization curve. For slow scan rates, accurate performance measurements are obtained at low-current densities; however, high-current densities yield poor agreement, and when scan rates are high, poor agreement is found throughout the polarization curve. In Figures 7.6 and 7.7, current sweep profiles and their corresponding polarization curves are shown.

To achieve good agreement between V_{0+} polarizations and polarizations from current sweep methods, two effects need to be considered—the double-layer discharge and the depletion of oxygen at the catalyst layer due to consumption. At low loads, slow current sweep rates are needed to allow the cathode double layer to discharge. This process can require several seconds and while the current ramps up, oxygen is being consumed. In order to ensure that the oxygen concentration does not change, constant

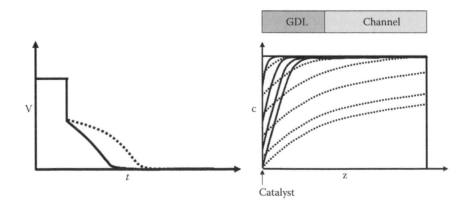

FIGURE 7.5 Expected voltage versus time and O_2 concentration versus time for high- (dotted lines) and low-diffusion coefficient electrodes (solid lines) at high load.

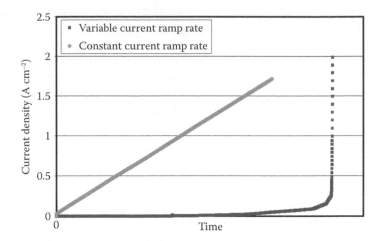

FIGURE 7.6 Examples of cathode discharge current profiles for constant ramp rate and variable ramp rate.

air flow at constant pressure is needed. Diffusion gradients in oxygen concentration between channel and catalyst can be neglected at low-current densities. Once higher currents are reached in the current sweep, diffusion effects start to create concentration gradients between channel and catalyst. Therefore, high current sweep rates are needed to minimize mass transport losses.

The variable sweep rate, constant oxidant flow approach has more sophisticated boundary conditions than the V_{0+} polarization. The current sweep profile needs to be optimized for individual tests. However, once optimized and carefully executed, current sweep cathode discharge methods enable quicker measurement of mass transport-free polarizations (seconds) than V_{0+} measurement methods.

7.3 Instruments and Measurements

In this section, practical aspects of performing cathode discharge experiments are treated in more detail. The selection of specific instrumentation for a particular experiment depends on the type of cathode discharge experiment, stack size, active area size, requisite spatial resolution, and other details

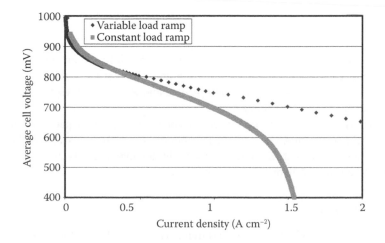

FIGURE 7.7 Polarization curves measured at constant ramp rate and variable ramp rate.

of the fluid systems and test station setup. The focus here shall be on the general requirements regarding test equipment and setup to execute cathode discharge experiments.

Cathode discharge experiments can be performed on most fuel cells, regardless of cell active area and stack size, often with little modification to test setups. For many tests, little change to the hardware, fluid streams, and connections is required. The art of well-executed cathode discharge experiments lies in the control and measurement of the electrical currents and voltages, and can require special equipment. The test station needs to be able to provide the following capabilities.

7.3.1 Fluid Streams

Oxidant and fuel need to be provided with flow rates and humidification levels that are typical for steady-state fuel cell operation. For some discharge experiments, particularly for the measurement of reactant contents, shut-off valves are required in close proximity to the fuel cell. The reactant and oxidant flows need to be switched off rapidly in some cases.

The coolant system needs to be able to keep the test cell at the target temperature during the experiment, which spans a few seconds.

7.3.2 Load

Control and real-time measurement of the applied load is critical for accurate cathode discharge measurements, as transient current changes are the basis for the methods described in this chapter. Typical challenges for load banks include very rapid and large current steps, as well as high currents at low cell or stack voltages.

The load bank should be capable of changing the actual current between any requested levels in less than 1–2 ms. Within the transition time, overshoots in load must be minimized. The required accuracy in the measurement of the current depends on the purpose of the experiment. Accuracy better than 0.1–1% of the full scale of the load bank is practically achievable and generally desired.

In cathode discharge experiments on stacks with low cell count, particularly single cells, the low voltage/high current demand poses a challenge to load banks. Situations can occur in which currents of hundreds of amperes are produced at stack voltages of <0.5 V.

The load requirement for some measurements can be controlled manually, for example with an on switch. However, for most practical applications and more complicated load ramp profiles (e.g., FTP), software-controlled load profiles, and simultaneously measured currents and voltages improve the ease of experimentation and the time needed for test execution.

7.3.3 Stack Size

One of the advantages of the cathode discharge method is the ability to quickly measure interesting characteristics of fuel cells of any stack size and power level. The nature of the experimental question determines what fuel cell stack size should ideally be used and what spatial resolution is required in the instruments measuring current and voltage.

In cathode discharge measurements on stacks, current and voltage information is typically gathered for the cells within the stack. Unless the current distribution within the active area of the cells is known, only cell average values for current and voltage are available for analysis. This is often sufficient to obtain critical information on stack, cell, or MEA characteristics.

7.3.4 Data Collection

The key information for cathode discharge measurements is derived from the time-dependent voltage and current signals, and needs to be recorded sufficiently rapidly. For other operational parameters such as pressure and temperature, single recorded values are often sufficient, as they remain quasiconstant during cathode discharge measurements.

For V_{0+} and FTP measurements, logging rates for voltage and for currents of 1000 Hz or higher are required to capture effects that occur in the 10–20 ms timescale.

For cathode discharge experiments, multichannel oscilloscopes or equivalent equipment are practical and useful devices.

7.3.5 Voltage Sensing

The accuracy requirement for voltage and current measurements can be derived from the magnitude of the expected effects. A measurement error of 1 mV is typical and acceptable.

In first order, the voltage measurement location on the cell is chosen to be representative of the average of the whole cell. This approach is useful for test situations where uniform MEA properties and conditions are expected throughout the cell. For nonuniform performing cells, such as those at the end of lifetime tests, the cell voltage may not be uniform and the cathode discharge approach is only favorable if the magnitude of the degradation effect is large. An alternative approach for those cases is to choose experimental setups that allow the simultaneous measurement of localized cell voltages and localized currents throughout the whole cell. Setups similar to the stack MEA resistance and electrode diffusivity (MRED) tool enable such experiments (Stumper et al., 2009, 2010).

7.3.6 Preconditioning

The operation of the cell prior to cathode discharge measurement has substantial influence on the subsequent cathode discharge results. This enables the investigation of how operating conditions affect performance loss characteristics. For example, the performance loss of a cell or stack can be measured at nearly any particular point in time of operation by abruptly stopping the test and the instantaneous measurement of V_{0+} or FTP.

If properties of uniform MEAs are being studied in cells or stacks, conditions that leave the MEA in a uniform state of hydration and oxide coverage are preferable to simplify analysis, as a one-dimensional situation is present during measurement. Operation with oversaturated conditions and low-temperature gradients along the cell is one option that leaves the MEA hydration state uniform.

7.3.7 Procedures

Cathode discharge experiments are typically executed seconds after regular fuel cell operation is stopped, and the duration of the actual discharge measurement lasts in the order of seconds. Often, little change in the state of conditions and MEA hydration is expected during these timescales, especially if the gas flows are stopped. As the specific test purpose varies, so does the test setup to perform discharge measurements. Therefore, only generic procedures to perform cathode discharge experiments are described below, to a level of detail that allows users to tailor procedures for their own needs.

In experimental test execution, safety considerations need to include the following concerns:

- The pressure in the reactant volumes can change substantially during discharge experiments. Cell and MEA damage may occur, which can lead to safety concerns during and after the experiments.
- Negative cell voltages can occur, which may damage the MEA and cell. Damaged MEAs may pose a safety risk during the tests or upon restart of regular operation.

Procedure to measure cathode reactant contents:

1. Operate fuel cell under conditions of interest.
2. Turn off load current.

3. Close cathode valves at inlet and outlet of stack.
4. Apply target load (typically low load) and record cell voltages until voltage drops to near 0 V or the load cannot be sustained.

Procedure to measure V_{0+} and reactant diffusivity:

1. Operate fuel cell at steady state under conditions of interest.
2. Turn off load.
3. Close cathode valves at inlet and outlet of stack.
4. Apply target load rapidly (within 1 ms) and record cell voltages with 1000 Hz or more for approximately 500 ms.

For measurement of transient polarizations, the load is ramped at varying rates. The exact profile depends on the cell and MEA properties, and needs to be uniquely tailored for the specific application.

Procedure for cathode discharge to measure FTP:

1. Operate fuel cell under conditions of interest.
2. Turn off load.
3. Continue fluid flows.
4. Once OCV is reached, apply load ramp profile and record cell voltages with 1000 Hz or more.

To dial in a FTP load profile is an iterative process. Initially, a load ramp profile is chosen; an example could be the one in Figure 7.6. If deviation between the V_{0+} and FTP is found, then parts of the load profile are adjusted. Generally, if the performance deviation is large at low-current densities, slower ramp rates at low loads are required. If deviations at high loads are observed, faster ramp rates at medium and high loads can reduce the gap between V_{0+} polarization and FTP.

7.3.8 Data Analysis of V_{0+}

The extraction of V_{0+} from cathode discharge measurements is not always straightforward. Once the load is turned on, several transient processes start occurring in parallel. The two most dominant processes are the fuel cell reaction and the discharge of the double layer. The discharge current can be a significant fraction of the total current; therefore, the fuel cell current is not known very accurately until the cathode electrode is substantially discharged. The fuel cell current created by the oxygen reduction reaction is typically of interest in these types of experiments. As the fuel cell current is not known in the initial stages of cathode discharge experiments, no meaningful point on the polarization curve can be measured during that time. Figure 7.8 shows voltage versus time during a cathode discharge experiment. During the first milliseconds after the load is applied, the cell voltage does not drop as rapidly as the total load ramps up, indicating that the fuel cell current is only a fraction of the total current.

After the double-layer discharge process is completed (a few milliseconds, depending on the load), the fuel cell consumes some of the reactant in the cathode. The concentration of reactant above the catalyst surfaces is therefore lower than the concentration prior to the application of the load. The V_{0+} performance is thus strictly speaking not directly measurable, and needs to be derived from the measured data, that is, by fitting the data, after complete discharge, to the time of V_{0+}. This is graphically shown in Figure 7.8, where the double-layer discharge appears complete after $t = 370$ ms. The linear curve fit in the time window from 370 to 440 ms extrapolated to $t = 345$ ms is one approach to obtain V_{0+}, as shown in Figure 7.8. This suggests the voltage that should be measured if there were no double-layer discharge effect.

7.4 Applications

In this section, several circumstances in which the cathode discharge method is applied are described.

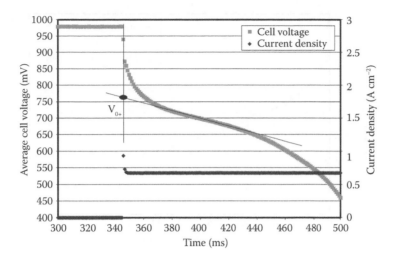

FIGURE 7.8   An example of data analysis: a real curve to obtain V_{0+}.

7.4.1 Performance Loss Diagnostics of New Test Articles

Measurement of the mass transport-free polarization is generally a quick and easy way to assess the state of a fuel cell stack. After initial cell conditioning is completed, analysis of the mass transport-free performance can be used to assess if the stack is built as intended.

In the development of new cell and hardware architectures, the mass transport-free performance measurement is a powerful tool to determine whether the cell achieves performance to full potential. If known membrane electrode designs with known mass transport-free performance characteristics are measured in new architecture, one can quickly assess if there are losses due to nonoptimized cell design or improper stack assembly. Figure 7.9 shows mass transport-free polarization curves of the same MEA type that are measured in different architecture types, two stacks of significantly different design and one small single cell. Good agreement is observed, confirming proper cell design and assembly.

7.4.2 Mass Transport-Free Polarization Curves for Model Input

Cathode discharge experiments have proven to be advantageous in the area of fuel cell performance modeling. Versions of the Butler–Volmer equation are typically used in computational models to calculate voltage as function of kinetic, ohmic, and reactant concentration terms. The challenge in fuel cell performance model development is the determination of the kinetic terms (exchange current density).

In single-crystal experiments with catalysts such as Pt, exchange current densities and symmetry factors can be measured quite accurately or are known. However, in real-world electrodes, these measurements prove to be challenging due to the sophistication of the boundary conditions in state-of-the-art porous electrode and MEA structures.

An alternative, empirical approach to obtain good inputs for cell performance models is based on accurate curve fits of mass transport-free polarization curves. In Figure 7.10, a measured mass transport-free polarization is shown along with the corresponding empirical curve fit. The fit variables used are resistance, exchange current density, and symmetry factor (oxygen concentration is known assuming performance is mass transport free). The excellent agreement between fit and experiment allows the determination of accurate empirical model input parameters.

The resulting kinetic parameters are empirical fit constants for electrodes in fuel cells rather than electrochemical material constants. Their value as model input parameters lies in the area of cell performance modeling of cells and stacks to optimize cell geometry and gas diffusion layers.

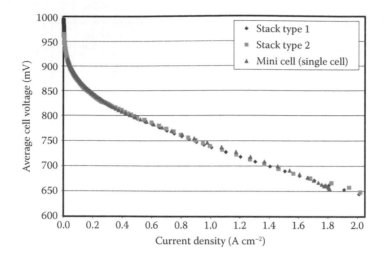

FIGURE 7.9 Mass transport-free performance of the same MEA type measured in two different stack architectures and in a small single cell.

7.4.3 Loss Analysis of Low-Performing Cells in Stack

An area where discharge methods can provide valuable information for test engineers is the diagnostics of low-performing cells within stacks. Some effects leading to low performance of individual cells in stacks include:

- Undersupply of reactants to individual cells or cell regions
- High-diffusion resistance of electrode (GDL and catalyst layer)
- Inactive cell regions due to liquid water or inert gas accumulation
- Below average electrode kinetics or excessive ohmic resistance (MEA or cell), throughout cell or in localized regions (i.e., degradation, MEA damage, dehydration, contamination)

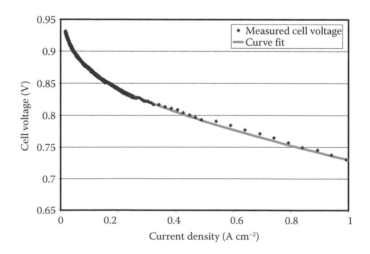

FIGURE 7.10 Measured FTP and corresponding curve fit using resistance, exchange current density, and transfer coefficient as fit variables.

This section discusses how some of these effects can be detected by discharge experiments on stacks.

By simultaneous measurement of the mass transport-free polarization of all cells in the stack, kinetic or ohmic effects can be separated from all mass transport-related effects. If the mass transport-free performance of an in-operation low-performing cell is comparable to those of well performing cells, then nonuniform fluid distribution or gas diffusion limitations between channel and catalyst are likely the cause for the low performance. If below-average mass transport-free performance is measured, then excessive kinetic or ohmic losses are present, either throughout the cell or in localized regions.

Figure 7.11 shows the cell scan of a 19-cell stack operated at high load. The cell voltage of cell 17 is approximately 30 mV below the stack average.

An FTP measurement is performed on the stack to obtain the mass transport-free polarization for each cell. The results are shown in Figure 7.12.

As can be seen from Figure 7.12, the mass transport-free polarization of cell 17 is below the polarizations of all other 18 cells. It appears that throughout the whole range in the polarization curve, cell 17 performs below average. Kinetic or ohmic effects are typically present when such behavior is observed. In this case, for experimental purposes, some sections within the active area of cell 17 are manufactured to be inactive during operation. The reduced active area of cell 17 caused the mass transport-free performance to be below stack average.

If low-performing cells in stacks of the same MEA type are caused by cathode mass transport limitations (insufficient flow or diffusive losses in electrodes), all individual mass transport-free performance curves typically overlay. To separate effects by insufficient oxidant flow from diffusive losses in the cathode electrode, step current measurements from OCV to high-current densities can be used (see also Section 7.4.6).

Low-performing cells caused by anode mass transport limitations (insufficient flow, diffusive losses, inert gas, or liquid water accumulation) typically exhibit overlaying mass transport-free polarizations of all individual cells. Discharge experiments can aid in the isolation of the different effects. Figure 7.4 shows an example in which after operation at low load, the fuel reactant supply is stopped, the fuel valves at the inlet and outlet of the stack are closed, and a low load is drawn. The cells with the lowest reactant content decay to low voltages first. Figure 7.13 shows a repeat of the experiment in Figure 7.4 but with

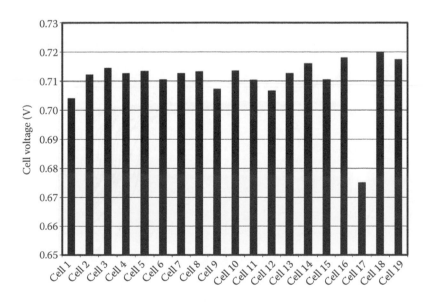

FIGURE 7.11 Cell voltage scan of stack containing low-performing cell (cell 17) during high-current density operation.

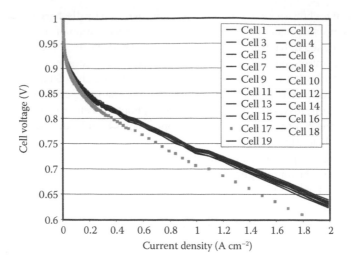

FIGURE 7.12 Mass transport-free polarization of each individual cell of Figure 7.11.

one deviation. The same stack is operated with the same operating conditions, but before the reactant volumes are closed, the stack is purged with very low fuel flows for 20 s at OCV. The observed cell voltage behavior is quite different between Figure 7.4 and Figure 7.13, caused by a fuel purge that has insufficient momentum to remove liquid water from the cell.

In the experiment of Figure 7.4, a significant fraction of the cell volume is occupied by nonreactant containing species in some cells. This species is unlikely liquid water as it can be removed by a 20 s low flow purge, pointing to inert gases. Typical sources of nitrogen or other inert gases that can accumulate in localized cell regions are from the fuel supply or nitrogen that diffuse from the cathode to the anode through the membrane over extended periods of time. In the case earlier described in Figure 7.4, nitrogen that originates from the cathode is believed to have accumulated in the low-performing cells in fuel-stagnant regions.

If liquid water would be the cause for the original void volume reduction of Figure 7.4, little difference in the discharge behavior with and without 20 s low flow fuel purge would be expected.

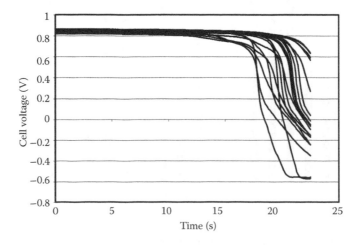

FIGURE 7.13 Cell voltage versus time at low load after 20 s low flow purge at OCV.

If low-performing cells in stacks show high sensitivity to fuel flow and discharge results similar to Figure 7.13 are observed without low flow purge, then diffusive losses rather than dilution effects are likely to be present.

As this section illustrates, various effects leading to low-performing cells in stacks can be isolated by discharge experiments.

7.4.4 Detection of Low MEA Hydration

An example of V_{0+} measurements being used to detect the hydration state of an MEA is described in a US Patent (St-Pierre et al. 2003). A stack is purged and V_{0+} periodically measured every few seconds. Once dehydration of the membrane gets more severe, the voltage drops become larger.

7.4.5 Effect of Electrode Oxide Coverage on Performance

The formation of oxides on the cathode catalyst surfaces can impact the performance of MEAs. This effect can be studied using FTPs. An experiment was performed in which a stack was operated in steady state at high load. Then, for a few seconds, the air flow was reduced such that the cell voltages dropped to near 0 V (air starvation). This state was held for 20 s to remove oxides from the catalyst surfaces. Subsequently, the load was removed and the cell voltages recovered to OCV, which was held for varying amounts of time to allow oxide buildup on the catalyst surfaces, followed by a fast transient measurement. Figure 7.14 shows the mass transport-free polarizations after 10 s, 100 s, and 1000 s of OCV. The longer the time spent at OCV, the lower the performance, caused by losses due to oxide buildup on the cathode catalyst surfaces. Significant performance losses can occur due to oxide effects in real-world applications. FTP experiments can help to quantify the loss at certain points during operation.

7.4.6 *In Situ* Measurement of Electrode Diffusivity in a Stack

The measurement of cathode electrode diffusivity is of great interest in the development of fuel cells. Higher performance and less mass transport loss are desirable in many fuel cell applications to reduce cost and increase power density. Electrode diffusivity is a key metric to achieve higher current density,

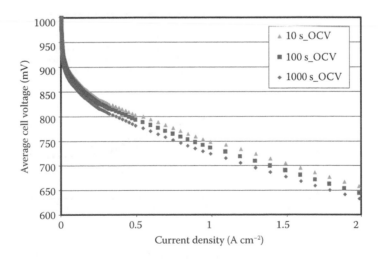

FIGURE 7.14 Mass transport-free performance after 10 s, 100 s, and 1000 s at OCV after air starvation.

yet *in situ* measurement techniques to quantify that metric are limited. One electrode diffusivity characterization technique is the measurement of limiting currents on small mini-cells (St-Pierre et al., 2007), but alternative test methods are desired, ideally for larger cells and stacks. To date, no method has been demonstrated for diffusivity measurements on stacks. The principles of Figure 7.5 theoretically apply to measure the diffusivity of electrodes—closed reactant and load step from OCV to high-current density. An experiment was designed to determine whether differences in the cathode discharge behavior can be seen for low- and high-diffusivity electrodes. A 20-cell stack is built with four different MEA types. Cells 1–5 contain type 1, cells 6–10 contain type 2, cells 11–15 contain type 3, and cells 16–20 contain type 4. A current step from 0 A cm^{-2} to 4.17 A cm^{-2} is applied and the voltage response measured. Figure 7.15 shows the average voltages for each MEA type.

A difference in cathode discharge behavior is observed for each type. The cell-to-cell variability within each type is small in comparison to the differences between the types. All MEA types contain very similar catalyst loadings, but the electrode structure and GDL vary from type to type. During the first few milliseconds after the load is applied, double-layer discharge currents can be expected. In the range 340–355 ms, diffusion of reactant gas to the catalyst is believed to influence the cathode discharge curve based on the principles described in Section 7.2.3. To gain further evidence that the diffusive properties are indeed measured, limiting current measurements are performed on individual MEAs from each type. Mini-cell measurements are used to determine the limiting current (St-Pierre et al., 2007). As a quantitative measure of diffusivity in Figure 7.15, the slope of the curves between 340 and 355 ms was taken. When plotting these slopes versus the limiting currents, a correlation is observed, as shown in Figure 7.16.

The results in Figure 7.16 confirm that by load step measurements from OCV to high-current density, the effective electrode diffusivity of electrodes can be measured *in situ*, as postulated in Section 7.2.3.

7.5 Advantages and Limitations

The family of cathode discharge experiments provides a number of useful techniques for diagnostic purposes. Some key advantages of the methods include:

- The setup and execution of most types of cathode discharge experiments are rapid.
- There is high flexibility in the experimental setups—little test stand modification is typically required.

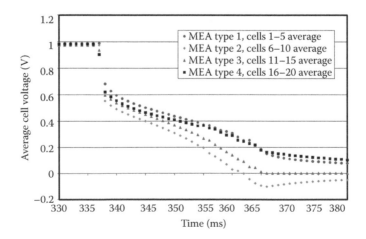

FIGURE 7.15 Cathode discharge on a 20-cell stack with 4 MEA types and 5 cells per MEA type (average cell voltage for each cell type; load step from 0 to 4.17 A cm^{-2}).

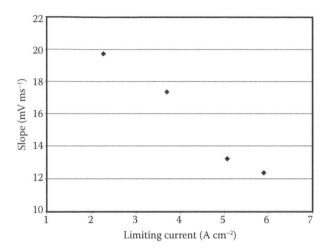

FIGURE 7.16 Slope of cell voltage drops in Figure 7.15 versus limiting current measured on individual cells for each MEA type.

- The nature of the experiments allows performance loss analysis snapshots at any point in time (e.g., during transient operation).
- The boundary conditions of the experiment are typically well defined.
- One can measure MEA properties one dimensionally, even in large cells (assuming that the MEA state is uniform and the conditions are uniform).
- The state of the MEA and cell (temperature, hydration, water distribution, catalyst oxide coverage, and so on) does not substantially change during measurement.
- Information can be obtained that either no other method can provide or where significantly higher sophistication in experimental setup would be required.

Some key limitations of discharge experiments are as follows:

- Obtaining experimental equipment that meets the requirements to control and measure current and voltage in short timescales requires investment.
- Spatial resolution within the cell can only be obtained with setups that have current mapping capability or MRED. MEAs that have nonuniform properties, such as locally degraded MEAs, cannot be investigated in detail by global stack voltage and current measurements.
- Determination of V_{0+} (Figure 7.8) requires skill and experience.
- When cathode discharge experiments are executed under states of low RH MEA hydration, rehydration effects can occur during short timescales, leading to error in V_{0+} determination and the resulting mass transport-free polarizations.
- The methods are not widely used and hence lack standards and comparative data.
- In cell/stack designs that do not allow reactant volumes to be sealed off, certain discharge experiments cannot be executed.
- The pressure in the reactant volumes changes during closed reactant discharge experiments. The cells need to be capable of tolerating cross-pressures.
- Experimentally, fast steps in load over wide load ranges are difficult to control. High load requests at low voltages can be experimentally challenging.
- Information on electrode diffusivity can be obtained with cathode discharge experiments. A breakdown into contributions by GDL, microporous layer, pores in the catalyst layer, and diffusion through the ionomer is currently not possible.

7.6 Outlook

Despite the rather recent approach of using cathode discharge for fuel cell diagnostics, several methods have been discussed here and in the references. Cathode discharge methods offer a wide range of possible experiments in fuel cell diagnostics. They are interesting, both for quick and semiquantitative investigations to test hypotheses on stacks, as well as for scientific purposes to study physical effects in detail.

7.6.1 Quick and Semiquantitative Methods

Cathode discharge methods are emerging as valuable techniques for quick diagnostics, which is desirable during stack development, validation, and manufacturing, as well as possibly for fuel cell system control strategies.

The development of robust and, ideally, standardized methods for mass transport-free polarizations and closed reactant consumption measurements would support the distribution of the method.

Methods to measure the diffusivity of electrodes in stacks are highly desired to enhance the diagnostic toolbox in stack development. One method has been experimentally demonstrated (Stumper et al., 2005a,b) using simultaneous discharge voltage and pressure measurements on a single cell. In Section 7.4.6, an alternative approach to obtain electrode diffusivity information has been conceptually demonstrated. Further development of these methods to obtain accurate and reliable data on electrode diffusivity within stacks would substantially simplify the diagnostics of diffusion-related questions.

The translation of test results obtained on small cells to observations on large cells and stacks is not always straightforward and limits the applicability of small-cell testing. The results in Figure 7.9 appear to be a step toward a method that allows better comparison of results that are obtained in different hardware and different laboratories. It appears that the mass transport-free performance measurements on small MEAs suffice—given proper cell design—as a predictive measure for large stacks.

The application of cathode discharge measurements during transient fuel cell operation can yield valuable insights. For example, during subzero temperature fuel cell operation, the temperature and hydration within the MEA can change rapidly, which could be investigated by cathode discharge experiments. Both the ohmic and the mass transport characteristics are expected to change with time.

7.6.2 Scientific Outlook

One avenue of research is the combination of discharge experiments with cell and stack current mapping, allowing the investigation of stack level effects and neighbor cell interactions (Stumper et al., 2010).

To date, the main step functions in load and specific current sweep patterns have been investigated. By using more sophisticated current versus time profiles, other effects may become measurable (i.e., diffusivity vs. z-location in the electrode). Theoretical investigations could open avenues to measure these or other properties.

Acknowledgments

Many individuals have contributed to the development of these methods over the years. Some of the key contributors are Ali Izadi-Najafabadi, Yeng Lim, Kelvin Fong, Cara Startek, Chris Richards, Michael Loehr, Stephen Hamada, Juergen Stumper, Mark Wolters, Reza Rahmani, John Kenna, Radu Bradean, Alex Szakaly, and Kyle Norris.

Thanks to Stephanie Westwood for execution of the tests that led to the experimental results in this chapter.

References

Lim, C.-Y. and Haas, H. R. 2006. *Diagnostic method for an electrochemical fuel cell and fuel cell components.* US Patent 2006/0051628 A1.

St-Pierre, J., Jia, N. Y., Van Der Geest, M., Atbi, A., and Haas, H. R. 2003. *Methods and apparatus for improving the cold starting capability of a fuel cell.* US Patent 2003/0186093 A1.

St-Pierre, J., Wetton, B., Kim, G.-S., and Promislow, K. 2007. Limiting current operation of proton exchange membrane fuel cells. *J. Electrochem. Soc.* 154(2): B186–B193.

Stumper, J., Haas, H., and Granados, A. 2005a. *In situ* determination of MEA resistance and electrode diffusivity of a fuel cell. *J. Electrochem. Soc.* 152: A837–A844.

Stumper, J., Löhr, M., and Hamada, S. 2005b. Diagnostic tools for liquid water in PEM fuel cells. *J. Power Sources* 143: 150–157.

Stumper, J., Rahmani, R., and Fuss, F. 2009. *In-situ* diagnostics for cell performance and degradation. *Electrochem. Trans.* 25(1): 1605–1615.

Stumper, J., Rahmani, R., and Fuss, F. 2010. Open circuit voltage profiling as diagnostic tool during stack lifetime testing. *J. Power Sources* 195: 4928–4934.

8

Water Transfer Factor Measurement

Tak Cheung Yau
University of British Columbia

Pierre Sauriol
University of British Columbia

Xiaotao Bi
University of British Columbia

Jürgen Stumper
Automotive Fuel Cell Cooperation Corporation

8.1 Introduction

8.1.1 Definition of Water Transfer Factor

Water management has been an important issue in the operation of hydrogen PEM fuel cells (Murahashi et al., 2006; Lu et al., 2007). If the water content in the membrane is too low, membrane proton conductivity and hence cell performance can be adversely affected. On the other hand, when the water produced from the cell reaction at the cathode cannot be removed efficiently, water fills up the pores in the electrode, blocking access of oxygen to the cathode catalyst. This is referred to as flooding of the electrode. Understanding how water is transferred in a fuel cell can provide insights in proper water management to avoid flooding and dry-out conditions.

A useful parameter to study in this context is the water transfer rate, which is defined as the rate of water transported between anode and cathode through the membrane. In the literature there are different sign conventions, but here we define the flow of water from the anode to the cathode as positive. The water transfer flux is defined as the water transfer rate over a certain area of the membrane. The water transfer flux can vary from one section to another in the same membrane. Experimentally, only the average water transfer rate over the membrane active area can be measured. For simplicity, the measured average water transfer rate and average water transfer flux will be referred to as water transfer rate and water transfer flux later in the chapter unless otherwise specified, but readers should keep the above distinctions in mind. A closely related parameter, the water transfer factor (or water transfer coefficient) α is defined as the water transfer rate divided by the rate of water generation from electrochemical reactions (Springer et al., 1991; Berg et al., 2004). As for the transfer rate and flux, the water transfer factor can vary within the cell, for simplicity throughout the remainder of the chapter, the term water transfer factor and α will represent the average water transfer factor, unless otherwise specified.

8.1.2 Importance of Water Transfer Factor Measurements and Potential Applications

8.1.2.1 Water Transfer Factor and Modeling

Even in simple fuel cell models, the importance of water transfer is already addressed. Water transfer affects the water balance between anode and cathode and thereby the reactant vapor pressures or the amount of liquid water present in the flow channels. In more advanced models, water content in the channels plays a significant part in determining the polymer electrolyte membrane resistance. As suggested by Springer et al. (1991), the water content in the channel affects the equilibrium water content of the membrane, which in turn affects the diffusivity of water (Springer et al., 1991) and proton conductivity (Zawodzinski et al., 1993) or membrane resistance (Springer et al., 1991).

8.1.2.2 Membrane Property Characterization

In the models for water transport in membranes, four important phenomena are suggested to occur simultaneously. These are electroosmotic drag, hydraulic permeation, diffusion, and thermoosmotic drag. Measurements of water transfer in general only provide an overall result from these processes. However, these phenomena could be studied individually through suitably designed experimental configurations.

In this chapter, various quantitative methods of measuring water content and hence water transfer is introduced along with their advantages and disadvantages. Then the design requirements in terms of sensitivity and response time are discussed at a system level. Examples are shown to illustrate the experimental results that can be expected from these systems, followed by other potential uses that can be derived from water transfer measurements.

8.2 Measurement Methods and Equipment

In this section, the key methods of measuring the water transfer flux are presented, covering their fundamentals, advantages, disadvantages, and their adaptation in PEMFC water transfer studies. However, these methods do not directly measure the water flow rate across the membrane but rather measure the water flow rate at the outlet of the fuel cell, or at a certain point in a channel. Mass balance calculations are necessary to determine the total water flow rate through the membrane.

We consider first the flow of water going into a fuel cell at both anode and cathode for the purpose of humidification. Between the two electrodes there is a water transfer rate $\dot{n}_{H_2O,X}$ which is our desired quantity to measure, and also a water generation rate at the cathode catalyst layer if the oxygen reduction reaction is taking place. On the anode side, between the inlet of the anode channel to the point of water flow measurement, the change in water flow rate along the channel can only be due to water transfer. Therefore,

$$\dot{n}_{H_2O,X,A} = \dot{n}_{H_2O,in,A} - \dot{n}_{H_2O,out,A} \tag{8.1}$$

On the cathode side, the change in water flow rate along the channel is the result of both water transfer and water generation. This can be written as

$$\dot{n}_{H_2O,X,C} = \dot{n}_{H_2O,out,C} - \dot{n}_{H_2O,in,C} - \dot{n}_{H_2O,gen} \tag{8.2}$$

Either one of the above two equations would suffice in determining the water transfer flow rate, and theoretically $\dot{n}_{H_2O,X,A}$ and $\dot{n}_{H_2O,X,C}$ should agree. The average water transfer factor α can then easily be calculated based on its definition in Section 8.1.

8.2.1 Collection Methods

Water transfer rate can be obtained by the collection methods in which the water vapor or liquid water coming out from the anode and cathode channels is collected and weighed. By collecting the water in the gas streams over a certain period of time, the water molar flow rate is found by dividing the mass of water collected by the time for collection and the molar mass of water.

8.2.1.1 Cold Traps

The simplest method to collect water vapor in the outlet streams is to condense it by cooling. The concentration of water vapor a gas can hold depends on both temperature and pressure. Assuming that there is only gas phase in a stream, the partial pressure of water vapor in the stream is defined as (Perry and Green, 2008):

$$p_{H_2O} = x_{H_2O}P = \frac{\dot{n}_{H_2O}}{\dot{n}_{total}}P \tag{8.3}$$

The saturation vapor pressure of water is related to the temperature by the empirical correlations. The most common one is the Antoine equation (Poling et al., 2007):

$$\log_{10}P_{sat}(T) = 7.11564 - \frac{1687.537}{T + 230.17} \tag{8.4}$$

This equation is applicable for temperatures from 0°C to 200°C. A similar alternative is the Magnus formula, applicable for temperatures from 0°C to 100°C with an error of 0.15% (Buck, 1981):

$$P_{sat}(T) = 0.61121\exp\left(\frac{17.123T}{T + 234.95}\right) \tag{8.5}$$

The dew point temperature is the temperature at which the saturation pressure equals the partial pressure of water in the stream, which could be obtained by solving for T when substituting p_{H_2O} for $P_{sat}(T)$ in the above correlations.

If the temperature of the stream is lower than its dew point, the partial pressure of water vapor in a given gas stream would be greater than the equilibrium vapor pressure and the excess water would condense into liquid. The flow rate of water at the outlet, in moles per second, would be

$$\dot{n}_{H_2O,out,A} = \dot{n}_{H_2O,collected,A} + \dot{n}_{drygas,out,A}\frac{P_{sat}(T_{trap})}{\left(P_{trap} - P_{sat}(T_{trap})\right)} \tag{8.6}$$

Atiyeh et al. (2007) verified the saturation of the gas stream leaving the cold trap in their setup.

Condensers by chilling water are commonly used. To obtain the value of T_{trap}, some researchers controlled the temperature of the condensed water (Husar et al., 2008; Colinart et al., 2009), while measuring the temperature of air leaving the condenser has also been practiced (Atiyeh et al., 2007; Dai et al., 2008), as shown in Figure 8.1. An alternative is to replace chilled water condensers by ice traps (Ge et al., 2005, 2006), so that the temperature of the air leaving the trap is kept at 0°C due to the high latent heat of fusion of ice. However, the amount of ice has to be adequate in order to collect enough water right at 0°C for accurate weighing.

Some authors, for the sake of simplicity, have assumed that all water vapor could be condensed by the cold trap, essentially replacing $P_{sat}(T_{trap})$ by 0 (Janssen and Overvelde, 2001; Murahashi et al., 2006; Ye and Wang, 2007; Adachi et al., 2009; Colinart et al., 2009). To achieve good accuracy, the mass of the collected water is recommended to be 1000 times of the error of the balance (Ye and Wang, 2007). The typical amount of water to be collected ranged from 3 g (Adachi et al., 2009) to 10 g (Ye and Wang, 2007); and the typical amount of time for collection ranged from 3 h to 12 h (Ge et al., 2006). Operating the condensers at elevated pressures favors the collection of liquid water and thus minimizes the impact of the water leaving with the gas stream on the overall balance.

Although α could be calculated using the water transfer rate obtained from Equation 8.6, cross-checking the results obtained from the anode-side and cathode-side measurements is a common practice to ensure accuracy. Assuming a zero $P_{sat}(T_{trap})$, Janssen and Overvelde (2001) obtained a deviation of

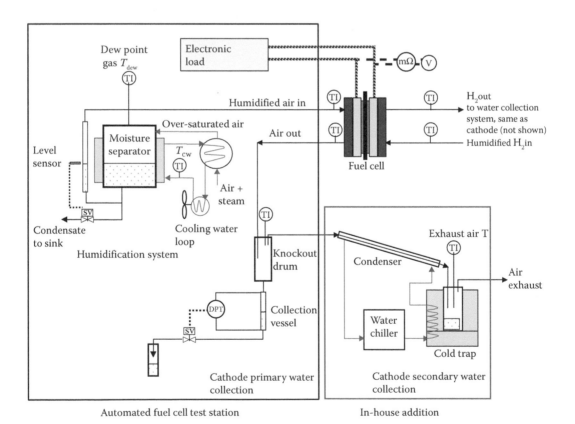

FIGURE 8.1 Condenser system used by Atiyeh et al. (2007). In addition to the air-cooled knock-out vessel built in the test station, they had a water-chilled condenser to collect the water vapor from the cell outlet. Note the use of thermocouples to measure the exhaust temperature (T_{trap}). (Reprinted from *J. Electrochem. Soc,* 156, Adachi, M. et al. Correlation of *in situ* and *ex situ* measurements of water permeation through Nafion NRE211 proton exchange membranes, B782–B790, Copyright (2009), with permission from Elsevier.)

4–7%, and 2.7–8.5% by Murahashi et al. (2006). Accounting for the amount of water left in the exhaust, Atiyeh et al. (2007) and Colinart et al. (2009) achieved deviations less than 5%. Murahashi et al. (2006) used the mean value of α from both sides in subsequent analysis. As discussed by Atiyeh et al. (2007), the accuracy of the humidification system is also important for obtaining reliable results.

The greatest advantage of using cold traps is the setup simplicity and low cost. Only condensers and balances are required. However, because of the constraints in the accuracy of balances and potential losses of liquid water after collection, a very long time is required to collect enough water to achieve an accuracy of <5% in water flow rate. Also, the measurement of mass is a batch process; cold traps are thus not suitable for transient measurements.

8.2.1.2 Sorbents

The water from the outlet of the cell could also be collected by sorbents prior to weighing (Yan et al., 2006). A typical sorbent for water is anhydrous calcium sulfate ($CaSO_4$), while silica gel is also employed for such purposes (Rajalakshmi et al., 2004). Readers are referred to Mallinckrodt Baker Inc. (2000) for a sorbent selection guide.

The maximum amount of water that can be sorbed depends on the sorption isotherm of the sorbent chosen. For example, at 32°C, the maximum mass of water that could be sorbed is around 5% of the mass of the $CaSO_4$ sorbent itself (Jury et al., 1972). When the sorbent can no longer take up additional moisture, a breakthrough is said to have occurred and the experimental results become erroneous. By weighing the anhydrous $CaSO_4$ before and after contact of the gases from the cell, the molar sorption rate of water is obtained by dividing the increase in mass by the time of sorption and the molar mass of water. Sorbents usually perform better at elevated pressures. Most spent sorbents can be regenerated by drying at low pressure and elevated temperature.

The advantage of the water sorption method over the cold traps is that a very low chilling temperature is not required. However, the ratio of mass of collected water to the error of the balance is much reduced because the water contributes only less than 5% of the measured mass (for $CaSO_4$). The accuracy on this adsorption method has not been reported for the measurement of water transport in running PEM fuel cells (Yan et al., 2006).

8.2.2 Dew Point/ Relative Humidity Methods

The partial pressure of water in a wet stream can be determined from dew point and relative humidity (RH) measurements. As shown above, the dew point of a stream of wetted gas reflects the partial pressure of water in that gas stream. In addition, at a certain temperature, the relative humidity of a wetted gas is given by

$$RH = \frac{p_{H_2O}}{P_{sat}(T)} \tag{8.7}$$

So the partial pressure of water can also be measured through RH. The molar flow rate of water can be determined from the partial pressure through Equation 8.3.

It should be noted that these methods apply to single-phase gases only.

8.2.2.1 Chilled Mirrors (Dew Point)

Chilled mirror sensors measure the temperature of the gas at which the first drop of water condenses (i.e., the dew point of the gas) by means of cooling a solid surface exposed to the sample gas. Unlike cold traps, the recorded quantity is the temperature rather than the mass of water, and the cooling is done gradually from the sample temperature to the dew point. Measurement is done when the first drop of condensed water is observed.

A chilled mirror sensor consists of a chamber in which a clean and highly polished mirror is placed for temperature and humidity measurement. The mirror is cooled by a Peltier cooler in which an electrical current is passed through a junction of two dissimilar conductors and the temperature of the mirror is closely monitored. The condensation of water vapor scatters a light beam originally reflected by the mirror to a detector. Then the mirror is heated by reversing the electrical current to the cooler. This is repeated until the mirror temperature and the thickness of dew is stabilized as the equilibrium is reached (Fraden, 2004). So this mirror temperature would be recorded as the dew point of the gas (Bentley, 1998).

8.2.2.2 Piezoelectric Hygrometers (Dew Point)

Alternatively, quartz crystals can be used in place of mirrors, and optical sensing of condensation can be replaced by the change in resonance frequency. This type of sensors is termed as oscillating or piezo-electric hygrometers (Ito, 1987; Fraden, 2004).

The advantage of using chilled mirrors is that they are highly accurate. The uncertainty of chilled mirrors can be as low as 0.1°C in dew point (Bentley, 1998; Sauriol et al., 2009), which translates into less than 1.4% in terms of RH (Bentley, 1998). However, thermal equilibrium has to be achieved before readings could be taken, which translates into a penalty in response time (Sauriol et al., 2009). Piezoelectric sensors can have response times of 5 min from high dew point to low dew point, and 1 min from low dew point to high dew point (Ito, 1987). If there is liquid water in the stream, the dew point is above the temperature of the gas, so it is impossible to trace the dew point by means of cooling. Therefore, preheating the gas prior to the sampling point of the sensor becomes necessary (Jung et al., 2007).

8.2.2.3 Capacitive Sensors (RH)

There are two main types of RH sensors: capacitive and resistive. In capacitive RH sensors, a hygroscopic polymer changes its dielectric constant upon adsorption of moisture from the environment. When sandwiched between porous electrodes, the capacitance of the resulting capacitor increases linearly with RH (Fraden, 2004). The uncertainty in the measurement is 2% RH at an RH range of 5–90% (Fraden, 2004), while the response time is a few seconds to about 4 min for temperature and humidity.

8.2.2.4 Resistive Sensors (RH)

In resistive sensors, the impedance of conductive polymers (Wilson, 2005) or a layer of porous aluminum oxide (Fraden, 2004) is measured. The impedance of such films could vary from 10 MΩ to 1 kΩ when the RH is increased from 20% to 90%. On account of the nonlinearity of the impedance response, the signal generated has to be linearized. The uncertainty of this type of sensors is 2% RH at an RH range of 15–85%, and it takes a few minutes for the sensor to respond when the RH changes from 100% to 30% (Wilson, 2005).

From Equation 8.7, a temperature reading is required to convert an RH value to the vapor pressure of water and hence water concentration. Therefore, RH sensors have to be coupled with temperature sensors in order to make the reading useful (Hinds et al., 2009).

One of the advantages of using capacitive and resistive RH sensors is that they are readily available commercially at reasonable costs (Sauriol et al., 2009). Also, because of their small size and minimally disruptive nature, it is possible to place a number of these sensors in a segmented cell to investigate the changes in RH along the channel (Nishikawa et al., 2006) (Figure 8.2). However, as encountered by the dew point method, RH sensors have to operate in single-phase flows without liquid water present. Heating up the gas samples may thus be necessary. In addition, when the RH of the sample is less than 10%, the error increases to 10%, rendering readings unreliable. To mitigate this situation, temperature-controlled RH sensors can be installed so that the RH can be maintained well above 10% (Majsztrik et al., 2008).

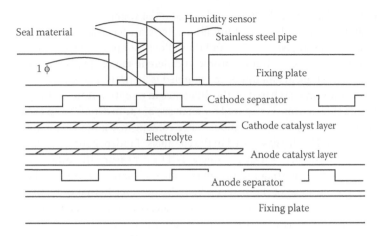

FIGURE 8.2 Humidity sensors installed in a cell to study the RH distribution along the channel. (Reprinted from *J. Power Sources*, 155, Nishikawa, H. et al. Measurements of humidity and current distribution in a PEFC, 213–218, Copyright (2006), with permission from Elsevier.)

8.2.3 Composition Methods

Composition methods can directly measure the mole fraction x_{H_2O} of water vapor in a gas sample. The flow rate of water can be directly obtained using Equation 8.3, without requiring $P_{sat}(T)$ as in the dew point and RH methods. The dew point of the sample fed to the analyzer has to be lower than the analyzer temperature to ensure all water stays vaporized.

8.2.3.1 Gas Chromatography-Thermal Conductivity Detector (GC-TCD)

In chromatography methods, the gas samples are first separated into different fractions, followed by measuring and identifying the fractions (Dean, 1995). The most straightforward implementation is through gas chromatography (GC). Samples from a fuel cell are mixed with a carrier gas (usually helium) and fed into a sorbent-filled column where certain gases in the sample are more readily adsorbed on the sorbent and hence stay longer in the column. To separate small molecules like water from hydrogen, nitrogen and oxygen, columns filled with molecular sieves were commonly used (Mench et al., 2003). To counter the problem of liquid water damage to columns and detectors, the gas fed to the column is preheated to vaporize all droplets, and the column is regularly back purged to remove water in the column (Mench et al., 2003).

The separated components (together with the carrier gas) are fed into a detector. On account of the different column retention times, they arrive at the detector at different times. To identify the gases, standard pure gases could be injected into the column and the retention time could serve as fingerprints of the gases. The most common detector is the thermal conductivity detector (TCD) (Dean, 1995; Dimitrova et al., 2002; Mench et al., 2003). In a TCD cell, there are two chambers: reference and sample. The reference chamber contains carrier gas only, while the samples are fed to the sample chamber. In the sample chamber, a filament is heated by current passing through it. While the thermal conductivity of the environment of the filament decreases with increasing sample content, the resistance of the filament in the sample chamber would be greater for higher sample concentrations due to a higher temperature. By measuring this resistance using a Wheatstone bridge, the concentration of a component in the sample could be found by area integration of the corresponding peak and correlating to a calibration curve (Dean, 1995). The measured water molar fraction is reported to be within 4.5% error in the temperature range of 40°C–85°C using helium as a carrier gas (Lu et al., 2007).

The advantage of using GC is that it is possible to obtain concentrations of other species than water in the same setup and same sample. Therefore, phenomena such as hydrogen crossover (Endoh et al., 2004), methanol crossover (Dimitrova et al., 2002), and oxidation of water at the anode under low fuel supply (Knights et al., 2004) can be studied in addition to water transfer. Also only a very small amount of sample would suffice for the analysis by GC, and thus allows sampling of gases at specific points along the channels to measure the concentration profile of different species (Mench et al., 2003; Yang et al., 2005; Lu et al., 2007). The typical amount of gas sampled is 20 μL (Lu et al., 2007). However, since the separation of gases in the sample is based on the difference in retention times of the gases in the column, it is inevitable that significant time is required for analysis, of the order of 5 min (Wu et al., 2008).

The TCD alone could be used for continuous water content measurements without separation. The reference in this case would be dry gases used in the channels (Fraden, 2004). Alternatively, Sauriol et al. (2009) considered the possibility of using a single TCD to measure the difference in water content between the inlet and outlet gas streams. This may reduce systematic errors in the measurement (see Section 8.3 for details). The ratio of thermal conductivity of water to that of hydrogen is 1:10 but only 1:1.4 for the water/nitrogen or water/oxygen pair. This makes the analysis on the cathode side more difficult (Sauriol et al., 2009). Also, at the cathode the molar ratio between nitrogen and oxygen is different between the inlet and outlet, depending on the load to the cell. Fortunately the thermal conductivity difference between nitrogen and oxygen is only 1.2% (Chemical Rubber Company, 2008), so that conductivity is barely load dependent.

8.2.3.2 Optical Methods

In optical methods, the concentration of water in a given gas sample is measured through the absorbance of electromagnetic radiation. A source of the radiation is passed through the sample gas, and the transmitted energy intensity is detected by photon detectors. For a certain wavelength the absorbance of water is defined as

$$\Gamma = \log \frac{\Phi_{blank}}{\Phi_{sample}} \tag{8.8}$$

which is dependent on the molar absorptivity, length of the optical path, and the molar concentration of water vapor. The absorption band for water is around 2.4 μm. Although light in the band around 123.6 nm is also strongly absorbed by water, oxygen also shows absorption at this range, making the readings inaccurate (Auble and Meyers, 1992).

Sauriol et al. (2009) used infrared gas analyzers to measure water concentrations at the cell outlet. The analyzer they used was designed for greenhouse applications (low pressure and temperature up to 50°C), and the outlet streams had to be diluted and depressurized rather than heated to vaporize any liquid water. Baseline measurements using dry gas were conducted to tackle drifts in the equipment. Also, frequent calibration against a known inlet water injection rate eliminated systematic offsets of the sensor, flow controllers and other system components in the sensor reading. By averaging values over 30 min, they were able to achieve an uncertainty level of 0.01 in α (Sauriol et al., 2009).

Basu et al. (2006a) passed a tunable diode laser beam of 1.491–1.492 μm in wavelength through a section of the flow channel to detect water. The widths of the fitted peaks in the absorption profile were used as a criterion for water vapor pressure measurement (see Figure 8.3). The error level is found to be around 10% in p_{H_2O}. The measurement time for a sample is reported to be 0.4 s. The total area of the peak is found to be weakly dependent on water vapor pressure but strongly dependent on temperature; hence, temperature and water pressure could be found simultaneously (Basu et al., 2006b). An improvement was made using a wavelength modulated laser beam. The ratio between the second and first harmonics of the resultant signal was used instead of the raw absorption profile as the fitting criteria for width and area determination. The error was found to be 2.5% in p_{H_2O} and 3°C in temperature. Also the ratio of the harmonics was independent of the incident laser intensity and system gain settings, so that the use of a background measurement could be eliminated (Sur et al., 2010).

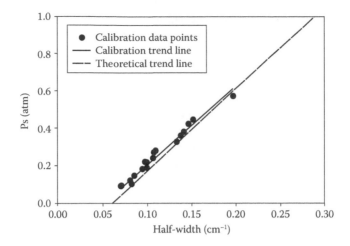

FIGURE 8.3 Relationship between water vapor pressure and the half-width of the Lorentz curve fitted to the broadened transmitted spectrum through the sample. (Reproduced from Basu, S. et al. 2006a. *J. Fuel Cell Sci. Technol.* 3: 1–7. With permission from ASME.)

The biggest advantage of optical methods is that it provides quick responses at a few milliseconds from the moment the sample is introduced into the measurement apparatus (Bentley, 1998). In general, the measurements are highly repeatable, but the variations in the cleanliness of the optical bench, its temperature, pressure, and flow rate may have an impact on the overall water transfer factor measurement. Frequent calibration (Sauriol et al., 2009), or referencing to a database (Basu et al., 2006b; Sur et al., 2010) are the solutions proposed to obtain accurate readings.

8.2.3.3 Mass Spectrometry

In mass spectrometry the gas samples are ionized and fragmented by electron beam bombardment. The ions then passed through an analyzer, most commonly a quadrupole, where an AC–DC superimposed voltage is varied to allow the passage of ions only with the desired mass-to-charge ratio. By scanning through different mass-to-charge ratios, the entire mass spectrum could be produced. It is capable of scanning up to a mass-to-charge ratio of 750 atomic mass unit per second. Ions passing through the quadrupole strike a conversion dynode which produces electron beams on detecting ions. This is the basis of the determination of the molar ratio of a particular species in the sample (Dean, 1995).

Dong et al. (2005) used a mass spectrometer (in the form of a real-time gas analyzer) to detect the molar fraction of water. The samples were heated to 150°C to have all the liquid water vaporized. The ion count from the spectrometer was calibrated using a stream of gas at a dew point of 50°C. The average error in molar fraction was 0.5% and the response time was around 1 s.

Similar to GC, the advantage of using mass spectrometry is the ability to study the concentration of species other than water, but with a higher degree of accuracy (Wu et al., 2010). In mass spectrometry a pure standard is not even necessary to recognize the species since molar mass information can be obtained from the spectrometer. It is minimally invasive to use mass spectroscopy in studying species distribution because of the small amount of sample required (Dong et al., 2005).

8.2.4 Mass Flow Methods

The total mass flow of a stream can be used to find out the corresponding water content, when the flow rates of components other than water are known. This is possible when we know the inlet gas flows, the current load of the cell, and the dry gas flow rate which could be calculated in terms of stoichiometry. Coriolis mass flow meters are good candidates. In a Coriolis mass flow meter the sample gas passes through a U-shaped horizontal tube and the mass flow rate within the tube is inversely proportional to the resonant frequency (Sparks et al., 2004).

8.3 Key Design Aspects

In this section, the key design aspects of a water transfer factor measurement apparatus are discussed. A selected number of existing and proposed designs are analyzed for their accuracy in measuring the water transfer factor based on typical operating conditions of PEMFCs and expected performance of key measurement devices. Finally, some design considerations affecting the capabilities of dynamic analysis will be presented.

8.3.1 Design Considerations

The analysis will attempt to estimate the accuracy of various water transfer factor measurement apparatus on a running PEMFC. The most important attributes of a water transfer factor measurement apparatus were identified as operating range, accuracy, spatial resolution, response time, engineering costs, and risks.

8.3.1.1 Operating Range

The operating range of existing fuel cell systems is extremely broad, covering several orders of magnitude in volumetric flow rates (small single-cell to large multicell stacks). Table 8.1 describes a wide range of operating conditions representative of a 300 cm² single-cell PEMFC. As the conditions inside a running fuel cell may change drastically along the length of the cell, so may the water transfer factor. All components in the measurement setup should have good accuracies in the desired operating range.

8.3.1.2 Accuracy

An accurate measurement of the water transfer factor is particularly difficult to obtain since the amount of water transported across the membrane usually represents a very small fraction of the water found in the outlet. Hence, small measurement errors in terms of water flow rate may yield large errors in the water transfer factor measurement. This is especially true for the cathode-based measurements where the net water transfer to outlet water flow ratio is typically an order of magnitude smaller than the anode. It is thus important to estimate the accuracy of all designs to assist in the selection of a water

TABLE 8.1 PEMFC Operating Conditions Based on a 300 cm² Fuel Cell

Load (A)	30	80	135	190	250	300
Anode (100% H_2, dry basis)						
Pressure (kPa)	210	230	250	270	295	320
Stoic.	2.4	1.6	1.6	1.6	1.6	1.6
Dew point (°C)	63	63	63	63	63	63
Cathode (air: 21% O_2, N_2 balance, dry basis)						
Pressure (kPa)	190	210	230	250	275	300
Stoic.	3.5	1.8	1.8	1.8	1.8	1.8
Dew point (°C)	67	67	67	67	67	67
Coolant (flowing in the same direction as the cathode)						
Inlet temperature (°C)	65	65	65	65	65	65
Outlet temperature (°C)	67.5	69	70.5	72.5	74	75.5

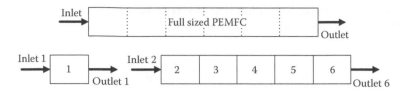

FIGURE 8.4 Representation of the subcell approach for localized water transfer measurement.

transfer factor measurement apparatus and its operations, as well as the identification of the measurands that play a determining role on the overall water transfer factor accuracy.

8.3.1.3 Spatial Resolution

Spatial resolution can be improved by positioning sampling points (using RH sensors as in Section 8.2.2.4 or GC as in Section 8.2.3.1) directly in the flow field. For other approaches in which only the outlet water content is measured, a subcell approach was proposed and demonstrated by Sauriol et al. (2005). The 300 cm^2 fuel cell is conceptually divided into a series of subcells as depicted in Figure 8.4. The subcells are to be operated and monitored one at a time with the outlet conditions of the current subcell serving as inlet conditions of the next one until all the subcells are resolved. The water transfer factor profile within the full-sized cell can then be determined by collating the subcell water transfer factors.

There is a relation between accuracy in water transfer factor and spatial resolution. As will be illustrated in Section 8.3.2, when the spatial resolution is increased, the accuracy in water transfer factor decreases in general.

8.3.1.4 Response Time

As discussed in Section 8.2.1.1, existing water balance techniques that rely on water collection usually require long time averages (several hours) to achieve acceptable accuracy levels. These techniques are incapable of tracking changes in the water transfer factor under unsteady operations, an important capability given the ongoing development efforts in the area of water management. While instantaneous changes in operating parameters such as load, pressure, or flow rate may induce a nearly instantaneous change in the local water transfer factor, the capacitive nature of the system will attenuate and delay the manifestation of these changes on the outlet stream; therefore, millisecond time response may not be necessary to allow for satisfactory dynamic analysis.

8.3.1.5 Engineering Costs and Risks

The cost of any individual component must be evaluated against its benefits. Off-the-shelf technologies should be favored in order to minimize the overall cost and risks of developing a new measurement instrument. If possible, the measurement instruments should be used in their known range of operation. Deviations from their known range will involve a risk associated with their out-of-range operation, and requires the development of a specific calibration for an instrument or conditioning of the measured subject so that it can fall within the known range of operation (e.g., heating stream to evaporate water when limited to single-phase measurement).

8.3.2 Sensitivity Analysis

8.3.2.1 Calculations

As described in Section 8.2, the determination of the water transfer factor from water balance calculations can be achieved by combining various measurands describing global quantities such as inlet and outlet flow rates (dry or wet), inlet and outlet compositions (molar fraction, dew point, RH), and subcell operating conditions (load) (Equation 8.9).

$$\overline{\alpha}_i = \frac{\dot{n}_{i,A,H_2O,in} - \dot{n}_{i,A,H_2O,out}}{(\nu_{H_2O}/\nu_{e^-})(I_i/F)} = \frac{\dot{n}_{i,C,H_2O,out} - \dot{n}_{i,C,H_2O,in}}{(\nu_{H_2O}/\nu_{e^-})(I_i/F)} - 1 \tag{8.9}$$

Alternatively, the calculations of the water transfer factor can be achieved on a total molar (Equation 8.10), total mass basis (Equation 8.11), or a molar composition difference basis (Equation 8.12). For simplicity, these equations are presented for the anode only.

$$\overline{\alpha}_i = \frac{\dot{n}_{i,A,tot,in} - \dot{n}_{i,A,tot,out} + (\nu_{H_2}/\nu_{e^-})(\overline{I}_i/F)}{(\nu_{H_2O}/\nu_{e^-})(\overline{I}_i/F)} \tag{8.10}$$

$$\overline{\alpha}_i = \frac{\dot{m}_{i,A,tot,in} - \dot{m}_{i,A,tot,out} + (\nu_{H_2}/\nu_{e^-})(\overline{I}_i/F)MW_{H_2}}{(\nu_{H_2O}/\nu_{e^-})(\overline{I}_i/F)MW_{H_2O}} = \frac{-\Delta\dot{m}_{i,A,tot} + (\nu_{H_2}/\nu_{e^-})(\overline{I}_i/F)MW_{H_2}}{(\nu_{H_2O}/\nu_{e^-})(\overline{I}_i/F)MW_{H_2O}} \tag{8.11}$$

$$\overline{\alpha}_i = \frac{-\Delta x_{i,A,H_2O}\dot{n}_{i,A,tot,in} - \left(x_{i,A,H_2O,in} + \Delta x_{i,A,H_2O}\right)(\nu_{H_2}/\nu_{e^-})(\overline{I}_i/F)}{\left(1 - x_{i,A,H_2O,in} - \Delta x_{i,A,H_2O}\right)(\nu_{H_2O}/\nu_{e^-})(\overline{I}_i/F)} \tag{8.12}$$

The molar flow rate may be replaced by other equivalent quantities such as volumetric flow rate, mass flow rate, wet composition, dry composition, relative humidity, dew point, and total volumetric flow rate depending on the measurement instruments available, referring to Equations 8.13 through 8.15.

$$\dot{n}_{H_2O} = \frac{\dot{v}_{H_2O}\rho_{H_2O}(T)}{MW_{H_2O}} = \frac{\dot{m}_{H_2O}}{MW_{H_2O}} = x_{H_2O}\dot{n}_{total} = y_{H_2O}\dot{n}_{drygas} \tag{8.13}$$

$$x_{H_2O} = \frac{y_{H_2O}}{1 + y_{H_2O}} = \frac{P_{sat,H_2O}(T)RH}{P} = \frac{P_{sat,H_2O}(T_{dew})}{P} \tag{8.14}$$

$$\dot{n}_{total} = \dot{v}_{total}\frac{P}{ZRT} \tag{8.15}$$

For the integration nature of collection methods, $\overline{\alpha}_i$ is given by Equation 8.16. As a first estimate, the gas leaving the collection vessel can be assumed to have reached equilibrium (condenser) or to be nearly dry (sorbent). For both types of collection methods, the analysis considers the collection of 10–200 g of water, and the sorbent capacity can be varied from 0.025 to 0.2 g of water per gram of sorbent.

$$\overline{\alpha}_i = \frac{n_{i,A,H_2O,out} - \int\limits_{time} \dot{n}_{i,A,H_2O,in}dt}{(\nu_{H_2O}/\nu_{e^-})\left(\int\limits_{time} I_i dt/F\right)} \tag{8.16}$$

To study the effect of spatial resolution on water transfer factor accuracy, the conditions of Table 8.1 and the model proposed by Berg et al. (2004) were used to generate representative current density and water transfer factor profiles of the full-sized PEMFC. The profiles were then integrated over various subcell sizes and positions within the full-cell profiles according to Equations 8.17 and 8.18 to give the average subcell water transfer factor.

$$I_i = \int_{S_i} j \, ds \tag{8.17}$$

$$\bar{\alpha}_i = \frac{\int_{S_i} \alpha j \, ds}{I_i} \tag{8.18}$$

The sensitivity analysis aimed at quantifying the impact of specific measurement instruments on the determined water transfer factor. The error in α caused by each measurand Y at subcell i is estimated by

$$\Delta\bar{\alpha}_{i,Y} = \frac{\partial\bar{\alpha}_i}{\partial Y} \Delta Y \tag{8.19}$$

where ΔY represents the standard deviation of the instruments. Considering that $|\Delta\bar{\alpha}_{i,Y}|$ is proportional to the standard deviation of a normally distributed function, the overall measurement accuracy (standard deviation) of the considered apparatus would be given by

$$\Delta\bar{\alpha}_i = \sqrt{\sum_Y \left(\Delta\bar{\alpha}_{i,Y}\right)^2} \tag{8.20}$$

TABLE 8.2 Summary of Measurement Instruments and Typical Accuracies

Measurand	Equipment	Typical Accuracies
Volumetric water flow rate ($\dot{v}_{H_2O}$)	Syringe pump Microannular gear pump HPLC pump	0.1–1% of set point
Gas molar flow rate ($\dot{n}_{tot}$ or $\dot{n}_{dry}$)	Thermal mass flow meters	1% of full scale
Water content (x_{H_2O})	Infrared sensors	1%
(Section 8.2.3.1)	Thermal conductivity detector	
Dew point (T_{dew}) (Sections 8.2.2.1 and 8.2.2.2)	Chilled mirrors	0.1°C
Relative humidity (RH)	Capacitive RH sensors	2% RH
(Sections 8.2.2.3 and 8.2.2.4)	Resistive RH sensors	
Saturation pressure (P_{sat,H_2O}). (Section 8.2.1.1)	Equations in Buck (1981) and Yaws (1999)	1% (from spread between selected correlations)
Pressure (P)	Pressure transducers	0.08% of full scale
Temperature (T)	Thermocouples Thermistors RTDs	1°C 0.1°C 0.03°C
Voltage (V_{shunt})	Data acquisition	<0.009%
Shunt resistance (R_{shunt})	Laboratory calibration	0.05%
Mass (m)	Electronic balance	0.001% of full scale
Volumetric gas flow rate ($\dot{v}_{total}$)	Turbine flow meter	0.6%
Mass flow rate ($\dot{m}_{total}$) (Section 8.2.4)	Coriolis mass flow meter	0.5%
Differential mass flow rate ($\Delta\dot{m}_{total}$)	Coriolis differential mass flow meter (Pradelli, 1998)	Unknown, 1–50% assumed
Differential water content (Δx_{H_2O})	Dual-channel infrared sensors	1%
(Section 8.2.3.1)	Thermal conductivity detector	
Compressibility factor (Z)	NIST data	N/A

8.3.2.2 Measurands

For each subcell, the possible measurands required in Equations 8.10 through 8.15 are briefly listed along with their typical accuracies in Table 8.2.

8.3.2.3 Results

The results of the sensitivity analysis are divided according to the main equation used and subdivided into contributing terms to facilitate the presentation and analysis of the results. The net accuracy can be estimated by applying Equation 8.20 to the contributing terms.

By combining instrument-specific accuracies, the impact of instruments combination on the water transfer factor accuracy is assessed across a wide range of subcell conditions. Figures 8.5 through 8.9 express the range of accuracy contributions of the selected measurement instruments combination that can be expected on both anode and cathode sides. The results are also discretized as a function of the number of subcells (i.e., the size of the sub-cell). In general, only the data from the instrument with the best accuracy is presented. Whenever the contribution of a specific measurand is lower than the overall accuracy, lower-accuracy measurement instruments could be considered without significant impact on the overall accuracy. In general, it is found that the use of thermistors would be preferable to thermocouples and that accounting for compressibility did not significantly improve the measurement accuracy.

Water molar balance (Equation 8.9): The overall accuracy is shown in Figure 8.5. Overall, the accuracy of the water transfer factor is nearly a full order of magnitude smaller on the anode side than on the cathode side. Considering the typical ranges of water transfer factors, the implementation of such a water transfer measurement procedure would only be recommended on the anode side with cell sizes corresponding to 1 and 6 subcells, or the cathode side with cell sizes corresponding to 1 subcell. Improvements to the determination of $\dot{n}_{i,H_2O,out}$, which is mostly comprised of the contributions of

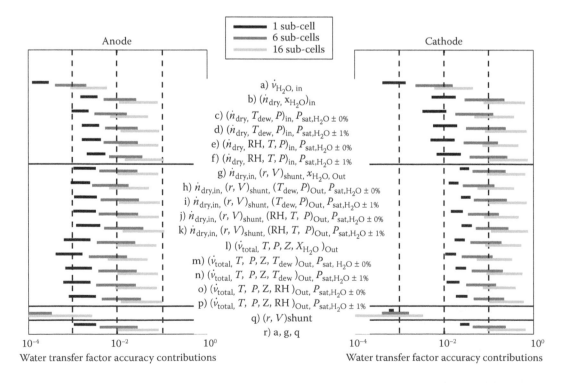

FIGURE 8.5 Water transfer factor accuracy contributions based on water molar balance (Equation 8.9).

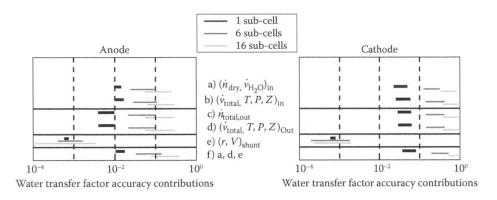

FIGURE 8.6 Water transfer factor accuracy contributions based on total molar balance (Equation 8.10).

$\dot{n}_{i,\text{drygas,in}}$ and $x_{i,\text{H}_2\text{O,out}}$, would be required for this approach to yield significantly more accurate measurement results. Readers are referred to Figure 8.5 for the effect of other measurands (e.g., T, P, RH) on the accuracy in water transfer factor.

Total molar balance (Equation 8.10): As shown in Figure 8.6, both molar flow rates terms $\dot{n}_{i,\text{H}_2\text{O,in}}$ and $\dot{n}_{i,\text{total,out}}$ contribute significantly to the overall accuracy, resulting in a nearly impractical measurement concept which can, at best, be used for the determination of the water transfer factor on the anode side for cell sizes corresponding to 1 subcell.

Total mass balance (Equation 8.11): The overall accuracies are presented in Figure 8.7. Both $\dot{m}_{i,\text{total,in}}$ and $\dot{m}_{i,\text{total,out}}$ contribute nearly equally to the overall accuracy. At best, this approach can be used on the anode side of cells with sizes corresponding to 1 and 6 subcells. The water transfer factor accuracy is more than one order of magnitude better on the anode side than on the cathode side, and thus cathode-side measurements would not be recommended. In differential mass flow measurements, anode and cathode yield identical accuracies. Considering the typical ranges of the water transfer factor, this approach may be applied for either electrode at any of the subcell sizes for differential mass flow rate with accuracies up to 10%.

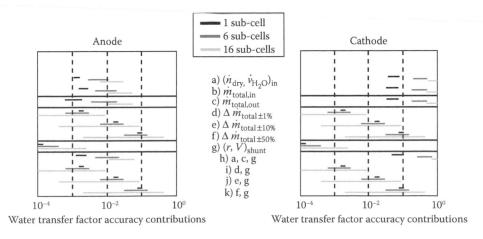

FIGURE 8.7 Water transfer factor accuracy contributions based on total mass balance (also includes differential mass flow measurement) (Equation 8.11).

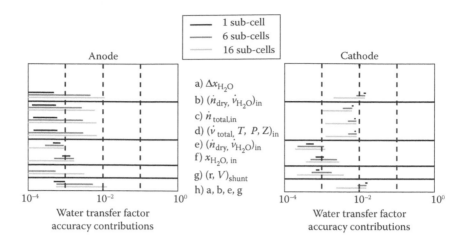

FIGURE 8.8 Water transfer factor accuracy contributions based on water molar balance with differential water composition measurement (Equation 8.12).

Water molar composition difference (Equation 8.12): As presented in Figure 8.8, on the anode side, $\Delta x_{i,H_2O}$, $\dot{n}_{i,total,in}$, and I_i contribute most to the water transfer factor accuracy of small-sized cells (corresponding to 6 and 16 subcells); while $x_{i,H_2O,in}$ contributes mostly to the larger-sized cells (corresponding to 1 subcell). On the cathode side, $\Delta x_{i,H_2O}$ is the term most influential on the overall water transfer factor accuracy. Based on the expected accuracies, this approach would lend itself well for measurement of the water transfer factor across all the conditions considered both on the anode and the cathode.

Collection-based water molar balance (Equation 8.16): In general, the anode-side accuracy is better than the cathode side by a factor of 3, as shown in Figure 8.9. At 10 g collection, a 0.1 g offset—the result of 2–3 water droplets remaining in the condenser/sampling lines or of gases other than water trapped by the sorbent—is critical on the overall accuracy. As the collected mass is increased, so does the overall accuracy. At 200 g collected mass, there is nearly no influence of the offset on the overall accuracy. In general, this approach could be implemented on the anode side of fuel cells of any size as long as the combined accuracy of the scale ($\sqrt{2}$ times the accuracy of scale to account for the difference) and the mass offset is less than 0.1% of the collected mass. At 1% offset, the measurements on the anode side should be limited to cell sizes corresponding to 1 and 6 subcells. On the cathode side, at 0.1% offset the approach would be limited to implementation on cell sizes corresponding to 1 and 6 subcells, and only to sizes corresponding to 1 subcell at 1% of the collected mass of water.

8.3.4 Monte Carlo Analysis

Sauriol et al. (2009) opted to adapt a water molar balance method (based on Equation 8.9) and implemented a self-reference step (a one point calibration). This configuration is depicted in Figure 8.10. By relying on the same instrument to calculate the difference of two close quantities, the error associated with the difference becomes more dependent on the repeatability rather than on the accuracy of the measurement instruments (as long as instruments are not affected by hysteresis). Equation 8.9 can be rewritten as Equation 8.21.

$$\bar{\alpha}_1 = \frac{\dot{n}_{i,A,H_2O,in} - \dot{n}_{i,A,H_2O,ref}\left(x_{i,A,H_2O,out} / x_{i,A,H_2O,ref}\right)\left(\dot{n}_{i,A,total,out} / \dot{n}_{i,A,total,ref}\right)}{(\nu_{H_2O} / \nu_{e^-})(\bar{I}_i / F)} \tag{8.21}$$

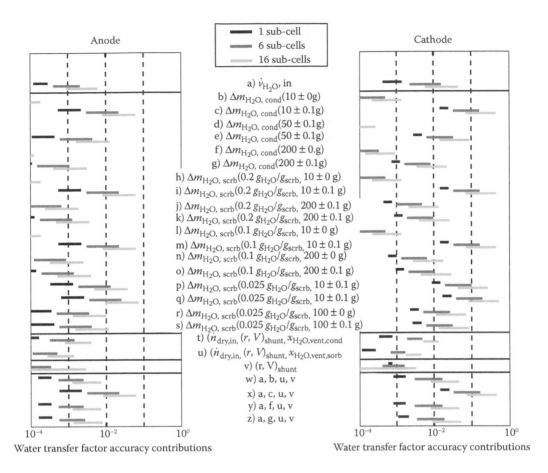

FIGURE 8.9 Water transfer factor accuracy contributions based on water molar balance with collection in a condenser or a sorbent bed (Equation 8.16).

To validate the proposed strategy, the measurement concept was evaluated by a Monte Carlo simulation which, as opposed to the sensitivity analysis, also accounted for instrument limitations and requirements. This enabled the anticipation of potential practical challenges while in the design phase. The classic measurement concept using similar measurement devices (refer to Figure 8.10a) but based on Equation 8.9 is also included for comparison purposes.

For every subcell condition, the measured water transfer factor is computed 100,000 times. The measured water transfer factors were compared against the expected value in order to verify the presence of a bias. For both approaches tested, the measured water transfer factors did not show evidence of a bias. The standard deviation of the measured water transfer factor was thus taken as the error of the measurement approach. The results of the Monte Carlo simulations are shown in Figure 8.11. The error of the measurement approach was found to correlate well with the "water ratio," which is the ratio between the water flow rate exiting a subcell to the water produced within the subcell $\left(\dot{n}_{i,\mathrm{H_2O,out}}/\dot{n}_{i,\mathrm{H_2O,rx}}\right)$. In general, the approach with self-reference reduces the expected water transfer factor error by a factor of 8 in comparison with the classic approach. In general, any condition with water ratios of 10 or less should yield acceptable water transfer factor accuracies (±0.03) within the typical ranges, thus permitting the use of the approach to every subcell conditions on the anode side and to subcell sizes corresponding to 1 and 6 subcell on the cathode side. Further details on the development of the measurement apparatus and the Monte Carlo simulations were reported by Sauriol et al. (2009).

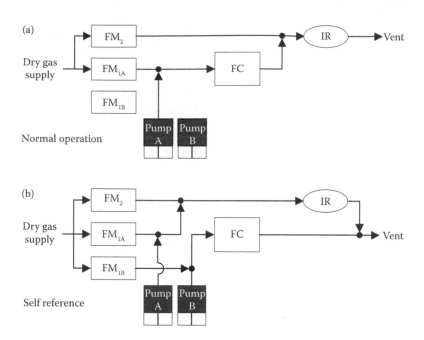

FIGURE 8.10 Configuration of the measurement concept proposed by Sauriol et al. (2009). (a) during normal operation the fuel cell is fed using the main pump (Pump A) and flow meter (FM_{1A}) and its outlet is combined with enough dry gas (FM_2) to allow complete evaporation of the water; (b) during the self-reference step, a backup pump (Pump B) and flow meter (FM_{1B}) are used to maintain operation of the fuel cell under the desired condition while the main pump (Pump A), main flow meter (FM_{1A}), and dilution gas stream (FM_2) are used to generate the self-reference data.

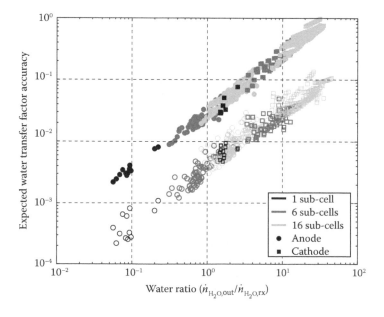

FIGURE 8.11 Expected water transfer accuracy from Monte Carlo simulations. Solid symbols represent the classic approach based on Equation 8.9; hollow symbols are for the self-reference approach based on Equation 8.21.

8.3.5 Further Design Considerations for Dynamic Analysis

Despite being capable of fast responses to sudden changes in flow rates or stream compositions, the dynamic analysis of the water transfer factor is often limited by the time required for the transferred water to make its way from the membrane interface to the measurement device. The following will deal with the configuration of the tubing/fittings from the outlet of the PEMFC to the measurement device.

In general, PEMFCs run at conditions where a significant amount of water is found in the outlet streams and visual observations of running PEMFCs at the outlet often reveal the presence of two-phase flow in the form of slugs of water which appear periodically. If the instantaneous water transfer measurements were achieved under such conditions, the results would have shown a succession of high and low values that would not reflect the true state of the PEMFC at the time of the measurement. To avoid these occurrences or decrease their impact it is important to understand how water slugs are formed and how they can be prevented.

Gas–liquid two-phase flow can take a number of flow patterns—stratified, wavy, annular, plug, slug, and bubble flows (Mandhane et al., 1974). It is generally recognized that the various flow patterns are a function of the superficial velocity of each phase (Mandhane et al., 1974; Chen et al., 2002). Recent studies have shown that the various flow patterns exhibit a complex behavior. For a given set of conditions (flow rates), more than one flow pattern could be observed over the course of time (e.g., slug/annular) and the size and shape of the cross-section affected the flow pattern transitions (Ekberg et al., 1999). Despite the complexity and the difficulty in predicting and describing the exact flow pattern in gas–liquid flow systems, it can be generalized that as the superficial velocity of both gas and liquid phases is increased, slug flow will be replaced by annular flow. A practical way to increase superficial velocities without modifying the flow rates is to reduce the cross-section for given flow rates. As a general rule of thumb, the Weber number (Equation 8.22) can be used to estimate the minimum stable liquid droplet size that occurs at Weber numbers between 10 and 20 (Eddingfield and Evers, 1981). Assuming that water slugs remain nearly static while forming ($U_g \gg U_l$), a gas (nitrogen) velocity between 50 m/s (1 atm and We = 10) and 100 m/s (2 atm and We = 20) would be sufficient to break up the water slugs into droplets of less than 1 mm in diameter.

$$\mathrm{We} = \frac{\rho_g \left(\left| U_g - U_l \right| \right)^2 D_{\mathrm{droplet}}}{\sigma} \tag{8.22}$$

Proper sizing is not limited to the tubing selection between the PEMFC and the measurement apparatus. Special attention should also be given to the transition zones where fitting assemblies can result in very large cross-sectional (Figure 8.12) zones with horizontal and upward flows, which are more prone to water accumulation and flooding, leading to slug formation.

Once water slugs are eliminated or broken into small water droplets, it is important to quickly evaporate the small droplets in order to have single-phase flow entering the measurement apparatus. Although providing energy to the water is necessary for its evaporation, the authors have found from several design iterations that ensuring that the gas phase had a high-enough temperature was even more important to avoid recondensation of water onto colder surfaces in the sampling lines.

8.4 Experimental Results and Uses

In this section, the experimental results obtained by various researchers are briefly reviewed to illustrate the information that could be obtained from the techniques discussed in Section 8.2. Some groups used water molar fraction in their analysis, while some used water transfer factor α. It should be noted that

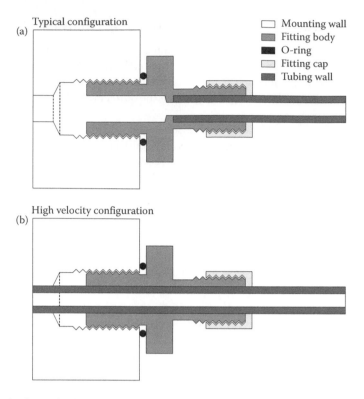

FIGURE 8.12 Tube/fitting body/mounting wall a) Typical configuration (low velocity) b) High velocity configuration.

in some articles α is referred to as the net electroosmotic drag coefficient, which is the water transfer rate divided by the proton generation rate. Shall we denote this quantity by β, we have

$$\beta = \frac{\alpha}{2} \tag{8.23}$$

where α is what was defined in Section 8.1.

8.4.1 Effects of Operating Conditions

There is an abundance of results in the literature on the effect of operating conditions on water transfer. An early work by Choi et al. (2000) reported that when the current density was increased, β decreased. For humidified cathode this trend ceased at current densities greater than 0.2 A cm^{-2}. Also by a mass balance on the nonhumidified cathode side they observed a decrease in water flow into the membrane as the current density increased. Janssen and Overvelde (2001) used a cold trap for water collection and showed that using dry anodes, β could be negative for counter-current gas flows. This meant that the diffusion of water from cathode to anode dominated over the electroosmotic drag that moved water from anode to cathode by means of proton conduction.

Various researchers studied the effect of inlet relative humidity on water transfer. In general, higher inlet humidity at the cathode side decreased β and also the anode-to-cathode water transfer factor, while higher anode inlet humidity would do the reverse (Colinart et al., 2009). This is a direct consequence of the water concentration gradient between the two sides of the membrane. Net electroosmotic drag coefficient β also decreased with increasing cathode pressure (Yan et al., 2006) or decreasing cathode temperature (Murahashi et al., 2006). This could be explained by the increased downstream cathode humidity due to the produced water. A higher cathode pressure increased the humidity for the same water vapor mole fraction (Yan et al., 2006), while a lower cell temperature would do the same (Colinart et al., 2009).

The overall water transfer is not the only property that can be studied using the techniques described in Section 8.2. The different components of water transfer, namely electroosmotic drag, diffusion, and hydraulic permeation, can be studied *in situ* using the same setup under special experimental conditions. For example, Husar et al. (2008) isolated the effect of hydraulic pressure by feeding liquid water with different pressure through the two channels. Diffusion is isolated by setting the same pressure for anode and cathode, while electroosmotic drag is eliminated by simply applying zero current density. An oversaturated wet side and unhumidified dry side were used to maximize diffusion gradients to determine diffusion coefficient. The electroosmotic drag is isolated in a running fuel cell by feeding fully humidified gas streams at the same pressure, therefore eliminating diffusion and hydraulic permeation effects.

An alternative method to measure the electroosmotic drag coefficient is to feed hydrogen to both anode and cathode channels, and a current is withdrawn by applying a potential difference across the electrodes (Ye and Wang, 2007). This "hydrogen pump" technique oxidizes hydrogen to produce protons at the anode, while protons are reduced at the cathode side. Thick membranes (200 μm) were used and the same inlet relative humidities at both sides of the membrane were set to minimize the rate of diffusion due to humidity gradients. Braff and Mittelsteadt (2008) used dead-ended anodes to minimize the humidity gradients along the channel. They searched for the feed ratio between water and hydrogen required for a metastable operation between dryout and flooding, with that ratio corresponding to the electroosmotic drag coefficient. Motupally et al. (2000) used water transfer results to demonstrate the relationship between literature-reported self-diffusion coefficients and Fickian diffusion in Nafion membranes. Kim and Mench (2009) used anode and cathode flow field plates with different coolant temperatures to apply a temperature gradient across the membrane, and found that water is transported from the cold side to the hot side of the membrane. This transfer rate increased with temperature gradient as well as the average temperature of the two sides of the membrane.

Adachi et al. (2009) used different combinations of liquid and vapor feeds at the anode and cathode sides and measured the permeation of water. They explained the low rate for liquid–liquid permeation by the low chemical potential gradient involved, while the much higher liquid–vapor permeation

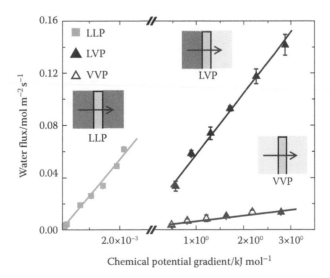

FIGURE 8.13 Water transfer flux measured using different feeds to the channels: LLP (liquid water on both sides), LVP (liquid water on one side and humidified gas on the other), VVP (humidified gas on both sides). Only LVP gave the same order of magnitude of flux compared to running fuel cell results. (Reprinted from Adachi, M. 2009. *J. Electrochem.* Soc. 156: B782–B790. With permission from the Electrochemical Society.)

compared to vapor–vapor permeation is explained by interfacial effects. They also compared these rates to the water transfer rates determined from running fuel cells, and found that only liquid–vapor permeation could match the water transfer rates. As a result, they suggested that liquid water is present on the cathode side of a running fuel cell (see Figure 8.13).

8.4.2 Effects of MEA Construction

Atiyeh et al. (2007) experimentally measured the water transfer factor from MEAs with and without microporous layers. They found that the addition of a microporous layer at the cathode did not change the overall water transfer significantly. The same results were obtained by Janssen and Overvelde (2001) at a current density of 0.4 A cm^{-2}. At 0.6 A cm^{-2}, the difference was more pronounced, regardless of humidification conditions. The water transfer was found to be less if microporous layers were applied.

Dai et al. (2009) used a water-filled anode and a dry-air cathode channel to measure water transfer to the cathode channel through the GDL. Contrarily, they found that the addition of a microporous layer could reduce water transfer to the cathode channel at dry air flow rates greater than 7.44 × 10^{-4} mol s^{-1}. The authors claimed that at higher air flow rates more water was driven out of the membrane which caused a higher capillary pressure in micropores and hence decreased the transport efficiency through the GDL. Murahashi et al. (2006) used hydrophobic microporous layers both in the anode and the cathode. Although they did not study directly the effects of the layer on water transfer, they found that when capillary pressure was included in their calculations, a lower overestimate in β was observed (see Figure 8.14). They claimed that capillary action increased the activity of water vapor at the membrane interface which in turn increased electroosmotic drag and back-diffusion coefficients and decreased water transfer from anode to cathode.

8.4.3 Trends from the Polarization-Type Tests with the Fitting of the Model Parameters

Apart from observing the trends in water transfer with current density, more information could be extracted from tests which measured water transfer along with polarization data.

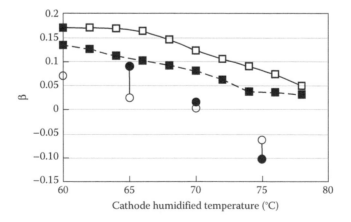

FIGURE 8.14 Model predictions of net electroosmotic drag coefficient β compared to experimental results. Hollow squares: predictions without capillary action at the microporous layer; solid squares: predictions with capillary action at the microporous layer; solid circles: experimental results obtained from the cathode; hollow circles: experimental results obtained from the anode. Note the disagreement between the circles. (Reprinted from *J. Power Sources*, 162, Murahashi, T., Naiki, M., and Nishiyama, E., Water transport in the proton exchange-membrane fuel cell: Comparison of model computation and measurements of effective drag, 1130–1136, Copyright (2006), with permission from Elsevier.)

Using an infrared sensing system with reference measurements, Yau et al. (2010) showed that the intercept of the relationship between water transfer flux and current density could be used to obtain the interfacial transport coefficient of water in the membrane; while the slope of the graph indicated the distribution of generated water between the anode and cathode channels. The intercept was not affected by cathode pressure but by temperature and the inlet humidity difference. The slope is not affected significantly by changes in inlet humidification, but temperature and cathode pressure did greatly change the slope with the same trends as described in Section 8.4.1.

Ge et al. (2005, 2006) used mass transfer models and curve fitting to determine diffusion and sorption coefficients and electroosmotic drag of Nafion membranes. Furthermore, Yau et al. (2010) showed that complementing polarization with water transfer data would yield a better set of fitted model parameters.

8.4.4 Local Alpha Profiles

Mench et al. (2003) measured water mole fraction profiles along the channel direction using a GC-TCD. The water mole fraction acquired increased along the channel at both anode and cathode, and is greatly dependent on inlet humidity. The current density did not significantly affect water mole fraction at the anode outlet but at the cathode outlet with dry feed increasing current densities yielded higher water mole fraction. They determined that the increase is due to generated water but not electroosmotic drag. Dong et al. (2005) obtained similar results by using a mass spectrometer.

Nishikawa et al. (2006) put RH sensors inside various points in the flow field on the cathode side. They also obtained an increasing trend along the channel, but the distribution became more and more even when cathode humidification increased. They also calculated β at the outlet, which became smaller and smaller in magnitude with increasing current densities.

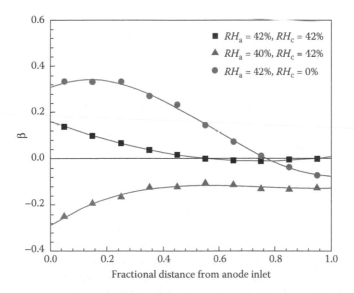

FIGURE 8.15 Net electroosmotic drag coefficient β decreased all the way from the inlet to the outlet using dry cathode, changing from positive (anode to cathode) at the inlet to negative (cathode to anode) close to the outlet because of the large driving force for back diffusion caused by water generation and humidity accumulation at upstream. (Reprinted from *J. Membr. Sci*, 287, Liu, F., Lu, G., and Wang, C. Water transport coefficient distribution through the membrane in a polymer electrolyte fuel cell, 126–131, Copyright (2007), with permission from Elsevier.)

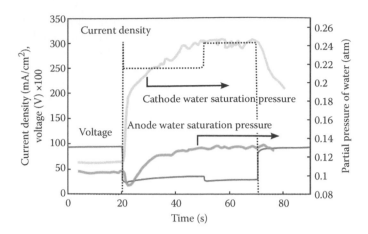

FIGURE 8.16 An increase in current density led to a sudden drop in water vapor pressure at the anode that became steady after 20 s. The authors explained the trends using the interaction among electroosmotic drag, advection of gases before reaching the sensor and back diffusion because of water production. (Adapted from Sur, R. et al. 2010. *J. Electrochem. Soc.* 157: B45–B53. Reproduced by permission of the Electrochemical Society.)

Lu et al. (2007) used an approach similar to that of Mench et al. (2003). At a voltage of 0.5 V for a fully humidified anode and partially humidified cathode, β decreased all the way from the inlet to the outlet. The current density decreased from the inlet and a small peak was observed near the outlet. The decrease in current density was due to the decrease in local stoichiometry, while near the outlet an increase in anode water mole fraction retained water in the membrane to give a lower resistance. When the anode is unsaturated, the decreasing trend of β along the channel did not change, despite the fact that water moved from the cathode side to anode side at the latter half of the channel. The trend in current density distribution did not change significantly when changing from a fully humidified to a partially humidified anode. Liu et al. (2007) expanded this work and found that even when dry cathode feed was used water still moved from the cathode to the anode near the outlet as a result of back diffusion of product water as shown in Figure 8.15.

8.4.5 Experimental Results from Dynamic Measurements

Dong et al. (2005) used a mass spectrometer with a time resolution of 1 s to study a fully humidified anode and a cathode with 50% RH. At the cathode inlet, the decrease in voltage did not change the water mole fraction significantly though the current density changed from 0.9 to 0.2 A cm^{-2}. At the midpoint and close to the outlet of the cathode channel, decreasing the voltage yielded more frequent formation of water droplets.

Sur et al. (2010) used a tunable diode laser system with a time resolution of 0.2 s. They showed that their fuel cell required approximately 20 s to reach steady state under a sudden change in current. At the onset of current increase, there is a dip in water partial pressure at the anode outlet, which is explained by the interaction among electroosmotic drag, advection of gases before reaching the sensor and back diffusion because of water production (see Figure 8.16). When the inlets were drier, more time would be required to humidify the membrane, leading to a drop in transient voltage.

8.5 Outlook

In this section, other possible uses of the water transfer factor measurement tool is presented using the one presented by Sauriol et al. (2009) as a model setup.

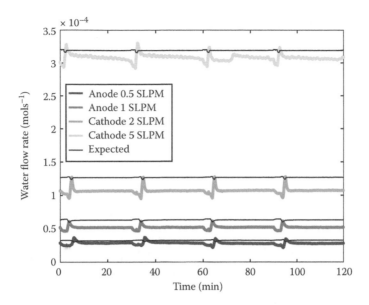

FIGURE 8.17 Contact humidifier performance under typical fuel cell operating conditions. Anode is run on dry hydrogen and cathode on dry air. Anode and cathode pressure 200 kPa, contact humidifier setpoint 55°C (dew point). The thick lines are the experimentally determined water flow rates, and the thin lines are the expected water flow rate based on the humidifier dew point and the measured flow rates and pressure.

8.5.1 Water Balance and Humidification Systems

Collection-based approaches have been widely used for the calibration of less accurate humidification techniques such as bubbler and contact humidifiers that rely on the water saturation of the gas phase under controlled conditions (temperature, pressure, flow rate) (Hyun and Kim, 2004). However, methods with shorter sampling times are highly recommended because substantial time required by collection methods does not permit transient changes that may occur during the humidification operation.

To configure a humidifier calibration setup, the PEMFC in Figure 8.10 was simply replaced by the contact humidifier under test. The injection pump was only used to provide the reference point. Equation 8.21 was updated to the more practical Equation 8.24 that yielded the effective amount of water fed by the contact humidifier.

$$\dot{n}_{H_2O,\text{humidifier}} = \dot{n}_{H_2O,\text{ref}} \left(\frac{x_{H_2O,\text{out}}}{x_{H_2O,\text{ref}}} \right) \left(\frac{\dot{n}_{\text{total,out}}}{\dot{n}_{\text{total,ref}}} \right) \tag{8.24}$$

Figure 8.17 depicts a contact humidifier calibration example. This exhibited a notable cyclic pattern. The refilling of the contact humidifier with fresh water and a subsequent heater overshoot every 30 minutes was characterized by a drop followed by a peak in the instantaneous measured water flow rate. Also in this example, it was found that the actual delivered water flow rate could be up to 45% lower than expected.

8.5.2 Water Transfer in Nonreactive Systems

As was the case with the water transfer factor, measurement of the water diffusion/convection on nonreactive systems (e.g., membranes) is made difficult from the low amount of water that transfers from one side to the other. Water permeable membranes can be quantified using the water transfer factor

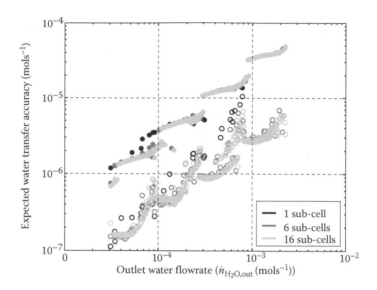

FIGURE 8.18 Expected water transfer accuracy from Monte Carlo simulations. Solid symbols represent the classic approach based on Equation 8.9; hollow symbols are for the self-reference approach based on Equation 8.25.

measurement apparatus, as long as the membranes do not allow significant convection or diffusion of other species. For the model setup, Equation 8.21 can be rewritten as

$$\dot{n}_{H_2O,X} = \dot{n}_{H_2O,in} - \dot{n}_{H_2O,ref} \left(\frac{x_{H_2O,out}}{x_{H_2O,ref}} \right) \left(\frac{\dot{n}_{total,out}}{\dot{n}_{total,ref}} \right) \tag{8.25}$$

Monte Carlo simulations were repeated using the inlet and water transfer fluxes in Section 8.3.2, but this time without load or reaction to assess the accuracy of the measurement for nonreactive systems. The results from this simulation are shown in Figure 8.18. It was found that the water transfer accuracy correlated well with the water flow rate in the outlet stream. As was the case with the water transfer factor accuracy, the classic approach is generally less accurate than the modified approach, which includes the self-reference point. The variability can be observed in Figure 8.18, which is mostly due to the selection of reference points which only reflects the amount of water fed.

8.5.3 Hydrogen Crossover on a Running Fuel Cell

By performing simultaneous water transfer factor measurements on the anode and cathode of a running PEMFC under a wide range of operating conditions, it has been found that there is a systematic difference between the anode- and cathode-side measurements (Yau et al., 2010). The difference always showed higher water transfer factors on the cathode-side measurements and decreased as the load increased. Once converted into a net water transfer flow rate or flux, this difference was nearly constant and ranged between 7×10^{-6}–1.1×10^{-5} mol s^{-1} (Figure 8.19). Several possible causes for this difference were considered (e.g., leaks, scaling, instrument offsets) but they failed to consistently explain the observations. It was speculated that this difference was the result of hydrogen crossover (i.e., hydrogen molecules that crossed the membrane from the anode to the cathode by convection, diffusion, or drag). Hydrogen reacted with oxygen on the cathode catalyst layer to form water, thus increasing the amount of product water found on the cathode side without affecting the load. This led to the strategy that only anode side results could be taken as representing water transfer alone and that the difference between the cathode and anode water transfer flow rates gave the hydrogen transfer flow rates.

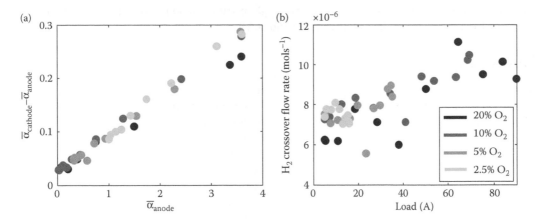

FIGURE 8.19 (a) Difference in measured water transfer factors between cathode and anode without accounting for hydrogen crossover and (b) corresponding hydrogen crossover flow rate. Anode inlet conditions: 4 SLPM (dry hydrogen), 382 µL min⁻¹ water injection, 230 kPa; cathode inlet conditions: 5 SLPM (dry, O2/N2 mixture), 163 µL min⁻¹ water injection, 200 kPa; cell temperature ~70°C (69°C low load up to 74°C high load).

In order to establish the reliability of the determined hydrogen crossover flow rates, Monte Carlo simulations were repeated following the earlier described procedure, only this time accounting for a small hydrogen crossover. Once it was determined that the approaches tested were unbiased, the standard deviation about each simulated condition was used as a measure of the expected accuracy. The results are shown in Figure 8.20. In general, the modified approach (based on Equation 8.21) yields expected hydrogen crossover uncertainties that generally lie between 10^{-6} and 3×10^{-6} mol s⁻¹ while the classic approach (based on Equation 8.9) usually yields uncertainties one order of magnitude greater. Considering the range of observed hydrogen crossover flow rates, the observations are likely to be real

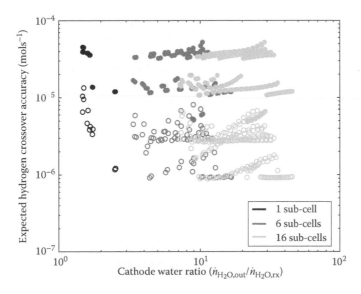

FIGURE 8.20 Expected hydrogen crossover flow rate accuracy from Monte Carlo simulations. Solid symbols represent the classic approach based on Equation 8.9; hollow symbols are for the self-reference approach based on Equation 8.21. (Reprinted from Sur, R. et al. 2010. *J. Electrochem. Soc.* 157: B45–B53. With permission from the Electrochemical Society.)

and not just related to instrumentation error, although instrumentation error is likely to be important on the hydrogen crossover values (between 10% and 50%).

Nomenclature

A	m^2	Area
C	mol m^{-3}	Molar concentration
C'		Neutron count
D	m	Diameter
F	C	Faraday constant
f	Hz	Resonant frequency
I	A	Current load
j	A m^{-2}	Current density
MW	g mol^{-1}	Molar mass
m	kg	Mass
$\dot{m}$	kg s^{-1}	Mass flow rate
$N\sigma$	cm^{-1}	Attenuation length
$\dot{n}$	mol s^{-1}	Molar flow rate
P	kPa	Total pressure
p	kPa	Partial pressure
RH		Relative humidity
s	m^2	Surface area
S		Surface
T	°C	Temperature
t	cm	Water film thickness
U	m s^{-1}	Superficial velocity
$\dot{v}$	m^3 s^{-1}	Volumetric flow rate
We		Webber number
X		Mole fraction (wet basis)
Y		Measurand
A		Water transfer factor
Γ		Absorbance
N		Stoichiometric coefficient
P	kg m^{-3}	Density
Σ	N m^{-1}	Surface tension
Φ	W	Optical power

Subscripts

A	Anode
Atm	Atmospheric
C	Cathode
Cal	Calibration
Collected	Collected
cond	Condensed
Droplet	Droplet of water
Drygas	Dry gas
e$^-$	Electron
Evap	Evaporation
Experimental	Experimental

final	Final
g	Gas
gen	Generation
H_2	Hydrogen
H_2O	Water
humidifer	Humidifer
i	Running index
in	Inlet
initial	Initial
l	Liquid
out	Outlet
ref	Reference mode
rx	Reaction
sat	Saturation
sensor	Sensor
sorb	Sorbed
total	All gases
trap	Cold trap
vent	Vented
X	Water Transfer (anode to cathode)
0	Reference value

References

Adachi, M., Navessin, T., Xie, Z., Frisken, B., and Holdcroft, S. 2009. Correlation of *in situ* and *ex situ* measurements of water permeation through Nafion NRE211 proton exchange membranes. *J. Electrochem. Soc.* 156: B782–B790.

Atiyeh, H., Karan, K., Peppley, B., Phoenix, A., Halliop, E., and Pharoah, J. 2007. Experimental investigation of the role of a microporous layer on the water transport and performance of a PEM fuel cell. *J. Power Sources* 170: 111–121.

Auble, D. L. and Meyers, T. P. 1992. An open path, fast response infrared-absorption gas analyzer for H_2O and CO_2. *Bound.-Lay. Meteorol.* 59: 243–256.

Basu, S., Renfro, M.W., and Cetegen, B. M. 2006b. Spatially resolved optical measurements of water partial pressure and temperature in a PEM fuel cell under dynamic operating conditions. *J. Power Sources* 162: 286–293.

Basu, S., Xu, H., Renfro, M.W., and Cetegen, B. M. 2006a. *In situ* optical diagnostics for measurements of water vapor partial pressure in a PEM fuel cell. *J. Fuel Cell Sci. Technol.* 3: 1–7.

Bentley, R. 1998. *Temperature and Humidity Measurement*. Singapore: Springer.

Berg, P., Promislow, K., St Pierre, J., Stumper, J., and Wetton, B. 2004. Water management in PEM fuel cells. *J. Electrochem. Soc.* 151: A341–A353.

Braff, W. and Mittelsteadt, C. 2008. Electroosmotic drag coefficient of proton exchange membranes as a function of relative humidity. In *ECS Transactions*. Honolulu, Hawaii 2008: 309–316.

Buck, A. 1981. New equations for computing vapor pressure and enhancement factor. *J. Appl. Meteor.* 20: 1527–1532.

Chen, W. L., Twu, M. C., and Pan, C. 2002. Gas–liquid two-phase flow in micro-channels. *Int. J. Multiphase Flow* 28: 1235–1247.

Choi, K., Peck, D., Kim, C., Shin, D., and Lee, T. 2000. Water transport in polymer membranes for PEMFC. *J. Power Sources* 86: 197–201.

Colinart, T., Chenu, A., Didierjean, S., Lottin, O., and Besse, S. 2009. Experimental study on water transport coefficient in proton exchange membrane fuel cell. *J. Power Sources* 190: 230–240.

Chemical Rubber Company 2008. *CRC Handbook of Chemistry and Physics*. Cleveland, OH: CRC Press.

Dai, W., Wang, H., Yuan, X., Martin, J., Luo, Z., and Pan, M. 2008. Measurement of the water transport rate in a proton exchange membrane fuel cell and the influence of the gas diffusion layer. *J. Power Sources* 185: 1267–1271.

Dai, W., Wang, H., Yuan, X., Martin, J., Shen, J., Luo, Z. et al. 2009. Measurement of water transport rates across the gas diffusion layer in a proton exchange membrane fuel cell, and the influence of polytetrafluoroethylene content and micro-porous layer. *J. Power Sources* 188: 122–126.

Dean, J. 1995. *Analytical Chemistry Handbook*. New York, NY: McGraw-Hill.

Dimitrova, P., Friedrich, K. A., Vogt, B., and Stimming, U. 2002. Transport properties of ionomer composite membranes for direct methanol fuel cells. *J. Electrochem. Soc.* 532: 75–83.

Dong, Q., Kull, J., and Mench, M. 2005. Real-time water distribution in a polymer electrolyte fuel cell. *J. Power Sources* 139: 106–114.

Eddingfield, D. and Evers, J. 1981. Techniques for the measurement of the air–water distribution in the flowfield of a high velocity water jet. In *First US Water Jet Conference*, Golden, CO, pp. 57–65.

Ekberg, N. P., Ghiaasiaan, S. M., Abdel-Khalik, S. I., Yoda, M., and Jeter, S. M. 1999. Gas–liquid two-phase flow in narrow horizontal annuli. *Nucl. Eng. Des.* 192: 59–80.

Endoh, E., Terazono, S., Widjaja, H., and Takimoto, Y. 2004. Degradation study of MEA for PEMFCs under low humidity conditions. *Electrochem. Solid-State Lett.* 7: A209–A211.

Fraden, J. 2004. *Handbook of Modern Sensors—Physics, Designs and Applications*. New York, NY: Springer.

Ge, S., Li, X., Yi, B., and Hsing, I. 2005. Absorption, desorption, and transport of water in polymer electrolyte membranes for fuel cells. *J. Electrochem. Soc.* 152: A1149–A1157.

Ge, S., Yi, B., and Ming, P. 2006. Experimental determination of electro-osmotic drag coefficient in Nafion membrane for fuel cells. *J. Electrochem. Soc.* 153: A1443–A1450.

Hinds, G., Stevens, M., Wilkinson, J., de Podesta, M., and Bell, S. 2009. Novel *in situ* measurements of relative humidity in a polymer electrolyte membrane fuel cell. *J. Power Sources* 186: 52–57.

Husar, A., Higier, A., and Liu, H. 2008. *In situ* measurements of water transfer due to different mechanisms in a proton exchange membrane fuel cell. *J. Power Sources* 183: 240–246.

Hyun, D. and Kim, J. 2004. Study of external humidification method in proton exchange membrane fuel cell. *J. Power Sources* 126: 98–103.

Ito, H. 1987. Balanced adsorption quartz hygrometer. *IEEE Trans. Ultrason. Ferroelectr. Freq. Control* UFFC-34: 136–141.

Janssen, G. and Overvelde, M. 2001. Water transport in the proton-exchange-membrane fuel cell: Measurements of the effective drag coefficient. *J. Power Sources* 101: 117–125.

Jung, S. H., Kim, S. L., Kim, M. S., Park, Y., and Lim, T.W. 2007. Experimental study of gas humidification with injectors for automotive PEM fuel cell systems. *J. Power Sources* 170: 324–333.

Jury, S., Pollock, M., and Mattern, J. 1972. The activated calcium sulfate–water vapor sorption therm. *AIChE J.* 18: 48–51.

Kim, S. and Mench, M. M. 2009. Investigation of temperature-driven water transport in polymer electrolyte fuel cell: Thermo-osmosis in membranes. *J. Membr. Sci.* 328: 113–120.

Knights, S. D., Colbow, K. M. St-Pierre, J., and Wilkinson, D.P. 2004. Aging mechanisms and lifetime of PEFC and DMFC. *J. Power Sources* 127: 127–134.

Liu, F., Lu, G., and Wang, C. 2007. Water transport coefficient distribution through the membrane in a polymer electrolyte fuel cell. *J. Membr. Sci.* 287: 126–131.

Lu, G., Liu, F., and Wang, C. 2007. An approach to measuring spatially resolved water crossover coefficient in a polymer electrolyte fuel cell. *J. Power Sources* 164: 134–140.

Majsztrik, P., Bocarsly, A., and Benziger, J. 2008. Water permeation through Nafion membranes: The role of water activity. *J. Phys. Chem. B* 112: 16280–16289.

Mallinckrodt Baker Inc. 2000, Dessicant Selection Guide, http://www.mallbaker.com/techlib/documents/americas/3045.html Online June 23, 2010.

Mandhane, J. Gregory, G., and Aziz, K. 1974. A flow pattern map for gas-liquid flow in horizontal pipes. *Int. J. Multiphase Flow* 1: 537–553.

Mench, M., Dong, Q., and Wang, C. 2003. *In situ* water distribution measurements in a polymer electrolyte fuel cell. *J. Power Sources* 124: 90–98.

Motupally, S., Becker, A., and Weidner, J. 2000. Diffusion of water in Nafion 115 membranes. *J. Electrochem. Soc.* 147: 3171–3177.

Murahashi, T., Naiki, M., and Nishiyama, E. 2006. Water transport in the proton exchange-membrane fuel cell: Comparison of model computation and measurements of effective drag. *J. Power Sources* 162: 1130–1136.

Nishikawa, H., Kurihara, R., Sukemori, S., Sugawara, T., Kobayasi, H., Abe, S. et al. 2006. Measurements of humidity and current distribution in a PEFC. *J. Power Sources* 155: 213–218.

Perry, R. and Green, D. 2008. *Perry's Chemical Engineers' Handbook*. New York, NY: McGraw-Hill.

Poling, B., Prausnitz, J., and O'Connor, J. 2007. *The Properties of Gases and Liquids*. New York, NY: McGraw-Hill.

Pradelli, A. 1988. Differential mass flowmeter. US Patent #4781068.

Rajalakshmi, N., Jayanth, T. T., Thangamuthu, R., Sasikumar, G., Sridhar, P., and Dhathathreyan, K. S. 2004. Water transport characteristics of polymer electrolyte membrane fuel cell. *Int. J. Hydrogen Energy* 29: 1009–1014.

Sauriol, P., Bi, X., Stumper, J., Nobes, D., and Kiel, D. 2005. A water transfer factor measurement apparatus for polymer electrolyte membrane fuel cells. In *Fuel Cell and Hydrogen Technologies: Proceedings of the First International Symposium on Fuel Cell and Hydrogen Technologies*, Calgary, AB, pp. 35–50.

Sauriol, P., Nobes, D., Bi, X., Stumper, J., Jones, D., and Kiel, D. 2009. Design and validation of a water transfer factor measurement apparatus for proton exchange membrane fuel cells. *J. Fuel Cell Sci. Technol.* 6: 041014.1–041014.13.

Sparks, D., Smith, R., Massoud-Ansari, S., and Najafi, N. 2004. Coriolis mass flow, density and temperature sensing with a single vacuum sealed MEMS chip. In *Solid-State Sensor, Actuator and Microsystems Workshop*. Hilton Head Island, South Carolina .

Springer, T., Zawodzinski, T., and Gottesfeld, S. 1991. Polymer electrolyte fuel-cell model. *J. Electrochem. Soc.* 138: 2334–2342.

Sur, R., Boucher, T. J., Renfro, M. W., and Cetegen, B. M. 2010. *In situ* measurements of water vapor partial pressure and temperature dynamics in a PEM fuel cell. *J. Electrochem. Soc.* 157: B45–B53.

Wilson, J. 2005. *Sensor Technology Handbook*. Boston: Elsevier.

Wu, J. F., Yuan, X. Z., Martin, J. J., Wang, H. J., Yang, D. J., Qiao, J. L. et al. 2010. Proton exchange membrane fuel cell degradation under close to open-circuit conditions Part I: *In situ* diagnosis. *J. Power Sources* 195: 1171–1176.

Wu, J. F., Yuan, X. Z., Wang, H. J., Blanco, M., Martin, J. J., and Zhang, J. J. 2008. Diagnostic tools in PEM fuel cell research: Part II–Physical/chemical methods. *Int. J. Hydrogen Energy* 33: 1747–1757.

Yan, Q., Toghiani, H., and Wu, J. 2006. Investigation of water transport through membrane in a PEM fuel cell by water balance experiments. *J. Power Sources* 158: 316–325.

Yang, X. G., Burke, N., Wang, C. Y., Tajiri, K., and Shinohara, K. 2005. Simultaneous measurements of species and current distributions in a PEFC under low-humidity operation. *J. Electrochem. Soc.* 152: A759–A766.

Yau, T., Sauriol, P., Bi, X., and Stumper, J. 2010. Experimental determination of water transport in polymer electrolyte membrane fuel cells. *J. Electrochem. Soc.* 157: B1310–B1320.

Yaws, C. 1999. *Chemical Properties Handbook*. New York, NY: McGraw-Hill.

Ye, X. and Wang, C. 2007. Measurement of water transport properties through membrane-electrode assemblies. *J. Electrochem. Soc.* 154: B676–B682.

Zawodzinski, T., Derouin, C., Radzinski, S., Sherman, R., Smith, V., Springer, T. et al. 1993. Water uptake by and transport through Nafion(R) 117 membranes. *J. Electrochem. Soc.* 140: 1041–1047.

9

Current Mapping

Dilip Natarajan
Experimental Station,
Du Pont de Nemours

Trung Van Nguyen
The University of Kansas

9.1 Introduction

The various transport processes that occur in a PEM fuel cell using a conventional gas distributor are visualized in Figure 9.1. The fuel and oxidant supplied at the inlet are consumed by the electrochemical reaction along the channel length (z-direction). Hence, a variation in the reactant and product concentrations is expected along the channel length. Such concentration variations along the length can be significant when air is supplied to the cathode as the oxygen source (use of reformate hydrogen can also lead to concentration variations along the channel at the anode).

Moreover, the fuel and oxidant need to diffuse through the GDL to reach the catalyst layer (CL). Within the three-dimensional CL, the reactive species need to dissolve in the ionomer phase and diffuse over a short distance (compared to transport distances within the GDL) to reach the electrochemically active sites. The cross-sectional area available for reactant transport and the diffusion distance within the diffusion layer and the CL are strongly influenced by the presence of liquid water (as product in the cathode or due to humidification at the anode) and the morphology of the porous components. Hence, it is intuitively evident that there exists a variation in the reactant and product distribution along the thickness of the electrode (y-direction) that affects the reaction rate within the CL. Also, different transport distances are encountered within the electrode along the x-direction due to the presence of ribs or shoulders (for current collection). Hence, variations in the reactant and product concentrations are also expected in the x-direction.

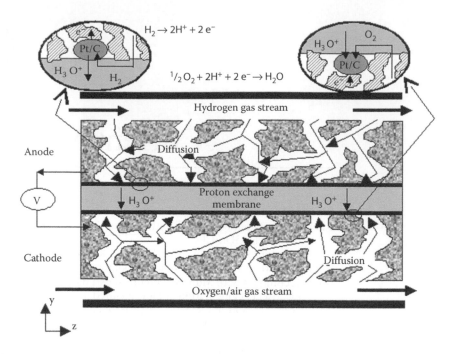

FIGURE 9.1 Visualized transport processes in a PEM fuel cell.

Thus, a nonuniform distribution of reactants and product is expected across the entire three-dimensional CL volume. Hence, the local reaction rate or current density at any point in space within this CL volume that is dependent on the local available reactant and product concentrations is expected to be nonuniform. Also, the local current density depends on the local solid and ionic phase potentials that are, in turn, dependent on the local electronic and ionic conductivities. In the case of electron transport, anisotropic electrical properties of the various layers along with differences in path lengths (due to alternating channels/shoulders) result in a spatial variation of current flux components. The proton conducting membrane needs to be properly hydrated with water to maintain sufficient ionic conductivity. The ionic conductivity of the proton exchange membrane fuel cell can vary over orders of magnitude depending on the hydration state. Usually PEM fuel cell operation temperatures are limited to values below 100°C to sustain proper membrane humidification. During operation, water is transported across the membrane from the anode to the cathode while it is also generated by the electrochemical reaction at the cathode. The transport of water across the membrane during operation tends to dry the anode CL and anode side of the membrane while in combination with the generated liquid water floods the cathode catalyst and diffusion layers. This results in higher ohmic resistance across the membrane and severe mass transport restrictions at the cathode. Thus, water activity also manifests through membrane conductivity on the spatial variation in the electrochemical reaction rates.

The voltage available from a single PEM fuel cell is usually too low for most practical applications. To obtain greater voltages, single cells are stacked in series through bipolar plates to form multicell fuel cell stacks. The bipolar plates are again made mostly of graphite with flow fields machined on both the major planar surfaces. Hydrogen is supplied to one side making it the anode of a given single cell in a stack while oxygen or air is supplied to the other side that acts as the cathode for the next cell. Other than the issues considered at the single-cell level, a major factor when it comes to stacks is thermal management. Fuel cell stacks can generate significant amounts of heat that demand special heat removal mechanisms employing ambient air or liquid coolants. These thermal gradients, design features, and variation in physical properties of the various components will render the electrochemical reaction rate map over the x–z plane in a given cell, a function of the position of the cell in the stack.

Traditional fuel cell experimentation does not entail the measurement of this nonuniform local reaction rate or current density distribution. For example, under potentiostatic operation, the current density measured at the current collector leads is only the average of the local reaction rates in the CL reaction volume. In other words, only the total current is measured and the density is calculated dividing by the entire active area. The potential difference measured is at the buss plates and not the solid phase potential at the CLs.

As evidenced from the history of fuel cell research, the overall average current and/or voltage response of a fuel cell can indeed be used as the primary criteria for evaluating performance and attempts to improve it. However, the local current density distribution that is a direct measure of reaction rates along the length (along the flow channel), width (across channels and shoulders), and thickness of the CL is of paramount interest. The ability to predict and/or measure the local current density distribution that ultimately decides the overall performance is essential to optimize fuel utilization that dictates the final efficiency. For a prescribed performance requirement, optimal utilization of the geometric reactive area (or the reactive volume) that determines fuel cell size and hence the power density is essential for cost optimization.

This need for the ability to measure and predict local current density distribution rates has prompted significant research over the past two decades. Traditional experimental research on PEM fuel cells were limited to average cell current density measurements due to lack of sufficient technological sophistication to measure local current density distribution. Also, experimental measurement of local reactant and product concentrations was not viable owing to the small thickness of the components involved in a PEM fuel cell (GDL and CL) and their extreme aspect ratios. Hence, PEM fuel cells researchers in the past resorted to first-principles-based mathematical models to determine the species distribution within the fuel cell (Bernardi and Verbrugge, 1992; Fuller and Newman, 1993; Nguyen and White, 1993; Springer et al., 1993; Gurau et al., 1998; He et al., 2000; Natarajan and Nguyen, 2001; Natarajan and Nguyen, 2003). However, these models still need reliable experimental techniques that provide accurate local electrochemical reaction rates distribution to validate them and transform them into design tools. This vital need has prompted significant research into experimental techniques that yield local current density measurements. While significant strides have been made in resolving the current map along the length and width of the CL, the variation across the CL thickness is beyond reach of current state of the art and presents the final frontier in current mapping. The ideal goal of any contemporary current mapping technique is to accurately measure the current (hence the reaction rate) emanating from a given unit area (decided by the resolution of the method) when a known solid-phase potential is imposed at the CL across that area or vice versa. However to get the complete picture, knowledge of the corresponding ionic potential in that prescribed unit area is needed because the driving force for the reaction is the difference between the solid- and liquid-phase potentials otherwise known as electrode over potential.

In this chapter, the approaches developed by various researchers in the PEM fuel cell community are presented along with their perceived advantages and disadvantages. Brief descriptions of the experimental setup, measuring technique, and in some cases representative results are provided. For further detail, the reader is referred to the original references. This review is categorized into four groups. The reader is cautioned that the sequence of articles in this review does not reflect the chronology of publication.

1. Techniques involving the use of a single potentiostat/load for cell control with/without electrical segmentation of flow field/GDL/CL
2. Techniques involving the use of multiple potentiostat/load for cell control with/without electrical segmentation of flow field/GDL/CL
3. Other indirect and noninvasive techniques
4. Reference electrode techniques to measure electrode over potential

9.2 Single Potentiostat/Load with/without Segmentation of Flow Field/GDL/CL

9.2.1 Pioneering Work by Stumper et al. (1998)

In their pioneering work, Stumper et al. (1998) analyzed three methods for current density distribution mapping namely the partial membrane electrode assembly (MEA) approach, the "subcell" technique, and the current mapping technique. The first approach involved the use of several different MEAs with a catalyzed active area of varying fractions of the total flow field area as seen in Figure 9.2a. In the subcell approach the authors used a number of "subcells" at various locations along the gas flow channel that were electrically insulated from the main active MEA and controlled by a separate load. The subcells

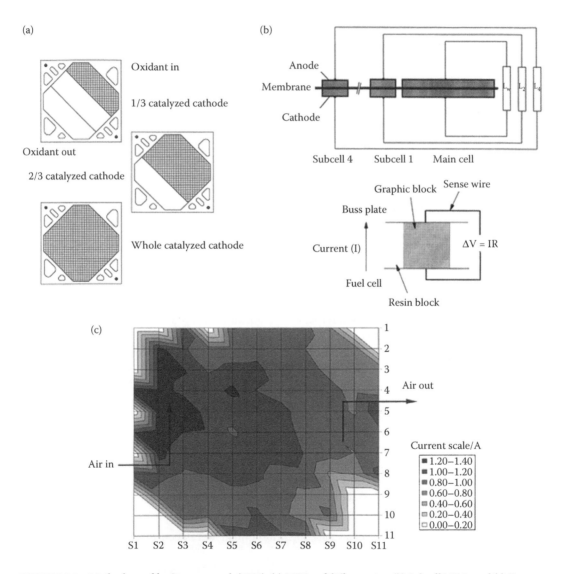

FIGURE 9.2 Methods used by Stumper et al. (1998). (a) MEAs of different size, (b) Subcell MEA, and (c) Contour plot of current density measured using the subcell method. (Reprinted from *Electrochim. Acta* 43(24), Stumper, J. et al. *In situ* methods for the determination of current distribution in PEM fuel cells. 3773–3783, 1998, with permission from Elsevier.)

were created by removing parts of the anode and cathode on the main MEA and replacing them with smaller electrodes, as seen in Figure 9.2b. While this configuration used multiple electrical loads it was included here for the sake of keeping Stumper et al.'s work together. In the third approach, a network of passive graphite resistors were placed between the flow field plate and the current collecting buss plate, while the potential drop across these resistors were monitored to establish the current flowing through them, shown in Figure 9.2c. The partial MEA approach does not provide sufficient spatial resolution and significant errors can arise due to inherent variations in electrical, transport, and kinetic properties between different MEAs. The "subcell" approach is plagued by the difficulty in properly isolating the "subcells" from the main electrode and achieving perfect alignment of the anode and cathode sides. The approach of passive resistor network is the most popular of the three that spurred significant follow-up refinement by other researchers.

9.2.2 Resistor Network Approach

Ghosh et al. (2006) demonstrated a novel approach in which they used a partially segmented expanded graphite to create a passive resistor network such as that of Stumper et al.'s in a single monolithic unit. The corners of each segment were left connected to the adjacent graphite segment as shown in Figure 9.3a to maintain mechanical integrity. The anisotropic nature of expanded graphite in terms of electrical conductivity was exploited to spatially realize the network and achieve a measurable potential drop in the through-plane direction. The potential difference was measured at the center of each resistor segment by thin copper wires. A set of 40 isolated copper wires was connected to both sides of the 20 segments to detect the potential difference as shown in Figure 9.3a. The wire diameter was kept to a minimum and cell compression ensured wire embedding in the soft expanded graphite. This unit was placed between the end plate and the bipolar plate on the cathode side of the fuel cell as shown in Figure 9.3b.

The passive resistor network technique was also demonstrated on a multicell stack by Geske et al. (2010). They developed an integrated sensor unit that can be inserted between cells in a stack. The sensor unit consisted of a shunt matrix with 112 elements and the measured potential drop across the sensors was used to calculate the current flowing through them. Active multilayer (AML®) integration technology was used to integrate the sensor elements inside of a multilayered printed circuit board (PCB). The sensors were low ohmic shunt resistors which were embedded in a solid epoxy resin layer. Buried vias connected the sensors with the segments outside which were in contact with the adjacent bipolar and cooling plates. The sensor unit was placed between the 2nd and 3rd cell in a PEMFC system with five cells.

A major issue with the resistor network approach is the loss of resolution because of in-plane currents in the bipolar plate, GDL and CL. This issue can be further exaggerated by nonuniform contact between the resistor network and the flow-field plate. Another disadvantage is the complexity of the experimental setup arising from the incorporation of a multitude of potential sensing wires (two per resistor). It is also worthwhile to mention here that the potential applied or measured in these techniques were at the buss plates. This necessitates postexperimental theoretical estimates to relate to the solid potential map across the CL that depends on design and operating parameters.

9.2.3 Segmented Flow Field

The issue of current spreading in the flow field plates was addressed by segmenting the graphite flow field plates and electrically isolating them. Wieser et al. (2000) authored the original article addressing this approach. They were also pioneers in incorporating Hall sensors that measure the magnetic field due to a current vector as opposed to passive resistor networks. They explicitly segmented the flow fields to avoid lateral currents as shown in Figure 9.4. From the magnetic field measurements, the current map was calculated. It should be noted that the hall sensors were built into the fuel cell assembly unlike

(a)

(b)

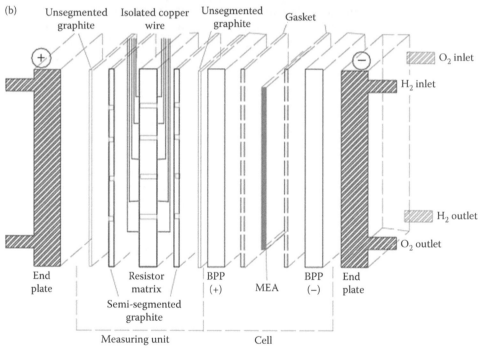

FIGURE 9.3 Ghosh et al.'s (2006) fixture of semisegmented resistor matrix on graphite plate with the isolated copper wire and schematic diagram of measuring unit inside the fuel cell. (Reprinted from *J. Power Sources* 154(1), Ghosh, P. C. et al. *In situ* approach for current distribution measurement in fuel cells. 184–191, 2006, with permission from Elsevier.)

noninvasive techniques like magnetotomography discussed in latter sections. Hwnag et al. (2008) also used Hall sensors with segmented flow fields on the cathode of a PEM fuel cell. Their fuel cell design is illustrated in Figure 9.5a. They used a 16 channel multiplexer to sequentially measure the current in each segment as configured in Figure 9.5b. They studied the effect of four different flow fields at the cathode namely, parallel, serpentine, interdigitated, and biomimic. Figure 9.6 shows an example of the current density distribution profiles obtained by them. Here the effect of two different types of flow field at the

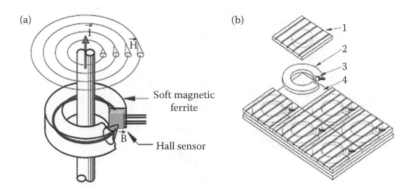

FIGURE 9.4 Schematic drawings of fixtures used in Wieser et al.'s (2000) study. (a) A current sensor with magnetic loop and hall sensor shown with magnetic field generating electric current, (b) current sensor in current collector with flow-field segment: (1) flow field segment, (2) annular ferrite, (3) Hall sensor, and (4) current collector. (With kind permission from Springer Science+Business Media: *J. Appl. Electrochem.*, A new technique for two-dimensional current distribution measurements in electrochemical cells, 30(7), 2000, 803–807, Wieser, Ch., Helmbold, A., and Gulzow, E.)

cathode on the transient development of the current density distribution is demonstrated. The serpentine flow field provides a more uniform current distribution than the parallel design. The reader is referred to the original article for further details on the experimental conditions.

Noponen et al. (2002) applied the segmented flow field in conjunction with a single potentiostat/load for control on free-breathing PEM fuel cells. They applied the segmentation to the cathode side. Segmentation of current collection at the flow field was achieved by embedding conductive gold-coated stainless-steel strips in a plastic PVC block. These strips acted as the shoulders while the gas channels were formed by the gaps between these strips. They incorporated elaborate set-screw assemblies to even the contact of these strips to the GDL. The strips were connected to a resistor in series and the potential drop across the resister provided a current measurement. Figure 9.7a provides a picture of the segmented cell used by Noponen et al. An example of the current density distribution measurements in the free-breathing PEM cells is provided in Figure 9.7b clearly demonstrating the effect of oxygen depletion from the entrance to the exit (arrow points the direction of air flow from bottom to top caused by thermally induced convection and diffusion).

While segmentation of the flow field helps mitigate current spreading in the graphite plates, significant in-plane currents can exist in the GDL and the CL. This phenomenon is further exaggerated by the anisotropic electronic conductivity of the GDL wherein the in-plane conductivity is more than an order of magnitude greater than the through-plane conductivity due to fiber orientation. Also the issue of contact resistance variations still exists at the interface between the segmented flow field and GDL. The effect of these contact variations and their drastic influence is addressed in detail in latter sections. The use of Hall sensors does have some disadvantages. They can significantly complicate the experimental setup making experimentation expensive and tedious. Moreover, interference from neighboring segments is also possible with the use of Hall sensors. The novel approach by Noponen et al. (2002) significantly alters the flow pattern of the electronic current. The through-plane primary current from the CL and GDL is completely redirected into an in-plane current along the shoulder in the flow field. This can artificially induce solid potential variations along the channel length (in the flow field that in turn reflects in the CL) that are not innate to a typical operating fuel cell hence influencing the current distribution.

9.2.4 Summary

A major drawback to all the approaches discussed so far is the use of single potentiostat or load control. When using a single potentiostat, if the voltage sensing and control point is after the resistors, the potential

(a)

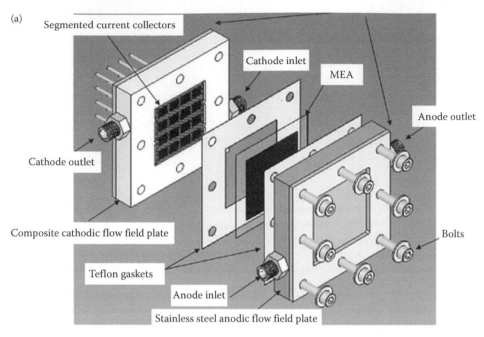

(b)

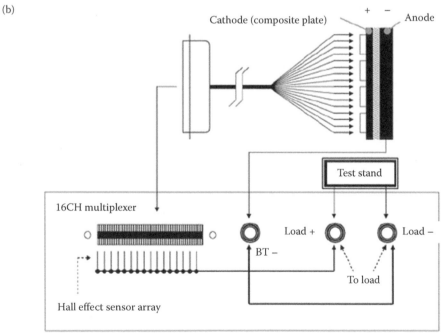

FIGURE 9.5 Test fixture of local current measurement used in Hwnag et al.'s (2008) study. (a) Photo of the test fixture, and (b) assembly diagram of the test fixture. (Reprinted from *J. Hydrogen Energy* 33(20), Hwnag, J. J., et al. Experimental and numerical studies of local current mapping on a PEM fuel cell. 5718–5727, 2008, with permission from Elsevier.)

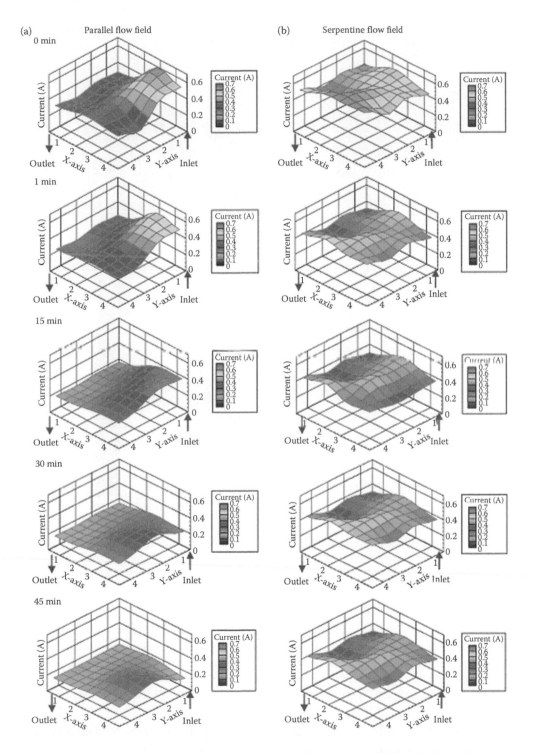

FIGURE 9.6 Transient developments of the local current distribution. (a) Parallel flow field and (b) serpentine flow field, E_{cell}: 0.5 V, T_{cell}: 40°C, anode stoich: 1.5 and cathode stoich: 2.5. (Reprinted from *J. Hydrogen Energy* 33(20), Hwnag, J. J., et al. Experimental and numerical studies of local current mapping on a PEM fuel cell. 5718–5727, 2008, with permission from Elsevier.)

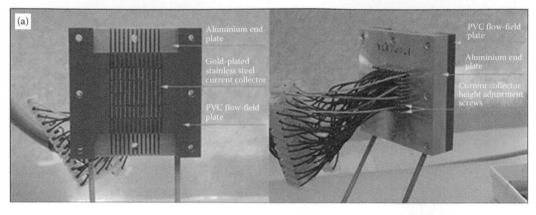

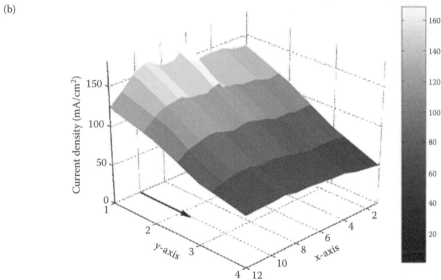

FIGURE 9.7 Segmented free-breathing PEM fuel cell used in Noponen et al.'s (2002) study, (left) the flow-field plate and the current collectors, (right) the end plate and the adjustment screws. (Reprinted from *J. Power Sources* 106(1–2), Noponen, M., et al. Measurement of current distribution in a free-breathing PEMFC. 304–312, 2002, with permission from Elsevier.)

at the electrode will not be uniform because of the differences in the current flowing through the resistors. In the case of segmented flow fields, by using a single control, the desired potential can be applied to a single point or segment while the potential at the other segments are decided by the design features, material properties, and possibly operating conditions. Thus, the actual solid potential at the CL corresponding to the other segments is unknown (can be estimated theoretically postexperimentation).

9.3 Techniques with Multiple Potentiostats/Loads with/without Electrical Segmentation of Flow Field/GDL/CL

9.3.1 Pioneering Work by Cleghorn et al. (1998)

Around the time of Stumper et al.'s work, Cleghorn et al. (1998) published some ground-breaking approaches in current density mapping. They pioneered the use of PCB technology (Geske et al.'s work

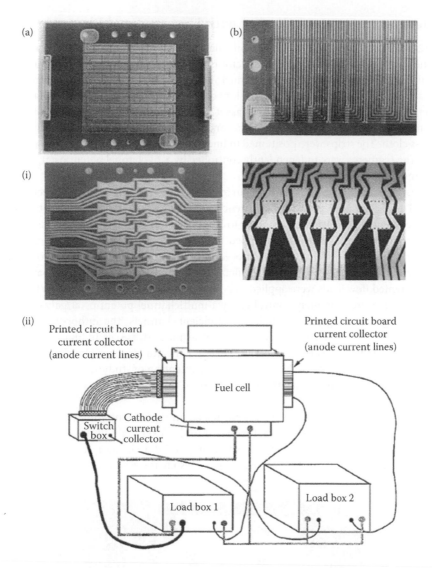

FIGURE 9.8 Photographs of the segmented printed circuit board flow-field/current collector used on the anode side of the cell and diagram of the electronics hardware used in Cleghorn et al.'s (1998) study. (With kind permission from Springer Science+Business Media: *J. Appl. Electrochem.*, A printed circuit board approach to measuring current distribution in a fuel cell. 28(7), 1998, 663–672, Cleghorn, S. J. C. et al.)

was an adaptation of this work) to create segmented flow fields shown in Figure 9.8(i). A PCB with isolated copper segments was fabricated and coated with gold to prevent corrosion. Serpentine flow channels were then machined into the face. Gold plated-through holes linked the top side of the board to the back side, where two current traces connected each segment to the fuel cell test equipment. One line was used as a voltage sense and the other to carry current. The GDL and the CL were also segmented to match the flow-field segments. The segmented assembly was applied to the anode. They used a combination of two load controls to run the experiment as depicted in Figure 9.8(ii). Any given segment of interest was controlled by one load while the other segments were maintained by the second load while transitioning was achieved through a switch box.

9.3.2 Independent Segment Control with Unsegmented Electrodes

Brett et al. (2001) adopted the PCB technique introduced by Cleghorn et al. (1998) to a single straight channel cathode where each of the current collector ribs (10 in all) on the PCB board was controlled by an individual load. They used a typical multichannel anode with hydrogen flowing perpendicular (cross-flow) to the air stream. The electrodes on both sides of the MEA were not segmented. Sun et al. (2006) developed a novel approach where they used a gasket between the flow field and the GDL that created a segmentation current path. They placed very thin conductive strips of copper in a gasket made of epoxy resin and glass cloth. The strips were positioned to line up with the shoulders on the flow field as shown in Figure 9.9. The segments were independently controlled by a 24 channel potentiostat. The electrodes, that is, GDL and the CL, were not segmented.

Some interesting steady-state and transient current density distribution data in direct methanol fuel cells and PEM fuel cells were published by Mench and Wang (2003) and Mench et al. (2003). The authors used segmented flow fields very similar to Noponen et al.'s design wherein the shoulders were made of gold-coated stainless-steel strips embedded in a polycarbonate block to complete the flow fields. However, the current collection was achieved through stainless-steel rods welded to the back of the strips that circumvented the problem of artificial change in the current direction pattern described above. The segmented flow fields were applied to both the anode and cathode sides. While most of the individual segments were separately controlled by a multichannel potentiostat/galvanostat, some of the segments were grouped together due to lack of sufficient channels. The authors employed commercially available MEAs where the GDL and CLs were not segmented, as seen in Figure 9.10. Bender et al. (2003) segmented the electrodes along with the current collectors for studying local current density distribution in an operating fuel cell. However, with the intention of studying the effect of feed stream impurities and poisoning of the anode, the authors applied the segmented assembly only to the anode and not the cathode.

In Cleghorn et al.'s approach, the use of a couple of load units with a multiplexer (i.e., voltage control is quickly switched from one electrode to another by the multiplexer) only allows analysis of steady-state behavior. Switching between electrodes gives rise to temporal double-layer charging and discharging currents and could induce undesired transient artifacts. Despite these issues, they were the first to carry

FIGURE 9.9 A photograph of the current distribution measurement gasket used by Sun et al. (2006). (Reprinted from *J. Power Sources* 158(1), Sun, H., A novel technique for measuring current distributions in PEM fuel cells. 326–332, 2006, with permission from Elsevier.)

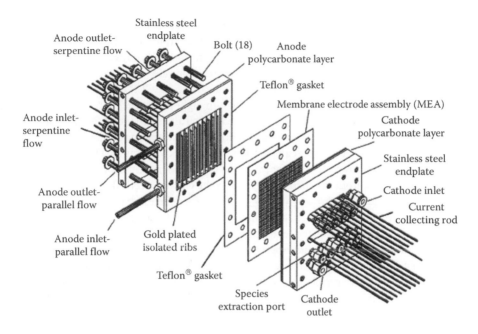

FIGURE 9.10 Schematic of the 50 cm² instrumented transparent fuel cell with segmented gold-plated current collector ribs used in the study by Mench et al. (2003). (From Mench, M. M., and Wang, C. Y. 2003. *J. Electrochem. Soc.* 150(1): A79–A85. With permission from Electrochemical Society.)

the segmentation all the way to the CL. The implications of the extent of segmentation are analyzed more in detail below. Sun et al.'s method has the disadvantage of artificially changing the current pattern as discussed above regarding Noponen et al.'s work.

Mench et al. (2003) were also the first to apply segmentation of both the anode and cathode simultaneously. In all the other literature, segmentation was predominantly applied to the cathode. While the electrode of interest was still the cathode, Cleghorn et al. (1998) segmented the anode side assuming that the facile kinetics of anode did not influence the overall current density measurement. Bender et al. (2003) and Stumper et al. (1998) applied segmentation to the anode to mainly study CO poisoning. During regular operation, depending on the operating conditions and load transients either electrode can be limiting and be the dominant influence on the current density distribution. Also, even if one of the electrodes is the focus, it is very difficult to ensure uniformity on the other electrode in terms of catalyst loading material properties and uniformity in fuel/water concentrations. Hence, the prudent approach to eliminate uncertainties arising from such assumptions is to segment both sides of the fuel cell. This way, any nonuniformity arising from the electrode other than the one of interest can be removed from the measured data. While the use of multichannel potentiostats/loads by Brett et al. (2001), Sun et al. (2006), and Mench et al. (2003) addresses the issue of independent control of the consequences at each segment, a common GDL and CL in terms of current spreading and interference between segments can still be an issue. Mench et al. (2003) made a significant effort to address this phenomenon. They compared *in situ* measured in-plane resistivity of the GDL with the average measured contact resistance and made the argument that the significantly greater in-plane resistance as compared to contact resistance mitigates cross-talk. However, if there are significant variations in contact resistances from segment to segment cross-talk may not be insignificant. Methods to avoid this contact resistance variation can be quite elaborate, like individual set-screws for each segment limiting their implementation. Hence, the major question to be addressed in current density distribution studies is whether the GDL and the CL that makes up the electrode need to be segmented along with the current collector.

Current density distribution studies are aimed at obtaining information on the local reaction rate distribution within the CL which is a strong function of the local electronic and ionic potentials, reactant and product concentrations, and membrane hydration state. Performance improvement and optimization of fuel cells directly relate to improving the electrochemical reaction rate within the CL. Hence, an accurate picture of the local current density distribution as it emanates from the CL is of interest to researchers from a fundamental point of view. However, another school of thought is that the current emanating from the unsegmented electrode represents real life operation of a fuel cell and segmenting the electrodes artificially disrupts the conductive paths. It then becomes a question of what information the experimenter is interested in to determine the suitable level of segmentation. In the case of a common electrode, the question of whether one sees the same current distribution, as in the CL, at the current collectors depends on: (1) the ratio of the normal distance from the current collector to the CL and electrode width along the channel, (2) the ratio of in-plane and through-plane area, (3) the ratio of in-plane and through-plane conductivities of the GDL material, and finally, (4) the relative contact resistances between the GDL and the current collectors. The above-mentioned parameters determine the extent of current spreading within the GDL. Mench et al. (2003) implemented elaborate experimental techniques like the use of pressure indicating film to minimize contact resistance. Another concern with the use of a common GDL is determining the active area associated with each current collector. Noponen et al. (2002) tried to address this by developing a simple model to estimate the errors arising from assigning equal areas to all the current collectors.

9.3.3 Independent Segment Control with Segmented Electrodes

Natarajan and Nguyen (2004) specifically addressed the effect of nonuniform contact resistance between the GDL/CL and adjacent segmented sections of the flow field. The influence of segmenting the GDL and the CL along with the flow field was also tackled. Figure 9.11 shows a schematic of the top and side view of the current collector and flow field block and MEAs used in this work. The flow-field block was made of acrylic with imbedded graphite current collectors. A single gas channel was machined to represent a

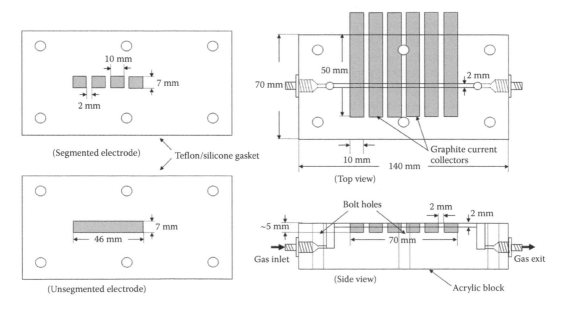

FIGURE 9.11 Schematic of the current collector/flow-field block and MEA used by Natarajan and Nguyen (2004). (Reprinted from *J. Power Sources* 135(1–2), Natarajan, D. and Nguyen, T. V. Effect of electrode configuration and electronic conductivity on current density distribution measurements in PEM fuel cells. 95–109, 2004, with permission from Elsevier.)

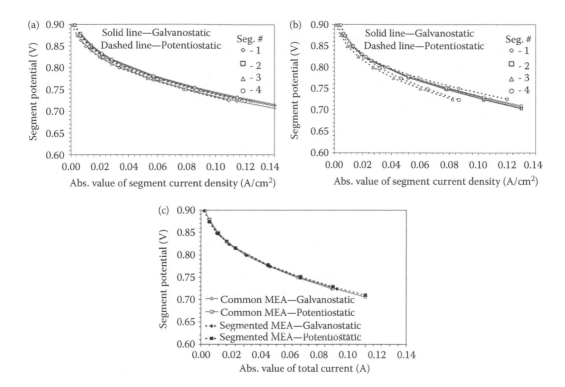

FIGURE 9.12 Comparison of common and segmented MEAs under galvanostatic and potentiostatic mode. (a) segmented, (b) unsegmented, and (c) total cell current. $F_{H2} = 2$ or 2.3 A cm^{-2}, $F_{O2} = 2$ or 2.3 A cm^{-2}, $T_{H2\ humidifier} = 70°C$, $T_{O2\ humidifier} = 25°C$ and $T_{cell} = 30°C$ (Natarajan and Nguyen, 2004). (Reprinted from *J. Power Sources* 135(1–2), Natarajan, D. and Nguyen, T. V. Effect of electrode configuration and electronic conductivity on current density distribution measurements in PEM fuel cells. 95–109, 2004, with permission from Elsevier.)

section of a conventional (or serpentine) flow field. Custom electrodes fabricated by catalyst coating on SIGRACET® GDL 30 BC from SGL-CARBON, Inc. were used to fabricate by hot bonding MEAs that had segmented or unsegmented electrodes on both the anode and cathode. Though the fuel cell assembly was capable of handling six segments, MEAs were prepared to correspond to four current collector segments for the sake of clarity of experimental data. The experiments on both types of MEA were conducted in excess supply of humidified fuel and oxidant at 30°C. After conditioning, the MEAs were subjected to galvanostatic and potentiostatic discharges.

Figure 9.12 provides the experimental data in the form of polarization curves based on the current and voltage responses from the four segmented current collectors, using an MEA with segmented (Figure 9.12a) and unsegmented (Figure 9.12b) electrodes. Figure 9.12c provides the total current from the entire cell versus the average voltage at the current collectors. It is clear that in the MEA with segmented electrodes the polarization curves of the individual segments were very close to each other irrespective of the discharge mode. On the other hand, in an unsegmented MEA, while the polarization curves from the four collectors were quite close to each other in the constant current discharge mode (within 5 mV at 100 mA/cm²), there existed significant spread in performance from segment to segment under constant voltage mode (40 mA/cm² at 0.725 V). Moreover, the polarization curves obtained in the galvanostatic or constant current discharge mode for two types of MEAs matched quite satisfactorily, indicating consistent catalytic properties in both the MEAs. The overall performances were also identical for both the segmented and unsegmented MEAs, irrespective of the discharge mode, suggesting minimal differences in terms of overall activation, ohmic and mass transport properties between the two types of MEAs (slightly greater area between the current collectors in the common MEA seemed to have a negligible effect probably due to lack of electronic contact). The fact that the common MEA showed very similar segment performances under the galvanostatic mode and not under the potentiostatic mode suggests that the interactions could be electronic and not kinetic in nature. Electronic interactions arise mainly due to differences in contact resistances between the various current collector segments and the GDL. However, the lack of contact influence in the segmented electrodes was intriguing and the authors developed a simple model to tease out the influence of contact resistance in interpreting the experimental results.

The two-dimensional mathematical model developed was to map the solid-phase potential distribution. The model domain is presented in Figure 9.13a along with the various boundaries and dimensions. The model was set up to accommodate both galvanostatic and potentiostatic simulations. The results confirmed that the discrepancy observed between the galvanostatic and potentiostatic discharges was caused by the difference in the contact resistance between the GDLs and flow-field/current collectors. The model results in terms of potential and electronic flux vector distributions within the common GDL are shown in Figures 9.13b and 9.13c for the two types of discharges. Figure 9.13b indicates a symmetric potential and electronic flux distribution within the GDL for the case of the galvanostatic simulation. The potentiostatic simulations shown in Figure 9.13c depicts a drop in the potential within the GDL from the region above current collector B toward the region over current collector A indicating the existence of in-plane current flow within the GDL. The electronic flux vector plot confirms this phenomenon indicating flow of electrons from current collector A toward the regions over current collector B. Thus despite the significant geometric size differences associated with the in-plane and through-plane directions, significant interactions between segments existed in the experiments with common electrodes. This was probably due to the inherent higher in-plane conductivity compared to through-plane values of typical GDL material arising from carbon fiber orientation. Noponen et al. (2002) used the through-plane resistance of the GDL in all directions in their model and hence did not observe a similar phenomenon.

One of the major concerns among the fuel cell research community regarding segmenting electrodes is achieving uniform comparable contact and performance from segment to segment. The experimental and model results from this article suggest that small differences in contact resistances do not significantly manifest themselves when the electrode is segmented. On the other hand, despite similarity in kinetic and transport properties the same difference in contact resistances significantly affects the results in the case of the common MEA, especially in the constant voltage mode.

As a follow-up to this publication, the authors conducted local current density distribution measurements using a similar segmented flow field and electrode setup (six operating segments) to study the influence of various operating parameters including transient behavior (Natarajan and Nguyen, 2005a,b). The experimental details are omitted here for the sake of brevity and only some sample data are shown. Figure 9.14 provides the results from two experiments that were conducted at a cell temperature of 30°C and 45°C with a corresponding anode sparger temperature of 40°C and 60°C. These experiments were conducted at a low air flow rate equivalent to 1 A. Increasing cell temperature resulted in a significant increase in the current densities at the first three segments while the latter segments registered a decrease. Sufficient oxygen was available at segments 1, 2, and 3 to reflect the increase in reaction rate due to enhanced kinetics and greater water removal capacity achieved through temperature rise. However, the performance limiting criteria for latter segments being the availability of oxygen, the enhancement in kinetics, and water evaporation at higher temperature was offset by the more pronounced lack of oxygen especially given the higher reaction rates near the channel entrance. This actually led to a drop in the current densities with increase in temperature at segments downstream.

Application of segmentation to the anode side from this work (Natarajan and Nguyen, 2005a,b) is demonstrated in Figure 9.15. In these experiments, five out of the six segments were subjected to a three-step potentiostatic discharge. Segment 6, which was left on open circuit, was used as an in-line voltage sensor to measure the effect of local gas composition. Figure 9.15a corresponds to experiment with humidified gas streams while neither gas streams were humidified in the experiments depicted in

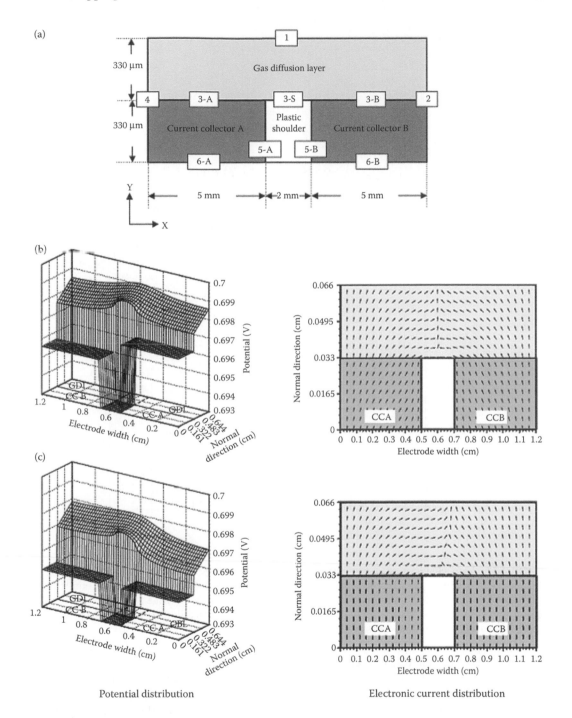

FIGURE 9.13 Schematic of the model domain and results on the effect of contact resistance between the GDL and flow-field/current collector (Natarajan and Nguyen, 2004). (Reprinted from *J. Power Sources* 135(1–2), Natarajan, D. and Nguyen, T. V. Effect of electrode configuration and electronic conductivity on current density distribution measurements in PEM fuel cells. 95–109, 2004, with permission from Elsevier.)

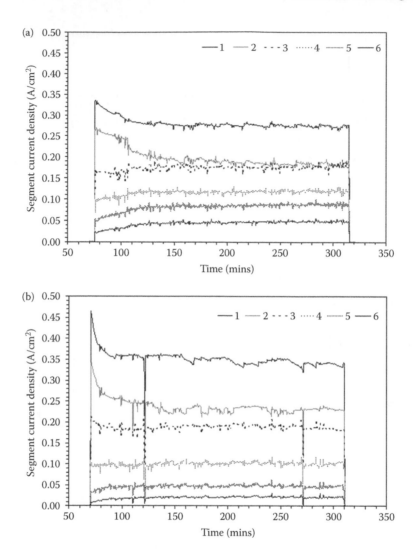

FIGURE 9.14 Individual segment current response to an imposed voltage of 0.45 V, (a) $T_{cell} = 30°C/T_{H2\ humidifier} = 40°C$ and (b) $T_{cell} = 45°C/T_{H2\ humidifier} = 60°C$. $F_{H2} = 48\ cm^3\ min^{-1}$ (6.9 A or 1.1 A cm^{-2}), $F_{air} = 16.6\ cm^3\ min^{-1}$ (1.0 A or 0.16 A cm^{-2}), $T_{air\ humidifier} = 30°C$ (Natarajan and Nguyen, 2005). (Reprinted from *AIChE J.* 51(9), Natarajan, D. and Nguyen, T. V. Spatiotemporal local current density distribution in PEM fuel cells: Part 1—Oxygen operation and hydrogen starvation studies. 2599–2608, 2005a, with permission from Elsevier.)

Figure 9.15b. Hydrogen flow rates equivalent to 1.47 A and 1.3 A were employed in these experiments respectively. In Figure 9.15a, when the segments were held at 0.7 V, the hydrogen supply was in excess of the reaction demand. Moreover, the gas streams were sufficiently humidified. Hence, the local current densities were uniform along the channel length as shown in the inset. Once the reaction driving force was increased by dropping the segment potentials to 0.6 V, starvation effects were observed in the performance of segments 4 and 5. Segment 5 being the furthest from the inlet was the most affected. Upon further decrease in the operating potential, most of the hydrogen was consumed at segments 1 and 2. This resulted in starvation effects being observed as close to the inlet as segment 3. Segments 4 and 5 were completely deprived of fuel and their current responses were close to 0. In the results shown in Figure 9.15b, at a segment potential of 0.7 V, fuel flow rates were sufficiently higher than reaction demand and the current density distribution along the channel was dominated by membrane dehydration (see

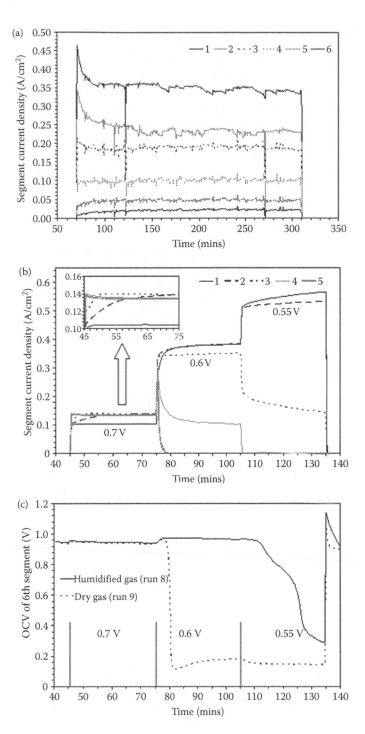

FIGURE 9.15 Individual segment current response at various applied segment potentials under fuel-starving conditions, $F_{H2} = 5.1$ cm^3 min^{-1} (1.47 A or 0.28 A cm^{-2}), $F_{O2} = 19$ cm^3 min^{-1} (5.5 A or 1.05 A cm^{-2}), $T_{cell} = 28°C$, (a) $T_{H2 \, humidifier} = 40°C$ and $T_{O2 \, humidifier} = 30°C$, (b) Dry H$_2$ and O$_2$, and (c) Sixth-segment open-circuit potential response during dry and humidified runs (Natarajan and Nguyen, 2005). (Reprinted from *AIChE J.* 51(9), Natarajan, D. and Nguyen, T. V. Spatiotemporal local current density distribution in PEM fuel cells: Part 1—Oxygen operation and hydrogen starvation studies. 2599–2608, 2005a, with permission from Elsevier.)

enlarged inset). Segment 1 being the closest to the inlet, was subjected to completely dry gases and showed the lowest current density. Along the length of the channel, the gases were humidified over a period of time by the product water generated at the cathode. This is clearly seen from the current responses of segments 2 and 3 that started out quite low and got better over a period of time. Segments 5 and 6 benefited from humidification byproduct water right from the beginning of the experiment. Once the segment potentials were dropped to 0.6 V, a transition was observed in the current density distribution from being dependent on membrane hydration to fuel availability. Due to the higher rate of water generation by reaction, segments 1 and 2 were quickly humidified within a few minutes and their current performances were the maximum due to greater fuel availability. The effect of hydrogen starvation was exhibited by segments 4 and 5. Further increase in the reaction driving force resulted in complete fuel depletion at segments 4 and 5 causing the current density to drop to 0.

The potential measured at segment 6 for the two experiments are plotted in Figure 9.15c. At high-segment potentials, when fuel was in excess, segment 6 registered typical open-circuit voltage values close to 1.0 V. As the fuel cell was taken to starvation regimes, the potential at segment 6 reflected the complete consumption of fuel. In the case of the dry hydrogen experiment with lower flow rate (1.3 A), the voltage signals from segment 6 for potential steps 0.6 and 0.55 V match the zero-current responses observed at segment 5 in Figure 9.15b indicating minimal availability of hydrogen past segment 5. On the contrary, when humidified hydrogen stream with higher flow rate (1.47 A) was used, the current density at segment 5 goes to 0 only after about 20 min into the 0.55 V potential step that is also captured in the voltage response of segment 6. The lag in the current response of segment 5 at 0.55 V can be attributed to the time for consumption of hydrogen available in the pore spaces of the electrode and the channel.

9.4 Other Indirect and Noninvasive Techniques

All the literature reviewed in the previous sections ultimately involved the measurement of current emanating from the fuel cell through modified architecture to map the current density distribution. In this section, three different techniques that are based on fundamentally different methodology for current mapping are reviewed. While the first two techniques can be considered as indirect/minimally invasive techniques, the last method of magnetotomography is a truly noninvasive approach to current mapping.

9.4.1 Indirect/Minimally Invasive Techniques

Recognizing the experimental complexities associated with traditional approaches of measuring current through artificially segmented pathways, Wilkinson et al. (2006) chose to measure the temperature distribution within an operation fuel cell to relate to the current density distribution. Their rationale for the approach was the relative ease of integrating thermocouples in the flow fuel cell assembly. In their method, the thermocouples were located in the landing area of the flow field plates in contact with the anode gas diffusion of the MEA as shown in Figure 9.16. Despite the cathode being the dominant source of heat generation, the thermocouples were incorporated on the anode side due to practical difficulties associated with the coolant channels on the backside of the cathode plate in the commercially procured single fuel cell assembly. The authors believed that the anode temperature was a good approximation of the cathode temperature since the temperature difference between the anode and cathode was found to be less than 1°C. The authors stabilized the performance of the fuel cell at a chosen operating point for 30 min at the end of which they stopped the coolant flow and purge the coolant chamber for 15 s. Following this, they recorded the temperature rise indicated by the thermocouples for 25 s. The rate of change of temperature with respect to time was then calculated from the measured data. The authors assumed that all the heat generated by the reaction resulted in this temperature change in the region around the thermocouple sensing tip. The above experimental procedure was devised to justify this

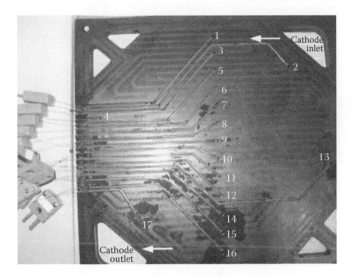

FIGURE 9.16 Photograph of the anode flow field used by Wilkinson et al. (2006) showing the locations of thermocouples embedded on the anode plate. (From Wilkinson, M. 2006. *Electrochem. Solid-State Lett.* 9(11): A507–A511. With permission from Electrochemical Society.)

assumption. The authors also conducted a heat transfer analysis to account for errors at an average current load of 100 mA cm^{-2} corresponding to a cell temperature of about 30°C. They calculated a heat loss of 1%, 2%, and 11% to latent heat of water evaporation, sensible heat loss to the gas streams, and convective/radiation losses to the ambient. The rate of temperature rate was used to show local current density trends for three different average current density set points.

While this technique is indeed innovative, some potential drawbacks exist. The major issue is their assumption that all the heat generated goes to heating the solid surface around the sensing point, that is, the region around the sensing point is adiabatic. Obviously, this assumption needs to be examined more carefully. Although they conducted a thermal balance at 100 mA cm^{-2} indicating a net 14% error, the choice of this set point might be misleading. At this average current density the cell temperature is 30–32°C where the vapor pressure of water is quite low evidently leading to a minimal error of 1%. However, the nonlinear dependence of water vapor pressure on temperature will significantly increase this error at higher cell operating temperatures/current densities. For instance, the vapor pressure of water at 32°C is about 32 mmHg while at 80°C it is 356 mmHg. This potential issue can also invalidate their assumption of reasonable uniformity of temperature across the membrane (and over the active area) thus rendering sensing at the anode not truly representative of the conditions at the cathode. Finally, as the authors duly recognized, the 15 s gap between shutting down the coolant and temperature measurement can also be a source of significant error.

Freunberger et al. (2006) have come up with a very interesting and unique technique to measure the current density distribution across the electrode width or in other words across the channel–shoulder pair while all the other techniques are focused along the channel. As explained in the introduction, significant current variations can be expected in this direction due to the different transport lengths for all the charged and neutral specie. This hypothesis has been supported by numerous modeling attempts cited previously. Figure 9.17(i) provides a schematic of their experimental setup. The authors sandwiched very thin gold wires between the GDL and the CL to directly measure the CL solid-phase potential with respect to the potential at the copper buss bar at various points along the width of the channel and shoulder. These data were mathematically correlated to local current density values by assuming that electron conductors in a plane perpendicular to the channel direction are two-dimensional resistors and

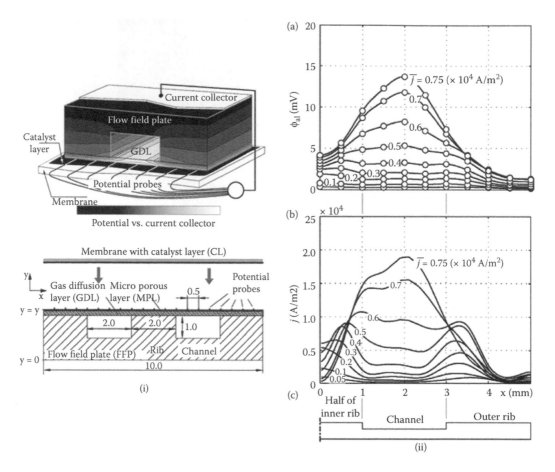

FIGURE 9.17 Schematic of the measurement principle employed and results in the study by Freunberger et al. (2006) that uses the potential drop over FFP and GDL for gaining information on the corresponding current distribution. The potential field is illustrated by means of iso-potential contours. Its value at the GDL-CL interface is probed using thin gold wires. (From Freunberger, S. A. 2006. *J. Electrochem. Soc.* 153(11): A2158–A2165. With permission from Electrochemical Society.)

the current density is obtained from the solution of Laplace's equation with the potentials at current collector and reaction layer as boundary conditions. Recognizing the strong dependence of GDL in-plane and through-plane conductivities and contact resistance on the compression pressure, the authors developed a model to map the compressive stresses across a channel and shoulder. The stress values were converted to relevant conductivities and contact resistances based on *ex situ* measurements of the mentioned variables with respect to compressive stress.

Figure 9.17(ii) provides an example of the results obtained by the authors. The flow of both reactants was set to a value to ensure virtually homogeneous conditions in the along-the-channel direction. The cell was operated at 65°C with minor gas humidity to prevent electrode flooding. The results clearly capture the theoretically predicted shift in the peak local current density from the center of the rib at low loads toward the center of the channel as the load is increased. At lower loads the oxygen concentration across the CL is more uniform and the performance is dominated by electron transport while at higher loads the slower oxygen transport dominates the current density distribution. The authors ran the gold sensing wires along the entire length of the channel. Hence, this setup cannot be used to look at the distribution along the channel length. A potential improvement to this experimental setup would be to have multiple sensing points along the channel which of course would make it more complicated. Also the accuracy and interpretation of results can be compromised by local liquid water flooding.

9.4.2 Noninvasive Approach

The final article reviewed here is a truly noninvasive technique. Hauer et al. (2005) adapted the method of magnetotomography to map the current density distribution in a working fuel cell. The authors built an elaborate four-axis computer-controlled system containing two magnetic sensors that measure the magnetic flux as shown in Figure 9.18. These sensors were programmed to measure the magnetic flux at different points in space. These data involving sensor position coordinates and special components of the magnetic flux were then processed to relate to a current density map. The mathematical treatise involved discretizing the fuel cell volume into components representing a current vector through it. This provided a set of unknowns that equals number of volume elements times three, representing the three spatial components of the current vector. The measurements were conducted to provide known values (spatial components of magnetic flux) at least equal to the number of unknown current components. The authors also investigated the current distribution using a segmented cell to qualitatively validate the measurements using the noninvasive techniques. Each measurement took about 15 min.

The long measurement time, which makes it unsuitable for measurement of transient behavior, is one of the major drawbacks of this technique. Other issues include the complexity of the experimental technique, the complexity of the mathematical exercise to glean information, and sensitivity of the analysis to minute errors in measuring the magnetic flux. Also the presence of cables and tubing can be a practical hindrance to measure the flux at desired location. The technique in its current form is only capable of mapping differences in the current distribution between two operating set points. Despite the stated drawbacks, the noninvasive nature of this technique warrants significant merit and needs to be refined further.

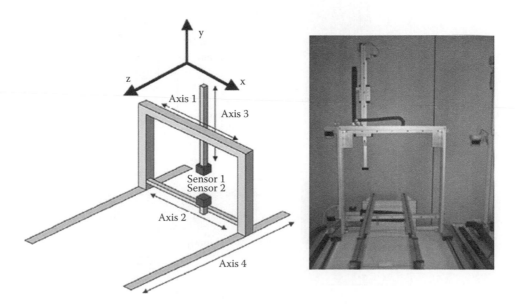

FIGURE 9.18 Positioning system and the location of the sensors in the setup used by Hauer et al. (2005). The physical dimensions are $L = 1.6$ m, $W = 1$ m, $H = 2$ m. (Reprinted from *J. Power Sources* 143(1–2), Hauer, K. et al. Magnetotomography—A new method for analyzing fuel cell performance and quality, 67–74, 2005, with permission from Elsevier.)

9.5 Reference Electrode Techniques to Measure Electrode Overpotential

While one can map the current distribution over the current collectors of a PEM fuel cell as illustrated by the methods discussed here, only the method that uses segmented electrodes (CLs and GDLs) and segmented flow fields can provide the current densities at the CLs. When only the current collectors are segmented and the electrodes are not segmented, depending on the conductivity of the porous support, the measured current density may not represent the current density at the CL because of in-plane flow of current in the porous GDL. Furthermore, to determine the root causes (e.g., activation, ohmic, and transport resistances) of local current distribution in a PEM fuel cell, electronic and ionic potentials at the CL of each segmented electrode are needed. An appropriately placed reference electrode can provide the ionic potential at the CL. With the ionic and electronic potentials and the local current density, one can then determine the voltage losses at each segmented electrode and in the membrane.

9.5.1 Electrode Edge Effects on Reference Electrode Measurements

Owing to the very low thickness of membrane electrolytes used in PEM fuel cells (10–50 μm) it is essentially impossible to place a reference electrode in the electrolyte in the main current path between two electrodes. Many researchers have studied the use of an external reference electrode placed in contact with the membrane electrolyte as a way to determine the ionic potential in the electrolyte, as seen in Figure 9.19. However, because of the edge effects and the difference in kinetic resistances at the electrodes,

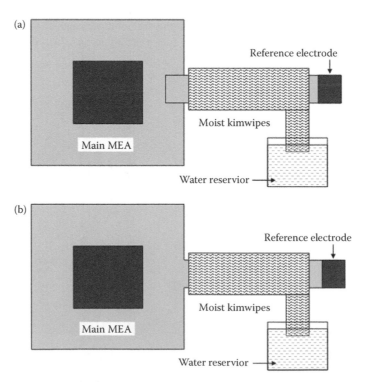

FIGURE 9.19 Schematic of membrane electrode assembly with an external reference electrode used in the study by He and Nguyen (2004). (a) Detachable version and (b) integrated version (From He, W. and Nguyen, T. V. 2004. *J. Electrochem. Soc.* 151(2): A185–A195. With permission from Electrochemical Society.)

accurate measurements of the ionic potentials at the electrodes could not be obtained with this method (He and Nguyen, 2004; Liu et al., 2004).

9.5.2 Reference Electrode Placement and Modification to Overcome Edge Effects

Recently, Piela et al. (2007) investigated three different reference electrode configurations (see Figure 9.20) for PEM fuel cells. In the first configuration (a), the reference electrode, which is made of a platinum black-coated wire covered with Nafion, is placed in direct contact with a region of the membrane surface that is not covered by the CL. The reference electrode is assumed to be in equilibrium with the gas phase in the flow channel where the reference electrode is mounted. The ionic potential measured by the reference in this configuration is found to be similar to that of a reference electrode placed in contact with the membrane surface outside of the electrode area, the case investigated by He and Nguyen (2004) and Liu et al. (2004) shown in Figure 9.19. In the second configuration (b), the reference electrode is placed in direct contact with the CL through an opening in the GDL. The ionic potential measured by configuration 'b' is also found to be unreliable. The reason is attributed to the different electrochemical performance of the CL area below the reference electrode caused by the lack of access to the reactant gas in the flow channel. In the third configuration (c), the reference electrode is placed in contact with the outer surface of the GDL. An ionic bridge or conduction pathway to the CL is established by coating the

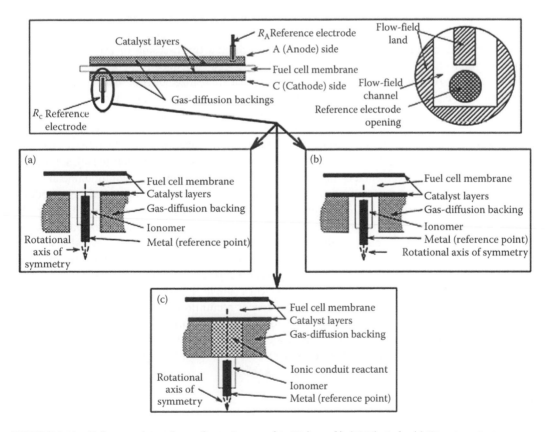

FIGURE 9.20 Reference electrode configurations used in Piela et al.'s (2007) study. (a) Direct contact to membrane, (b) direct contact to CL, and (c) ionic bridge through gas diffusion backing. (Reprinted with permission from Piela, P. et al. 2007. Direct measurement of iR-free individual-electrode overpotentials in polymer electrolyte fuel cells. *J. Phys. Chem.* 111(17): 6512–6523. Copyright 2007 American Chemical Society.)

surface of the GDL in the contact area with Nafion ionomer. This configuration is found to provide the correct ionic potential in the membrane electrolyte phase in the CL. Since the thickness of the electrolyte phase over the catalyst surface in the CL is very small (submicron), the potential difference across this layer is also very small. So, not knowing the exact location of the measured ionic potential in this layer is not a big concern.

The reason that the reference electrode in configuration "c" works is because the reaction and transport processes in the CL region in contact with the reference electrode are not affected. The other reason may be that the reference electrode is sensing the ionic potential of a very small region in the CL, where the specific current density is very small, and not in the main membrane electrolyte where interference from currents converging from all the area in the CL is much more significant. Furthermore, the fact that the reference electrode in configuration "c" measures the local ionic potential in the membrane electrolyte phase in the CL instead of at the membrane and CL interface is considered to be an added advantage by Piela et al. (2007). The information can be used to determine the ionic conductivity of the electrolyte phase in the the CL as a function of the CL thickness, Nafion loading in the CL, and local gas and humidity conditions. One could use a reference electrode similar to that in configuration "c" in each segmented electrode to determine the ionic potential in the electrolyte phase in the CL in each segment. This approach will allow one to map the three-dimensional potential distribution in a membrane electrode assembly.

9.6 Conclusions

In this chapter, several articles regarding current density mapping have been summarized. It is evident that no one technique can be declared as a clear choice for current mapping. All the techniques have pros and cons, and it is up to the researcher to select the technique that best suits their needs and means. The pioneering work by Cleghorn et al. (1998) and Stumper et al. (1998) can still be suitable as long as the limitations are recognized. Passive resistor networks that are realized by several means described in this review can be easily incorporated and are valuable tools for quick and cost-effective current mapping estimates. The use of a single potentiostat/load control, while being simple and cost effective does not provide accurate control of the potential across the entire CL surface. While the use of a couple of controls as in Cleghorn's work mitigates this issue, the preferred solution would be to have the same number of controls as the number of artificially created electron pathways to be able to look at transient behavior. The necessity of segmentation of the GDL and CL is discussed in detail. For fundamental studies aimed at understanding the kinetic and transport processes in the CL, segmenting the electrodes is recommended. Investigation of other components design, fabrication, and operating conditions can be conducted with unsegmented electrode. In either case special attention is needed in minimizing and equalizing contact resistances between the segmented and unsegmented elements. Several new minimally or completely noninvasive have been proposed in the recent years that show a lot of promise. All of them involve significant mathematical modeling to correlate the measured data to a current map that leaves the door wide open for further innovation to yield a simple and accurate noninvasive technique. Finally, to obtain a better understanding of the causes of nonuniform current distribution over the whole membrane electrode assembly, in addition to electronic potential distribution in the electrode local ionic potentials in the electrolyte at the electrode reactive surface will also be needed. The reference electrode configuration developed by Piela et al. (2007) that allows the ionic potential in the electrolyte phase in the CL to be measured may make this goal possible.

References

Bender, G., Wilson, M. S., and Zawodzinski, T. A. 2003. Further refinements in the segmented cell approach to diagnosing performance in polymer electrolyte fuel cells. *J. Power Sources* 123(2): 163–171.

Bernardi, D. M. and Verbrugge, M. W. 1992. A mathematical model of the solid–polymer–electrolyte fuel cell. *J. Electrochem. Soc.* 139(9): 2477–2491.

Brett, D. J. L., Atkins, S., Brandon, N. P., Vesovic, V., Vasileiadis, N., and Kucernak, A. R. 2001. Measurement of the current distribution along a single flow channel of a solid polymer fuel cell. *Electrochem. Commun.* 3(11): 628–632.

Cleghorn, S. J. C., Derouin, C. R., Wilson, M. S., and Gottesfeld, S. 1998. A printed circuit board approach to measuring current distribution in a fuel cell. *J. Appl. Electrochem.* 28(7): 663–672.

Freunberger, S. A., Reum, M., Evertz, J., Wokaun, A., and Büchi, F. N. 2006. Measuring the current distribution in PEFCs with sub-millimeter resolution. *J. Electrochem. Soc.* 153(11): A2158–A2165.

Fuller, T. F. and Newman, J. 1993. Water and thermal management in solid–polymer–electrolyte fuel cells. *J. Electrochem. Soc.* 140(5): 1218–1225.

Geske, M., Heuer, M., Heideck, G., and Styczynski, Z. A. 2010. Current density distribution mapping in PEM fuel cells as an instrument for operational measurements. *Energies* 3: 770–783.

Ghosh, P. C., Wüster, T., Dohle, H., Kimiaie, N., Mergel, J., and Stolten, D. 2006. *In situ* approach for current distribution measurement in fuel cells. *J. Power Sources* 154(1): 184–191.

Gurau, V., Liu, H., and Kakac, S. 1998. Two-dimensional model for proton exchange membrane fuel cells. *AIChE J.* 44(11): 2410–2422.

Hauer, K., Potthast, R., Wüster, T., and Stolten, D. 2005. Magnetotomography—A new method for analyzing fuel cell performance and quality. *J. Power Sources* 143(1–2): 67–74.

He, W. and Nguyen, T. V. 2004. Edge effects on reference electrode measurements in PEM fuel cells. *J. Electrochem. Soc.* 151(2): A185–A195.

He, W., Yi, J. S., and Nguyen, T. V. 2000. Two-phase flow model of the cathode of PEM fuel cells using interdigitated flow fields. *AIChE J.* 46(10): 2053–2064.

Hwnag, J. J., Chang, W. R., Peng, R. G., Chen, P. Y., and Su, A. 2008. Experimental and numerical studies of local current mapping on a PEM fuel cell. *Int. J. Hydrogen Energy* 33(20): 5718–5727.

Liu, Z., Wainright, J. S., Huang, W., and Savinell, R. F. 2004. Positioning the reference electrode in proton exchange membrane fuel cells: Calculations of primary and secondary current distribution. *Electrochim. Acta* 49(6): 923–935.

Mench, M. M. and Wang, C. Y. 2003. An *in situ* method for determination of current distribution in PEM fuel cells applied to a direct methanol fuel cell. *J. Electrochem. Soc.* 150(1): A79–A85.

Mench, M. M., Wang, C. Y., and Ishikawa, M. 2003. *In situ* current distribution measurements in polymer electrolyte fuel cells. *J. Electrochem. Soc.* 150(8): A1052– A1059.

Natarajan, D. and Nguyen, T. V. 2001. A two-dimensional, two-phase, multi-component, transient model for the cathode of a proton exchange membrane fuel cell using conventional gas distributors. *J. Electrochem. Soc.* 148(12): A1324–A1335.

Natarajan, D. and Nguyen, T. V. 2003. Three-dimensional effects of liquid water flooding in the cathode of a PEM fuel cell. *J. Power Sources* 115(1): 66–80.

Natarajan, D. and Nguyen, T. V. 2004. Effect of electrode configuration and electronic conductivity on current density distribution measurements in PEM fuel cells. *J. Power Sources* 135(1–2): 95–109.

Natarajan, D. and Nguyen, T. V. 2005a. Spatiotemporal local current density distribution in PEM fuel cells: Part 1—Oxygen operation and hydrogen starvation studies. *AIChE J.* 51(9): 2599–2608.

Natarajan, D. and Nguyen, T. V. 2005b. Spatiotemporal local current density distribution in PEM fuel cells: Part 2—Air operation and temperature case studies. *AIChE J.* 51(9): 2587–2598.

Nguyen, T. V. and White, R. E. 1993. A water and heat management model for proton-exchange-membrane fuel cells. *J. Electrochem. Soc.* 140(8): 2178–2186.

Noponen, M., Mennola, T., Mikkola, M., Hottinen T., and Lund, P. 2002. Measurement of current distribution in a free-breathing PEMFC. *J. Power Sources* 106(1–2): 304–312.

Piela, P., Springer, T. E., Davey, J., and Zelenay, P. 2007. Direct measurement of iR-free individual-electrode overpotentials in polymer electrolyte fuel cells. *J. Phys. Chem.* 111(17): 6512–6523.

Springer, T. E., Wilson, M. S., and Gottesfeld, S. 1993. Modeling and experimental diagnostics in polymer electrolyte fuel cells. *J. Electrochem. Soc.* 140(12): 3513–3526.

Stumper, J., Campbell, S. A., Wilkinson, D. P., Johnson, M. C., and Davis, M. 1998. *In situ* methods for the determination of current distribution in PEM fuel cells. *Electrochim. Acta* 43(24): 3773–3783.

Sun, H., Guangsheng, Z., Lie-Jin, G., and Hongtan, L. 2006. A novel technique for measuring current distributions in PEM fuel cells. *J. Power Sources* 158(1): 326–332.

Wieser, Ch., Helmbold, A., and Gulzow, E. 2000. A new technique for two-dimensional current distribution measurements in electrochemical cells. *J. Appl. Electrochem.* 30(7): 803–807.

Wilkinson, M., Blanco, M., Gu, E., Martin, J. J., and Wilkinson D. P. 2006. *In situ* experimental technique of temperature & current distribution in proton exchange membrane fuel cells. *Electrochem. Solid-State Lett.* 9(11): A507–A511.

10

Transparent Cell

Chun-Ying Hsu
Yuan-Ze University

Fang-Bor Weng
Yuan-Ze University

10.1 Introduction

Heat and water management in a fuel cell occurs due to several coupled processes, such as liquid and gas flows, as well as heat and mass transfers. Improper thermal and water management may cause severe mass transport limitations. In a PEMFC, the cathode is the performance limiting component due to the sluggish oxygen reduction kinetics. Other limitations in mass transport are imposed by the liquid water generated by the electrochemical reaction and brought to the cathode through electroosmotic (EO) drag. Proper water management ensures that the membrane remains fully hydrated (Su et al., 2006). A major obstacle preventing a fuel cell from realizing its theoretical current density is the reduction in effective porosity due to the accumulation of liquid product water inside the flow channels, the gas diffusion layer (GDL), or the catalyst layer (CL). Therefore, information regarding the amount of liquid water inside the fuel cell is of great technical significance.

In order to maintain membrane conductivity, the polymer electrolyte must be sufficiently hydrated. However, excess water causes the electrodes, which are bonded to the electrolyte, to flood, blocking the pores of the GDL. This problem worsens when designing larger cells and stacks. The importance of this water balance and management arises from the fact that the polymer membrane used as the electrolyte in PEMFCs requires full hydration in order to maintain good performance and durability. Furthermore, water transport also occurs due to local gradients of pressure, concentration, and surface tension. Consequently, flooding occurs if the water is not adequately removed from the cell (the cathode in particular). Water flooding makes the PEMFC performance unpredictable and unreliable under nominally identical operating conditions, and elaborate experimental diagnostics and schemes must be implemented to achieve consistent results with much lower performance and restricted current densities. Maintaining a perfect water balance during the dynamic operation process has therefore posed a significant challenge for PEMFC design.

Weng et al. (2010) designed a PEMFC that was divided into eight segments to study the aging phenomenon under transient loading conditions. The results suggested that the membrane might experience mechanical degradation in the downstream segments. The surface area of the catalyst was also seen to decrease due to downstream water flooding, which during operation might block mass transfer and result in fuel starvation, carbon corrosion, and catalyst degradation. Litster et al. (2007) incorporated an EO pump into a 25 cm² PEMFC using a new pump integration strategy. This EO pump actively removed excess water from the channels and diffusion layer, and was hydraulically coupled to an internal wick, with an electrically conductive structure that simultaneously served as current collector, a flow-field/channel structure, and an actively controlled wick. Ge et al. (2005b) described the water transport process through the membrane. The mass-transfer coefficients for the absorption and desorption of water and the water diffusion coefficients were measured using steady-state permeation under both vapor and liquid water conditions. The micro porous layers (MPLs) were prepared with varying polytetrafluoroethylene (PTFE) contents on carbon-fiber substrates. The mercury porosimetry and water permeation experiments showed that the addition of PTFE increased the resistance to water flow through the GDL due to a decreased MPL porosity and increased volume fraction of hydrophobic pores (Park et al., 2008).

The use of SGL carbon paper with an MPL gave the best fuel cell performance even at low air stoichiometries (Lin and Nguyen, 2005). Adding PTFE to the GDL enhanced gas and water transport when operating under flood conditions; however, excessive PTFE loading could lead to flooding in the CL (Lin and Nguyen, 2005). The resistance of the thin membrane varied slightly when the anode humidity was changed from saturated to dry. However, the humidity at the cathode was found to have more serious implications on the cell performance. Water balance experiments show that the best water transport coefficient is negative, even when the anode is humidified and liquid water exists both in the cathode and anode. High cathode humidity is a disadvantage for the removal of water from both electrodes (Cai et al., 2006). The effect of compression on cell performance is found to be the greatest when the gas diffusion backings (GDB) have poor water management properties or low permeability (Ohonen et al., 2004).

The water transfer coefficient—the number of moles of water transported per Faraday through the membrane—is governed by two characteristics: electrostatic interaction between ions and water dipoles, and the size of the cation. It has been shown that an increase in water content improves the transport characteristics of the membrane by increasing the size of hydrophilic domains through which the ions move (Okada et al., 1998). At high-current densities, excess water is generated, which condenses and fills the pores of the electrodes and limits the transport of reactants to the active catalyst. Liquid water transport across the GDL has also been found to be controlled by capillary forces resulting from the gradient in phase saturation (Pasaogullari and Wang, 2004). An optimized MPL should provide a small surface pore size on the CL side, which can efficiently confine the interfacial droplets formed on the surface of the CL and reduce their size and saturation level (Nam et al., 2009). The net water transport coefficient through the membrane, defined as the ratio between the net water flux from the anode to cathode and the protonic flux, is used as a quantitative measure of water management in a PEMFC. The local current density is dominated by the membrane hydration and the gas RH has a large effect on water transport through the membrane (Liu et al., 2007b).

10.2 Transparent Cell Design

10.2.1 Design Concept

Weng et al. (2006) designed a transparent fuel cell to investigate water flooding in a PEMFC. The membrane electrode assembly (MEA) is sandwiched between two bipolar plates, represented schematically in Figure 10.1a.

The bipolar plate, which is placed between two transparent acrylic cover plates, is shown in Figure 10.1b. The extension areas become fins to aid in temperature reduction through free or forced air convection with the environment. When the temperature becomes too low, a heater can be attached to the extension area to increase the cell temperature.

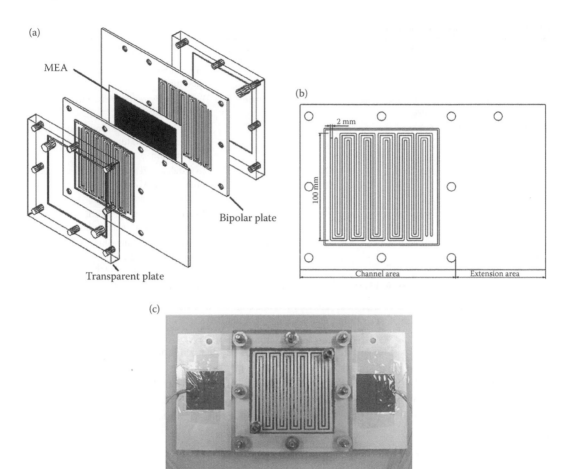

FIGURE 10.1 (a) Schematic of a transparent PEMFC; (b) schematic of a bipolar plate (Reprinted from Weng, F.-B. et al. 2006. *J. Power Sources* 157: 674–680.); and (c) picture of the transparent PEMFC. (Reprinted from Weng, F.-B., Su, A., and Hsu, C.-Y. 2007. *Int. J. Hydrogen Energy* 32: 666–676.)

An alternate approach is to use a transparent cathode to observe the transport of liquid water in the GDL and flow fields. However, this method has limited capability in terms of quantifying the amount of liquid water in the fuel cell. The most common technique for observing flow-field flooding is the use of a transparent fuel cell. The study of different phenomena within fuel cells has led to a number of specific designs for transparent fuel cells. Table 10.1 summarizes the designs from several research groups, including relevant channel dimensions, materials, and flow-field design (Anderson et al., 2010).

10.2.2 Instruments and Measurements

The experimental setup shown in Figure 10.2 consists of a gas-supply unit, a fuel cell, a closed-circuit digital (CCD) camera, an electronic load, and a data-acquisition system for analysis. The flow rates of air, oxygen, and hydrogen are regulated using mass flow controllers (Model #1179, MKS). All gases are humidified at 70°C. The fuel cell is connected to an electronic load (890B, SAI) that is operated in a constant-current discharge mode.

Flooding phenomena were observed at different current densities. Cathode gas stoichiometries of 2, 4, 6, 10, and 20 were used for each value of current density and the hydrogen stoichiometry was kept constant at 1.2. The cycle time of each experiment was 1 h for each cathode gas stoichiometry. Images

TABLE 10.1   Transparent Fuel Cell Designs

Author	Flow-Field Material	Transparent Plate Material	Subject/Application
Tüber et al. (2003)	Stainless steel	Plexiglas	To explain the phenomenon of water flooding
Yang et al. (2004)	Gold coated; stainless steel	Polycarbonate	To observe water droplets emerged from the GDL surface
Hakenjos et al. (2004)	Graphite	Zinc selenide	To perform a simultaneous evaluation of current, temperature, and water distribution
Sugiura et al. (2005)	NA	Polycarbonate	To install a water absorption layer (WAL). The new "WAL type" design is to use the visualization technique.
Weng et al. (2006)	Brass	Acrylic	To use a transparent fuel cell to observe the flooding inside the cathode channel.
Theodorakakos et al. (2006)	Plexiglas	Plexiglas	To investigate the detachment of water droplets under the influence of an air stream flowing around them.
Spernjak et al. (2007)	Stainless steel 316	Polycarbonate	To investigate the influence of the MPL on water transport.

Source: Adapted from Anderson, R. et al. 2010. *J. Power Sources* 195: 4531–4553.

were taken by the CCD in 1 s intervals and the current, voltage, fuel cell, and humidifier temperatures were recorded with a data-acquisition system.

For every test, the MEA undergoes an activation process to ensure that the fuel cell has reached a stable state. This process involves running the fuel cell at: (i) 0.6 V for 15 min; (ii) 0.4 V for 14 min; (iii) open-circuit potential for 1 min. This cycle is repeated several times until the MEA is stable.

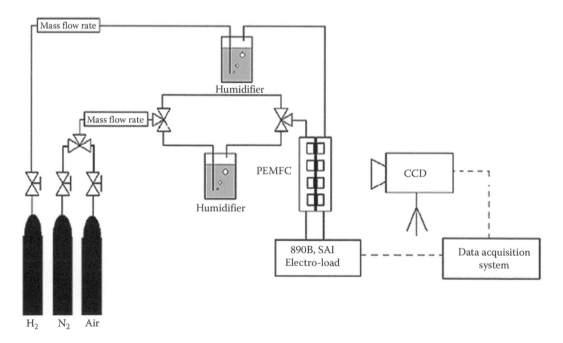

FIGURE 10.2   Schematic of experimental equipments. (Reprinted from *J. Power Sources*, 157, Weng, F.-B. et al. Study of water-flooding behaviour in cathode channel of a transparent proton-exchange membrane fuel cell, 674–680, Copyright (2006), with permission from Elsevier.)

10.2.3 Advantages and Limitations

Water produced at the cathode can evaporate into the cathode stream and leave the PEMFC. Therefore, water exhibits a two-phase flow in the cathode of PEMFCs, that is, in vapor and liquid phases. Understanding the dynamics of this two-phase flow allows for the optimization of PEMFC design, and has been an important subject of research in this field. Many techniques have been developed for this purpose. The transparent fuel cell allows for the observation of flooding phenomena using a camera that can then be analyzed to find correlations with operational parameters. However, a major limitation of this experimental method is the difficulty in quantifying the water flooding inside the fuel cell. Because of this, the images of water flooding obtained are explained qualitatively. Furthermore, the collection plates in transparent PEMFCs are gold-coated metals, resulting in lower performance due to greater surface resistance when compared with conventional graphite plates. Metal surfaces can also be easily corroded, producing ions that can contaminate the CL, and degrade the PEMFC performance. A new, low-resistance collection plate material is required for future experiments to better match the performance of real PEMFCs.

Table 10.2 summarizes the state-of-the-art spatial and temporal resolutions of a number of methods for liquid water visualization reported by Bazylak (2009) such as nuclear magnetic resonance spectroscopy, neutron imaging, synchroton x-ray radiography, microtomography, optical photography, and fluorescence microscopy. Table 10.2 also shows the merits and challenges of each method. For example, neutron imaging is used to visualize spatial water accumulation and distribution in PEMFCs. The neutron intensity measured by the detector changes in proportion to the amount of water present in the fuel cell, revealing the spatial accumulation and transport of water in the fuel cell. However, this method requires expensive hardware and is challenging to calibrate for reliable data quantification.

Table 10.3 summarizes the advantages and shortcomings of the water measurement methods reported. Solid-state humidity sensors provide an alternative to conventional measures for the relative humidity (RH) of the PEMFC anode and cathode output streams. These sensors measure the change in a stream capacitor dielectric constant and temperature to quantify the RH. Unfortunately, when used with a saturated stream, water condensation on these sensors gives rise to unreliable results.

For direct measurement of two-phase water transport in PEMFCs (Niroumand and Saif, 2010), magnetic resonance imaging (MRI) is used to study the spatial water distribution. In this method, an array of radiofrequency (RF) transmitters and receivers is used to provide a two-dimensional slice from the

TABLE 10.2 Summary of Reported Resolution Capabilities for Liquid Water Visualization

Method	Spatial Resolution (μm)	Temporal Resolution (s)	Merits	Challenges
Nuclear magnetic resonance spectroscopy	50	50	Compatible with operating fuel cell	Incompatible with carbon materials
Neutron imaging	25	5.4	Compatible with operating fuel cell and carbon materials	Limited spatial and temporal resolutions
Synchroton x-ray	3–7	4.8	Compatible with operating fuel cell and carbon materials	Limited temporal resolution
Microtomography	10	0.07	Through-plane resolution available	Is yet to be demonstrated with an operating fuel cell
Optical photography	10	0.06	Compatible with operating fuel cell	Transparent window requires substitution materials for operating fuel cell
Fluorescence microscopy	5.38	0.3	High spatial and temporal resolutions	Is yet to be demonstrated with an operating fuel cell

Source: Adapted from Bazylak, A. 2009. *Int. J. Hydrogen Energy* 34: 3845–3857.

TABLE 10.3   Water Measurement Techniques Used for PEM Fuel Cells

Methods	Strength	Weakness
Magnetic resonance imaging	Membrane water distribution	Not compatible with carbon; expensive
Transparent cathode	Phenomenological study	Limited quantification
Neutron imaging	Nonintrusive, spatial water distribution	Measure anode+ cathode; limited accuracy; expensive
Laser spectroscopy	Measure spatial distribution	Cannot measure liquid water
Gas chromatography	Spatial species distribution	Work only for gasses
Condensation	Total water balance	Small temporal resolution
Humidity sensor	Accurate, cheap	Measure only water vapor

Source: Data from Niroumand, A. M. and Saif, M. 2010. *J. Power Sources* 195: 3250–3255.

volume of interest. However, MRI requires expensive hardware and is difficult to calibrate such that the amount of water in the PEMFC can be quantified.

10.3 Applications

10.3.1 Optimizing Fuel Cell Operating Conditions

During the experiment, the voltage is controlled between 0.4 and 0.6 V, and the humidified gas temperature is fixed at 70°C. The operating temperature of the fuel cell is fixed at 50°C. According to the polarization curve, the fuel cell operates in the mass-transfer region. In this region, the performance is affected by both flooding within the flow channel and excess water content in the MEA.

10.3.1.1 Effects Arising from Humidified Oxygen Stoichiometry

The power and resistance curves of the PEMFC operating under various oxygen stoichiometries as a function of time are shown in Figure 10.3a and b, respectively. It can be seen that at a stochiometry of 2, the cell power and resistance are unstable and are characterized by large and unsteady fluctuations. These fluctuations reduce as the stoichiometry increases. At a high electronic load specified for this experiment, a large concentration of H+ ions is required to retain a constant output voltage. It is then reasonable to conclude that an oxygen stoichiometry of 2 gives insufficient humidity at the cathode, causing large fluctuations in output voltage. When the rate of gas flow increases, the output power also increases due to the increase in voltage, reaching a constant level at a stoichiometry of 6, beyond which no change in output power is observed. The steady performance at stochiometries above 6 may be related to stable water content within the fuel cell.

Images of the cathode side under operating conditions are shown in Figure 10.4a and b. It can be seen that the cathode channels do not become clogged even at an oxygen stoichiometry of 20. The status of the channels for stoichiometries of 2 and 20 are very similar. This explains why the cell performance remains stable at higher stoichiometries. No channel flooding is observed under such operating conditions.

10.3.1.2 Effects Arising from Humidified Air Stoichiometry

The power and resistance curves for varying stoichiometries of humidified air are shown in Figure 10.5a and b. As expected, the cell power increases with stoichiometry; however, the resistance of the cell for all stoichiometries remains similar: the difference between the maximum and minimum resistance is only 0.3 mΩ. This result suggests that the water content in the membrane is sufficient to retain stable operation. The fluctuation in the output power is also more stable than discussed in the previous section. It appears that the effects associated with the water content required for normal operation are overwhelmed by those associated with the oxygen concentration required for the reaction. The vapor carried

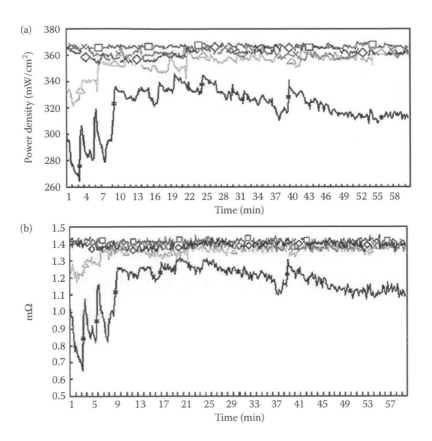

FIGURE 10.3 The power (a) and resistance (b) curves at a constant electronic load of 700 mA cm^{-2} with humidified oxygen at a stoichiometry of: (∗) 2; (△) 4; (×) 6; (◇) 10; (□) 20. (Reprinted from *J. Power Sources*, 157, Weng, F.-B. et al. Study of water-flooding behaviour in cathode channel of a transparent proton-exchange membrane fuel cell, 674–680, Copyright (2006), with permission from Elsevier.)

FIGURE 10.4 Images of cathode gas-flow channels, operated with humidified air at a constant electronic load of 700 mA cm^{-2} and a stoichiometry of (a) 2, and (b) 20. (Reprinted from *J. Power Sources*, 157, Weng, F.-B. et al. Study of water-flooding behaviour in cathode channel of a transparent proton-exchange membrane fuel cell, 674–680, Copyright (2006), with permission from Elsevier.)

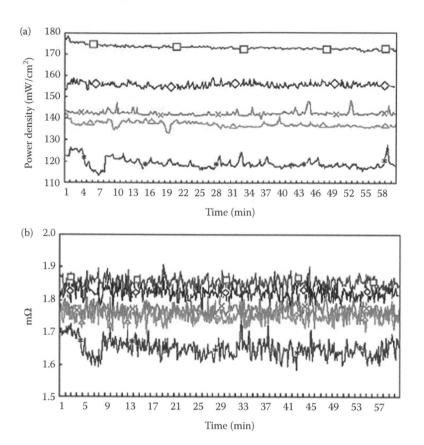

FIGURE 10.5 The power (a) and resistance (b) curves at a constant electronic load of 300 mA cm^{-2} with humidified air at a stoichiometry of: (⋆) 2; (△) 4; (×) 6; (◇) 10; (□) 20. (Reprinted from *J. Power Sources*, 157, Weng, F.-B. et al. Study of water-flooding behaviour in cathode channel of a transparent proton-exchange membrane fuel cell, 674–680, Copyright (2006), with permission from Elsevier.)

by nitrogen is sufficient to support the reaction in the fuel cell. Liquid water can be removed using large stoichiometries (e.g., 20), and the amount of vapor in the air can support the cell reaction in the mass-transfer region. In contrast to the previous case, at a stoichiometry of 2, some of the cathode flow channels are flooded with water, as shown in Figure 10.6a. On the other hand, liquid water is pushed along the flow channels at an air stoichiometry of 20 (Figure 10.6b). These results suggest that it is better to operate the fuel cell with a humidified air stoichiometry greater than 15.

10.3.1.3 Effects Arising from Nonhumidified Oxygen Stoichiometry

The power curves for operation under different stoichiometries of nonhumidified oxygen are shown in Figure 10.7a. These data show that the cell behavior is the reverse of that found in cases 1 and 2 above: the cell performance deteriorates when the stoichiometry of the cathode gas increases. The resistance of the cell measured for all stoichiometries less than 20 is steady, as shown in Figure 10.7b. It is known that a large gas flow rate can remove too much water, dehydrating the membrane and causing instabilities in the resistance. The fluctuations observed in output power at a stoichiometry of 20 become severe due to insufficient water content in the membrane.

Images of the cathode under operating conditions are shown in Figure 10.8a and b. Water flooding is obvious at the bottom of the image at a stoichiometry of 2, which disappears at a stoichiometry of 20. As described above, the membrane becomes dehydrated due to the large stoichiometry of nonhumidified gas, which causes an unsteady output voltage that decays at large stoichiometries.

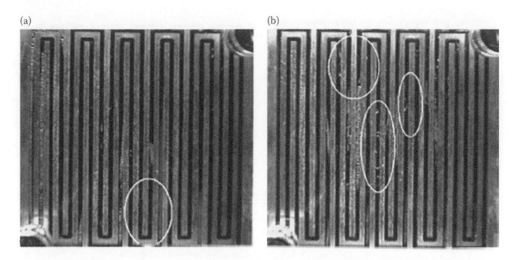

FIGURE 10.6 Images of cathode gas-flow channels, operated with humidified air at a constant electronic load of 300 mA cm^{-2} and a stoichiometry of (a) 2 and (b) 20. (Reprinted from *J. Power Sources*, 157, Weng, F.-B. et al. Study of water-flooding behaviour in cathode channel of a transparent proton-exchange membrane fuel cell. 674–680, Copyright (2006), with permission from Elsevier.)

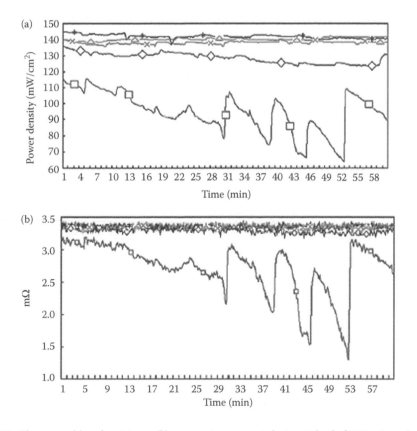

FIGURE 10.7 The power (a) and resistance (b) curves at a constant electronic load of 300 mA cm^{-2} with nonhumidified oxygen at a stoichiometry of: (⋆) 2; (△) 4; (×) 6; (◇) 10; (□) 20. (Reprinted from *J. Power Sources*, 157, Weng, F.-B. et al. Study of water-flooding behaviour in cathode channel of a transparent proton-exchange membrane fuel cell, 674–680, Copyright (2006), with permission from Elsevier.)

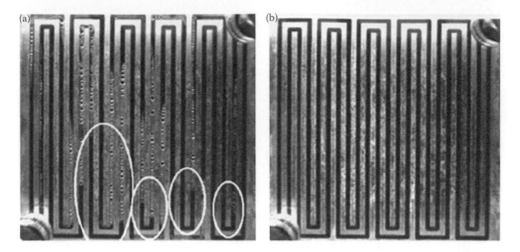

FIGURE 10.8 Images of cathode gas-flow channels operated with nonhumidified oxygen at a constant electronic load of 300 mA cm^{-2} and a stoichiometry of (a) 2, and (b) 20. (Reprinted from *J. Power Sources*, 157, Weng, F.-B. et al. Study of water-flooding behaviour in cathode channel of a transparent proton-exchange membrane fuel cell, 674–680, Copyright (2006), with permission from Elsevier.)

10.3.1.4 Effects of Concentration and Humidification on Cell Voltage Fluctuation

The effect of gas concentration and humidification on fuel cell performance under three different conditions (humidified oxygen at 70°C, humidified air at 70°C, and nonhumidified oxygen) is shown in Figure 10.9. The current remains constant in each experiment, meaning increased voltage corresponds to improved overall cell performance. The output voltage of the humidified oxygen is slightly higher than the other two conditions, as shown in Figure 10.9. The behavior for all conditions is similar when the stoichiometry is below 10 as the cathode gas cannot easily remove water. Because of this, water is still present in the MEA to support the reaction. According to this result, water removal has little influence on the performance of the fuel cell at certain stoichiometries (a value of 10 in the present study). This reveals that at stoichiometries of less than 10, the control of parameters such as concentration and humidification is important.

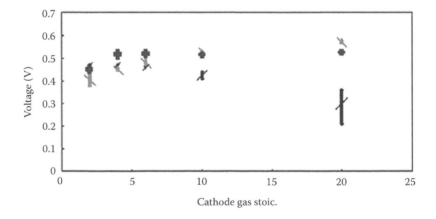

FIGURE 10.9 Fluctuation of voltage under different conditions at a constant electronic load of (\\) 300 mA cm^{-2} with humidified air; (/) 300 mA cm^{-2} with nonhumidified oxygen; (■) 700 mA cm^{-2} with humidified oxygen. (Reprinted from *J. Power Sources*, 157, Weng, F.-B. et al. Study of water-flooding behaviour in cathode channel of a transparent proton-exchange membrane fuel cell, 674–680, Copyright (2006), with permission from Elsevier.)

10.3.2 Optimizing Components and Flow Channels

A long, serpentine oxidant channel layout, along with proper design of the channel dimensions, has been employed as an effective strategy for removing liquid water. The combination of convective forces and high-pressure drop along the channel length drives out excess liquid water (Knobbe et al., 2004; Pasaogullari and Wang, 2005). Most studies are limited to traditional flow channel systems such as serpentine and straight channels; however, interdigitated flow channels have also been used (Weng et al., 2008a). This novel flow channel resembles veins in a leaf, allowing the nonuniform pressure distribution found in conventional interdigitated flow channels to be overcome.

In addition to channel layout strategies, the incorporation of special hydrophilic wicking structures into cathode flow channels have also been proposed to redistribute liquid water (Eckl et al., 2004; Ge et al., 2004; Ge et al., 2005a). Further research into flow-field design has been demonstrated by UTC fuel cells (Miachon and Aldebert, 1995; Meyer et al., 1996; Yi et al., 2004) for fuel cell stacks, where a porous bipolar plate is used as the water transport plate (WTP). This plate is not permeable to gas, preventing ingestion into the coolant streams while allowing excess liquid water to be wicked out through the coolant passage network due to the predetermined pressure difference between the gas and the coolant flow field. Other strategies regarding flow-field design involve the use of a porous gas distributor plate (Eckl et al., 2004; Ge et al., 2005a), and the inclusion of several sequential regions across the WTP that have different functions related to water removal and humidification (Vanderborgh and Hedstrom, 1990).

Water flooding in the GDL under ribs is usually more serious than elsewhere at the cathode. Therefore, enhancing the convection through the GDL by optimizing the flow field is an effective way to reduce water flooding, enhance the mass transport, and improve the cell performance and operating stability (Xu and Zhao, 2007). Kim et al. (2005) found that the undershoot of the final steady-state current that follows the resulting overshoot was observed with a standard triple-path serpentine flow field (SFF) but not with a single-path SFF. The results also show that the dimensionless peak current and the percentage of the overshoot current depend on both the starting cell voltage and the range of the voltage change. Both channel geometry and surface properties were seen to have appreciable effects on the volume of accumulated water and on the morphology of water droplets retained in the flow-field channels. Both for rectangular and triangular channels with the same cross-sectional area, channel-level water accumulation is seen to reduce with a PTFE coating, which provides a static contact angle of 95°. For a given flow-field surface energy, the triangular channels retained less water (Owejan et al., 2007). Wang et al. (2007) proposed a novel serpentine-baffle flow-field (SBFF) design to improve the cell performance compared to that of a conventional SFF. At high operating voltages both the conventional and the baffled design perform similarly; however, at low operating voltages, the baffled design displays improved performance. Analyses of the local transport phenomena in the cell indicate that the baffled design induces larger pressure differences between adjacent flow channels over the entire electrode surface, enhancing under-rib convection through the porous electrode layer, and improving the limiting current density and overall cell performance.

An active water management system that decouples water removed from oxidant delivery uses a porous carbon flow-field plate as an integrated wick that can passively redistribute water within the fuel cell. The system also employs an external EO pump that actively removes excess water from channels and the GDL (Litster et al., 2007).

The effect of an added microporous layer at the cathode on the net water transport in a PEMFC was also investigated. A key observation was that the inclusion of an MPL on the cathode improves the overall performance and durability of cells (Karan et al., 2007). Water accumulates at the membrane/electrode interface until sufficient hydrostatic pressure builds up and overcomes the water/GDL interfacial tension. After a water drop detaches, it moves freely along the surfaces in the gas flow channel, combining with other drops. Flooding occurs when the slugs of liquid water hinder the gas transport downstream and result in the starvation of reactants to the membrane/electrode interface. The motion of the liquid slug is gravity dependent (Kimball et al., 2008). With a combination of fluorescence microscopy and liquid pressure drop measurements, it was found that with an initially dry GDL and gas channel apparatus, the emergence and detachment of individual droplets is followed by slug formation and channel flooding (Bazylak et al., 2008). Fluorescence microscopy is applied to provide *ex situ* visualizations, showing that the compressed GDL provides preferential pathways for water transport and breakthrough. SEM images clearly illustrating the degradation of the GDL under varying compression pressures were presented by Bazylak (2009).

The water distribution in the GDL strongly depends on the water production rate and up to two diffusion blockages caused by liquid water were detected at high-current densities. To maintain a satisfactory water content and distribution within the MEA, a novel GDL was designed by inserting a water management layer (WML) between the traditional GDL and the CL in the PEMFC. A simulator was also developed to optimize the GDL, in which the water distribution in the electrode and profile of the water transport in the polymer membrane can be predicted (Chen et al., 2004).

10.3.3 Aging Phenomena under Transient Loading Conditions

Hsu et al. (2009) studied the transient loading of a PEMFC by step switching the current or voltage to understand the fuel cell's response rate under different humidified gas temperatures and flow rates. Two mechanisms were found to retard the response rate: obstruction of the reaction path due to the hydrated GDL, and mass transfer polarization rendering the gas supply insufficient when the gas stoichiometry was low.

Figure 10.10 shows the transient loading curve with a 5 cm^2 transparent PEMFC switching between 0.8 and 0.6 V. The total experiment time was 60 h, with 36,000 cycles. Each cycle lasted a total of 6 s, comprising 3 s at 0.8 V and 3 s at 0.6 V. This experiment focused on observing the water flooding patterns in the flow channels. The accelerated aging was also studied using electrochemical diagnostic methods, which were recorded during the experiment. The PEMFC performance was steady for the first 15 h, but the stability was then lost.

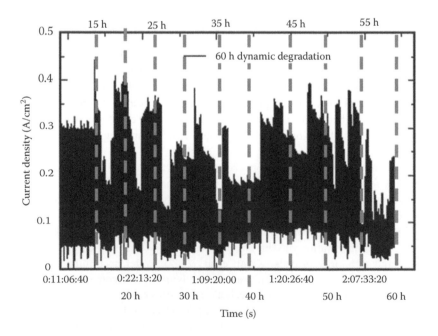

FIGURE 10.10 Dynamic loading curves from 0 to 60 h (0–36,000 cycles).

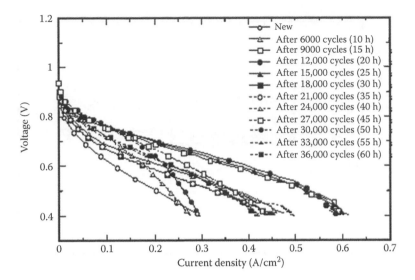

FIGURE 10.11 Polarization curves at different experiment times.

The initial conditions gave rise to the lowest performance, which then increased with time, as shown in Figure 10.11. However, the polarization curves can be divided into three distinct regions. The cell performance peaks after 15 h, then decreases.

Figure 10.12 shows the linear sweep voltammeter (LSV) curves. The limited current does not display an obvious increase due to the advantageous membrane state, suggesting that the membrane does not have large pores. This result shows that transient loading has a lesser negative effect on the membrane.

Figure 10.13 shows the cyclic voltammeter (CV) curves for different cycles, in which the hydrogen adsorption area becomes obviously small after 15 h. This supports the suggestion that catalyst particles

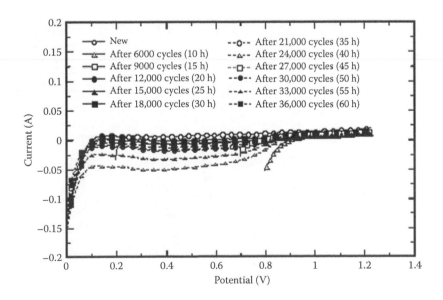

FIGURE 10.12 LSV curves at different experiment times.

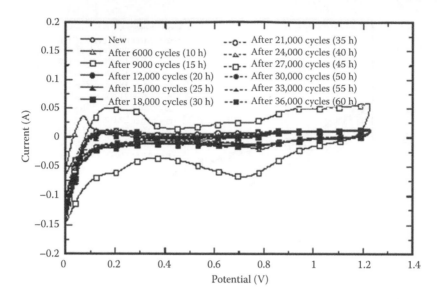

FIGURE 10.13 CV curves at different experiment times.

will cluster as a result of transient loading, decreasing the catalyst surface area, and giving rise to the decay seen in the polarization curves in Figure 10.10.

Figure 10.14 shows the flooding pattern in the flow channel after 25 h. Obvious traces of liquid water cannot be found in the flow channel, showing that the GDL remains dehydrated and can still remove liquid water easily. This accounts for the steady performance shown in Figure 10.10, especially before 15 h. As experimental time increases, the performance becomes unsteady and declines due to dehydration in the GDL. Figure 10.15 shows that the water droplets (shown as white particles) move within the flow channels, implying poor GDL hydration that results in more water in the GDL. The cell performance becomes increasingly unsteady as shown in Figure 10.10, especially after 60 h.

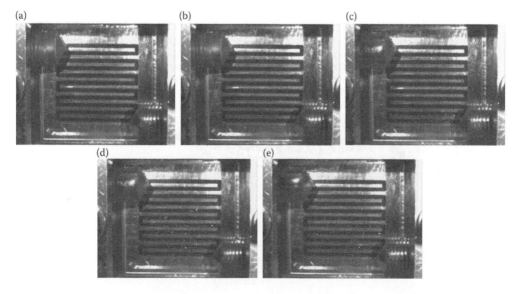

FIGURE 10.14 Snapshots of the flooding process after 25 h: (a) 0; (b) the 4th; (c) the 8th; (d) the 12th; and (e) the 14th second.

FIGURE 10.15 Snapshots of the flooding process after 60 h: (a) 0; (b) the 2nd; (c) the 4th; (d) the 6th; and (e) the 8th second.

10.3.4 Transparent Cells Integrated with Other Diagnostic Tools

The measurement of local currents in PEMFCs provides an important tool for the diagnosis and development of fuel cells. A segmented cell was developed to serve as an essential instrument to investigate the different operating conditions in the cells and stacks of technical relevance. Regional-averaged current through each cell segment can be determined using a Hall-effect sensor. By probing the high-frequency internal resistance and performance of this cell, the effects of fuel and oxidant flow rates, RH, and directional channel flows were investigated to analyze the performance and stability of local segmented regions. The results of these experiments demonstrate that the local current distribution is strongly influenced by the RH of fuel, the stoichiometry of the processed air, and the cell operational mode (Weng et al., 2008a,b).

Local current distribution in a PEM fuel cell has been mapped experimentally using a custom-designed single-cell fixture. An array of 16 individual conductive segments was distributed on the composite plate. Then, the effects of flow-field patterns, dew points for the cathodic feedings, and cathodic stoichiometrics on the local current distribution were examined. The transient variation of the local current distribution on the cathode under supersaturated conditions was further visualized to illustrate the flooding phenomena in different flow patterns (Hwang et al., 2008). For local current density measurements, there are two different feasible approaches: (1) construction of a model cell, optimized for local current density measurements to ensure accuracy and resolution, and (2) adopting a real cell with the necessary instrumentation (Büchi et al., 2005).

Anode water removal can be used as a diagnostic technique to remove water in the cathode without directly affecting other conditions. A water management technique has been developed for solid polymer electrolyte fuel cells, demonstrating both performance and diagnostic benefits. Using this technique a substantial proportion of the water in the cathode can be removed through the anode fuel stream (Voss et al., 1995).

Another novel diagnostic test method provides an insight into the distribution of liquid water in an operating fuel cell. In this method, the amount and distribution of liquid water in a fuel cell is quantified indirectly through the measurement of certain physical and chemical properties (Stumper et al., 2005).

An electrode flooding monitoring device designed for PEM fuel cells with interdigitated flow distributors has also been proposed and tested. The pressure drop between the inlet and outlet channels can be used as a diagnostic signal to monitor the liquid water content in the porous electrodes due to its strong influence on electrode gas permeability (He et al., 2003). Ma et al. (2006) used a transparent PEMFC to study the correlation between liquid water removal and the pressure drop between the channel inlets and outlets (ΔP). Pressure-drop measurements have been established as a suitable diagnostic tool to determine the effects of gas velocity on liquid water removal in a straight channel.

Hartnig et al. (2009) investigated the cross-sectional transport of liquid water in porous gas diffusion materials as employed in low-temperature fuel cells by means of synchrotron x-ray radiography with a spatial resolution of 3 μm and a time resolution of 5 s.

Extensive neutron imaging experiments were performed using several different flow-field geometries with varying landing-to-channel (L:C) ratios. In these methods, the liquid water accumulation and residual water content in the fuel cell were analyzed with relatively dry and fully humidified inlet conditions. The results indicate: (1) for an L:C ratio of one, the liquid water tends to preferentially accumulate under the landings rather than in, or under the channels, (2) as the L:C ratio reduces, the liquid stored in the cell decreases, with optimal conditions reached when the L:C ratio becomes smaller than 2:3, and (3) under dry operation, a high L:C ratio can be helpful, whereas at higher humidities, a low L:C is preferred (Turhan et al., 2008). The total pressure drop in the flow channels mainly depends on the resistance of the liquid water in the flow channels to gas flow, and different flow patterns distinguish the total pressure drop in the flow field. Clogging by water columns results in a high-pressure drop in the flow channels, particularly in the cathode, where the amount of liquid water in the cathode flow channels becomes much greater than in the anode flow channels and thus exhibits a higher pressure drop (Liu et al., 2007a). Hakenjos et al. (2004) designed a special fuel cell that combined three methods of spatially resolved measurements. This experimental setup enables a simultaneous evaluation of current, temperature, and water distribution in an operating PEMFC. Additionally, infrared (IR) thermography was used to record the temperature distribution on the active area, and digital photography allowed for the optical surveillance of condensed water within the flow field.

Increasing cell temperature can enhance the overall performance by improving the electrochemical kinetics and activity of the catalyst, and reducing the condensation of liquid water to retain mass transport performance (Liu et al., 2008). The liquid water distribution on the GDL surface and inside the gas channel can be quantified in an operating transparent PEMFC. Liquid droplet formation and emergence from the GDL surface are characterized and two modes of liquid water removal from the GDL surface are identified: droplet detachment by the shear force arising from the core gas flow, followed by a mist flow in the gas channel and capillary wicking onto the more hydrophilic channel walls, and annular film flow and/or liquid slug flow in the channel (Zhang et al., 2006). Experimental results indicate that the liquid water columns accumulating in the cathode flow channels can reduce the effective electrochemical reaction area by limiting mass transfer, which results in cell performance loss. When water flooding begins, an increase in the cathode flow rate can remove excess water and prevent this performance loss (Liu et al., 2006).

10.4 Recent Advances and Future Outlooks

Water management is of key importance in order to ensure high performance and durability for PEMFCs. Liquid water has been found to accumulate and block the flow channel, and is referred to as channel flooding or channel clogging. Due to excellent spatial and temporal resolution, optical visualization is a powerful and convenient method that is particularly suited for studying two-phase flow regimes in flow channels.

However, little work has been done on liquid water behavior inside the GDL and CL. Ge and Wang (2006, 2007) used a transparent PEMFC to study liquid water and ice formation during start-up from subzero temperatures. It was found that at a current density of 0.02 A cm^{-2} and a start-up temperature of −5°C, water in the CCL existed in solid and gas phases. At startup temperatures of greater than −3°C, water droplets were found on the CL surface and the cold-start operation was significantly prolonged. *In situ* observations revealed that liquid water in the anode channels resulted from condensation on the cooler and more hydrophilic channel walls, with the water vapor coming either from the cathode through membrane transport or from hydrogen consumption. No water droplets were found on the anode GDL surface, in sharp contrast with the cathode side. Oxygen consumption and H_2O_2 formation, as well as the local catalyst have been investigated and visualized in a solution with a scanning electron microscope.

Despite considerable recent advances, existing fuel cell technology still has drawbacks, such as the kinetic limitations on the oxygen reduction reaction, and the instability of Pt catalysts and polymer membranes. An oxygen-sensitive porphyrin, tetrakis porphyrinatoplatinum, is used in the transparent fuel cell system. This complex dye is dispersed in an oxygen-permeable polymer matrix, poly (1-trimethylsily-1-propune), to create a thin, water-insoluble dye film. By combining other dye materials sensitive to temperature, CO_2 or infrared absorption for water vapor, simultaneous visualization of these parameters can be achieved in further in-depth diagnosis of fuel cell performance (Inukai et al., 2008).

In the future, the transparent cell will still be an important diagnostic tool to investigate the effects of water flooding on cell performance and durability. Utilizing this tool in conjunction with optical methods to study the CL and GDL with advanced spatial and temporal resolution will be important to improve fuel cell durability. In addition, these methods can be integrated with other diagnostic tools, such as electrochemical methods and micro electromechanical systems (MEMS) humidity sensors to quantitatively analyze flooding in transparent cells.

References

Anderson, R., Zhang, L., Ding, Y. et al. 2010. A critical review of two-phase flow in gas flow channels of proton exchange membrane fuel cells. *J. Power Sources* 195: 4531–4553.

Bazylak, A. 2009. Liquid water visualization in PEM fuel cells: A review. *Int. J. Hydrogen Energy* 34: 3845–3857.

Bazylak, A., Sinton, D., and Djilali, N. 2008. Dynamic water transport and droplet emergence in PEMFC gas diffusion layers. *J. Power Sources* 176: 240–246.

Bazylak, A., Sinton, D., Liu, Z. S. et al. 2007. Effect of compression on liquid water transport and microstructure of PEMFC gas diffusion layers. *J. Power Sources* 163: 784–792.

Büchi, F. N., Geiger, A. B., and Neto, R. P. 2005. Dependence of current distribution on water management in PEFC of technical size. *J. Power Sources* 145: 62–67.

Cai, Y., Ma, J. H., Yi, B. et al. 2006. Effect of water transport properties on a PEM fuel cell operating with dry hydrogen. *Electrochem. Acta* 51: 6361–6366.

Chen, J., Matsuura, T., and Hori, M. 2004. Novel gas diffusion layer with water management function for PEMFC. *J. Power Sources* 131: 155–161.

Eckl, R., Zehtner, W., Leu, C. et al. 2004. Experimental analysis of water management in a self-humidifying polymer electrolyte fuel cell stack. *J. Power Sources* 138: 137–144.

Ge, S. and Wang, C. Y. 2006. *In situ* imaging of liquid water and ice formation in an operating PEFC during cold start. *Electrochem. Solid-State Lett.* 9: A499–A503.

Ge, S. and Wang, C. Y. 2007. Liquid water formation and transport in the PEFC anode. *J. Electrochem. Soc.* 154: B998–B1005.

Ge, S., Li, X., and Hsing, I. M. 2005a. Internally humidified polymer electrolyte fuel cells using water absorbing sponge. *Electrochem. Acta* 50: 1909–1916.

Ge, S., Li, X., Yi, B. et al. 2005b. Absorption, desorption, and transport of water in polymer electrolyte membranes for fuel cells. *J. Electrochem. Soc.* 152: A1149–A1157.

Ge, S. H., Li, X. G., and Hsing, I. M. 2004. Water management in PEMFCs using absorbent wicks. *J. Electrochem. Soc.* 151: B523–B528.

Hakenjos, A., Muenter, H., Wittstadt, U. et al. 2004. A PEM fuel cell for combined measurement of current and temperature distribution, and flow field flooding. *J. Power Sources* 131: 213–216.

Hartnig, C., Manke, I., Kuhn, R. et al. 2009. High-resolution in-plane investigation of the water evolution and transport in PEM fuel cells. *J. Power Sources* 188: 468–474.

He, W., Lin, G., and Nguyen, T. V. 2003. Diagnostic tool to detect electrode flooding in proton-exchange-membrane fuel cells. *AIChE J.* 49: 3221–3228.

Hsu, C.-Y., Weng, F.-B., Su, A. et al. 2009. Transient phenomenon of step switching for current or voltage in PEMFC. *Renewable Energy* 34: 1979–1985.

Hwang, J. J., Chang, W. R., Peng, R. G. et al. 2008. Experimental and numerical studies of local current mapping on a PEM fuel cell. *Int. J. Hydrogen Energy* 33: 5718–5727.

Inukai, J., Miyatake, K., Takada, K. et al. 2008. Direct visualization of oxygen distribution in operating fuel cells. *Angew. Chem. Int. Ed.* 47: 2792–2795.

Karan, K., Atiyeh, H., Phoenix, A. et al. 2007. An experimental investigation of water transport in PEMFCs. *Electrochem. Solid-State Lett.* 10: B34–B38.

Kim, S., Shimpalee, S., and Zee, J. W. V. 2005. Effect of flow field design and voltage change range on the dynamic behavior of PEMFCs. *J. Electrochem. Soc.* 152: A1265–A1271.

Kimball, E., Whitaker, T., Kevrekidis, Y. G. et al. 2008. Drops, slugs, and flooding in polymer electrolyte membrane fuel cells. *AIChE J.* 54: 1313–1332.

Knobbe, M. W., He, W., Chong, P. Y. et al. 2004. Active gas management for PEM fuel cell stacks. *J. Power Sources* 138: 94–100.

Lin, G. and Nguyen, T. V. 2005. Effect of thickness and hydrophobic polymer content of the gas diffusion layer on electrode flooding level in a PEMFC. *J. Electrochem. Soc.* 152: A1942–A1948.

Litster, S., Buie, C. R., Fabian, T. et al. 2007. Active water management for PEM fuel cells. *J. Electrochem. Soc.* 154: B1049–B1058.

Liu, X., Guo, H., and Ma, C. 2006. Water flooding and two-phase flow in cathode channels of proton exchange membrane fuel cells. *J. Power Sources* 156: 267–280.

Liu, X., Guo, H., Ye, F. et al. 2007a. Water flooding and pressure drop characteristics in flow channels of proton exchange membrane fuel cells. *Electrochem. Acta* 52: 3607–3614.

Liu, X., Guo, H., Ye, F. et al. 2008. Flow dynamic characteristics in flow field of proton exchange membrane fuel cells. *Int. J. Hydrogen Energy* 33: 1040–1051.

Liu, F., Lu, G., and Wang, C. Y. 2007b. Water transport coefficient distribution through the membrane in a polymer electrolyte fuel cell. *J. Membr. Sci.* 287: 126–131.

Ma, H. P., Zhang, H. M., Hu, J. et al. 2006. Diagnostic tool to detect liquid water removal in the cathode channels of proton exchange membrane fuel cells. *J. Power Sources* 162: 469–473.

Meyer, A. P., Scheffler, G. W., and Margiott, P. R. 1996. Water management system for solid polymer electrolyte fuel cell power plants. *US Patent* Ch.5, 503,994.

Miachon, S. and Aldebert, P. 1995. Internal hydration H_2/O_2 100 cm^2 polymer electrolyte membrane fuel cell. *J. Power Sources* 56: 31–36.

Nam, J. H., Lee, K. J., Hwang, G. S. et al. 2009. Microporous layer for water morphology control in PEMFC. *Int. J. Heat Mass Transfer* 52: 2779–2791.

Niroumand, A. M. and Saif, M. 2010. Two-phase flow measurement system for polymer electrolyte fuel cells. *J. Power Sources* 195: 3250–3255.

Ohonen, J., Mikkola, M., and Lindbergh, G. 2004. Flooding of gas backing in PEFCs. *J. Electrochem. Soc.* 151: A1152–A1161.

Okada, T., Xie, G., Gorseth, O. et al. 1998. Ion and water transport characteristics of Nafion membrane as electrolytes. *Electrochem. Acta* 43: 3741–3747.

Owejan, J. P., Trabold, T. A., Jacobson, D. L. et al. 2007. Effect of flow field and diffusion layer properties on water accumulation in a PEM fuel cell. *Int. J. Hydrogen Energy* 32: 4489–4502.

Park, S., Lee, J. W., and Popov, B. N. 2008. Effect of PTFE content in microporous layer on water management in PEM fuel cells. *J. Power Sources* 177: 457–463.

Pasaogullari, U. and Wang, C. Y. 2004. Liquid water transport in gas diffusion layer of polymer electrolyte fuel cells. *J. Electrochem. Soc.* 151: A399–A406.

Pasaogullari, U. and Wang, C. Y. 2005. Two-phase modeling and flooding prediction of polymer electrolyte fuel cells. *J. Electrochem. Soc.* 152: A380–A390.

Spernjak, D., Prasad, A. K., and Advani, S. G. 2007. Experimental investigation of liquid water formation and transport in a transparent single-serpentine PEM fuel cell. *J. Power Sources* 170: 334–344.

Stumper, J., Löhr, M., and Hamada, S. 2005. Diagnostic tools for liquid water in PEM fuel cells. *J. Power Sources* 143: 150–157.

Su, A., Weng, F.-B., Hsu, C.-Y. et al. 2006. Studies on flooding in PEM fuel cell cathode channels. *Int. J. Hydrogen Energy* 31: 1031–1039.

Sugiura, K., Nakata, M., Yodo, T. et al. 2005. Evaluation of a cathode gas channel with a water absorption layer/waste channel in a PEFC by using visualization technique. *J. Power Sources* 145: 526–533.

Theodorakakos, A., Ous, T., Gavaises, M. et al. 2006. Dynamics of water droplets detached from porous surfaces of relevance to PEM fuel cells. *J. Colloid Interface Sci.* 300: 673–687.

Tüber, K., Pócza, D., and Hebling, C. 2003. Visualization of water buildup in the cathode of a transparent PEM fuel cell. *J. Power Sources* 124: 403–414.

Turhan, A., Jeller, K., Brenizer, J. S. et al. 2008. Passive control of liquid water storage and distribution in a PEFC through flow-field design. *J. Power Sources* 180: 773–783.

Vanderborgh, N. E. and Hedstrom, J. C. 1990. Fuel cell water transport. US Patent Ch. 4,973,530.

Voss, H. H., Wilkinson, D. P., Pickup, P. G. et al. 1995. Anode water removal: A water management and diagnostic technique for solid polymer fuel cells. *Electrochem. Acta* 40: 321–328.

Wang, X. D., Duan, Y. Y., and Yan, W. M. 2007. Novel serpentine-baffle flow field design for proton exchange membrane fuel cells. *J. Power Sources* 173: 210–221.

Weng, F.-B., Hsu, C.-Y., Chan, S.-H. et al. 2008a. Experimental investigation of polymer electrolyte membrane fuel cells with ramification flow fields. *Proc. IMechE Part A: J. Power Energy* 22: 771–779.

Weng, F.-B., Jou, B.-S., Li, C.-W. et al. 2008b. The effect of low humidity on the uniformity and stability of segmented PEM fuel cells. *J. Power Sources* 181: 251–258.

Weng, F.-B., Su, A., and Hsu, C.-Y. 2007. The study of the effect of gas stoichiometric flow rate on the channel flooding and performance in a transparent fuel cell. *Int. J. Hydrogen Engery* 32: 666–676.

Weng, F.-B., Su, A., Hsu, C.-Y. et al. 2006. Study of water-flooding behaviour in cathode channel of a transparent proton-exchange membrane fuel cell. *J. Power Sources* 157: 674–680.

Weng, F.-B., Su, A., Hsu, C.-Y. et al. 2010. Experimental investigation of PEM fuel cell aging under current cycling using segmented fuel cell. *J. Power Sources* 35: 3664–3675.

Xu, C. and Zhao, T. S. 2007. A new flow field design for polymer electrolyte-based fuel cells. *Electrochem. Commun.* 9: 497–503.

Yang, X. G., Zhang, F. Y., Lubawy, A. L. et al. 2004. Visualization of liquid water transport in a PEFC. *Electrochem. Solid-State Lett.* 7: A408–A411.

Yi, J. S., Yang, J. D., and King, C. 2004. Water management along the flow channels of PEM fuel cells. *AIChE J.* 50: 2594–2603.

Zhang, F. Y., Yang, X. G., and Wang, C. Y. 2006. Liquid water removal from a polymer electrolyte fuel cell. *J. Electrochem. Soc.* 153: A225–A232.

11

Magnetic Resonance Imaging

Ziheng Zhang
Yale University School of Medicine

Bruce J. Balcom
University of New Brunswick

11.1 Introduction

Water plays a crucial role in the operation of all polymer electrolyte membrane (PEM) fuel cells. Water is the main byproduct of power generation in the fuel cell and water is the basic carrier of protons through the membrane. Water distribution across the membrane, Nafion for our purposes in this chapter, in an operational fuel cell varies with the rate of water formation and removal, and directly determines the cell's reliability and efficiency (Eikerling, 2006; Springer et al., 1991; Zawodzinski et al., 1993). Low relative humidity in the membrane will exacerbate degradation of the fuel cell (Shim et al., 2007). Thus, direct monitoring of the water content distribution, spatially and temporally, in an operating fuel cell is critical to understanding PEM fuel cell behavior, and further developing fuel cell technology (US Department of Energy, 2003). Many imaging techniques have been proposed for water content visualization in fuel cells (US Department of Energy, 2003).

Magnetic resonance imaging (MRI) is among the most promising such techniques because it directly detects ^{1}H nuclei in the water phase, which in principle permits direct noninvasive visualization and measurement of water content behavior inside an operating fuel cell. Measurements can be designed for one-, two-, or three-dimensional studies, with a variety of contrasts depending on the specific research question (Haacke et al., 1999). Such measurements are not limited to steady-state examination, since measurements are rapid enough in many instances to time resolve dynamic features associated with fuel cell operation (Zhang et al., 2008a,b). MRI has principally been employed in fuel cell studies to probe in-plane and through-plane water content in the Nafion membrane in addition to water dynamics in the gas flow channels.

Water in the Nafion membrane is transported through two primary mechanisms: electroosmotic drag of water by protons transported from the anode to the cathode, and diffusion of water along water concentration gradients (Springer et al., 1991; Zawodzinski et al., 1993). Electroosmotic drag and the oxidation reaction tend to create an excess of water at the cathode, which causes deleterious "flooding." The accumulation of excess liquid water in the gas diffusion layer (GDL) and gas flow channels may block the channels and locally poison the cell. In addition to these mass transport limitations, excess liquid water can also lead to nonhomogeneous current density and membrane swelling (Wielstich et al., 2003). All the above phenomena can be examined using MRI, and have been investigated by a number of research groups worldwide. The most active groups have been Tsushima and coworkers (Ikeda et al., 2008; Shim et al., 2007, 2009; Teranishi et al., 2002; Tsushima et al., 2005a,b, 2007; Tsushima and Hirai, 2009) and Wayslishen and coworkers (Feindel et al., 2004, 2007a,b; Wang et al., 2010). Representative publications for each group are referenced in this chapter but the publication list for each is incomplete.

Water management aims at balancing the water content across the membrane electrode assembly (MEA) by controlling the relative humidity of the feed gases and thereby improving the fuel cell performance. Water management relies on understanding the proton conductivity, water diffusion coefficient, and electroosmotic drag coefficient—all as functions of water content in the membrane. Numerous phenomenological models of PEM behavior in fuel cells have been created (Springer et al., 1991; Zawodzinski et al., 1993). It is critical to develop an efficient measurement to directly characterize the water distribution and/or transport across the PEM in an operational PEM fuel cell, in order to validate and guide these models. While MRI seems a natural technique for such studies, it does have several important limitations, not always fully appreciated, as will be outlined in this chapter.

After introducing the basic principles and hardware of magnetic resonance and MRI, this chapter will describe specific PEM fuel cell MRI measurements including cell design, construction, and materials selection. A literature review of MRI applications to fuel cells is provided. A summary of the advantages and disadvantages of MRI for PEM fuel cell studies concludes the chapter.

11.2 Theory of Magnetic Resonance and Magnetic Resonance Imaging

Magnetic resonance is a quantum mechanical phenomenon that arises from quantization of spin angular momentum, and thereby quantization of the magnetic moment of individual nuclei. A detailed explanation of the quantum mechanics of magnetic resonance is however beyond the scope of this chapter. A large number of reference books cover the subject (Abragam, 1961; Levitt, 2001; Slichter, 1990).

In this chapter, we will introduce, building on magnetic resonance basics, the salient features of MRI for PEM fuel cell studies. The MRI background provided is not meant to be an exhaustive survey of the subject. A large number of reference sources describe MRI in detail, and the interested reader is referred to one of the many available texts (Callaghan, 1991; Haacke et al., 1999; Vlaardingerbroek and den Boer, 1999).

11.2.1 Concepts of Magnetic Resonance

11.2.1.1 Magnetic Resonance

Nuclei such as ^1H in a macroscopic sample of water (H_2O) exposed to a static magnetic field B_0 tend to align their individual magnetic moments either partially parallel or partially antiparallel to the direction of the static field. The population of these two states differ according to a Boltzmann distribution with the lower-energy state preferred (parallel) but the high-energy state (antiparallel) populated to a near equivalent extent. The very similar populations result from the thermal energy kT, at realistic temperatures, being large compared to the interaction energy of individual nuclei with the static field.

If the total population of ¹H nuclei is N then N_p is the number of nuclei with a parallel orientation while, N_a, is the number antiparallel. The number of excess spins oriented parallel to the static field B_0 is given by $\Delta N = N_p - N_a = N(\gamma hB_0 / 2kT)$, where k is Boltzmann's constant, and T the temperature. The equilibrium sample magnetization M_0 is

$$\vec{M}_0 = N\frac{\gamma^2 h^2 I(I+1)}{4\pi kT}\vec{B}_0. \tag{11.1}$$

In Equation 11.1 the spin quantum number I appears, $I = 1/2$ for ¹H, as does the gyromagnetic ratio γ which is characteristic of the nuclei under study. The constant h in Equation 11.1 is Planck's constant. Note that equilibrium sample magnetization is directly proportional to the quantity of material in the sample through N the number of ¹H nuclei.

The energy difference ΔE between the parallel and antiparallel states, illustrated in Figure 11.1, is given by Equation 11.2 where μ is the magnetic moment of nuclei in the parallel and antiparallel states.

$$\Delta E = \left(\vec{\mu}_{N_a} - \vec{\mu}_{N_p}\right)\cdot\vec{B}_0 = \gamma\left(\frac{1}{2}\hbar - \left(-\frac{1}{2}\hbar\right)\right)B_0 = \gamma\hbar B_0 = \hbar\omega_0 \tag{11.2}$$

The relation $\omega_0 = \gamma B_0$, from Equation 11.2, is the Larmor equation. With reference to Figure 11.1 it suggests that electromagnetic radiation of frequency ω_o is required to cause transitions of nuclei between the two energy states. For typical magnetic field strengths of a few Tesla (T), the resonance frequencies are radio frequency band, with 2.4 T yielding a frequency of 100 MHz.

11.2.1.2 Spin Excitation and Relaxation

The quantum mechanical view of Equation 11.2 is less convenient in terms of understanding a magnetic resonance experiment than a more classical approach where we think of excitation and manipulation of the sample magnetization from Equation 11.1 by alternating magnetic fields. Application of an appropriately phased radio frequency magnetic field B_1, at the Larmor frequency, results in rotation of the sample magnetic moment away from the z-axis, defined by the direction of the static field B_0, toward the transverse plane as illustrated in Figure 11.2. The extent of rotation, flip angle α, is determined by the duration of the radio frequency pulse excitation and the strength of the associated magnetic field.

The pulsed RF excitation is generated by a radio frequency probe, which is a tuned electrical circuit that generates an alternating magnetic field in the sample space. It is commonly but mistakenly assumed that the radio frequency probe generates an electromagnetic wave with orthogonal magnetic and electric field components. The radio frequency probe is a near-field device where the sample space experiences solely an alternating magnetic field (Hoult, 1978).

The transverse sample magnetization created by the radio frequency excitation pulse will precess about the z-axis due to the interaction between the sample magnetic moment and the static field B_0. The

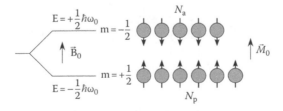

FIGURE 11.1 The Zeeman energy levels with potential energies of $\pm(1/2)\hbar\omega_0$ for a spin 1/2 system. The difference in spin population between N_p and N_a in an external magnetic field $\vec{B}_0$ yields the equilibrium magnetization $\vec{M}_0$.

radio frequency probe switched to a receive mode detects the precessing sample magnetization, inducing a voltage in the probe which is the experimental magnetic resonance signal.

The sample magnetization rotated toward the transverse plane in Figure 11.2 is nonequilibrium and in the fullness of time the system must return to equilibrium. We can distinguish longitudinal and transverse components of the sample magnetization. These are $\vec{M}_z$ and $\vec{M}_{xy}$, respectively and they return to equilibrium values by relaxation processes characterized by different time constants. Recovery of equilibrium $\vec{M}_z$ magnetization occurs by a process known as spin–lattice relaxation with a time constant T_1. Decay of transverse sample magnetization occurs by a process known as spin–spin relaxation with a time constant T_2. Equations 11.3 and 11.4 describe these relaxation processes for recovery of an idealized system after a 90° RF excitation pulse.

$$M_z(t) = M_0\left(1 - e^{-t/T_1}\right) \tag{11.3}$$

$$M_{xy}(t) = M_0 e^{-t/T_2} \tag{11.4}$$

11.2.1.3 Free Induction Decay and Spin Echo

The transverse magnetization, M_{xy}, begins to lose phase coherence after the RF pulse at a rate generally faster than suggested by the T_2 time constant and Equation 11.4. The time constant describing this decay, after a single RF excitation pulse, is T_2^* and the experimental signal is known as a free induction decay (FID). The T_2^* time constant is shorter than T_2 because inhomogeneities in the static field B_0 yield local precession frequencies in the sample space that vary. This causes dephasing of the sample magnetization M_{xy} at an enhanced rate. The two time constants are related by, $1/T_2^* = 1/T_2 + \gamma\left\langle\left|\Delta B_0\right|\right\rangle$, with the last term describing a distribution of the static magnetic field in the sample space. Signal loss in the FID due to this inhomogeneous field is reversible using a spin-echo (SE) measurement.

A typical SE measurement employs a 180° RF pulse, a chosen time after the initial 90° pulse, to invert the phase accumulated in the inhomogeneous static magnetic field. The result is a rephasing signal after the 180° pulse. The time between the 180° degree pulse and the rephasing signal, an echo, is equal to the time between the initial two RF pulses. The time from the first RF pulse to the echo is sensibly termed the echo time (TE). This procedure may be repeated if desired, with repeated echoes acquired. The decay of the echo amplitude with time is governed by Equation 11.4 with the time variable replaced by the TE or some multiple of the TE for multiple echoes. The echo signal decay is frequently multiexponential if multiple ^{1}H environments exist in the sample, but may average to a single exponential decay if there is rapid molecular exchange between environments.

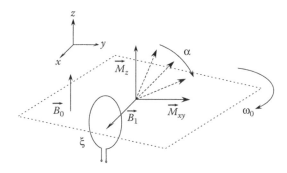

FIGURE 11.2 Longitudinal magnetization is rotated into the transverse plane by an RF pulse, flip angle, α. The transverse magnetization $\vec{M}_{xy}$, induces a voltage in a receiver coil tuned to the Larmor frequency. ξ is the voltage induced by the precessing $\vec{M}_{xy}$.

Fuel cell MRI experiments are almost entirely based on SE measurements since the heterogeneous environment of the fuel cell ensures an inhomogeneous static field. Molecular diffusion through these inhomogeneous fields, including the deliberately applied magnetic field gradients, may introduce an additional dephasing and signal loss mechanism in the SE experiment.

11.2.2 Magnetic Resonance Imaging

11.2.2.1 Concepts of MRI

From the Larmor equation, the resonance frequency of the ^{1}H MR experiment is proportional to the local magnetic field. A controlled linear magnetic field variation, termed a gradient $G(r)$, can be employed to spatially encode the spin frequency. The local magnetic field can be expressed as, $\vec{B} = \vec{B}_0 + \vec{G}(\vec{r})$. This manner of spatial encoding for MRI can be expanded into three dimensions and implicitly incorporates time as a variable,

$$\omega(\vec{r},t) = \omega_0 + \omega_G(\vec{r},t) = \omega_0 + \gamma \vec{r} \cdot \vec{G}(\vec{r},t), \tag{11.5}$$

where $\vec{r}$ and $\vec{G}(\vec{r},t)$ are the position vector and the magnetic field gradient, respectively. The frequency spectrum of the MRI signal obtained from the sample arises from ^{1}H resonance frequencies that vary linearly with r.

Assuming no MR relaxation effects, the total signal is comprised of the sum of the density of ^{1}H at each location, weighted by a spatially dependent phase term,

$$S(\vec{k}) = \iiint \rho(\vec{r}) \cdot e^{i2\pi \vec{k} \cdot \vec{r} + \Phi_0} d\vec{r}, \tag{11.6}$$

where Φ_0 is a phase offset and $\vec{k}$ is the reciprocal space vector, defined as

$$\vec{k} = \frac{\gamma \vec{G} t}{2\pi}. \tag{11.7}$$

This expression illustrates the Fourier transform relationship between the signal, $S(\vec{k})$ and the spin density, $\rho(\vec{r})$. The two domains r and k have a reciprocal relationship with r having natural units of cm while k has natural units of cm^{-1}.

In practice, k-space is sampled on a finite Cartesian grid with a sampling period $\Delta \vec{k}$, and maximum value $\vec{k}_{max}$ representing the highest spatial frequency. The k-space sampling interval is inversely related to the image field of view (FOV) FOV $= 1/\Delta k$ following Fourier transformation of the k-space data to reconstruct the image. The nominal image resolution is simply the FOV divided by the number of pixels. The number of pixels, in any dimension, is equal to the number of k-space data points acquired encoding that dimension. The number of data points is always a power of 2 with 64, 128, or 256 the most common.

The differential form of Equation 11.7 is

$$\Delta \vec{k} = \frac{\gamma}{2\pi} \left(\Delta \vec{G} \cdot t + \vec{G} \cdot \Delta t \right), \tag{11.8}$$

which suggests two possibilities for sampling k-space. The first, suggested by the term $\vec{G} \cdot \Delta t$, maintains the gradient constant while sampling the MR signal as a function of time, and is named frequency encoding. The second, suggested by the term $\Delta \vec{G} \cdot t$, keeps the encoding time constant while incrementing the gradient, and is named phase encoding.

Frequency encoding is typically faster than phase encoding but is more prone to image artifacts, particularly in the case of an inhomogeneous static magnetic field which may well result in nonlinear magnetic field gradients (Gravina and Cory, 1994). Phase encoding, while generally slower, in some instances will be more sensitive than frequency encoding, and it is very robust to image artifacts. Most importantly, it is largely insensitive to inhomogeneities in the static magnet field, which are a major concern in fuel cell MRI. The heterogeneous structure of even a simple PEM fuel cell, with substantial metal components, one anticipates will result in an inhomogeneous static field.

11.2.2.2 Two-Dimensional Spin-Echo MRI

Most conventional frequency encoding MRI techniques are based on echo measurements. Gradient echo measurements are common in clinical MRI (Reichenbach et al., 2005), but are very vulnerable to signal loss and image artifacts in systems with inhomogeneous static magnetic fields. All literature fuel cell MRI measurements are based on SE methods.

A representative two-dimensional spin-echo MRI measurement with frequency and phase encoding is illustrated in Figure 11.3. A slice selective 90° RF pulse is applied in conjunction with a slice selection gradient to make the three-dimensional problem immediately two dimensional. The 90° RF pulse rotates longitudinal magnetization into the transverse plane with precession occurring as described earlier. The pulsed magnetic field gradient in the frequency encode direction encodes spatial position in reciprocal space, k-space, as does the phase encode gradient, which is applied at different amplitudes in successive repetitions of the experiment. The 180° RF pulse makes this a SE measurement. The echo maximum occurs at the center of k-space, $k = 0$, to ensure the best signal-to-noise ratio (SNR). The two-dimensional data set that results from this experiment is Fourier transformed to generate a two-dimensional image. The local image intensity, for a single-echo image, is given by

$$S = M_0 e^{-TE/T_2} \left(1 - e^{-TR/T_1}\right). \tag{11.9}$$

The basic SE measurement may be implemented in many different variations. Minard et al. (2006) used a frequency selective initial 90° RF pulse to reduce image artifacts in their PEM fuel cell imaging, with slice selection implemented for the 180° RF refocusing pulse.

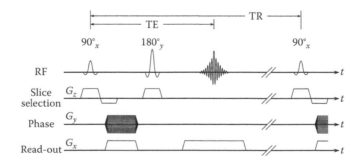

FIGURE 11.3 2D SE sequence. TR is the time between initial excitation pulses. TE is the time from the first excitation pulse to the peak of the first echo. Repetitive multiecho acquisition is possible. Echo rise and fall is exponential with the time constant T_2^*.

11.2.2.3 Nafion Depth Imaging and Pure Phase Encode Spin-Echo SPI

11.2.2.3.1 Nafion Depth Imaging

The perfluorinated polymer, Nafion, functions as both an electrolyte and gas separator in a PEM fuel cell. It consists of a polytetrafluoroethylene backbone and perfluorinated side chains with terminating sulfonic acid groups ($-SO_3H$) (Wielstich et al., 2003). Hydrophobicity of the backbone, with hydrophilicity of the sulfonic acid functional group, results in a constrained hydrophobic/hydrophilic nanoseparation. The sulfonic acid functional groups aggregate to form a hydrophilic domain that is hydrated upon absorption of water. It is within this continuous domain that ionic conductivity occurs: protons dissociate from their anionic counterion ($-SO_3^-$) and become solvated and mobilized by the hydration water.

The detailed mechanisms of H^+ ion transport in Nafion are not well established in the literature. It is generally agreed however that proton conduction occurs in the aqueous phase with rapid hydrogen bond breaking and forming processes, which permit long-range proton transport (Kreuer et al., 2004).

The Nafion microstructure is known to be anisotropic (Mauritz and Moore, 2004), which indicates the distribution of water clusters through the Nafion layer will not be a linear function of depth. It is also well known that Nafion absorbs water to different extents when exposed to liquid or water vapor. This is known as Schroeder's paradox (Futerko and Hsing, 1999) and suggests that the anode will always be much drier than the cathode, which is exposed to liquid water.

To understand the above phenomena is a core issue in water management, and requires quantitative measurement to directly characterize the water transport properties of Nafion. While Nafion membranes of several hundred microns thickness may be employed in model systems, the Nafion thickness for realistic fuel cells will be significantly reduced (Hoogers, 2000) for better cell performance. If one simultaneously demands high-temporal resolution to observe transient state behavior, traditional frequency and phase encode spin-echo MRI methods are ill suited to the problem.

To overcome these difficulties a one-dimensional pure phase encode method has been developed to permit rapid (several minutes) high-resolution (4 μm) imaging of water content in thin films. This method is termed spin-echo SPI (Ouriadov et al., 2004). The most quantitative version of the experiment for Nafion fuel cell measurements is described below.

11.2.2.3.2 Multiecho Spin-Echo SPI

High-resolution MRI is demanding on SNR. High resolution with frequency encoding methods requires high strength magnetic field gradients which impair the native SNR of the experiment (Ouriadov et al., 2004). In cases where acceptable image SNR with frequency encoding requires a number of signal averages that is greater than the number of pixels, a pure phase encoding approach to spatial encoding is inherently more sensitive (Ouriadov et al., 2004). The fuel cell requirement for high spatial resolution and rapid imaging argues strongly for a pure phase encode approach to spatial encoding. In addition, since one can reasonably anticipate that the water relaxation times T_1 and T_2 will vary with water content in the Nafion membrane, quantitative imaging requires that one must map the relaxation times and local water content in an MRI measurement. Imaging of the type suggested by Equation 11.9 will be nonquantitative unless great care is exercised in the measurement.

The pure phase encode fuel cell MRI methods developed by Zhang et al. (2008b) are inherently insensitive to T_1 relaxation. Multiple spin echoes are generated and spatially encoded in these methods, with an image generated for each echo. The resulting signal equation, Equation 11.10, is given below with the dependent variable y indicating depth through the Nafion membrane,

$$S(y) = \rho_0(y)\exp\left(-\frac{n\mathrm{TE}}{T_2(y)}\right). \tag{11.10}$$

In Equation 11.10 n is the echo number, TE is the echo time, and $\rho_0(y)$ is the spin density, which is the water content. By fitting each pixel in the profile series to Equation 11.10, it is possible to spatially resolve both T_2 and $\rho_0(y)$. High sensitivity with these methods is achieved in part by making the measurement one dimensional. It is also achieved through employment of a high-sensitivity RF probe to excite and detect the signal. In studies of isolated thin films a surface coil is employed (Ouriadov et al., 2004). In a model fuel cell, as described below, a new style of integrated RF probe is required.

11.2.3 MRI Hardware

An MRI scanner consists of four components: the magnet, the gradient coils, the RF resonator and receiver, and the console. The parameters affecting the SNR have been explored by Hoult et al. (Chen and Hoult, 1989; Hoult, 1978),

$$\text{SNR} \propto \left(B_0\right)^{7/4} \cdot K \cdot \eta \cdot \left(\frac{QV_c}{\Delta f}\right)^{1/2}, \qquad (11.11)$$

where η is the "filling factor" and is a measure of the fraction of the coil volume occupied by the sample, Q is the quality factor of the receiver coil, K is a numerical factor dependent on the receiving coil geometry, B_0 is the field strength, V_c is the volume inside the coil, and Δf is the bandwidth of the receiver in Hertz.

More importantly for this work, Hoult has also developed the principle of reciprocity, which states that the local sensitivity of an RF probe, for reception, is directly proportional to the local B_1 field strength of the probe, per unit current, generated for excitation (Chen and Hoult, 1989; Hoult, 1978).

A small RF probe generates a much stronger B_1 field per unit current than a large probe. Therefore, it is highly advantageous to customize the RF probe making it as small as possible and putting it in close proximity to the region of interest in the PEM fuel cell to be imaged.

11.2.3.1 MR Magnets and Gradient Coils

Three principle styles of magnet are employed for MRI, water-cooled resistive magnets, permanent magnets, and superconducting magnets. Increasing the static field strength increases the Larmor frequency and overall sensitivity of the experiment as described in Equation 11.11. The SNR $\propto (B_0)^{7/4}$ dependence is only true for nonconducting samples (Chen and Hoult, 1989). For systems with high conductivity, fuel cells for example, we anticipate instead SNR $\propto B_0$ (Hoult, 1978). The stability and homogeneity of B_0 are also critical for high-quality MRI. All fuel cell MRI studies in the literature have employed superconducting magnets with Larmor frequencies of hundreds of MHz. The lowest field strength employed is 2.4 T (Zhang et al., 2008a,b), with field strengths of 9.4 T more common.

Gradient coils are required to produce a pulsed linear variation in magnetic field, with high efficiency, low inductance, and low resistance to minimize the required current, dissipated heat, and gradient switching times. The slew rate, determined as the ratio between the gradient strength and the rise time, is an important determinant of performance. Increasing the slew rate by decreasing the inductance and the resistance of the gradient coil may introduce eddy currents, which distorts the gradient waveform. Compensation for eddy currents, currents generated in adjacent conductive structures, is usually achieved through pre-emphasis, in which the gradient waveform is modified to counter the distortion (Gach et al., 1998).

11.2.3.2 RF Probes

RF coils fall into two main categories, surface coils, the simplest kind of RF coil, and volume coils. The latter class, which fully encloses the sample, includes solenoidal coils, saddle coils, and birdcage coils (Chen and Hoult, 1989).

For a resonant RF probe circuit, the resistance of the RF coil should be matched to 50 Ω, the characteristic impedance of coaxial cable. The inductance and the resistance of the RF coil are used to tune and match the probe. The quality factor, Q, of the RF probe is determined by the pulse length, dead time, required excitation voltage, and the sensitivity of the probe:

$$Q = \frac{\omega_0 L}{R} = \frac{1}{\omega_0 R C},$$ (11.12)

where $\omega_0 = 1/\sqrt{LC}$. The quality factor of a circuit is affected by resistance along the leads of the coil and dielectric losses. An increase in the Q value of the probe will increase the sensitivity but will also increase the probe dead time, which limits ones ability to image species with short relaxation times.

11.2.3.3 Receiver and Signal Processing

If, as is usually the case, the RF coil is also the signal receiver, the RF circuit must be specifically designed to prevent interference between the high power RF pulse and the detected signal voltage, which is on the scale of microvolts. Once the sample signal voltage is amplified by the preamplifier, it is mixed in a phase sensitive detector to form a complex final signal. The magnetic resonance signal is filtered, digitized, and sent to an image processor for Fourier transformation or other processing.

As outlined above, the RF probe is a compromise and one should not anticipate a generic design for all possible measurements. For fuel cell MRI the principle of reciprocity, Section 11.2.3, is critical since one is seeking high resolution with rapid image acquisition. The RF probe should be customized for the fuel cell under study although this has not been the typical approach since most laboratories tend to utilize RF probes provided by the instrument vendor.

11.3 Instruments and Measurements

The PEM fuel cell is a complex electrochemical device, which substantially complicates the MRI measurement. The majority of problems are associated with metallic and conductive components of the fuel cell. Nonferromagnetic metals are required for MRI-compatible cells since otherwise a very strong force of attraction irretrievably pulls the cell into the magnet. Metal and conducting cell components must not fully enclose the PEM layer since RF excitation is required for signal excitation and detection. Magnetic susceptibility differences between the conductive (metal) and nonconductive elements of the fuel cell will spoil the static magnetic field homogeneity, which may lead to a wide variety of image artifacts. Electrical connections to the external load of the fuel cell, or to monitoring equipment, may introduce external RF noise into the MRI experiment introducing image artifacts or degrading the image SNR.

These ideas are discussed in more detail below.

11.3.1 MRI-Compatible PEM Fuel Cell Design

In most commercial PEM fuel cells, graphite plates, tightly held by a steel frame, contain the gas flow channels while the plates simultaneously function as the current collector. Their high conductivity shields the RF B_1 field required to excite and detect the ^{1}H signal inside the flow channels. The graphite plates also distort the local static magnetic field, $\vec{B}_0$, and may shield the PEM membrane from RF excitation as well. MRI fuel cell studies, therefore, rely on model MRI-compatible fuel cells constructed to minimize the usage of conductive and magnetic materials.

State-of-the-art MRI-compatible PEM fuel cells remain single-layer fuel cells due to constraints on materials. Nonconductive materials, with good mechanical properties, such as acrylic, Delrin, PEEK,

Teflon, and G10 fiberglass, have been employed to manufacture the flow fields and provide cell assembly support. These materials are either ^{1}H free or feature short T_2 spin–spin relaxation times, which ensure they do not appear in the experimental MRI images with finite TEs. The exact nature of the flow channels are variable and depend on specific research goals. The current collector can be fabricated from a thin layer of conductor machined into the proper shape and affixed to the flow-field plates.

The MEA, the most critical part of a PEM fuel cell, is generally unchanged except for increasing the thickness of the Nafion layer for experimental convenience. Catalysts are usually Pt or Pt–Ru with loadings from 0.3 to 3 mg cm^{-2} (Feindel et al., 2007a; Minard et al., 2006) dependent on the specific study. The GDL is usually either carbon paper or carbon cloth.

The size and layout of MRI-compatible PEM fuel cells is restricted by the inner diameter of the RF coil employed, the orientation of the magnet, and ancillary concerns such as gas supply and load circuit. Two typical designs are reproduced in Figure 11.4.

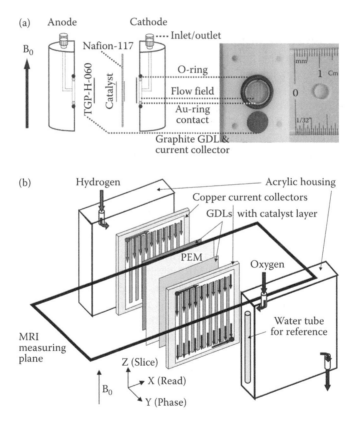

FIGURE 11.4 PEMFC designs: (a) Reported by Feindel et al. (2006). The body is composed of two halves of a vertically sliced Delrin cylinder, with gas inlet and outlet machined through the top of each half that leads to a combination of gas channels, which are slotted into the flat face of the cylinder half. The membrane electrode assembly is oriented parallel with $\vec{B}_0$. Gold wire, used as the electrical circuit contact, enters through the rounded side of each cylinder half and forms an unclosed ring around the flow field. The PEM is Nafion-117. (b) Reported by Tsushima et al. (2007). PEM was sandwiched with gas diffusion layers, in which fine platinum particles are dispersed as a catalyst facing the membrane. A parallel flow configuration from the top to the bottom is adopted, with a thin copper plate with a gold coating affixed against the GDL as a current collector. MRI visualization is conducted in a plane perpendicular to the static magnetic field in the MRI system. ((a) Reproduced with permission from Feindel, K. W. et al. 2006. *J. Am. Chem. Soc.* 128: 14192–14199. Copyright 2006 American Chemical Society; (b) With kind permission from Springer Science+Business Media: Tsushima, S. et al. 2007. *Appl. Magn. Reson.* 32: 233–241.)

11.3.2 RF Excitation and Detection

Except for the studies of Zhang et al. (2008a), which integrated the MR resonator into an operational fuel cell, all other reported studies were carried out using birdcage RF coils to excite and detect the ^{1}H signal from custom-designed fuel cells.

With traditional volume resonators, such as the birdcage coil, the Nafion film, typically 250 µm thick in a model PEM fuel cell, occupies a very small fraction of the measurement volume of the probe. This introduces three related difficulties: (1) the SNR is poor since the filling factor is low, (2) even if one minimizes the amount of conductor in the model PEM fuel cell, it will still contain significant electrical conductor which introduces thermal noise decreasing the SNR of the experiment, and (3) the RF magnetic field must be precisely aligned with the cell to permit the magnetic field to propagate through the membrane between the MEA electrodes. This alignment can be difficult to achieve in practice. The MEA electrodes will therefore partially shield the incident RF magnetic field resulting in unintentional heating of the electrode.

Minard et al. (2006) made a concerted effort to avoid significant RF screening effects in their studies. Similar effects are important in all MRI PEM fuel cell studies but these effects are usually not discussed. The best diagnostic for significant RF interaction with conductors in the PEM fuel cell is the 90° pulse length that is increased when significant pulse energy is absorbed by the conductor.

The parallel plate resonator, developed by the coauthors (Zhang et al., 2008a), is purpose designed for high-resolution PEM fuel cell imaging. This novel probe design largely avoids the above problems since it uses the cell electrodes to generate the RF field. The required field is thus naturally parallel to the Nafion plane and the filing factor is very good, yielding a very sensitive experiment. In the study reported by Zhang et al. (2008a), as described above, the MEA electrodes function as a fuel cell at DC but act as an RF probe at the Larmor frequency. As illustrated in Figure 11.5a, an outer electrical load was connected to complete the DC circuit. Capacitors, C_c, C_f, C_b, and C_m are soldered to the electrodes to complete the RF circuit and allow probe tuning and matching. The equivalent RF circuit incorporating the operating fuel cell is illustrated in Figure 11.5b.

11.3.3 Effects from Conductive Materials

Although, as explained above one seeks to minimize conductive materials in the MRI-compatible PEM fuel cell, the GDL and current collector must remain and must reside in close proximity to the Nafion layer. The major issue in these structures is the RF shielding effect, potentially impairing signal excitation and detection in the Nafion or other parts of the device.

The skin depth is the best way to collect sample and electromagnetic effects into one expression to evaluate potential shielding. The skin depth δ is defined as the distance that the wave has to travel for it to be attenuated to 37% of its original amplitude, with $\delta = \sqrt{2 / \omega\mu\sigma}$. The skin depth depends on frequency ω, conductivity σ, and permeability μ.

The skin depth explains why water content measurements within the GDL have not been possible with conventional high-frequency superconducting magnet-based instruments. The frequency is too high and shielding is too great. Low-frequency measurements with permanent magnet-based instruments have a greater probability of success.

11.3.3.1 Spatial Variation of Magnetic Susceptibility

Assuming the GDL and current collector, as a minimum, are paramagnetic materials the magnetic field within the fuel cell will be altered. The induced magnetic field in a material is proportional to the strength of the static magnetic field, $\vec{B}_0$:

$$\vec{B}_0^s = \left(1 + \chi_m\right)\vec{B}_0 \tag{11.13}$$

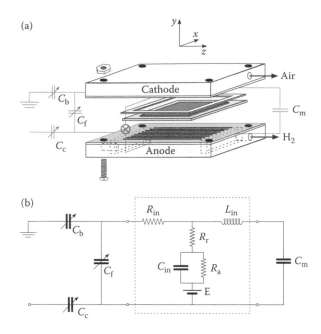

FIGURE 11.5 (a) Schematic of the homemade operating fuel cell reported by Zhang et al. (2008b). The Nafion layer is sandwiched between two catalyst-sprayed gas diffusion layers to comprise a MEA. Gold-plated PC board, with slots etched on the surface as flow channels, as the electrodes connects the outer load (⊗) and RF components. Two G10 fiberglass substrates, which support the PC boards, are machined, as shown by discrete lines in the figure, to generate paths for gas flow. Two homemade Teflon O-rings seal the cell. Only one, the frame shown above the Nafion, is illustrated. The entire apparatus is held together by four copper screws and tightened with a controlled torque. (b) The RF circuit diagram, with the operating fuel cell integrated into the circuit as the RF resonator. The equivalent circuit of the operating fuel cell consists of R_r, R_a, C_{in}, and E. R_{in} and L_{in} represent the parallel plate transmission line. C_c, C_f, C_b, and C_m, are capacitance. (Reproduced from *J. Magn. Reson.*, 194, Zhang, Z. et al. Spatial and temporal mapping of water content across Nafion membranes under wetting and drying conditions, 245–253, Copyright (2008b), with permission from Elsevier.)

where $\vec{B}_0^s$ is the mean magnetic field in the sample, while χ_m is the diamagnetic susceptibility (Callaghan, 1991). The magnetic field inside the conductive structures of the fuel cell will be higher than the magnetic field in surrounding space. Since lines of magnetic flux are continuous, increasing the field strength in some areas (conductors) will distort the magnetic field in surrounding areas (Nafion). The resulting inhomogeneous magnetic fields may distort the MR image, perturb image intensities, and cause slice-selection errors (Park et al., 1988).

The geometrical image distortion, due to field inhomogeneities, occurs primarily along the frequency-encode direction. It can be mitigated by applying stronger frequency-encode gradients (larger bandwidth), which will however decrease SNR.

Magnetic susceptibility effects also impair high-resolution imaging that will be detailed in Section 11.3.5.2.

11.3.3.2 Eddy Currents due to Switched Magnetic Field Gradients

The time rate of change of magnetic flux through a conductor induces a current in the conductor, an eddy current, which has an associated magnetic field that opposes the changing field. The eddy current, described by Lenz's law, alters the waveform of the switched magnetic field gradients inherent to spatial encoding in MRI. The GDL and current collector are once again the most obvious source of such currents in a PEM fuel cell, but all metallic components have the potential to support eddy currents.

Significant eddy currents yield geometric artifacts in an MRI image with potential signal loss. Orienting the fuel cell such that the GDL and current collector are largely parallel to the direction of the static field will significantly reduce eddy current effects in PEM fuel cell MRI experiments.

11.3.4 Quantitative MRI

Both qualitative and quantitative studies of water content distribution have been undertaken in PEM fuel cell MRI studies. Quantitative studies are the obvious end goal of the vast majority of studies undertaken. The simplest quantitative study would be mapping the water content either absolutely or relatively as a function of space and time in an operating PEM fuel cell. Valuable data of this type would permit investigators to validate and improve mathematical models of fuel cell operation and to observe trends in fuel cell behavior as a function of operating conditions or aging for example.

As described in Equation 11.9 the local image intensity is usually affected by the MR signal lifetimes. These lifetimes can be difficult and time consuming to measure independently and so careful experimental design is required for quantitative imaging.

11.3.4.1 Relaxation Contrast Effects in PEM Fuel Cell MRI

As the relaxation time of ^1H in Nafion varies with water content, it is clear that relaxation time weighting of the signal equation, for example, Equation 11.9, will be important since one anticipates variable water content in the Nafion membrane.

The dependence of the Nafion relaxation times on water content are well known in the literature with the first studies reported by MacMillan et al. (1999). MacMillan found that water in Nafion existed in both bound water and bulk-like water forms. The large population of bulk-like water exists in discrete clusters associated with acid sites and dominates the global relaxation time behavior. Fast magnetization exchange, through water exchange or magnetization cross-relaxation, of the water molecules from multiple sites yields single exponential magnetization decays (MacMillan et al., 1999).

A similar picture was introduced by Zhang et al. (2008b) in their *ex situ* Nafion study with controlled boundary conditions. Their explanation of relaxation time trends incorporated the physical model of Weber and Newman with multiple microstructural environments for water in the Nafion (Weber and Newman, 2003),

$$\frac{1}{T_2} = \sum_i p_i \cdot \frac{1}{T_{2,i}}$$

(11.14)

In Equation 11.14, i indexes the multiple water environments, such as channels and clusters, all of which exchange rapidly, each with an associated T_2. These environments must include bound and free water in both clusters and channels, with the probability of each subpopulation represented by P_i. The weighted average T_2 has short T_2 contributions from bound-water populations and long T_2 contributions from bulk-like water. Increasing or decreasing the water content alters P_i and potentially the associated $T_{2,i}$ in Equation 11.14.

Zhang's results showed that although the T_2 in Nafion is water content dependent, with a strong dependence on short-term sample age, it remained single exponential in accord with MacMillan's findings. This suggests that although T_2 should be measured in PEM fuel cell imaging experiments, the measurement is simplified because of the single exponential behavior.

Equation 11.9 shows that the local signal intensity in a simple SE imaging measurement should incorporate T_1 contrast in addition to T_2 contrast. If T_2 varies with water content one could reasonably assume that T_1 will have a similar dependency. The simplest way to ensure a more quantitative SE imaging experiment is to simply increase the repetition time (TR), of the experiment. The TR is the time interval between successive repetitions of the basic SE measurement. If TR is maintained much longer than the longest anticipated T_1 then the third term of Equation 11.9 will have a value of 1 in all cases. Increasing the repetition time is not however automatically the best strategy since a longer repetition time increases the duration of the imaging experiment.

Conventional wisdom holds that the repetition time should be chosen to be $5T_1$, which assures 99.3% recovery of longitudinal magnetization. However, 95% of M_z is recovered with a repetition time of $3T_1$, with substantial savings in image acquisition time.

11.3.4.2 Calibration of T_2-Weighted Image

Spin-echo MRI methods inevitably feature T_2 contrast in the resultant images unless the TE can be set shorter than T_2. Since T_2 decay is an exponential decay the TE must be dramatically shorter to remove entirely the T_2 contrast, which is difficult to achieve in practice.

From previous bulk measurements, T_2 is known to vary in the range of 6–50 ms, as the Nafion increases in moisture content to saturation (MacMillan et al., 1999; Sivashinsky and Tanny, 1981; Slade et al., 1983). These T_2 values are long enough to permit multiecho MRI experiments with the base TE shorter than or on the order of the shortest T_2.

Zhang et al. in their studies (Zhang et al., 2008a,b), used a multiecho pure phase encode spin-echo SPI measurement, as introduced in Section 11.2.2.3.2. This method generated a rapid series of T_2-weighted profiles according to Equation 11.10. By fitting each pixel in the profile series to Equation 11.10, it is possible to spatially resolve both T_2 and the local proton density, $\rho_0(y)$, which corresponds to the water content. In their work, the true water content variation across a Nafion membrane in an operating fuel cell was monitored by high-resolution MRI profiles. The T_2 fitting procedure was simplified because T_2 decay was observed to be single exponential in all cases.

Wang et al. (2010) have recently adopted a similar approach to quantifying T_2 in PEM fuel cells but with a more conventional multiecho frequency-encode approach.

These approaches to quantifying the MRI images quite naturally separate T_2 contrast from the density term, which is water content, but they generate relative water content maps after fitting, not absolute water content maps.

11.3.4.3 Image Calibration

It is possible to calibrate MRI images by simultaneously or sequentially imaging standard samples then using a simple ratio to assign quantity to the experimental image. This works best if the images do not have variable relaxation time weighting.

Dunbar and Masel used water-filled capillaries attached to the fuel cell to quantify the ^{1}H MRI signal for water condensed in flow fields (Dunbar and Masel, 2007, 2008). This work assumed that water vapor was too low in density to contribute to the image and so the image is dominated by liquid water with long T_2 lifetimes. In this case, the direct water calibration works well because of no relaxation time weighting of the water signal.

Tsushima and coworkers used a calibration curve obtained from ^{1}H MRI images of an MEA exposed to reactant gases with known relative humidity to quantify the through-plane water content distribution. This methodology works if T_2 is a function solely of local water content, and does not vary with depth during fuel cell operation. Direct calibration of images by this procedure assumes that the calibration sample accurately mimics the experimental sample, which can be a dubious assumption in some cases. Zhang et al. (2008b) showed that the Nafion T_2 value for nominally static samples continued to change for many hours after initial water exposure.

11.3.5 High-Resolution MRI

High resolution is the goal in most PEM fuel cell MRI studies. High resolution requires high SNR, hence the need for the most sensitive measurements. Hardware concerns, and the choice of pulse sequence, were introduced in previous sections. In this section, we discuss other factors which will limit the achievable resolution in an MRI experiment.

Unlike whole-body MRI scanners used for clinical imaging, scientific studies are usually undertaken on narrow bore scanners equipped with high-strength gradients that rapidly switch. The available hardware is therefore not a principal determinant of resolution, assuming eddy currents are minimal.

11.3.5.1 Sample Alignment

Misalignment of the spatial encoding gradient(s) orthogonal to the sample plane is the principal cause of resolution loss in high-resolution Nafion depth imaging measurements. Conventional SE imaging measurements employ a thick imaging slice with two-dimensional spatial encoding such that misalignment of the imaging axes with the sample plane are revealed in the two-dimensional image as sample rotation (Tsushima et al., 2005a). One can then always choose to examine a profile from the two-dimensional image that is orthogonal to the Nafion plane.

Zhang et al. (2008a,b) used linear combinations of the spatial encoding gradients to ensure, in their one-dimensional measurements, that the imaging axis was strictly orthogonal to the sample plane ensuring high-resolution images.

11.3.5.2 Line-Width, Susceptibility, and Diffusion Limits on Resolution

MRI image resolution with frequency encoding is limited by the natural line-width of the MR signal. If the natural spread of frequencies of a point source sample were 400 Hz then a pixel size of 200 Hz would show the point source sample spread over two pixels. This simple example shows that the nominal resolution of an MRI experiment, FOV divided by the number of pixels, can be a quite misleading statement of resolution. Phase encoding is often preferred for high-resolution imaging because the sample line width is not a restriction on resolution, due to the nature of the encoding (Gravina and Cory, 1994).

Any source of static field inhomogeneity in the sample space will act to increase the natural line-width of the sample at a point, and decreases the image resolution. Static field inhomogeneity due to magnetic susceptibility mismatch causes exactly the same line broadening and resolution loss. These features once again do not cause resolution loss in pure phase encode measurements which makes these techniques naturally high resolution.

Diffusion of water in the presence of a frequency-encoded magnetic field gradient causes signal attenuation and line broadening along the read-out direction. From the formula of Hahn and Torrey (Callaghan, 1991), the additional attenuation due to diffusion, resulting from the application of a steady gradient, is $\exp(-\gamma^2 G^2 D T_{acq}^3/3)$ where T_{acq} is the time during which signal is acquired. In the frequency domain, therefore, the image is broadened by the convolution of the spectrum with this function, with the resultant spread in frequency given by $\Delta f_D = 0.6(\gamma^2 G^2 D/3)^{1/3}$. If diffusion dominates the line width broadening, which will be the ideal regime for fuel cell MRI, the diffusion-limited resolution can be estimated by (Ahn and Cho, 1989),

$$\Delta x_D = \sqrt{\frac{2}{3} D T_{acq}} \tag{11.15}$$

Given the well-accepted water membrane diffusivity in Nafion, $5.80 \times 10^{-6}\,\mathrm{cm^2\,s^{-1}}$ (Zawodzinski et al., 1991), and a readout gradient for high-resolution imaging with a maximum of 60 Gauss cm^{-1}, a T_{acq} on the order of 8 ms results in a diffusion-limited resolution of 5.5 µm ignoring the influence of all other factors.

On the basis of Equation 11.15, we can predict that the pure phase encode spin-echo SPI approach will be less effected by molecular diffusion than frequency encoding. A theoretical diffusion-limited resolution was reported as 2 µm with a 4 µm nominal resolution readily achieved in previous work (Ouriadov et al., 2004).

The true resolution for high-resolution PEM fuel cell MRI is best revealed by imaging a sharp discontinuity in the sample. This reveals the true resolution of the image and automatically combines the effect of all possible resolution loss mechanisms. For this reason true experimental resolution is best evaluated for image profiles that show the edges of the Nafion membrane in the image. Displaying only the center of the profile does not permit evaluation of the image resolution and is a practice that should be discouraged.

11.4 Imaging Water in PEM Fuel Cell Using MRI

Fuel cell MRI is still a measurement and research area that is evolving. Many of the measurements reported in the literature are therefore best considered feasibility studies. A brief review of work in the literature is given below.

11.4.1 General Considerations for a Fuel Cell MRI Measurement

A general schematic of the experimental setup for fuel cell MRI measurement is illustrated in Figure 11.6. We amplify, however, a number of important experimental considerations.

Both the fuel cell and the RF probe should be carefully aligned such that the RF probe B_1 is parallel to the Nafion layer while the layer is carefully aligned with respect to the magnetic field gradients.

As the sample fuel cell is running during MRI data acquisition, it is usually connected to a load or a cell-testing device outside of the magnet. The electrical connection to the external load or testing circuit may introduce RF noise to the MRI signal. To minimize this electromagnetic interference, wire leads should be filtered, shielded, or tightly twisted together in a pair to cancel interference effects.

To maintain the cell temperature, which varies with operation, one can add a heater inside the PEM fuel cell system or blow air over the sample at a desired temperature to provide a consistent temperature.

11.4.2 Water in Flow Fields

Water in the PEM fuel cell flow fields has been investigated by several groups. Feindel et al. (2006) investigated the effect of gas flow configuration on the distribution of water in the PEM and cathode flow field and found that the counterflow configuration yielded a more uniform distribution of water throughout the PEM. They also observed that maximum power output from the PEM fuel cell, with a constant

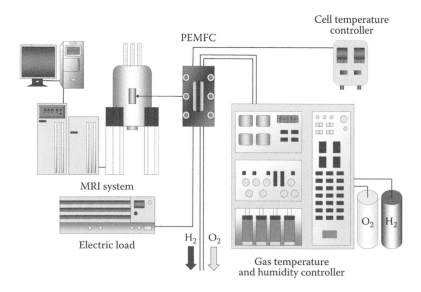

FIGURE 11.6 Schematic of a typical setup used to perform MRI studies of an operating fuel cell reported by Ikeda et al. (2008). (Reproduced from Ikeda, T. et al. 2008. *ECS Trans.* 16: 1035–1040. With permission from ECS.)

external load, occurred at the onset of water accumulation at the cathodes. The power declined with further water accumulation.

Minard et al. (2006) investigated liquid water behavior in fuel cells, with a single serpentine flow channel and a constant electrical load. They observed that after Nafion dehydration occurred, channels in the gas manifold began to flood on the cathode side.

Dunbar and Masel used MRI to examine and quantify the flow of water in both Teflon- and graphite-coated flow fields to investigate water transport and water accumulation in an operating PEM fuel cell (Dunbar and Masel, 2007, 2008). Some of their results are reproduced in Figure 11.7. The authors reported that water drawn away from the cathode GDL accumulated at the bottom of the flow field, even with hydrophobic Teflon flow fields. The authors also considered the significant effect of small defects in

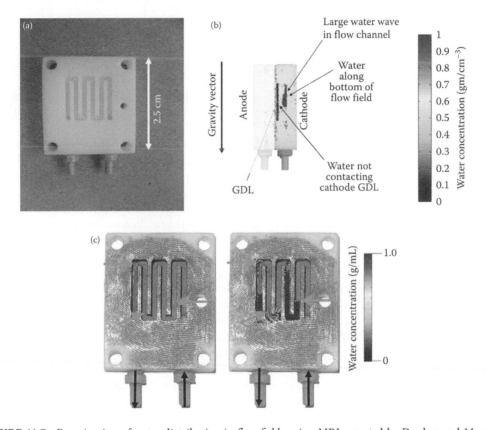

FIGURE 11.7 Examination of water distribution in flow fields using MRI reported by Dunbar and Masel (a) Frontal view of a Teflon flow field used in study (Dunbar and Masel, 2007). Channels are 1 mm wide and 3 mm deep, in a serpentine pattern. (b) Profile view of the water concentration in the cathode flow field at steady state, accompanied with 200 mA cm^{-2} constant current hold at all times. The anode gas is dry hydrogen, with a flow rate of 40 standard cubic centimeters per minute (sccm), while the cathode is dry oxygen, at 40 sccm. A large water wave occupies the bulk of one flow channel. (c) Another study executed by the same group (Dunbar and Masel, 2008), where the flow fields was sprayed a layer of graphite. The figure shows a top view of the cathode flow field illustrating the position of the water waves. The arrows indicate the direction of gas flow. The waves are stationary most of the time but periodically, a wave will "slip" from one location in the flow channel to another. ((a,b) Reproduced from *J. Power Sources*, 171, Dunbar, Z. W. and Masel, R. I. Quantitative MRI study of water distribution during operation of a PEM fuel cell using Teflon(R) flow fields, 678–687, Copyright (2007), with permission from Elsevier. (c) Reproduced from *J. Power Sources*, 182, Dunbar, Z. W. and Masel, R. I., Magnetic resonance imaging investigation of water accumulation and transport in graphite flow fields in a polymer electrolyte membrane fuel cell: Do defects control transport? 76–82, Copyright (2008), with permission from Elsevier.)

the wall of the flow field on water accumulation/blockage (Dunbar and Masel, 2008). Multislice SE pulse sequences were used to acquire T_2-weighted two-dimensional images. The resultant images were calibrated with a standard sample as detailed in Section 11.3.4.3.

Water flow in the flow-field plates would seem to be an obvious target for velocity imaging but no such studies have as yet been reported.

11.4.3 Water in the PEM

Measurement of water content in the PEM layer has been the target of most MRI studies reported in the literature. The obvious goal of measuring water content through the Nafion layer is difficult because of the demanding resolution required. Many studies have therefore emphasized lateral measurements of water content distribution in the plane of the Nafion membrane.

11.4.3.1 In-Plane Water Distribution

The in-plane or lateral distribution of water in a PEM is indicative of the uniformity of electro-chemical reactions in the PEM fuel cell. A major reason for heterogeneity in the lateral water content distribution is the variation of the partial pressures of hydrogen, oxygen, and water vapor along the flow channels. In-plane water measurement is helpful to examine the flow-field configurations, the gas flow rate, and PEM fuel cell operating conditions.

In their first fuel cell MRI paper, Feindel et al. (2004) measured the radial diffusion of water from the MEA into the surrounding Nafion membrane during operation. They found the integrated MRI image signal intensity from the region of the PEM between the catalyst layers correlated well with the power output of the fuel cell. They also examined the influence from flow configurations, flow rate of supplied oxygen, and the resistance of the external circuit to the integrated signal intensity. It was revealed that the counterflow configurations yielded greater integrated intensities than coflow, as illustrated in Figure 11.8, despite neither of them generating an isotropic water distribution. Hydrogen–deuterium (H–D) exchange was employed as a method to introduce contrast in MRI images and to investigate the dynamic distribution of water throughout an operating H_2/O_2 PEM fuel cell (Feindel et al., 2007a). Through cycling the D_2O/H_2O into the flow channels of an operating PEM fuel cell, the removal/recovery of the proton signal as H–D exchange occurred revealed where H_2O was contained in the PEM. At similar currents, the H–D exchange rate in the PEM during operation was faster when the PEM was saturated with water than when run under low relative humidity conditions, which indicated that the current per area of conductive hydrophilic domain was higher when the PEM operated under low relative humidities.

More recently, the same group reported (Wang et al., 2010) a measure of water content per acid site within a Nafion membrane between the catalyst stamps of an MEA. The signal intensity was observed with multiecho T_2 measurement to yield proton density-weighted images, thereby establishing the relationship between the density-weighted image intensity and the water/acid ratio.

Minard et al. (2006) carried out a time-series measurement of the in-plane water distribution in a PEM. Measurement every 128 s revealed the formation of a dehydration front that propagated slowly, starting from the gas inlets and progressing toward the gas outlets.

Tsushima et al. (2005b) examined the effects of the flow-field pattern on membrane hydration using a 3D MRI technique to visualize the in-plane water content distributions in a PEM in an operating PEMFC. The variation of lateral water content distribution in the PEM with a serpentine flow channel and a parallel flow channel, as illustrated in Figure 11.9, suggests that the flow channel pattern affects accumulation of liquid water.

11.4.3.2 Through-Plane Water Distribution

As detailed in Section 11.2.2.3.1, through-plane water distribution and transport is a fundamental issue of water management. It has been investigated by Tsushima and coworkers in addition to Zhang and coworkers.

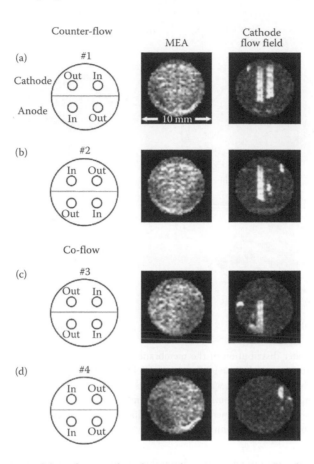

FIGURE 11.8 Investigation of the influence of gas flow configuration on water distribution reported by Feindel et al. (2007b). (a)–(d) illustrate four possible gas flow configurations for the PEM fuel cell and the region of interest in images acquired from a 500 μm slice containing the MEA and the cathode flow field of the operating PEMFC. The larger signal intensities for the counterflow configurations than the coflow configurations indicate that counterflow configurations maintain a greater amount of H_2O in cathode-side PEM than the coflow configurations. (From Feindel, K. W. et al. 2007b. The influence of membrane electrode assembly water content on the performance of a polymer electrolyte membrane fuel cell as investigated by 1H NMR microscopy. *Phys. Chem. Chem. Phys.* 9: 1850–1857. Reproduced by permission of The Royal Society of Chemistry.)

In 2002, Tsushima et al. (Teranishi et al., 2002) commenced investigations using ^1H NMR microscopy to measure water distribution in PEMs during fuel cell operation. The majority of their studies probed *in situ* water content through-plane within a PEM, for example, AciplexS-1 112 and Nafion 1110. The PEM thicknesses were chosen to be relatively thick for ease of experiment, 340 and 250 μm respectively. As illustrated in Figure 11.10, their spatial resolution was 25 μm. In more recent work they have reported increased resolution, to better than 10 μm (Shim et al., 2009).

They examined the performance of operating fuel cells under the influence of water supply (Tsushima et al., 2005a), and membrane thickness. In addition to examining water content variation with humidification (Tsushima et al., 2005b), they have found the water concentration gradient in the PEM, and the overall water content, decreased with an increase of the cell current. During critical periods such as "start-up" or with a rapid change of circuit resistance, this may help guide modification of gas flow rates or humidification to maintain optimal operating conditions.

They compared two types of polymer electrolyte membrane, Nafion and Flemion, in terms of diffusivity and compared the performance of two fuel cells (Tsushima et al., 2007), respectively, encased with

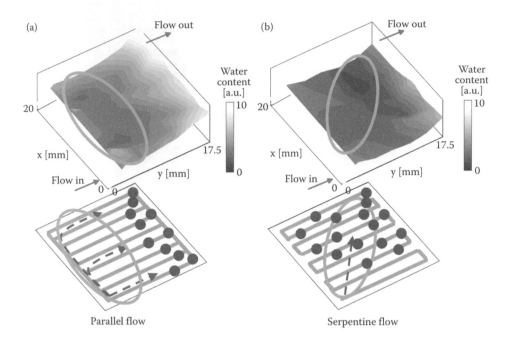

FIGURE 11.9 Lateral water distribution of the membrane in the operating fuel cell with different flow configurations, studied by Tsushima et al. (2005b). For the parallel flow pattern (a), the overall pressure gradient from the top to the bottom of the cell is uniform with water droplets plugging the channels. For the serpentine flow pattern (b), dehydration from the inlet to the outlet across the membrane is observed which suggests intensive electrochemical reaction in this portion of the membrane. (Reproduced from Tsushima, S. et al. 2005b. *ECS Trans.* 1: 199–205. With permission from Electrochemical Society.)

a hydrocarbon membrane and a perfluorinated membrane. They confirmed an improvement of fuel cell performance due to pressurization, through the observation that both activation polarization and ohmic polarization decrease, which leads to an increase in hydration with increase in the operating pressure of the fuel cell.

They also employed MR contrast from H–D exchange to examine fuel cell fueling dynamics, and the hydration path, and found that membrane hydration is dominated by water generated by chemical reaction at the cathode (Kotaka et al., 2007). Using the same method, they also analyzed water transportation and cell degradation (Hirai and Tsushima, 2008), and microporous layer effects on membrane hydration under low-humidity conditions.

To elucidate the water transport mechanism in fuel cell operation at various levels of relative humidity and current density, an environmentally controlled MRI measurement was undertaken (Ikeda et al., 2008). Back permeation of water generated by chemical reaction at the cathode was observed, as well as an increase of the water concentration gradient due to the electroosmotic drag increases with relative humidity.

One significant limitation of the studies undertaken by Tsushima and coworkers is their neglect of T_2 contrast effects, which make studies less quantitative than might otherwise be the case. They employed calibration methods as described earlier.

Using multiecho spin-echo SPI methods purpose designed for PEM fuel cell imaging, Zhang et al. (2008b) directly measured *ex situ* the water content profiles across a Nafion layer under wetting and drying conditions. High-resolution (<8 μm) depth images with an SNR greater than 20 were acquired with image-acquisition times of less than 2 min. Water content maps, at high resolution, were attained with T_2 mapping. It was also observed, as illustrated in Figure 11.11, that the membrane shrinks as it dries,

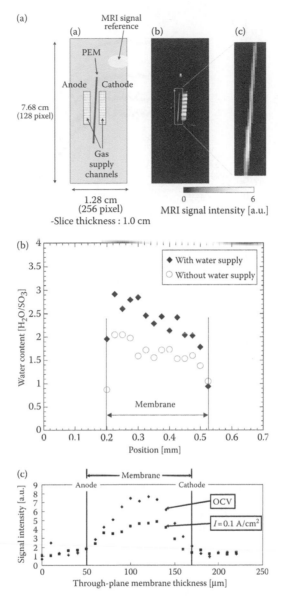

FIGURE 11.10 Investigation of water through-plane distribution in PEM reported by Tsushima et al. (a) Schematic of a typical fuel cell configuration, corresponding to Figure 11.4b, for high-resolution through-plane MRI (Tsushima et al., 2007). An MR image, positioned at the middle, covers a full field of view illustrated in Figure 11.4b, with its extended image, which corresponds to the part of membrane region, shown at right. (b) A representative 1D high-resolution image reported by Tsushima et al. (2005a), with a nominal resolution of 25 μm. (c) High-resolution 1D image recently reported by the same group (Shim et al., 2009), with two profiles under different currents. No edge apparent in the original profile images makes evaluating the resolution problematic. ((a) With kind permission from Springer Science+Business Media: *Appl. Magn. Reson.*, MRI application for clarifying fuel cell performance with variation of polymer electrolyte membranes: A comparison of water content of a hydrocarbon membrane and a perfluorinated membrane, 32, 2007, 233–241, Tsushima, S. et al. (b) Reproduced from *Magn. Reson. Imaging*, 23, Tsushima, S. et al., Water content distribution in a polymer electrolyte membrane for advanced fuel cell system with liquid water supply, 255–258, Copyright (2005a), with permission from Elsevier. (c) Reproduced from Shim, J. et al. 2009. *ECS Trans.* 25: 523–528. With permission from Electro-chemical Society.)

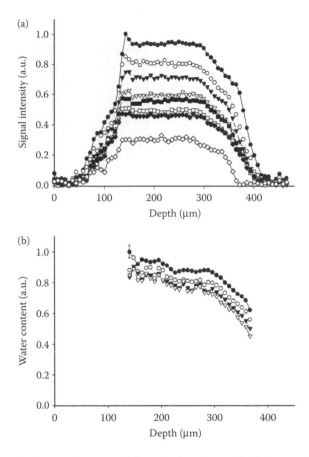

FIGURE 11.11 High-resolution one-dimensional through-plane image of a Nafion membrane in a mock fuel cell reported by Zhang et al. (2008a). (a) Water content across the Nafion membrane undergoing boundary conditions with wetting on one side while drying on the other. Experiments (●), (○), (▼), (▽), (■) (□), (◆), (◇), correspond to experimental times of 0.17, 1.9, 4.6, 8.1, 11.0, 14.0, 16.3, 18.5 h. The water reservoir, which feeds water to the membrane, is at the left in the profiles with the drying chamber at the right. At the right top corner, a discontinuity in signal is pronounced. The water reservoir emptied after 18.5 h with an obvious drop in signal intensity at that point. (b) 4 averaged water content profiles from 20 profiles over 3.4 h. Experiments (●), (○), (▼), (▽) correspond to the average from the experimental time range of 0–3.4, 3.4–6.8, 6.8–10.2, 10.2–13.6 h. The nonuniform profiles present two different regimes of water content. Water content decreased over 20 h, due to an aging effect of the membrane. A clear drying front is observed on the drying side. (Reproduced from *J. Magn. Reson.*, 194, Zhang, Z. et al., Spatial and temporal mapping of water content across Nafion membranes under wetting and drying conditions, 245–253, Copyright (2008a), with permission from Elsevier.)

without a water supply, due to the high mobility of water inside the Nafion membrane that yields fast water redistribution.

Zhang also undertook (Zhang et al., 2008a) direct measurements of water content across the Nafion membrane in an operational PEM fuel cell, using the same MRI method. To avoid the effects of RF screening, a novel parallel plate resonator was specifically developed, by employing the electrodes inherent to the fuel cell to create a resonant circuit at RF frequencies for MR excitation and detection, while still operating as a conventional fuel cell at DC. Three stages of fuel cell operation, activation, operation, and dehydration, were investigated, as illustrated in Figure 11.12. T_2-calibrated water content profiles were obtained, with 6 min acquisition time, 6 μm nominal resolution, and an SNR of better than 15. At room temperature, partial dehydration of the membrane at the anode side of the fuel cell was observed with high spatial resolution.

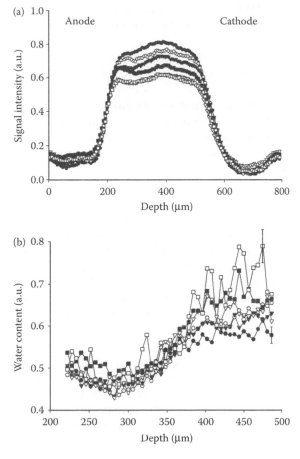

FIGURE 11.12 High-resolution one-dimensional through-plane MRI study of a Nafion membrane in an operational fuel cell reported by Zhang et al. (2008b). (a) Water content profiles across the Nafion during quasisteady state fuel cell operation, starting at approximately 1.2 h and lasting 6 h of subsequent operation. Ten first-echo MRI profiles were averaged over 1 h intervals to produce the profiles shown in (a). Profiles (●), (○), (▼), (▽), (■), and (□) correspond to time windows of, 1.1–2.1, 2.1–3.1, 3.1–4.1, 4.1–5.1, 5.1–6.1, 6.1–7.1 h. (b) True water content profiles determined by T_2 relaxation time mapping. The profiles displayed are also averaged. The profiles in (b) show stable water content close to the anode, with a higher water content near the cathode. (Reproduced from *J. Magn. Reson.*, 193, Zhang, Z. et al., Magnetic resonance imaging of water content across the Nafion membrane in an operational PEM fuel cell, 259–266, Copyright (2008b), with permission from Elsevier.)

The studies of Zhang et al., illustrate the importance of T_2 measurement because true water content profiles differ significantly from simple T_2-weighted images in the case of operating PEM fuel cells.

11.5 Advantages and Limitations

MRI is a natural technique for water content distribution imaging in fuel cells and has already provided valuable information to fuel cell researchers. However, the measurements are not routine and the advantages and disadvantages of the methodology should be considered objectively. We summarize the advantages and limitations below.

Advantages:

- *Noninvasive measurement:* MRI is a noninvasive measurement which permits *in situ* determination of water content spatially and temporally resolved in model fuel cells.

- *Flexible dimensionality:* MRI permits one-, two-, or three-dimensional imaging according to the research need. Spatial encoding is driven by the magnetic field gradient(s) employed, not sample orientation.
- *Measurement of water dynamics:* MRI is sensitive to molecular displacement and this may be exploited for measurement of local velocity and flow. This capability has been minimally exploited in fuel cell studies to this point.
- *Quantitative measurement:* MRI permits quantitative measurements of water content and water content changes to calibrate and/or develop numerical models for water management.
- *Contrast enhanced imaging:* The contrast inherent to most MRI measurements may be exploited to suppress or reveal features of interest in the fuel cell. This occurs most naturally through relaxation time weighting.

Limitations:

- *Fuel cell materials:* To the best of our knowledge, no commercial fuel cells have been studied with MRI methods. The major difficulty is material compatibility. Ferromagnetic materials cannot be employed. Use of conductors must be limited to avoid RF screening. Materials must be chosen to limit magnetic susceptibility mismatch. These limitations require measurements be undertaken in purpose-built model PEM fuel cells. One must also be careful of material compatibility of ancillary monitoring equipment and the electrical load in the vicinity of the MRI magnet.
- *Spatial resolution:* PEM fuel cells envisioned for commercial use will feature membranes less than 50 μm thick. The resolution of current MRI methods may not be sufficient for high-quality studies.
- *Undetectable water in GDL:* Due to RF screening effects, water in the GDL is not currently detectable with high-field MRI. Low-field measurements are more promising.
- *Limited sample size:* Small RF probes are required for rapid high-resolution PEM fuel cell imaging. The limited field of view limits the overall size of the fuel cell. Once more model systems are required and it is unlikely if anything larger than a single cell will ever be imaged.

11.6 Outlook

MRI studies of PEM fuel cells have been in progress for less than 10 years. In this time dramatic improvements in image quality have been achieved, but significantly more development is required before the goals initially envisioned for PEM fuel cell MRI are realized. It is still realistic to imagine high-resolution water content measurements in model systems, temporally resolved, that permit one to examine water behavior in the PEM as a function of a wide range of operating conditions. Measurements of lateral water distribution in the PEM are relatively easy since the resolution required is less demanding. Through-plane measurements will remain challenging and will require dedicated RF probes and purpose-designed MRI methods.

Future efforts should certainly include "hyphenated methods" wherein one monitors all aspects of fuel cell performance with MRI providing one critical piece of an array of measurements.

References

Abragam, A. 1961. *The Principles of Nuclear Magnetism*. London: Oxford University Press.

Ahn, C. B. and Cho, Z. H. 1989. A generalized formulation of diffusion effects in μm resolution nuclear magnetic resonance imaging. *Med. Phys.* 16: 22–28.

Basic Research Needs for the Hydrogen Economy: Report on the Basic Energy Sciences Workshop on Hydrogen Production, Storage, and Use, US Department of Energy, 2003.

Callaghan, P. T. 1991. *Principles of Nuclear Magnetic Resonance Microscopy*. Oxford: Clarendon Press.

Chen, C.-N. and Hoult, D. I. 1989. *Biomedical Magnetic Resonance Technology*. Bristol: Adam Hilger.

Dunbar, Z. W. and Masel, R. I. 2007. Quantitative MRI study of water distribution during operation of a PEM fuel cell using Teflon(R) flow fields. *J. Power Sources* 171: 678–687.

Dunbar, Z. W. and Masel, R. I. 2008. Magnetic resonance imaging investigation of water accumulation and transport in graphite flow fields in a polymer electrolyte membrane fuel cell: Do defects control transport? *J. Power Sources* 182: 76–82.

Eikerling, M. 2006. Water management in cathode catalyst layers of PEM fuel cells: A structure-based model. *J. Electrochem. Soc.* 153: E58–E70.

Feindel, K. W., Bergens, S. H., and Wasylishen, R. E. 2006. Insights into the distribution of water in a self-humidifying H_2/O_2 proton-exchange membrane fuel cell using ^1H NMR microscopy. *J. Am. Chem. Soc.* 128: 14192–14199.

Feindel, K. W., Bergens, S. H., and Wasylishen, R. E. 2007a. Use of hydrogen–deuterium exchange for contrast in ^1H NMR microscopy investigations of an operating PEM fuel cell. *J. Power Sources* 173: 86–95.

Feindel, K. W., Bergens, S. H., and Wasylishen, R. E. 2007b. The influence of membrane electrode assembly water content on the performance of a polymer electrolyte membrane fuel cell as investigated by ^1H NMR microscopy. *Phys. Chem. Chem. Phys.* 9: 1850–1857.

Futerko, P. and Hsing, I. M. 1999. Thermodynamics of water vapor uptake in perfluorosulfonic acid membranes. *J. Electrochem. Soc.* 146: 2049–2053.

Feindel, K. W., LaRocque, L. P.-A., Starke, D., Bergens, S. H., and Wasylishen, R. E. 2004. *In situ* observations of water production and distribution in an operating H_2/O_2 PEM fuel cell assembly using ^1H NMR microscopy. *J. Am. Chem. Soc.* 126: 11436–11437.

Gach, H. M., Lowe, I. J., Madio, D. P., Caprihan, A., Altobelli, S. A., Kuethe, D. O., and Fukushima, E. 1998. A programmable pre-emphasis system. *Magn. Reson. Med.* 40: 427–431.

Gravina, S. and Cory, D. G. 1994. Sensitivity and resolution of constant-time imaging. *J. Magn. Reson. A* 104: 53–61.

Haacke, E. M., Brown, R. W., Thompson, M. R., and Venkatesan, R. 1999. *Magnetic Resonance Imaging— Physical Principles and Sequence Design*. Toronto: John Wiley & Sons.

Hirai, S. and Tsushima, S. 2008. Water transport and degradation analysis in PEMFC by *in-situ* MRI visualization. *ECS Trans.* 16: 1337–1343.

Hoogers, G. 2000. *Fuel Cell Technology Handbook*. New York, NY: CRC Press.

Hoult, D. I. 1978. The NMR receiver: A description, and analysis of design. *Prog. NMR Spectr.* 12: 41–77.

Ikeda, T., Koido, T., Tsushima, S., and Hirai, S. 2008. MRI investigation of water transport mechanism in a membrane under elevated temperature condition with relative humidity and current density variation. *ECS Trans.* 16: 1035–1040.

Kotaka, T., Tsushima, S., and Hirai, S. 2007. Visualization of membrane hydration path in an operating PEMFC by nuclei-labeling MRI. *ECS Trans.* 11: 445–450.

Kreuer, K. D., Paddison, S. J., Spohr, E., and Schuster, M. 2004. Transport in proton conductors for fuel-cell applications: Simulations, elementary reactions, and phenomenology. *Chem. Rev.* 104: 4637–4678.

Levitt, M. H. 2001. *Spin Dynamics: Basics of Nuclear Magnetic Resonance*. Toronto: John Wiley and Sons.

MacMillan, B., Sharp, A. R., and Armstrong, R. L. 1999. An NMR investigation of the dynamical characteristics of water absorbed in Nafion. *Polymer* 40: 2471–2480.

Mauritz, K. A. and Moore, R.B. 2004. State of understanding of Nafion. *Chem. Rev.* 104: 4535–4585.

Minard, K. R., Viswanathan, V. V., Majors, P. D., Wang, L. Q., and Rieke, P. C. 2006. Magnetic resonance imaging (MRI) of PEM dehydration and gas manifold flooding during continuous fuel cell operation. *J. Power Sources* 161: 856–863.

Ouriadov, A. V., MacGregor, R. P., and Balcom, B. J. 2004. Thin film MRI—High resolution depth imaging with a local surface coil and spin echo SPI. *J. Magn. Reson.* 169: 174–186.

Park, H. W., Ro, Y. M., and Cho, Z. H. 1988. Measurement of the magnetic susceptibility effect in high-field NMR imaging. *Phys. Med. Biol.* 33: 339–349.

Reichenbach, J. R., Venkatesan, R., Yablonskiy, D. A., Thompson, M. R., Lai, S., and Haacke, E. M. 2005. Theory and application of static field inhomogeneity effects in gradient-echo imaging. *J. Magn. Reson. Imaging* 7: 266–279.

Shim, J., Tsushima, S., and Hirai, S. 2009. High resolution MRI investigation of transversal water content distributions in PEM under fuel cell operation. *ECS Trans.* 25: 523–528.

Shim, J. Y., Tsushima, S., and Hirai, S. 2007. Preferential thinning behaviors of the anode-side of the PEM under durability test. *ECS Trans.* 11: 1151–1156.

Sivashinsky, N. and Tanny, G. B. 1981. The state of water in swollen ionomers containing sulfonic acid salts. *J. Appl. Polym. Sci.* 26: 2625–2637.

Slade, R. C. T., Hardwick, A., and Dickens, P. G.1983. Investigation of H$^+$ motion in Nafion film by pulsed ^{1}H NMR and A.C. conductivity measurement. *Solid State Ionics* 9&10: 1093–1098.

Slichter, C. P. 1990. *Principles of Magnetic Resonance: Third Enlarged and Updated Edition*. Berlin: Springer-Verlag.

Springer, T. E., Zawodzinski, T.A., and Gottesfeld, S. 1991. Polymer electrolyte fuel cell model. *J. Electrochem. Soc.* 138: 2334–2342.

Teranishi, K., Tsushima, S., and Hirai, S. 2002. Measurement of water distribution in polymer electrolyte fuel cell membrane by MRI. *Therm. Sci. Eng.* 10: 59–60.

Tsushima, S. and Hirai, S. 2009. Magnetic resonance imaging of water in operating polymer electrolyte membrane fuel cells. *Fuel Cells* 5: 506–517.

Tsushima, S., Hirai, S., Kitamura, K., Yamashita, M., and Takase, S. 2007. MRI application for clarifying fuel cell performance with variation of polymer electrolyte membranes: a comparison of water content of a hydrocarbon membrane and a perfluorinated membrane. *Appl. Magn. Reson.* 32: 233–241.

Tsushima, S., Nanjo, T., Nishida, K., and Hirai, S. 2005b. Investigation of the lateral water distribution in a proton exchange membrane in fuel cell operation by 3D-MRI. *ECS Trans.* 1: 199–205.

Tsushima, S., Teranishi, K., Nishida, K., and Hirai, S. 2005a. Water content distribution in a polymer electrolyte membrane for advanced fuel cell system with liquid water supply. *Magn. Reson. Imaging* 23: 255–258.

Vlaardingerbroek, M. T. and den Boer, J. A. 1999. *Magnetic Resonance Imaging*, 2nd Edition. Berlin: Springer.

Wang, M., Feindel, K. W., Bergens, S. H., and Wasylishen, R. E. 2010. *In situ* quantification of the in-plane water content in the Nafion membrane of an operating polymer–electrolyte membrane fuel cell using ^{1}H micro-magnetic resonance imaging experiments. *J. Power Sources* 195: 7316–7322.

Weber, A. and Newman, J. 2003. Transport in polymer–electrolyte membranes I. Physical models. *J. Electrochem. Soc.* 150: A1008–A1015.

Wielstich, W., Gasteiger, H. A., and Lamm, A. 2003. *Handbook of Fuel Cell—Fundamentals, Technology and Applications*. New York, NY: John Wiley & Sons.

Zawodzinski, T. A., Derouin, C., Radzinski, S., Sherman, R. J., Smith, V. T., and Springer, T. E. 1993. Water uptake by and transport through Nafion® 117 membranes. *J. Electrochem. Soc.* 140: 1041–1047.

Zawodzinski, T. A., Neeman, M., Sillerud, L. O., and Gottesfeld, S. 1991. Determination of water diffusion coefficients in perfluorosulfonate ionomeric membranes. *J. Phys. Chem.* 95: 6040–6044.

Zhang, Z., Marble, A. E., MacMillan, B., Promislow, K., and Balcom, B. J. 2008b. Spatial and temporal mapping of water content across Nafion membranes under wetting and drying conditions. *J. Magn. Reson.* 194: 245–253.

Zhang, Z., Martin, J., Wu, J., Wang, H., Promislow, K., and Balcom, B. J. 2008a. Magnetic resonance imaging of water content across the Nafion membrane in an operational PEM fuel cell. *J. Magn. Reson.* 193: 259–266.

12

Neutron Imaging

Pierre Boillat
Paul Scherrer Institut

Günther G. Scherer
Paul Scherrer Institut

12.1 Introduction

12.1.1 Overview

Proper water management is usually recognized as a key topic in the *proton exchange membrane fuel cell (PEMFC)* technology. Either too dry or too wet operating conditions can have a direct impact on cell performance, in particular limiting the maximal power density. Reducing these detrimental effects to their minimum is therefore a requisite to maximize the power which can be obtained out of a defined fuel cell size, or inversely, reduce the cell size—and subsequently its cost—for a given target maximal power. Besides the direct effect on power density, the topic of water management is highly relevant in terms of improving fuel cells durability. On one side, the durability of a fuel cell can be affected by the distribution of water (as a liquid or as vapor) in different ways. The chemical degradation of Nafion, a widely used polymer electrolyte, has been reported to be much faster in low-humidity conditions (Endoh et al., 2004; Sethuraman et al., 2008; Chen and Fuller, 2009). Dry conditions combined with sudden load changes can lead to localized temperature elevations in the membrane leading to the formation of pinholes (Schneider et al., 2008). Excess of water can also have negative impact on durability, as flooding can induce reactant starvation, which in turn favors degradation processes (Yousfi-Steiner et al., 2009). The amount of residual water present after cell shutdown is also critical. Carbon corrosion induced by start/stop cycles is reported to have a clear dependence on the relative humidity (Linse et al., 2009). When residual water is present in nonoperating fuel cell in regions subject to cold climatic conditions, damage can be induced by freeze/thaw cycles (Kim and Mench, 2007; Kim et al., 2008). On the other side, some degradation processes occurring in a fuel cell may disturb the water management, as, for example, a change in wetting properties of the porous media. Thus, part of the power losses of a degraded cell can for example have its origin in increased flooding. Proper water management should therefore be designed not only to be effective in the pristine cell, but also after several thousands of hours of operation. This later issue has not been addressed yet in the literature in the frame of visualization studies. Visualization techniques have an obvious interest in the context of water management studies, either with the idea of directly drawing conclusions from the observation of liquid water distribution, or with the aim of using liquid water distribution measurements in order to refine models of water transport. Therefore, these techniques have received an increasing attention during the last decade. Several different methods have been reported for the observation of liquid water in operating PEMFCs. These methods include optical imaging on transparent fuel cells (Zhang, F. Y. et al., 2006), nuclear magnetic resonance imaging (Tsushima et al., 2005; Minard et al., 2006; Dunbar and Masel, 2008; Zhang et al., 2008), neutron imaging (*cf.* Section 12.5), synchrotron x-ray imaging (Manke et al., 2007b; Büchi et al., 2008; Hartnig et al., 2008b), and Raman spectroscopy (Matic et al., 2005). This chapter, dealing with neutron imaging, is intended to explain the basics of this unique method, to demonstrate the results obtained out of it, and to motivate for further studies using this tool. In the examples and in the review being part of this chapter, the theme of durability will only scarcely be explicitly mentioned. But one should keep in mind, as cited above, that proper water management is in all cases relevant for fuel cell durability.

12.1.2 Basics of PEMFCs and Water Management

This chapter is not intended to give a description of the PEMFC technology, but to focus on the application of neutron imaging. However, for the sake of clarity, a few basics about the PEMFC technology and about water management will be reminded.

12.1.2.1 PEMFC Operation and Water Transport

The theory of operation of a PEMFC is simple: hydrogen gas (H_2) is oxidized at the anode and electrons are conducted through an external circuit, while protons are conducted though the membrane. At the cathode, electrons and protons combine with oxygen (O_2), supplied as a pure gas or from air, and the product of this reaction is water (H_2O). Water transport in a PEMFC implies several different physical phenomena. In the GDL and MPL, water can be transported as vapor by diffusion in dry conditions where gradients of relative humidity exist. In fully humidified conditions, water is transported by capillary flow. In the case of fully humidified conditions in the presence of temperature gradients, water can be transported by vapor diffusion, induced by the dependency on temperature of the water saturation pressure. In the flow channels, water is transported as vapor by convection and as droplets or slugs by viscous forces. Relevant phenomena for water transport across the polymer electrolyte membrane include diffusion induced by gradients of membrane hydration, permeation induced by gradients of hydraulic pressure, and electroosmotic drag, a process in which protons migrating through the membrane drag water molecules with them. Thermal gradients can also induce a water transport effect called thermoosmotic drag.

12.1.2.2 Effect of Water on Cell Performance

The effects of the lack or excess of water on cell performance can be separated into direct and indirect effects. *Direct effects* represent an impact on cell performance at a defined location, depending on the humidification state at this very location. They include the loss of membrane conductivity in the case of too dry conditions and the increase of diffusive mass transport losses due to accumulation of water in the porous media. They also include the total shutdown of the cell due to reactant starvation in the case of very severe flooding. All these local effects can be well studied in a small-scale *differential cell* such as defined in Section 12.4.2.1. In addition to these direct effects, accumulation of droplets or slugs in flow channels and in the cell manifolds can induce *indirect effects* on cell performance. At the scale of a cell, channels with a higher accumulation of water will exhibit a higher flow resistance, which induces an uneven flow distribution between the gas channels. The effective stoichiometric ratio for the regions with less flow will be reduced, leading to a loss of performance. On the single cell scale, this effect is somehow stabilized by the fact that the channels with higher water accumulation produce less current, and thus less water. At the scale of a fuel cell stack, the same indirect effects can appear. But, the cell being connected in series, a cell with lower gas flow will still produce the same current as all other cells, though with a lower voltage. The water production remains constant while the water removal reduces due to the reduced flow, potentially leading to the reversal of the cell. Such effects cannot be studied using single small-scale cell, but require hardware emulating a real stack.

12.2 Neutron Imaging Principles

12.2.1 Neutron Interaction with Matter

12.2.1.1 Types of Interaction

Neutrons, similarly as photons, have interaction with matter by different mechanisms such as absorption or scattering, the latter being either classified as incoherent or coherent scattering. The type of interaction is of high relevance for neutron scattering studies, in particular for the microstructural analysis of the atomic level (e.g., crystallography). On the contrary, for imaging studies, distinction is made in first order between neutrons transmitted through the sample and neutron interacting with the sample, regardless of the type of interaction. For the sake of simplicity, all interactions are considered as being equivalent to absorption, even though in many cases the dominating mechanism of interaction is scattering. For this simplified view to be reasonably accurate, the scattered neutrons must have a low probability of hitting the detector. Otherwise, the amount of scattered neutrons collected by the detector has to be either measured or calculated, and subsequently accounted for. Illustrations of a neutron imaging setup as compared to a neutron scattering setup is illustrated in Figure 12.1.

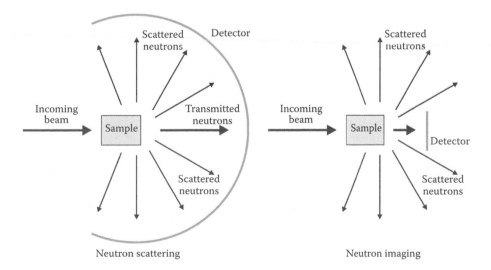

FIGURE 12.1 In a neutron scattering setup, the angular distribution of neutrons scattered in the sample is measured. In a neutron imaging setup, the spatial distribution of transmitted neutrons is measured.

12.2.1.2 Comparison with Photons

Photon beams such as x-rays interact with the electronic cloud of the atoms and neutrons interact with the atoms nuclei. Thus, while x-rays feature increasing sensitivity with the increase of the atomic number, making those mostly suited to detect relatively heavy elements, neutron sensitivity has no straightforward dependence on the weight of the elements. Interestingly, some light elements such as hydrogen, lithium, or boron, which are difficult to sense with x-rays, strongly interact with neutrons. This makes photons and neutrons complementary sensors for many applications. Additionally, the interaction with nuclei implies dependence not only on the count of protons of the atom, but also on its count of neutrons. In other terms, different isotopes of the same chemical element will exhibit different interaction with neutrons.

12.2.1.3 Microscopic Cross Section and Attenuation Coefficient

The neutron interaction with matter is described in a quantitative way by the *microscopic cross section* (σ), which has the dimension of an area and is usually expressed in *barns (b)* with the equivalence $1\ b = 10^{-28}\ m^2$. The microscopic cross section describes the probability of interaction of a neutron with a single atom, regardless of the concentration of atoms in the concerned material. The effective attenuation of a neutron beam will naturally depend on this concentration, and for practical use the quantity called *attenuation coefficient* (Σ) is introduced as the product of the microscopic cross section and the atomic density. The attenuation coefficient, which is sometimes called *macroscopic cross section*, has the unit of the inverse of a distance, usually expressed in cm^{-1}.

12.2.2 Imaging Configurations

12.2.2.1 Radiography

Radiographic imaging consists of a single exposure of the object. The intensity of the beam on a single point of the detector depends on the integral quantity and nature of materials in the sample along a single ray impinging on the considered point. Thus, only two-dimensional information is obtained. If this information is sufficient, radiography has the advantages of speed and setup simplicity.

12.2.2.2 Tomography

For tomographic imaging, a series of radiographs of the sample are acquired from different directions. In practice, this is done by rotating the sample around one axis (usually, around the vertical axis), as illustrated in Figure 12.2. Using a reconstruction algorithm, three-dimensional images of the sample are obtained.

More constraints are introduced than for single radiography: the sample must be free to rotate, and must be entirely contained in the field of view for what concerns the direction perpendicular to the rotation axis. Additionally, the acquisition of all images is time consuming (up to several hours), which limits the amount of experiments. For the particular case of fuel cells, meeting the constraint of sample stability during the acquisition time is not straightforward. The study of transient processes is also excluded.

12.2.2.3 Time-Resolved Imaging

Imaging can be considered as "time resolved" as long as the used acquisition time is low enough to catch dynamic processes occurring in a fuel cell. The limit between steady-state and time-resolved imaging is therefore arbitrary and depends on the kind of dynamic processes that are to be studied. For example, vibrations of droplets before their detachment from the GDL are processes involving dynamics in the millisecond scale. Several processes however occur on timescales of 1–5 min, as illustrated by the examples in this chapter. For these, acquisition rates of a few images per minute are well suited. A possibility to monitor reproducible processes is to conduct the experiment several times and cumulate the results

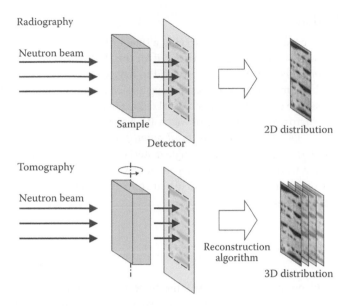

FIGURE 12.2 Acquisition of 2D and 3D data using radiography and tomography.

obtained for each sequence. This allows using short exposure times while obtaining a good image quality on the cumulated sequence. This technique was recently used successfully in fuel cell studies (Boillat et al., 2011; Schneider, I. A. et al., 2010).

12.2.3 Image Analysis and Processing

12.2.3.1 Referencing

Neutron imaging provides a unique combination of high transparency of fuel cell structural materials and high contrast for liquid water. However, even in these conditions, the attenuation contribution of the structural materials cannot be neglected. In order to distinguish between attenuation induced by the sample structure from attenuation due to water, a *referencing* process is done. In this process, a reference image is acquired, which represents the attenuation of the structure without liquid water. Typically, this is realized by flowing dry nitrogen into the cell. The attenuation due to the water only is then computed by dividing the image of the operating cell by the image of the dry cell pixel-wise, as illustrated in Figure 12.3. In some cases, structural modifications are induced by the change of humidity, due to the change in the swelling state of the membrane (Hussey et al., 2007). This problem can be solved by using other types of references, for example, a reference where the membrane is humidified with heavy water (*cf.* Section 12.4.4.1).

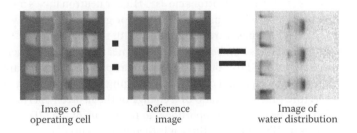

| Image of | Reference | Image of |
| operating cell | image | water distribution |

FIGURE 12.3 Extraction of the attenuation due to liquid water by a referencing process.

12.2.3.2 Quantification

Neutron imaging can be used not only for observing a qualitative distribution of water, but also for obtaining quantitative data. Two approaches can be used for this. In the first approach, called *segmentation-based quantification*, a pixel of the data is assumed to have only different discrete values (such as "full" or "empty"), and the number of pixel corresponding to each category is counted. Such an approach requires, in all but very simple cases, three-dimensional data as obtained from tomography, with a resolution higher than the microscopic structure of water. Both these conditions are usually not met for neutron imaging of fuel cells, and *attenuation-based quantification* is preferred. In this approach, all attenuation is considered as originating from a single material (in our case, water). This assumption is realistic for fuel cells when working on radiograms previously referenced such as described above. The accumulated thickness of water can then be calculated from the attenuation coefficient of water (Σ) using the Lambert–Beer relation:

$$\frac{I}{I_0} = e^{-\Sigma \times \delta} \tag{12.1}$$

where I_0 represents the incoming beam intensity, I the outgoing intensity, and δ the thickness of water. Due to the dependence of Σ on neutron energy, the effective value of the attenuation coefficient depends on the beam spectrum and on the sensitivity spectrum of the detector system. Usually, an empirical value of Σ is measured for a given setup, based on the measurement of water layers of known thicknesses.

12.3 Instruments and Measurements

12.3.1 Neutron Sources and Beam Lines

12.3.1.1 Neutron Source Types

Although other types of neutron sources exist, only two categories of them can provide a flux that is sufficient for performing imaging with reasonably low exposure times: nuclear fission reactors and spallation sources. Research reactors, with a typical nominal power of 10 MW, where neutrons are extracted from the moderator are optimized for the highest possible neutron output at the well-designed beam ports mostly linked to a specific moderator system. Spallation sources are based on the destructive collision of heavy nuclei (in the target) by high-energy particles like protons with energy in the order of 1 GeV. The emitted fast neutrons from the spallation act needs also to be moderated as in the case of reactor neutrons. Spallation neutron sources thus require to be "driven" by a proton accelerator. There are stationary sources (like SINQ, CH), but also pulsed ones (SNS, ISIS, or J-PARC).

12.3.1.2 Beam Line Characteristic Parameters

When considering a beam line for imaging purposes, the following parameters are to be taken into account:

- The *neutron flux* (Φ) is expressed in neutrons cm^{-2}s^{-1}. Higher neutron fluxes mean lower exposure times in order to obtain a sufficient count of neutrons for good image quality. Typical values of neutron fluxes for imaging purposes are in the range of 10^6–10^8 neutrons cm^{-2} s^{-1}.
- The *L/D ratio* is a dimensionless number representing a measure of the divergence of the beam at the position of the sample. High *L/D* values represent a quasiparallel beam and are usually required for high-resolution imaging. Typical *L/D* ratios for imaging beam lines range from 100 to 5000. Some beam lines feature tunable *L/D* ratio by a change of the neutron aperture. In this case, the neutron flux will usually be inversely proportional to the *square* of the *L/D* ratio. When comparing the neutron fluxes of different beam lines, it is therefore essential to compare them at similar values of *L/D* ratio.

- The *energy spectrum* represents the distribution of the neutrons energy. As the attenuation of materials is energy dependent, different spectra result in different effective attenuation coefficients. Imaging beam lines usually exhibit a spectrum in the thermal (12–100 meV) and cold (0.12–12 meV) neutron ranges. For fuel cell applications cold neutrons provide higher attenuations for water and are preferred, except for configurations with a long transmission path through water.
- The *beam size* at the position of the sample constitutes one of the limitations for the maximal sample size, although the detector itself can as well be a limiting factor. Typical beam diameters range from a few centimeters up to 40 cm. The largest beams thus allow imaging of a whole cell of technical size.

12.3.2 Detectors for Imaging Applications

12.3.2.1 Scintillator Screens with Optical Detectors

The combination of a scintillator screen with an optical detector can be considered as the standard detection setup for neutron imaging applications. In this configuration, the neutrons impinging on the scintillator screen are converted into visible light in a two step process: first, the neutrons are converted to a secondary radiation by a neutron capture material. The secondary radiation is then converted to visible light by the scintillating material itself. The visible light is recorded by a CCD camera chip. Some devices, based on amorphous silicon, can be placed in the beam directly next to the scintillator screen. Other devices are placed outside the beam, and the image of the scintillator screen is projected onto them using a mirror and an optical system.

12.3.2.2 Neutron-Sensitive Imaging Plates

Neutron-sensitive imaging plates in combination with a high-resolution scanner were up to recently the benchmark for high-resolution neutron imaging, featuring pixel sizes slightly bigger than 10 μm and effective resolutions under 100 μm. They are similar to x-ray imaging plates (e.g., for dental applications), with the addition of a converter material to make them sensitive to neutrons. The major drawback of the use of imaging plates is the multistep process for image acquisition: after the exposure of the sample, the plate must be placed in a scanner and finally be erased for further use. This excludes their use for applications such as tomography or dynamic imaging and strongly limits the throughput of experiments for stationary imaging. Furthermore, no pixel-wise referencing (*cf.* Section 12.2.3.1) is possible.

12.3.2.3 Microchannel Plates

Microchannel plates are plates made of glass, having cylindrical pores with diameters in the range of 10 μm (Siegmund et al., 2007; Tremsin et al., 2009). The glass is doped with neutron-absorbing elements such as ^{10}B or Gd. The secondary particles stemming from the capture of the neutrons produce the emission of electrons at the walls of the channels. These electrons are accelerated by a strong electric field and, if they hit the channel walls, result in the emission of more electrons, finally resulting in an avalanche effect. The resulting electron cloud is captured by a pixel array placed just behind the microchannel plate.

The use of microchannel plate as a detector for cross-sectional imaging of fuel cells was recently reported at NIST (Hussey et al., 2007).

12.3.2.4 Other Detector Types

Historically, other detector types have been used. For example, *track etch foils* are based on the conversion of neutrons to alpha particles using a ^{10}B based converter. The "scratches" produced by the alpha particles in the foils can be enlarged by chemical treatment in an alkaline bath. Although providing high resolution, the necessity of a chemical processing step renders them unpractical for experiments requiring a high throughput of data acquisition.

Recently, pixilated detectors (Pilatus, Eiger, Medipix), primarily designed for x-ray detection were made although sensitive for neutrons by adding Gd, ^{10}B, or ^{6}Li conversion layers. The potential of these systems is still under investigation. The major advantage of the systems is the low-noise level in the images.

12.3.3 High-Resolution Imaging

Over the last few years, cross-sectional imaging of polymer electrolyte fuel cells has been a strong driving force for improving the spatial resolution of neutron imaging. The major advances realized are presented here. The concept of resolution is sometimes falsely assimilated to the pixel size, which is only one of the resolution limiting factors. To avoid any confusion, we will speak of *effective resolution*, which represents the resolution of the setup including any unsharpness effects, and will be measured using the full-width at half-maximum (FWHM) definition.

12.3.3.1 Resolution Enhancements

Three major factors can be considered as limiting for the spatial resolution:

- The effective pixel size of the setup, which depends on the pixel size of the sensor, and on the conversion factor of the optical system
- The inherent unsharpness of the detector (e.g., originating from the scintillator screen)
- The geometric unsharpness introduced by the beam divergence

Realizing a high-resolution setup implies addressing all these factors. In the approach used at PSI (Lehmann et al., 2007), based on the conventional scintillator screen/CCD camera setup, the pixel size was reduced by using custom-made optics, providing a 1:1 conversion factor between the scintillator plane and the CCD plane, while keeping a very good light efficiency. The unsharpness introduced by the conversion process was addressed by using thin scintillator screens. Combining these, a pixel size of 13.5 μm, and an effective resolution of approximately 50 μm were obtained. The work realized at NIST (Hussey et al., 2007; Hickner et al., 2008b) is based on microchannel plates, as described in Section 12.3.2.3. An effective resolution of 25 μm was reported.

The so-called *geometric unsharpness* is introduced by the divergence of the beam, combined by the necessary distance between the sample and the detector, as illustrated in Figure 12.4. The divergence of the beam is characterized by the *L/D ratio*, defined as the ratio between the source aperture (*D*) and the distance from the source aperture to the sample (*L*).

The geometric unsharpness r_G can be calculated using the simple formula:

$$r_G = \frac{d}{L/D} \tag{12.2}$$

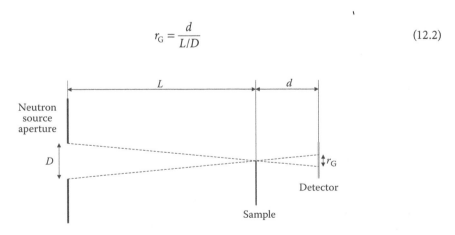

FIGURE 12.4 Unsharpness induced by the divergence of the neutron beam.

where *d* represents the distance between the sample and the detector. Reducing the geometric unsharpness thus implies either reducing *d*, which is for practical reasons not always possible, or increasing the *L/D* ratio. The latter can be done either by reducing the aperture size—some beam lines feature a tunable aperture—or using a measurement position as far as possible from the aperture. In both cases, the neutron flux at the position of the sample is reduced, and longer exposure times are required. If, as a first approximation, the neutron flux is considered proportional to the source aperture area, the flux reduction is proportional to the *square* of the increase in *L/D* ratio. The trade-off between temporal and spatial resolution also applies to the reduction of pixel size: by reducing the pixel area, a higher exposure time is needed in order to keep the same count of captures neutrons. Again, a reduction of temporal resolution scaling to the square of the improvement in spatial resolution is expected.

12.3.3.2 Anisotropic Setups

As mentioned in the previous section, enhancements in spatial resolution usually imply a reduction in temporal resolution. In specific applications—such as cross-sectional imaging of fuel cells—where a high spatial resolution is only required in a specific direction, this reduction can be mitigated by using anisotropic setups. Concerning the *L/D* ratio, an anisotropic setup means that the *L/D* ratio is different in horizontal and vertical directions. This can be realized in practice by using a slit aperture instead of a round aperture. The aperture size being reduced only in one direction, the scaling between resolution improvement and flux reduction is now linear. Besides the *L/D* ratio, the other limitations—pixel size and inherent blurring—can also be addressed by an anisotropic approach. By tilting the detector around a vertical axis, the image of the sample is projected on a skew surface, resulting in a magnification of the image in a specific direction. As this magnification can be considered as occurring "before" the image acquisition, the inherent blurring of the detector, as well as the pixel size, will be reduced when converting back to the real sample dimensions. Again, the reduction of temporal resolution will be moderate, as the effective pixel dimension is only reduced in one direction.

As already reported (Boillat et al., 2010) and illustrated in Figure 12.5, the combination of detector tilting and anisotropic beam collimation enabled reaching effective spatial resolution better than 10 μm.

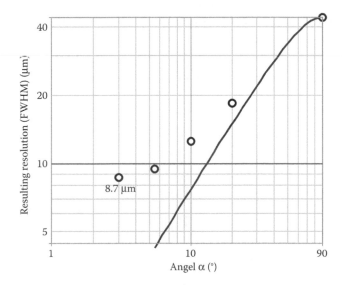

FIGURE 12.5 Measured spatial resolution (FWHM) as a function of the tilting angle. (Reproduced from Boillat, P. et al. 2010. *Electrochem. Solid-State Lett.* 13: B25–B27. With permission from the Electrochemical Society.)

Typical setups used for fuel cell imaging combine an effective spatial resolution of approximately 20 μm and exposure times of 10 s only. As a comparison, for measurements reported with a similar resolution but using an isotropic setup (Turhan et al., 2010), typical exposure times of 1 min are used.

12.4 Application to PEMFCs

This section presents a few examples of the application of neutron imaging to PEMFCs corresponding to original work of the authors. These examples are not meant to cover the whole field of applications, in particular for through-plane imaging, and the reader interested in a broad overview is invited to consult the review part of this chapter (Section 12.5).

12.4.1 Through-Plane Imaging

In through-plane imaging, the cell is placed such that the beam axis is perpendicular to the plane of the membrane, as illustrated in Figure 12.6. This offers the advantage of being able to observe the distribution of water over large areas, and can potentially be applied to cells of technical size, as illustrated by the numerous publications mentioned in Section 12.5. In the present examples, however, through-plane imaging is applied to a small-scale cell (0.5 cm^2 of active area).

12.4.1.1 Water Measurement in Flow Channels

The radiograms corresponding to the water distribution at different relative humidities of the cathode gas flow are shown in Figure 12.7. The relative humidity of the anode gas flow is kept constant at 90%.

At some specific locations, water droplets appear even with a nonsaturated gas flow (down to 60% relative humidity for one specific location). The apparition of a liquid water droplet in an unsaturated gas flow is somewhat counter intuitive. Possible mechanisms explaining this behavior are.

- Locally higher relative humidity due to diffusive uptake of water vapor by the gas flow
- Limitations in evaporation kinetics
- Lower temperature in the near vicinity of the droplet induced by water phase change

A further insight is given when observing the temporal evolution of a single droplet, as illustrated in Figure 12.8. For a relative humidity of 60–80%, the water grows until an equilibrium size, at which its evaporation and refilling rates equilibrate. At 90%, the droplet grows until it is detached, and a new

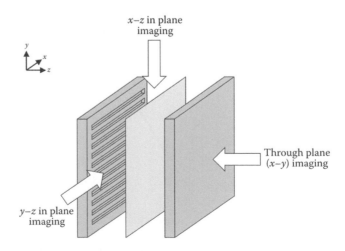

FIGURE 12.6 Definition of the imaging directions (through plane and in plane) relatively to the cell structure.

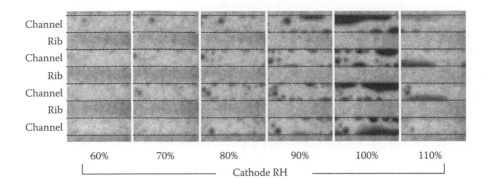

FIGURE 12.7 Referenced radiograms of water distribution for different relative humidities of the cathode gas flow. Gas flow is from left to right.

droplet starts to grow at the very same place. The reason of the large increase at 100% is a slug occupying the same position.

12.4.1.2 Water Measurement in Porous Media

An often cited issue in through plane neutron imaging is the inability to directly distinguish the different layers of the cell. Several strategies have been proposed to workaround this problem, as mentioned in Section 12.5.1.2. We will here propose a method based on histogram analysis, in order to distinguish water in the flow channels from water in the MEA (including GDL) and use it on the radiograms displayed in the previous section. The analysis is based on the following assumptions:

- The quantity of water in the MEA is considered as homogeneous over a local region (but with possible differences between rib and channel regions).
- Water in the channels is considered to be inhomogeneously distributed (e.g., as droplets or slugs).

In these conditions, when plotting a histogram of the neutron transmission intensity, channel water produces a broad and flat contribution, while the homogeneously distributed water in the MEA

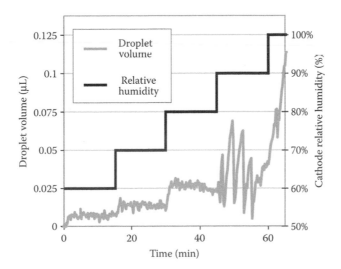

FIGURE 12.8 Temporal evolution of the size of a single droplet, when increasing the cathode flow relative humidity.

produces a sharp peak. The mean value of neutron transmission includes contribution of both channel and MEA water, while the *modal* value (the position of the highest peak of the histogram) represents MEA water only. This is represented in Figure 12.9, based on the data of the radiogram with a cathode relative humidity of 100%.

We can thus approximate the MEA water by the modal value of the histogram and the channel water by the difference between modal and mean value. The result of this analysis for different values of cathode relative humidity, with a separation of rib and channel areas, is illustrated in Figure 12.10.

As naturally expected, no "droplet water" can be measured in the rib area (but for a slight measurement error). At 100% of cathode relative humidity, a large amount of droplets is found in the channel area, but the amount of water measured in the MEA only slightly increases. This indicates that the proposed analysis successfully removed the major part of the channel water contribution. Additional verification experiments would however be needed in order to assess whether the slight increase of measured MEA water is real or corresponds to a residual error. With an over humidified flow, the amount of water in the channels decreases again (as it can be seen from the radiograms as well). A possible reason is that droplets brought externally from the gas flow help to "wipe out" the droplets forming in the gas channels.

A limitation of the method has to be cited: in the situation when a film of water covers the bottom of the channel the assumption that channel water is unevenly distributed will not hold, and this amount of water will falsely be attributed to the MEA. In order to assess the validity of the proposed analysis, a possibility would be to compare through-plane and in-plane images taken successively. The major difficulty is to ensure that the cell remains in the very same state between the two acquisitions.

12.4.2 In-Plane Imaging

12.4.2.1 Definition of x–z and y–z Imaging

The term "in-plane imaging" refers to the configuration where the beam axis is parallel to the membrane plane (*cf.* Figure 12.6). For flow field designs having a dominant channel orientation, such as parallel or serpentine flow fields, distinction can further be made between an imaging configuration where the beam axis is perpendicular or parallel to the channels direction. These two configurations are

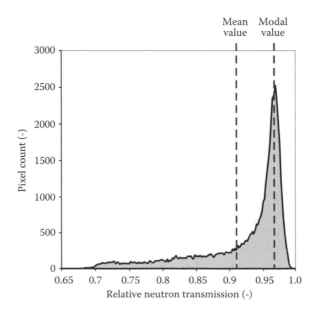

FIGURE 12.9   Difference between mean and modal value.

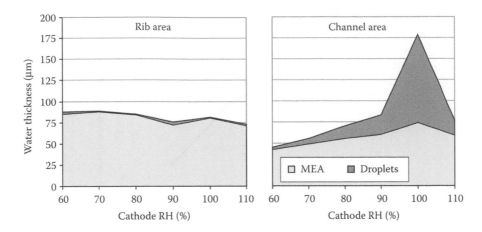

FIGURE 12.10 Measured amount of MEA and droplet water based on histogram analysis.

respectively referred to as *x–z imaging* and *y–z imaging*, according to the directions of Figure 12.6 which are visible in the image. According to this convention, through-plane imaging could as well be referred as *x–y imaging*.

The choice between *x–z* and *y–z* imaging depends on the type of cell used. In a cell of standard size operated with technically relevant stoichiometries, large inhomogeneities are expected along the channels, and thus *x–z* imaging is usually preferred. The major drawback is the loss of information about the distribution on the rib/channel scale. Often, *differential cells* are used for fuel cell research. These cells are of reduced dimensions but are operated with gas flow velocities similar to those of a technical cell, which results in very high stoichiometric ratios and a homogeneous distribution of parameters such as temperature, current density and humidity in the direction along the channels. In this case, the distribution in direction *x* is of reduced interest and the *y–z* configuration is usually preferred.

12.4.2.2 Measurement of Cross-Sectional Water Distribution

In order to illustrate the possibilities of cross-sectional imaging, the water distribution as a function of current density in a small-scale differential cell is presented hereunder. Radiograms are acquired at different positions during the acquisition of current–voltage curves. In Figure 12.11, the performance of the cell in two different conditions of local relative humidity is presented, while the corresponding radiograms are depicted in Figure 12.12.

In the high-current density range, the performance curves exhibit significant differences between the two operating conditions. By comparing the radiograms, a clear difference is observed. For the higher-humidity condition (90/90%), water accumulates in all regions of the cathode side GDL, as well as under the ribs of the anode side GDL. For the lower-humidity condition (100/0%), water accumulates exclusively under the ribs, but the amount of water there is similar to the one of the high-humidity condition. As it can reasonably be assumed that the accumulation of water on the anode side has no direct negative effect on performance, water in the GDL under the cathode channels seems to be responsible for the reduction of cell performance. This is not unexpected since this region is on the way of the diffusive gas flow for the entire active area.

Besides direct observations on its relation to the local cell performance, the observation of liquid water distribution as a function of operating conditions can provide very valuable information in order to understand water transport on a local scale, in order, for example, to validate numerical studies. An interesting observation from the radiograms acquired under the high-humidity condition is that the water distribution remains virtually unchanged in a wide range of current density, from approximately 0.2 up to 1.5 A cm^{-2}. This result suggests the existence of a hard limit of water saturation values under fuel cell operation, as an intrinsic property of the porous media.

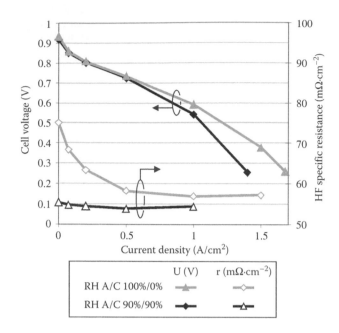

FIGURE 12.11   Cell IV performance in different conditions of relative humidity.

12.4.3 Transients of Water Distribution

In the previous example, only steady-state conditions were considered, resulting in a limited amount of information. If the imaging setup allows exposure times short enough, inclusion of the temporal dimension opens a wide field of new possibilities. The examples below illustrate two interesting aspects of water distribution transients.

12.4.3.1 Water Distribution Transients Following Local Condition Changes

As mentioned earlier, observation of liquid water in operating PEMFCs is not only of interest for making direct relations to local performance losses, but also to help understand the phenomena governing water transport. The operation of PEMFCs being highly dynamic, the exclusive study of equilibrium conditions is not sufficient. Typically, the reaction of the cell to different stimuli such as a change of current density or a change of local relative humidity is of interest. In Figure 12.13, the evolution of water content in the cathode GDL as the consequence of a change in local relative humidity is presented. The temporal evolution of the cell high-frequency resistance, representative of the humidification state of the membrane, is shown as well.

The beneficial process (humidification of the membrane) is faster than the detrimental process (accumulation of liquid water), and the cell performance reaches a peak during the transient state. This result illustrates the interest of understanding water transport in dynamic conditions for designing new control strategies.

12.4.3.2 Water Distribution Transients as a Diagnostic Tool

Apart from studying them to understand transport processes, water distribution transients can be used as a tool for identifying the relation between water accumulation and performance losses. An example is given in Figure 12.14, showing the temporal evolution of liquid water amount after a change in the gas relative humidity.

The area of the cathode GDL under the flow channels dries faster than the area under the ribs. The temporal evolution of liquid water in both regions can be compared with the temporal evolution of cell

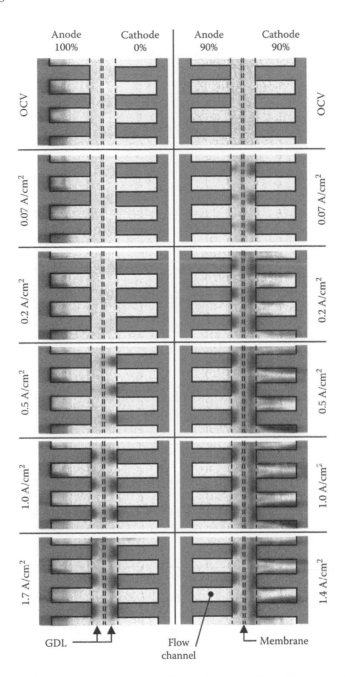

FIGURE 12.12 Measured distribution of liquid water for the operating points of Figure 12.12.

performance. The period corresponding to an increase in cell performance is well correlated with the drying of the GDL under channels, suggesting that the performance losses due to liquid water are dominated by the area under the channels.

12.4.4 ^{1}H–^{2}H Contrast Imaging

An interesting characteristic of neutron imaging is its sensitivity to different isotopes of the same chemical element. In particular, the neutron cross section of ^{2}H is almost one order of magnitude lower than

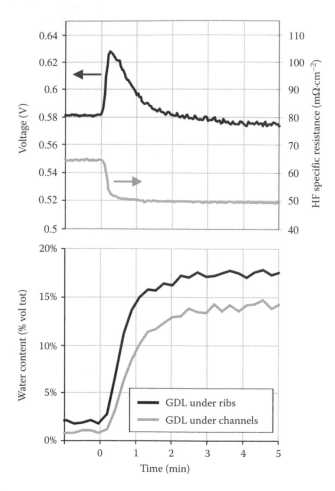

FIGURE 12.13 Evolution of cell performance and liquid water distribution following a dry => wet change of relative humidity. The cell is operated at 1 A cm^{-2}. The anode/cathode relative humidity is changed at $t = 0$ from 40/0% to 90/90%.

the one of ^{1}H, rendering heavy water (^{2}H$_2$O) almost invisible in comparison to light water (^{1}H$_2$O). This contrast opens new field of possibilities for fuel cell neutron imaging, among which two examples are given below. In this section, we refer to the hydrogen isotope as ^{1}H, and to the deuterium isotope as ^{2}H. When simply using the symbol H, this refers to hydrogen as a chemical element, meaning any of its isotopes, or any mixture of them.

12.4.4.1 Use of ^{2}H$_2$O Humidification for Reference Images

When realizing in-plane imaging of fuel cells, the dimensional changes related with the swelling of the membrane constitutes an issue (Hussey et al., 2007). In order to compensate for the attenuation of the structural material, a referencing process is needed, as described in Section 12.4.4.2. Typically, an image of the dry cell is used as reference. As the membrane swells, regions that were outside the membrane become part of the membrane and the structural attenuation is modified, introducing imaging artifacts. Moreover, the gas diffusion layers (GDLs) are further compressed by the swelling of the membrane, which increase their density and can induce errors in quantification. This problem is highly disturbing when making experiments using thick membranes. The use of ^{1}H–^{2}H contrast imaging can conveniently solve this problem. When gases humidified with ^{2}H$_2$O are fed to the cell, the membrane swells with heavy water.

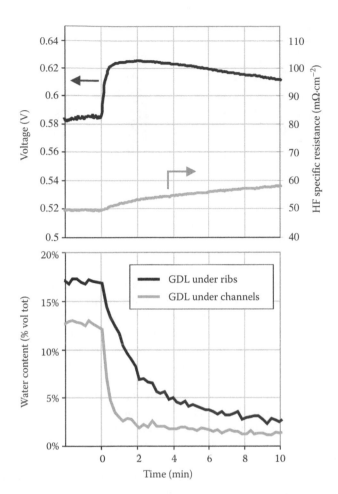

FIGURE 12.14 Evolution of cell performance and liquid water distribution following a wet ⇒ dry change of relative humidity. The cell is operated at 1 A cm^{-2}. The anode/cathode relative humidity is changed at $t = 0$ from 90/90% to 40/0%.

The structural changes will be the same as when humidified with light water, but the neutron attenuation related to ^{1}H atoms would have disappeared. The example in Figure 12.15 illustrates the potential of contrast imaging for solving the swelling problem. In this example, a single piece of Nafion 117 (approx. 180 μm thick) is compressed between a GDL and a PTFE sheet. The image of the membrane exposed to water vapor with a relative humidity of 80% is referenced once to an image of the dry membrane, and once to an image of the membrane exposed to heavy water vapor with the same relative humidity.

When using the dry reference, strong imaging artifacts are observed on the left side of the membrane, where it can expand due to the compressibility of the GDL. When using the ^{2}H$_2$O reference, these artifacts disappear completely. It can also be observed that the measured intensity is slightly lower (lighter shade of gray) when using the dry reference. The reason for this can be the higher density of the membrane structure in the dry state, or residual water present in the dry reference. Thus, references using heavy water are expected to give better quantitative results as well.

12.4.4.2 Use of ^{2}H Labeling for Studying Transport Parameters

The isotopic sensitivity of neutron imaging provides the potential to make observations beyond the distribution of liquid water, and give an insight into transport phenomena. In a simple representation,

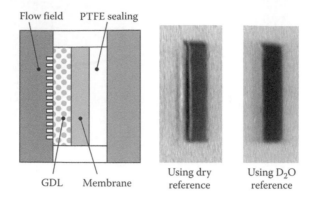

Flow field   PTFE sealing

GDL   Membrane

Using dry reference

Using D$_2$O reference

FIGURE 12.15 Result of referencing using a dry image as a reference, and using an image of the membrane swollen with heavy water as a reference.

when switching from normal operation to operation where a source of hydrogen atoms is labeled with ^{2}H, the water will progressively be replaced by heavy water following through the water transport paths. However, this representation is a crude oversimplification, as exchange processes involving H atoms can represent flows having the same order of magnitude than the water fluxes involved, even at the highest current densities. This was pointed out in a study on replacement of ^{1}H atoms in the membrane by ^{2}H atoms, following a change of the fuel gas from hydrogen (^{1}H$_2$) to deuterium (^{2}H$_2$). At low-current densities, the rate of replacement of H atoms is observed to be much higher than could be explained by the electrochemical production of water (Boillat et al., 2008b). This effect is easily understandable by considering a relatively high exchange current density of the hydrogen oxidation reaction (HOR). No consensus exists in the literature about the value of this exchange current density. Values over a wide span from 0.1 up to 200 A cm^{-2} (referring to the active surface of platinum) have been reported (Feindel, 2007; Neyerlin, 2007). For electrodes with a typical ratio of 200 between the platinum area and the geometrical area, these values represent a current density ranging from 20 to 40 A cm^{-2}. The upper range of this span is consistent with the observations reported above. The interpretation of *in situ* data collected by ^{2}H labeling experiments therefore needs a proper understanding of the exchange processes, among which the HOR exchange current density is only an example. In this optic, experiments were realized on a nonoperating fuel cell with hydrogen flown on both sides in order to isolate such effects. The results of these experiments were analyzed based on a simple transport model. In the bulk of the membrane, diffusive transport based on Fick's law is assumed, as in the representation described in previously published work (Boillat et al., 2008b). However, finite exchange rates between the membrane and the gas phase are considered as illustrated in Figure 12.16.

The exchange of H atoms between the membrane and the hydrogen gas phase is called k_H and is considered as originating from the HOR exchange current density. The exchange of H atoms between the membrane and the water vapor phase corresponds to a concurrent condensation/evaporation process, and is called k_W. A more detailed description of this representation is to be found in a forthcoming publication (Boillat et al., 2011). The interest of using ^{2}H labeling is the possibility to study these exchanges with a membrane humidification which is homogeneous and steady. The results of two types of experiments are presented here. In the first experiment, symmetric conditions are applied between the anode and cathode side, and either the hydrogen gas or the water vapor is labeled using ^{2}H. In these conditions, no transport across the membrane occurs and the fraction of ^{1}H isotope in the membrane depends only on the ratio between k_H and k_W. For a temperature of 70°C and a relative humidity of 80%, we measured a ^{1}H isotope fraction of approximately 0.5 in the membrane (*cf.* Figure 12.17), which indicates that for this sample the values of k_H and k_W are approximately identical.

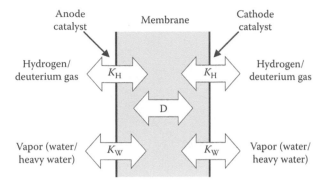

FIGURE 12.16 Illustration of the simple isotope transport model used for analyzing the experimental results.

In the second experiment, both the hydrogen and water vapor at the anode are labeled (meaning that we use deuterium gas humidified with heavy water), while normal hydrogen and water are used on the cathode side. The resulting profile of ^{1}H isotope fraction as shown in Figure 12.18 depends on the relation between the sum of interfacial exchanges and the H diffusivity in the membrane.

Thus, based on the previous knowledge of the k_H/k_W ratio, and using a value for the H self-diffusivity in Nafion 117 measured by NMR in similar conditions from the literature (Ochi et al., 2009), the value of k_H and k_W can be calculated. For the present experiment, we obtained a value of approximately 8×10^{-6} mol cm^{-2} s^{-1} for both k_H and k_W. As a comparison, the flow of H atoms equivalent to a current density of 1 A cm^{-2} is equal to ~10^{-5} mol cm^{-2} s^{-1}, in the same order of magnitude of our calculated values for k_W and k_H. This means that these exchange processes cannot be neglected when interpreting the results of labeling experiments. The measurement of k_W has further implications: in water transport modeling, the assumption is sometimes made that the interfacial exchange is infinitely fast for water, meaning that the water content in the membrane is in equilibrium with the water amount in the gas phase, even in the presence of a water flow across the membrane. Our calculated value of k_W indicates, on the contrary, that the interfacial resistance cannot be neglected. Beside the interest for studying water transport, the measurement of these interfacial exchanges can also be useful for assessing catalyst degradation, as discussed in Section 12.7.3.

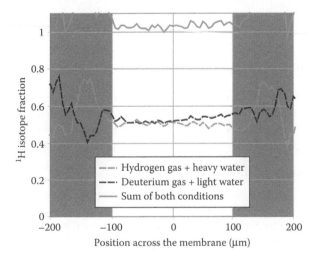

FIGURE 12.17 Profile of ^{1}H isotope fraction measured across the membrane when applying symmetric conditions.

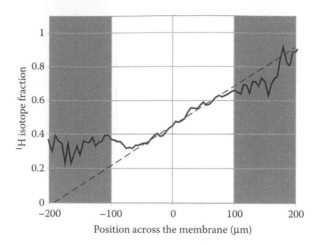

FIGURE 12.18   Profile of ¹H isotope fraction measure across the membrane. On the anode side (left), deuterium gas humidified with heavy water is used. On the cathode side (right), hydrogen gas humidified with normal water is used.

Some limitations of this method have to be mentioned. Differences in ¹H isotope fraction were observed between different positions along the membrane, which indicate that our sample may have suffered from inhomogeneities. The results presented here above need to be repeated, if possible with a more homogeneous sample. The assumption that the exchanges represented by k_H and k_W are independent from each other is yet to be verified, otherwise the model has to be improved to take into account a possible coupling. A clear drawback is the necessity of relying on *a priori* values for the diffusivity, based on published values measured on different samples. A possibility for solving this issue is to rely on transient experiments, from which the diffusivity in the membrane can be directly extracted (Boillat et al., 2011).

12.5   Literature Review/Recent Advances

This section is intended to give an overview on the reported activities involving neutron imaging of fuel cells over the past 10 years. This review is organized according to the different topics of study and/or techniques involved.

12.5.1   Through-Plane Imaging

Although the first report on the use of neutron imaging on fuel cell was using an in-plane imaging configuration on a very thick membrane (Bellows et al., 1999), the majority of the available literature concerns through-plane imaging. Due to the abundant literature on this topic, this review is further divided into different topics of study.

12.5.1.1   Evolution of the Method/Quantification

The use of neutron imaging of a fuel cell in through-plane imaging configuration was first reported by Geiger et al. (2003) and concerned the observation of gas clusters in a DMFC. Radiographies of operating PEMFCs were reported by Satija et al. (2003) including quantified values of water accumulation, although no detailed information on the quantification method is available. The use of neutron imaging on PEMFCs was also reported by Pekula et al. (2005). Kramer et al. (2005a) reported studies on DMFCs, including a quantification method based on segmentation. A detailed method based on statistics in order to avoid quantification errors in PEMFCs was then published (Kramer et al., 2005b). All studies cited above were published in the time frame between 2003 and 2005, after which the focus of published work is the study of fuel cells using neutron imaging, rather than the method itself.

12.5.1.2 Discrimination between Water in Different Layers

A frequently reported issue with through-plane imaging is the inability to directly separate the contributions from different layers of the cell (e.g., porous media and flow channels). Different approaches were proposed besides the use of in-plane imaging (*cf.* Section 12.4.2) or tomography (*cf.* Section 12.5.3). Chen et al. (2007) used a special design where the anode ribs face the cathode channels, and claimed that water would mainly be found under the ribs, thus allowing their approach to discriminate between the anode and cathode porous media. Owejan et al. (2007b) used a flow field with the anode and cathode channel perpendicular to each other, allowing discriminating water slugs in anode and cathode channels. The amount of water in the porous media under the channels was estimated using an algorithm extrapolating the amount of water measured in areas made of lands only. The main drawback of these methods is their potential effect on water distribution and transport, due to the change of vapor diffusion, capillary transport and thermal conduction pathways. More recently, Owejan et al. (2009b) proposed a method comparing the water purge rate in a previously operated cell with that of cells assembled with either an anode or cathode GDL partially saturated with water. This method has the advantage of requiring no design modification, but the drawback of being time consuming and of providing the distribution at a snapshot moment only. However, when not used as a stand-alone method but combined to water purge studies as the authors did, it remains an attractive option.

12.5.1.3 Effect of GDL Materials in PEMFCs

A first study including a comparison between different GDL materials was published by Zhang, J. et al. (2006). The water accumulation in three different types of porous media, all of them having an MPL, was compared. By observing significantly different performance losses in two materials having similar average water contents, but different channel/rib distributions, the authors pointed out that water accumulations under ribs and channels may have a different effect on cell performance. Owejan et al. (2007b) compared the amount of water accumulations in cells with three different carbon paper types, one of them without MPL. They measured higher water contents as well as a lower cell performance for the paper without MPL. Yoshizawa et al. (2008) compared water accumulation in cells having either carbon paper or carbon cloth as diffusion media. They measured a higher water amount in the cell with carbon cloth. In this study, the performance of the cell with higher water content was better, which the authors explained by the broader pore size distribution of the carbon cloth GDL, which provides different pathways for water and gas transport.

12.5.1.4 Effect of Flow Channels Shape/Wetting Angle in PEMFCs

A few studies deal with the effect on liquid water distribution of the shape and wetting angle properties of the flow channels. In the work published by Zhang J. et al. (2006), cells with flow fields made of either graphite or more hydrophilic gold-coated aluminum are compared. The authors observed the tendency of water to form small droplets at the edge of channels for the gold-coated aluminum, and larger droplets and slugs for the graphite. Owejan et al. (2007b) compared channels of triangular and rectangular shapes, and further examined the effect of coating gold-plated channels with PTFE. In both cases, an effect on the amount and morphology of water is observed: triangular channels held less water than rectangular ones, and PTFE coating tended to reduce the amount of water in the channels. Turhan et al. (2008) studied the effect of different land and channel sizes and as a general trend observed a lower amount of water for larger channel/land sizes.

12.5.1.5 Effect of Flow-Field Geometries of PEMFCs

A relatively high number of studies dealt with flow-field geometries and their effect on liquid water accumulation. A recurring observation is the effect of flow channel bents in serpentine flow fields, where several publications report a preferential accumulation of water before and at the bents (Satija et al., 2003; Pekula et al., 2005; Kramer et al., 2005b; Trabold et al., 2006; Zhang, J. et al., 2006; Hickner et al., 2010a,b). Kramer et al. (2005b) compared water accumulation in serpentine and interdigitated flow fields and observed that the removal of water from the latter is very poor. Owejan et al. (2006) studied interdigitated flow fields as well and measured a strong effect of water accumulation in the porous media on the pressure drop across the cell. Li et al. (2007) proposed a design method for the channels dimensions in order to optimize water removal, and verified its effectiveness using neutron imaging. Murakawa et al. (2009) compared water accumulation in single and threefold serpentine flow fields. They observed a similar amount of water, though the water accumulating in the threefold design was less constant over time, which correlated with variations of the pressure drop and of the cell voltage. Spernjak et al. (2010) compared three different geometries: one single serpentine, a parallel, and an interdigitated flow field. They observed the lowest water accumulation and best performance for the single serpentine and higher water content for the other flow fields, although a significant impact on performance was only observed for the cell with parallel channels. Finally, in a miniaturized cell with microchannels, Seyfang et al. (2010) compared a regular serpentine flow field with an optimized flow field with merging channels, and observed a positive effect on the removal of water.

12.5.1.6 Effect of Operating Conditions

Besides studies dealing with different cell constructions, several reports about the effect of operating conditions are found in the literature (Turhan et al., 2006; Zhang, J. et al., 2006; Hickner et al., 2008a). Typically varied parameters are the gas flows and the inlets relative humidity, the temperature, and the operating pressure. Hickner et al. (2006) observed an amount of water reaching a maximum at middle current densities, and decreasing at highest current densities, even in the case of a constant gas flow. Kim, T. J. et al. (2009) varied the pressure difference between the anode and cathode, but found no notable effect on liquid water distribution.

12.5.1.7 Study of Water Purge in PEMFCs

Recently, attention has been placed not only on the accumulation of water during cell operation, but also on the purge procedures which can be applied for removing the water after shutdown. As mentioned in the introduction, residual water after cell shutdown can have a negative effect on cell durability. Cho et al. (2009) studied the effect of using different gas flow humidification during the purge, and concluded that purging with dry gas on the anode and fully humidified gas on the cathode resulted in removal of liquid water without drying the membrane. Owejan et al. (2009a,b) studied the removal of water in a previously operated cell by a sequence including first, a rapid reduction of the pressure for dragging remaining water slugs out of the channels, followed by a purge with constant gas flow of dry nitrogen on the cathode only. They observed that the pressure reduction effectively removed slugs in the channels without affecting water in the GDLs, and subsequently analyzed the evolution of the drying front as a function of time.

12.5.1.8 Simulated Stack Operation

As described in Section 12.1.2.2, some indirect effects of liquid water can only be observed in real stack operation. Up to now, no neutron imaging measurements on a fuel cell stack were reported, except for the tomographic measurements mentioned in Section 12.5.3 of this review. The study of instabilities induced by indirect effects of water was however reported by Owejan et al. (2007a) in a so-called *simulated stack* mode. Instead of using a fixed gas flow as in usual single-cell measurements, the authors applied a constant differential pressure on the fuel cell cathode, in order to simulate the way this single cell would "view" the rest of a stack. They found that, from an initial value of more than 3, the cathode stoichiometric ratio is reduced to nearly 1 as water accumulates in the gas channels and manifolds, which eventually leads to complete cell failure. Another experiment approaching stack operation was reported by Siegel et al. (2008). In this case, a cell was operated with the anode in *dead end mode*, a way of controlling the hydrogen delivery that minimized the complexity of the external system and is therefore attractive for portable applications. The authors measured the water accumulation during this operation mode and drew correlations with the evolution of cell performance. They also studied the effect of short purges of the anode using a solenoid valve.

12.5.1.9 Study of DMFCs

Although one of the first applications of neutron imaging to fuel cells was the study of gas bubbles in DMFCs (Geiger et al., 2003), this topic has been much less studied than the accumulation of water in PEMFCs. Kramer et al. (2005a) compared serpentine and parallel flow-field geometries and established a relation between the methanol flow and the amount of gas-filled anodic channels. Recently, two studies by Schröder et al. (2009, 2010) were published about gas and water accumulations in DMFCs. In the first study, serpentine and grid-type flow fields were compared. In the second one, GDL with different wetting angles were compared. The authors observed no significant effect of the anode-side wetting angle, but observed a correlation between the cathode-side GDL wetting angle and the water accumulation and the cell performance.

12.5.2 In-Plane Imaging

The first report about performing in-plane imaging on fuel cell was published by Bellows et al. (1999). This study was focused on the water distribution across the membrane. To deal with the limited spatial resolution available at that time, the authors used a very thick membrane (500 μm). The observed water profiles were much flatter than expected from simulations. Due to recent improvements in detector technology, the use of in-plane imaging was increasingly reported in the last few years. Hussey et al. (2007) reported measurements on a cell with a thick Nafion 117 membrane and 500 μm GDLs, and pointed out the issue introduced by the swelling of the membrane. Hickner et al. (2008b) applied the same detector technology to a fuel cell with materials of state-of-the-art thicknesses, and studied the effect of current density and operating temperature on the cross-sectional water distribution. Boillat et al. (2008a) measured the water distribution at a constant current density but with different humidifications of the gas flows, and observed a negative impact on cell voltage of too humid conditions. Kim and Mench (2009) studied temperature-driven water flows in an *ex situ* configuration, where a temperature differential was applied between the anode and cathode end plates. The use of time-resolved imaging with an in-plane configuration and with exposure times of 10 s was reported by Boillat et al. (2010), demonstrating the ability of neutron imaging to resolve the dynamics of liquid water evolution in porous media following a step in gas flow humidification. Seyfang et al. (2010) observed the cross-sectional profile of water distribution in a miniaturized fuel cell without GDLs, and observed a strong accumulation of water near the catalyst layer. The issue of membrane swelling was addressed there by using images of a cell humidified with heavy water as a reference. Finally, Turhan et al. (2010) studied the effect of channel walls wetting angle on the water content in porous media. They observed that a PTFE coating on the cathode side resulted in a higher accumulation of water in porous media than a PTFE coating on the anode side.

12.5.3 Tomography

Only a few publications exist about neutron tomography for 3D imaging of fuel cells, the probable reason being the high amount of beam time required for imaging single steady-state conditions. A first report of applying neutron tomography to a fuel cell was published by Satija et al. (2003). In this study, only the structure of the nonoperating cell was imaged. Manke et al. (2007a) reported the use of neutron tomography with a so-called *quasi in situ* configuration, in which the cells were first operated and then stopped for imaging during a total acquisition time of approximately 5 h. This study used stacks made of 3 and 5 cells, and the tomography of the 5-cell stack showed higher water accumulation in the flow channels of end cells, which correlated with a lower potential of these cells during earlier operation. Sakata et al. (2009) performed neutron tomography on a small-scale single cell, reporting a total acquisition time of 20 min. Accumulation of water in the flow channel could be observed, but the ability of the setup to distinguish water in porous media was not demonstrated.

12.5.4 Comparison with Numerical Simulations

Despite the high potential of experimentally measured water distribution for validation of water transport models, the quantity of published work on this topic remains relatively scarce. Ju et al. (2007) compared the maximal amount of water predicted in porous media by their water transport model, and noted that it was in the same order of magnitude as the values measured by neutron imaging usually reported in the literature. Weber and Hickner (2008) intended to interpret the cross-sectional profiles of liquid water distribution measured by high-resolution in-plane neutron radiography, using a 1 + 1D transport model. Relatively large discrepancies were, however, found between the model and experimental data. Finally, Chen and Peng (2008) compared the distribution of liquid water measured by through plane neutron radiography over the area of a fuel cell, and compared it to the result of a numerical simulation. A relatively good agreement was found on the general trend of water distribution, though a significant number of model parameters were adjusted based on experimental data.

12.5.5 Combination with Other Methods

The use of neutron imaging for liquid water detection has increasingly been combined with other advanced characterization methods, in order to provide a better picture of the involved phenomena. As imaging gives access to the local quantities of liquid water, there is an obvious interest in obtaining local electrochemical data such as the current density distribution as well. The combination of neutron imaging with the measurement of current density distribution was first reported by Schneider et al. (2005), using a ninefold segmented cell. In this study, the authors measured as well local electrochemical impedance spectra, for a cell operated on hydrogen and pure oxygen. In dry regions, they observed a significant increase of the membrane resistance measured at high frequency, but also of the kinetic resistance. Hartnig et al. (2008a) measured the local current density in a 5×5 segmented cell and compared the temporal evolution of both liquid water and current density. Changes in current density synchronous to changes in water contents were found in certain regions. The same 5×5 segmented setup was used by Schröder et al. (2009, 2010) for the measurements on DMFC mentioned in Section 12.5.1.9. Gagliardo et al. (2009) combined neutron imaging with the measurement of local current density and high-frequency resistance. In regions in which channels clogged with water were observed, they found both a reduction of the local current and more surprisingly an increase in high-frequency resistance. They attributed this increase to the drying of the membrane by the flow on the nonflooded side. Spernjak et al. (2009) combined neutron and optical imaging using a quartz window transparent to both neutrons and visible light, which allowed them to image the amount of water in porous media under the lands of the flow fields while obtaining high-resolution images of channel water, though with the drawbacks in terms of invasiveness usually related to transparent cell setups. This setup was used in a further study of flow-field geometries (Spernjak et al., 2010), as mentioned in Section 12.5.1.5. Chen and Peng (2009) combined the use of neutron imaging with local measurements of relative humidity using miniature RH sensors embedded in the flow field. Finally, Iwase et al. (2009) reported the use of a setup combining neutron imaging with small-angle neutron scattering (SANS), in order to obtain quasisimultaneous information about liquid water in the flow fields and porous media from the imaging method, and about the membrane hydration state using the method based on SANS originally described by Mosdale et al. (1996).

12.5.6 Use of ^{2}H Labeling

The domain of ^{2}H labeling combined with neutron imaging is up to now scarcely studied. Manke et al. (2008) realized time resolved through-plane imaging of a fuel cell of technical size, and changed the fuel gas from hydrogen to deuterium. The reported idea is to measure the rate at which water is replaced by electrochemical water. However, the impact of the interfacial exchanges and diffusive processes as developed in Section 12.4.4.2 was not considered. These effects were reported by Boillat et al. (2008b) in a study using time-resolved imaging of $^1H_2$ to $^2H_2$ transitions as well, but with in-plane imaging. The observed replacement rate in the membrane was significantly higher than calculated by the electrochemical water production, indicating than additional exchange processes cannot be neglected.

12.6 Advantages and Limitations

12.6.1 Advantages

12.6.1.1 Noninvasive Locally Resolved Information

The advantage of providing locally resolved information is shared by all imaging methods (e.g., magnetic resonance imaging, x-ray imaging, optical microscopy, Raman confocal microscopy). All these methods can to some extent be considered as "noninvasive," as they do not imply the actual insertion of sensors into the fuel cell itself. However, they often require modifications of the cell structure and/or materials in order the get sufficient transparency. As developed in the following paragraph, neutron imaging is usually able to observe the liquid water distribution on unmodified fuel cell designs, providing the lowest possible level of invasiveness. Another aspect of "invasiveness" is the possible radiation damage induced to the cell, either by direct ionization or by the effect of absorbed heat. Such damage has been reported in certain conditions when using synchrotron x ray imaging (Schneider, A et al., 2010). To the best of our knowledge, no such effects were reported for other imaging methods. Cold and thermal neutron have energies lower than 100 meV, clearly in the nonionizing range. Despite this, damage induced by the gamma background, as well as by radiation generated by the activated cell housing cannot be totally excluded. Concerning the setup and cells used for the examples presented in this chapter, some samples were operated for more than 50 h with almost uninterrupted imaging, without any noticeable effect on performance.

12.6.1.2 High Transparency of Usual PEMFC Structural Materials

Materials used for fuel cell housings and flow fields (aluminum, steel, graphite), as well as sealings (PTFE) exhibit a good transparency to neutron rays. The typical mean free path values, corresponding to the distance through which the beam intensity is reduced to 33% of its initial value, are given in Table 12.1.

12.6.1.3 High Contrast for Liquid Water

On the contrary to structural materials, liquid water provides a high attenuation of neutron beams, having an attenuation coefficient of 3–5 cm^{-1} (depending on the beam spectrum), allowing to image even thin layers of liquid water. Combined to the high transparency of fuel cell structural materials, this high attenuation provides a unique characteristic of neutron imaging. Illustrating this, the attenuation of different materials, compared to that of water, is plotted in Figure 12.19. As a comparison, the same ratio is plotted for soft x-rays as well.

TABLE 12.1 Neutron Attenuation Coefficients for PEMFC Structural Materials

Material	Attenuation Coefficient Σ (cm^{-1})	Mean Free Path λ (cm)
Aluminum	0.1	10
Stainless steel	1	1
Graphite (incl. polymer filler)	1	1
PTFE	0.3	3.3

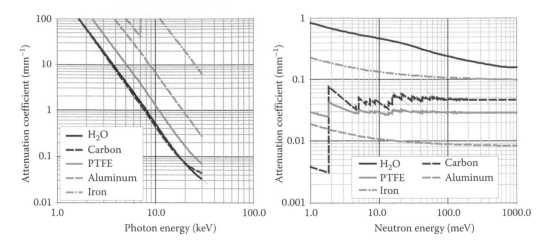

FIGURE 12.19   Attenuation of x-rays and neutrons as a function of energy for water and different PEFC structural materials.

12.6.1.4 Isotopic Sensitivity

As mentioned previously, neutron imaging provides sensitivity to different isotopes of the same chemical element, and, in particular, ^{1}H and ^{2}H isotopes have very different cross sections. As illustrated in Section 12.4.4, this characteristic is very useful for fuel cell research. When compared with MRI, another imaging method featuring isotopic sensitivity, neutron imaging is characterized by a better spatial and temporal resolution.

12.6.2 Limitations

12.6.2.1 Beam Line Availability

The use of neutron imaging with fuel cells implies the availability of a neutron source of sufficient flux. Due to the high construction cost of such facilities, they are usually not dedicated to fuel cell research but beam time is shared between numerous applications. Performing neutron imaging of fuel cell thus implies coping with a limited time for experiments, without the possibility to extend them. For these reasons, experiments concerning fuel cell research have been reported at only a few facilities over the world.

12.6.2.2 Activation of Materials

Exposing materials to neutron beams results in neutron capture by some of the nuclei. Different isotopes of the elements present in the sample are thus produced, some of which are radioactive. Care has to be thus taken in order to avoid the presence of materials with long-life radio-isotopes. The presence of some short-life isotopes also may limit the access to the fuel cell for manipulation during a short-time period after exposure.

12.6.2.3 Temporal and Spatial Resolution

The relatively low neutron fluxes (when compared, for example, to synchrotron radiation) imply limitations in the combination of temporal and spatial resolution. At the current state of the art, effective spatial resolution is limited to 10–20 μm.

12.7 Outlook

Neutron imaging has been increasingly used in the past years in the context of fuel cell management studies, and has proven to deliver important information for understanding water transport processes.

However, several fields of applications of neutron imaging are still unexplored. A short outlook on possible future development is therefore given here, organized in different topics.

12.7.1 Resolution Improvements

With application to fuel cells as a significant driving force, the spatial resolution reachable with neutron imaging has benefited from important improvement in the past years. Further improvements in this domain are of interest for fuel cell imaging, for example, in order to image the distribution of water across a catalyst layer or across a thin membrane. The approach using detector tilting presented in this chapter can potentially be pushed further, though some observed limitations are yet to be understood. An issue not to be neglected is the planarity of the fuel cell sample, which is an absolute requisite for high-resolution imaging. Another aspect of imaging improvement is the extension of the field of view, currently limited to relatively small sizes for high-resolution imaging (typ. 20 mm). This limitation is mainly related to the detector technology, as the flux areas of neutron beams are usually larger. First promising results were obtained at the ICON beam line at PSI with a detector based on the same technology than for the high-resolution example provided in this chapter, but with a vertical range extended to 120 mm. The potential of larger detector areas is to realize in-plane imaging on fuel cell samples closer to technical fuel cells or on several differential fuel cells at the same time, providing an improved efficiency of beam time use.

12.7.2 Validation of Modeling Studies

A few authors have reported comparisons done between water distributions as predicted by water transport models and as measured by neutron imaging (*cf.* Section 12.5.4). The potential of neutron imaging for modeling validation is however far from having been fully developed yet. A limitation is the discrepancy sometimes found between the model representation and the experimental object (e.g., simplification neglecting the rib/channel dimension). The comparison of complex fully three-dimensional models with cells of technical size also raises the issue of the uncertainties in material characteristics, which are sometimes adjusted for fitting the experimental data. If a large number of parameters are fitted on a limited set of experimental data, no valuable validation of the fuel cell model is provided. Therefore, on the view of the authors, validation studies realized on fuel cells of reduced complexity—such as differential cells—would be of highest interest, as a building block for studying more complex systems. The possibility also exists, thanks to imaging methods, to realize validation studies on partial models. An example of this would be the modeling of liquid water impact on cell performance, based on the experimental water distribution as obtained by neutron imaging. In such a comparison, relatively simple models can be used as they do not require predicting the water transport. Finally, inclusion of the temporal dimension in modeling is highly desired. As shown in a few examples in this chapter, temporal evolution of liquid water can be an effective way of separating different processes related to water transport.

12.7.3 Study of Degradation Processes

As mentioned in the introduction, water can have a significant effect on degradation processes. Degradation of the cell could also influence the water transport properties (e.g., by damaging the hydrophobic agent in GDLs) and induce increased mass transport losses. The comparison of water distribution in fresh cells and in previously aged cells has, to our knowledge, not been reported yet. Neutron imaging could additionally be a useful tool for studying the distribution of cell degradation without the need of cell disassembly. As illustrated in Section 12.4.4.2, isotope labeling can be used to study the magnitude of the exchange between the hydrogen gas phase and the membrane, which is related to the presence of an electrocatalyst. Concerning the study of local membrane degradation, the local amount of water uptake by the membrane at a given humidity in a nonoperating cell could, in principle, be used as a probe for the chemical degradation of the membrane. The ideas above are yet to be experimented and validated. In the case of positive results, the possibility would be open to measure the local degradation of the electrocatalyst and of the membrane in a totally noninvasive way, along the life time of a fuel cell.

References

Bellows, R. J., Lin, M. Y., Arif, M. et al. 1999. Neutron imaging technique for *in situ* measurement of water transport gradients within Nafion in polymer electrolyte fuel cells. *J. Electrochem. Soc.* 146: 1099–1103.

Boillat, P., Frei, G., Lehmann, E. H. et al. 2010. Neutron imaging resolution improvements optimized for fuel cell applications. *Electrochem. Solid-State Lett.* 13: B25–B27.

Boillat, P., Kramer, D., Seyfang, B. C. et al. 2008a. *In situ* observation of the water distribution across a PEFC using high resolution neutron radiography. *Electrochem. Commun.* 10: 546–550.

Boillat, P., Oberholzer, P., Seyfang, B. C. et al. 2011. Using ^{2}H labeling with neutron radiography for the study of polymer electrolytes water transport properties. *J. Phys.: Condens. Matter.* Accepted manuscript, in press.

Boillat, P., Scherer, G. G., Wokaun, A. et al. 2008b. Transient observation of ^{2}H labeled species in an operating PEFC using neutron radiography. *Electrochem. Commun.* 10: 1311–1314.

Büchi, F. N., Flückiger R., Tehlar D. et al. 2008. Determination of liquid water distribution in porous transport layers. *ECS Trans.* 16: 587–592.

Chen, C. and Fuller, T. F. 2009. The effect of humidity on the degradation of Nafion membrane. *Polym. Degrad. Stab.* 94: 1436–1447.

Chen, Y. S. and Peng, H. 2008. A segmented model for studying water transport in a PEMFC. *J. Power Sources* 185: 1179–1192.

Chen, Y. S., and Peng, H. 2009. Studying the water transport in a proton exchange membrane fuel cell by neutron radiography and relative humidity sensors. *J. Fuel Cell Sci. Technol.* 6: 31016–31028.

Chen, Y. S. Peng, H., Hussey, D. S. et al. 2007. Water distribution measurement for a PEMFC through neutron radiography. *J. Power Sources* 170: 376–386.

Cho, K. T., Turhan, A., Lee, J. H. et al. 2009. Probing water transport in polymer electrolyte fuel cells with neutron radiography. *Nucl. Instr. and Meth. A* 605: 110–122.

Dunbar, Z. W. and Masel, R. I. 2008. Magnetic resonance imaging investigation of water accumulation and transport in graphite flow fields in a polymer electrolyte membrane fuel cell: Do defects control transport? *J. Power Sources* 182: 76–82.

Endoh, E., Teranozo, S., Widjaja, H. et al. 2004. Degradation study of MEA for PEMFCs under low humidity conditions. *Electrochem. Solid-State Lett.* 7: A209–A211.

Feindel, K. W., Bergens S. H., and Wasylishen R. E. 2007. Use of hydrogen-deuterium exchange for contrast in ^{1}H NMR microscopy investigations of an operating PEM fuel cell. *J. Power Sources* 173: 86–95.

Gagliardo, J. J., Owejan, J. P., Trabold, T. A. et al. 2009. Neutron radiography characterization of an operating proton exchange membrane fuel cell with localized current distribution measurements. *Nucl. Instr. Meth. A* 605: 115–118.

Geiger, A. B., Tsukada., A., Lehmann, E. et al. 2003. *In situ* investigation of two-phase flow patterns in flow fields of PEFCs using neutron radiography. *Fuel Cells* 2: 92–98.

Hartnig, C., Manke, I., Kardjilov, N. et al. 2008a. Combined neutron radiography and locally resolved current density measurements of operating PEM fuel cells. *J. Power Sources* 176: 452–459.

Hartnig, C., Manke, I., Kuhn, R. et al. 2008b. Cross-sectional insight in the water evolution and transport in polymer electrolyte fuel cells. *Appl. Phys. Lett.* 92: 134106–134108.

Hickner, M. A., Chen, K. S., and Siegel., N. P. 2010a. Elucidating liquid water distribution and removal in an operating proton exchange membrane fuel cell via neutron radiography. *J. Fuel Cell Sci. Technol.* 7: 11001–11005.

Hickner, M. A., Siegel, N. P., Chen, K. S. et al. 2006. Real-time imaging of liquid water in an operating proton exchange membrane fuel cell. *J. Electrochem. Soc.* 153: A902–A908.

Hickner, M. A., Siegel, N. P., Chen, K. S. et al. 2008a Understanding liquid water distribution and removal phenomena in an operating PEMFC via neutron radiography. *J. Electrochem. Soc.* 155: B294–B302.

Hickner, M. A., Siegel, N. P., Chen, K. S. et al. 2008b. *In situ* high-resolution neutron radiography of cross-sectional liquid water profiles in proton exchange membrane fuel cells. *J. Electrochem. Soc.* 155: B427–B434.

Hickner, M. A., Siegel, N. P., Chen, K. S. et al. 2010b. Observations of transient flooding in a proton exchange membrane fuel cell using time-resolved neutron radiography. *J. Electrochem. Soc.* 157: B32–B38.

Hussey, D. S., Jacobson, D. L., Arif, M. et al. 2007. Neutron images of the through-plane water distribution of an operating PEM fuel cell. *J. Power Sources* 172: 225–228.

Iwase, H., Koizumi, S., Iikura, H. et al. 2009. A combined method of small-angle neutron scattering and neutron radiography to visualize water in an operating fuel cell over a wide length scale from nano to millimeter. *Nucl. Instr. Meth. A* 605: 95–98.

Ju, H., Luo, G., and Wang, C. Y. 2007. Probing liquid water saturation in diffusion media of polymer electrolyte fuel cells. *J. Electrochem. Soc.* 154: B218–B228.

Kim, S. and Mench, M. M. 2007. Physical degradation of membrane electrode assemblies undergoing freeze/thaw cycling: Micro-structure effects. *J. Power Sources* 174: 206–220.

Kim, S. and Mench M. M. 2009. Investigation of temperature-driven water transport in polymer electrolyte fuel cell: Phase-change-induced flow. *J. Electrochem. Soc.* 156: B353–B362.

Kim, S. Ahn, B. K., and Mench, M. M. 2008. Physical degradation of membrane electrode assemblies undergoing freeze/thaw cycling: Diffusion media effects. *J. Power Sources* 179: 140–146.

Kim, T. J., Kim, J. R., Sim, C. M. et al. 2009. Experimental approaches for distribution and behavior of water in PEMFC under flow direction and differential pressure using neutron imaging technique. *Nucl. Instr. Meth. A* 600: 325–327.

Kramer, D., Lehmann, E., Frei, G. et al. 2005a. An on-line study of fuel cell behavior by thermal neutrons. *Nucl. Instr. Meth. A* 542: 52–60.

Kramer, D., Zhang, J., Shimoi, R. et al. 2005b. *In situ* diagnostic of two-phase flow phenomena in polymer electrolyte fuel cells by neutron imaging: Part A. Experimental, data treatment, and quantification. *Electrochim. Acta* 50: 2603–2614.

Lehmann, E. H., Frei, G., and Kühne, G. 2007. The micro-setup for neutron imaging: A major step forward to improve the spatial resolution. *Nucl. Instr. Meth. A* 576: 389–396.

Li, X., Sabir, I., and Park, J. 2007. A flow channel design procedure for PEM fuel cells with effective water removal, *J. Power Sources* 163: 933–942.

Linse, N., Aellig., C., Wokaun, A. et al. 2009. Influence of operating parameters on start/stop induced degradation in polymer electrolyte fuel cells. *ECS Trans.* 25: 1849–1859.

Manke, I., Hartnig, C., Grünerbel, M. et al. 2007a. Quasi-*in situ* neutron tomography on polymer electrolyte membrane fuel cell stacks. *Appl. Phys. Lett.* 90: 184101–184103.

Manke, I., Hartnig, C., Grünerbel, M. et al. 2007b. Investigation of water evolution and transport in fuel cells with high resolution synchrotron x-ray radiography. *Appl. Phys. Lett.* 90: 174105–174107.

Manke, I., Hartnig, C., Kardjilov, N. et al. 2008. Characterization of water exchange and two-phase flow in porous gas diffusion materials by hydrogen–deuterium contrast neutron radiography. *Appl. Phys. Lett.* 92: 244101–244103.

Matic, H., Lundblad, A., Lindbergh, G. et al. 2005. *In situ* micro-Raman on the membrane in a working PEM cell. *Electrochem. Solid-State Lett.* 8: A5–A7.

Minard, K. R., Viswanathan, V. V., Majors, P. D. et al. 2006. Magnetic resonance imaging (MRI) of PEM dehydration and gas manifold flooding during continuous fuel cell operation. *J. Power Sources* 161: 856–863.

Mosdale, R., Gebel, G., and Pineri, M. 1996. Water profile determination in a running proton exchange membrane fuel cell using small-angle neutron scattering. *J. Membr. Sci.* 118: 269–277.

Murakawa, H., Ueda, T., Yoshida, T. et al. 2009. Effect of water distributions on performances of JARI standard PEFC by using neutron radiography. *Nucl. Instr. Meth. A* 605: 127–130.

Neyerlin, K. C., Gu, W., Jorne, J. et al. 2007. Study of the exchange current density for the hydrogen oxidation and evolution reactions. *J. Electrochem. Soc.* 154: B631–B635.

Ochi, S., Kamishima, O., Mizusaki, J. et al. 2009. Investigation of proton diffusion in Nafion®117 membrane by electrical conductivity and NMR. *Solid State Ionics* 180: 580–584.

Owejan, J. P., Gagliardo, J. J., Sergi, J. M. et al. 2009a. Water management studies in PEM fuel cells, Part I: Fuel cell design and *in situ* water distributions. *Int. J. Hydrogen Energy* 34: 3436–3444.

Owejan, J. P., Gagliardo, J. J., Falta, S. R. et al. 2009b. Accumulation and removal of liquid water in proton exchange membrane fuel cells. *J. Electrochem. Soc.* 156: B1475–B1483.

Owejan, J. P., Trabold, T. A., Jacobson, D. J. et al. 2006. *In situ* investigation of water transport in an operating PEM fuel cell using neutron radiography: Part 2—Transient water accumulation in an interdigitated cathode flow field. *Int. J. Heat Mass Transfer* 49: 4721–4731.

Owejan, J. P., Trabold, T. A., Gagliardo, J. J. et al. 2007a. Voltage instability in a simulated fuel cell stack correlated to cathode water accumulation. *J. Power Sources* 171: 626–633.

Owejan, J. P., Trabold, T. A., Jacobson, D. L. et al. 2007b. Effects of flow field and diffusion layer properties on water accumulation in a PEM fuel cell. *Int. J. Hydrogen Energy* 32: 4489–4502.

Pekula, N., Heller, K., Chuang, P. A. et al. 2005. Study of water distribution and transport in a polymer electrolyte fuel cell using neutron imaging. *Nucl. Instr. Meth. A* 542: 134–141.

Sakata, I., Ueda, T., Murakawa, H. et al. 2009. Three-dimensional observation of water distribution in PEFC by neutron CT. *Nucl. Instr. Meth. A* 605: 131–133.

Satija, R., Jacobson, D. L., Arif, M. et al. 2003. *In situ* neutron imaging technique for evaluation of water management systems in operating PEM fuel cells. *J. Power Sources* 129: 238–245.

Schneider, I. A., Bayer, M. H., Wokaun, A. et al. 2008. Transient response of the local high frequency resistance in PEFCs. *PSI Electrochemistry Laboratory Annual Report 2007, ISSN-1661-5379*: 18.

Schneider, I. A., Kramer, D., Wokaun A. et al. 2005. Spatially resolved characterization of PEFCs using simultaneously neutron radiography and locally resolved impedance spectroscopy. *Electrochem. Commun.* 7: 1393–1397.

Schneider, I. A., Von Dahlen, S., Bayer, M. H. et al. 2010. Local transients of flooding and current in channel and land areas of a polymer electrolyte fuel cell. *J. Phys. Chem.* 114: 11998–12002.

Schneider, A., Wieser, C., and Roth, J. 2010. Impact of synchrotron radiation on fuel cell operation in imaging experiments. *J. Power Sources* 195: 6349–6355.

Schröder, A., Wippermann, K., Lehnert, W. et al. 2010. The influence of gas diffusion layer wettability on direct methanol fuel cell performance: A combined local current distribution and high resolution neutron radiography study. *J. Power Sources* 195: 4765–4771.

Schröder, A., Wippermann, K., Mergel, J. et al. 2009. Combined local current distribution measurements and high resolution neutron radiography of operating Direct Methanol Fuel Cells. *Electrochem. Commun.* 11: 1606–1609.

Sethuraman, V. A., Weidner, J. W., Haug, A. T. et al. 2008. Durability of perfluorosulfonic acid and hydrocarbon membranes: Effect of humidity and temperature. *J. Electrochem. Soc.* 155: B119–B124.

Seyfang, B. C., Boillat, P., Simmen, F. et al. 2010. Identification of liquid water constraints in micro polymer electrolyte fuel cells without gas diffusion layers. *Electrochim. Acta* 55: 2932–2938.

Siegel, J. B., McKay, D. A., Stefanopoulou, A. G. et al. 2008. Measurement of liquid water accumulation in a PEMFC with dead-ended anode. *J. Electrochem. Soc.* 15: B1168–B1178.

Siegmund, O. H., Vallerga, J. V., Martin, A. et al. 2007. A high spatial resolution event counting neutron detector using microchannel plates and cross delay line readout. *Nucl. Instr. Meth. A* 579: 188–191.

Spernjak, D., Advani, S. G., and Prasad, A. K. 2009. Simultaneous neutron and optical imaging in PEM fuel cells. *J. Electrochem. Soc.* 156: B109–B117.

Spernjak, D., Prasad, A. K., and Advani, S. G. 2010. *In situ* comparison of water content and dynamics in parallel, single-serpentine, and interdigitated flow fields of polymer electrolyte membrane fuel cells. *J. Power Sources* 195: 3553–3568.

Trabold, T.A., Owejan, J. P., Jacobson, D. L. et al. 2006. *In situ* investigation of water transport in an operating PEM fuel cell using neutron radiography: Part 1—Experimental method and serpentine flow field results. *Int. J. Heat Mass Transfer* 49: 4712–4720.

Tremsin, A. S., McPhate, J. B., and Vallerga, J. V. 2009. High-resolution neutron radiography with microchannel plates: Proof-of-principle experiments at PSI. *Nucl. Instr. Meth. A* 605: 103–106.

Tsushima, S., Teranishi, K., Nishida, K. et al. 2005. Water content distribution in a polymer electrolyte membrane for advanced fuel cell system with liquid water supply. *Magn. Reson. Imaging* 23: 255–258.

Turhan, A., Heller, K., Brenizer, J. S. et al. 2006. Quantification of liquid water accumulation and distribution in a polymer electrolyte fuel cell using neutron imaging. *J. Power Sources* 160: 1195–1203.

Turhan, A., Heller, K., Brenizer, J. S. et al. 2008. Passive control of liquid water storage and distribution in a PEFC through flow-field design. *J. Power Sources* 180: 773–783.

Turhan, A., Kim, S., Hatzell, M. et al. 2010. Impact of channel wall hydrophobicity on through-plane water distribution and flooding behavior in a polymer electrolyte fuel cell. *Electrochim. Acta* 55: 2734–2745.

Weber, A. Z. and Hickner, M. A. 2008. Modeling and high-resolution-imaging studies of water-content profiles in a polymer-electrolyte-fuel-cell membrane-electrode assembly. *Electrochim. Acta* 53: 7668–7674.

Yoshizawa, K., Ikezoe, K., Tasaki, Y. et al. 2008. Analysis of gas diffusion layer and flow-field design in a PEMFC using neutron radiography. *J. Electrochem. Soc.* 155: B223–B227.

Yousfi-Steiner, N., Moçotéguy, P., Candusso, D. et al. 2009. A review on polymer electrolyte membrane fuel cell catalyst degradation and starvation issues: Causes, consequences and diagnostic for mitigation. *J. Power Sources* 194: 130–145.

Zhang, F. Y., Yang X. G., and Wang C. Y. 2006. Liquid water removal from a polymer electrolyte fuel cell. *J. Electrochem. Soc.* 153: A225–A232.

Zhang, J., Kramer, D., Shimoi, R. et al. 2006. *In situ* diagnostic of two-phase flow phenomena in polymer electrolyte fuel cells by neutron imaging: Part B. Material variations. *Electrochim. Acta* 51: 2715–2727.

Zhang, Z., Martin, J., Wu, J. et al. 2008. Magnetic resonance imaging of water content across the Nafion membrane in an operational PEM fuel cell. *J. Magn. Reson.* 193: 259–266.

II

Ex Situ
Diagnostic Tools

13

X-Ray Diffraction

Justin Roller
*Institute for Fuel Cell
Innovation*

13.1 Introduction

X-rays were originally discovered by German physicist Roentgen in 1895 and were used without a deeper fundamental understanding in the fields of radiography, which allowed shadow pictures of opaque objects to be taken so that their internal structure could be explored. It was discovered that they affected photographic film in the same way as visible light albeit much more penetrating than visible light. However, the limiting resolution of internal features was limited to 10^{-3} mm. It was not until 1912 that the phenomenon of XRD by crystals was established as well as the wave nature of x-rays. This discovery allowed for observation of structure of the order of 10^{-7} mm in size. Max von Laue, in 1912, discovered that crystalline substances act as three-dimensional diffraction gratings for x-ray wavelengths similar to the spacing of planes in a crystal lattice.

13.2 Principles and Measurements

13.2.1 X-Rays Generation and Detection

X-rays, like visible light, are formed by oscillations in the electro-magnetic field. Whereas the wavelength of visible light is of the order of 3900–7500 Å, x-rays are some 3000–7800 times shorter, of the order of 0.5–2.5 Å. X-rays are sandwiched between the ultraviolet and gamma-ray spectrums. X-rays are produced when any electrically charged particle of sufficient kinetic energy decelerates.

In an XRD experiment the incident beam x-rays are created by accelerating electrons between two metal electrodes maintained at a bias of several thousands of volts (10–20,000 V). On impact of the electrons onto the anode target (i.e., copper, molybdenum) the x-rays emanate in all directions. Most of the kinetic energy of the electrons is converted into heat with only a small percentage transformed into x-rays. The x-rays coming out of the target consist of a mixture of different wavelengths with the intensity dependent on the tube voltage. Due to the large amount of heat generated, the target must be externally cooled by a recirculating water bath to avoid melting or uneven production of x-ray intensity. Some x-ray generators utilize a rotating anode material so that the incident electron beam is never focused on the same spot for an appreciable amount of time.

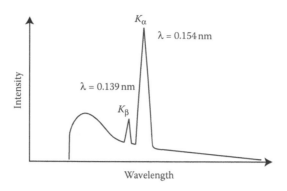

FIGURE 13.1   Distribution of x-ray intensity versus wavelength for a copper target anode.

Examination of the x-ray intensity versus wavelength for the emission from a copper target anode shows no intensity up to a minimum or cut-off wavelength. In Figure 13.1, this minimum energy is marked by a rapid increase in x-ray intensity at the lower end of the wavelength spectrum. The reason that there is no x-ray intensity lower than this cutoff is that the emitted energy can never be higher than the incident energy used to accelerate the electron into the target. This limit decreases as the accelerating voltage is increased. Immediately after the limit there is a sharp rise in the intensity followed by a gradual tailing off of intensity. This broadly distributed radiation is named polychromatic, continuous, *Bremsstrahlung* (braking radiation), or white radiation. This is analogous to white light in the visible spectrum being a combination of many wavelengths.

When the voltage on the x-ray tube is raised above a certain critical value, which is dependent on the anode material, a set of sharp spikes in intensity are seen to superimpose on the continuous spectrum. These spikes are referred to as characteristic lines since they are "characteristic" of the decelerating material used to produce x-rays and stand out prominently above the background continuous radiation. The lines are labeled in increasing order K, L, M, and so on. The letters refer to the principal quantum numbers $n = 1, 2, 3$, and so on. The K lines are of most interest to this discussion since the higher wavelength lines are more easily absorbed. A closer inspection of the K lines reveals that they consist of three lines close in energy $K\alpha_1$, $K\alpha_2$, and $K\beta_1$. An increase in accelerating voltage above a critical value increases the intensity, often 100× higher than the continuous background for a $K\alpha$ line, but does not change the wavelength. The narrowness of the characteristic lines is often less than 0.001 Å, and is critical to allowing for precise atomic spatial determination.

Many XRD experiments require radiation that is highly monochromatic (one defined wavelength). However, the beam from an x-ray tube operated at a voltage above the K excitation voltage contains not only the strong K_α line but also the weaker K_β line and some continuous spectrum (Cullity and Stock, 2001). The intensity of these undesirable wavelengths can be decreased relative to K_α line by passing the beam through a filter material (usually a thin metallic foil) with an atomic number one less than that of the target metal. This places the absorption edge of the foil between K_α and K_β of the target material and allows only the desired K_α radiation to pass with acceptable attenuation. Filtration is never perfect; however, a reduction of the intensity ratio of K_β/K_α to 1/500 of the transmitted beam is usually sufficient for most purposes.

Once the x-rays are sufficiently monochromatic they are then sent through a slit for further conditioning of the incoming beam. A smaller slit size allows for better resolution of the diffracted peaks but cuts some of the radiation and therefore decreases the signal-to-noise ratio. On the other hand, a larger slit allows more radiation through the instrument but because of beam spreading contributes to peak broadening.

13.2.1.1 Powder Diffraction versus Laue Diffraction

There are several possible geometries used for observation of XRD. The most commonly used geometry for PEM fuel cell catalyst degradation is called powder diffraction. This occurs when a large amount of sample of a polycrystalline specimen is illuminated by monochromatic (single wavelength) x-ray radiation. Polycrystalline materials are solids that are composed of many crystallites of varying size and orientation. The variation in direction can be random (called random texture) or directed, possibly due to growth on a substrate with a high degree of lattice orientation that acts as a template to constrain the growth along a particular direction. Almost all common metals and many ceramics are polycrystalline. The crystallites are often referred to as grains, it should be noted that powder grains are many times themselves composed of smaller polycrystalline grains, and this will be further discussed in relation to particle size determination. Polycrystalline is the structure of a solid material that, when cooled, form or nucleate crystallite grains at different points within it. The areas where these crystallite grains meet are known as grain boundaries. Since the sample is polycrystalline, all orientations of the diffracting lattice planes are present to the beam for simultaneous diffraction. Additional circular rotation of the sample may be employed to further randomize the exposure of every crystal plane.

This must be distinguished from Laue diffraction in which the x-ray beam, with a continuous range of wavelengths, impinges on a stationary single crystal; the resulting diffraction pattern is recorded in transmission and contains a very large number of spots, arising from reflections occurring due to the many different impinging wavelengths (Harding, 1991). A single crystal solid is a material in which the crystal lattice of the entire sample is continuous and unbroken to the edges of the sample, with no grain boundaries. Grain boundaries can have significant effects on the physical and electrical properties of a material. Due to the fact that entropy favors the presence of some imperfections in the microstructure of solids, such as impurities, inhomogeneous strain, and crystallographic defects such as dislocations, perfect single crystals of meaningful size are rare in nature, and can only be produced in the laboratory under controlled conditions. Crystal orientation is determined from the position of the spots in Laue diffraction. Each spot can be indexed, that is, attributed to a particular plane, using special charts. The Laue technique can also be used to assess crystal perfection from the size and shape of the spots. If the crystal has been bent or twisted in anyway, the spots become distorted and smeared out (Gene and Pang, 2009).

13.2.1.2 XRD Schematics

An x-ray diffractometer consists of the following setup as outlined in Figure 13.2.

The setup consists of an x-ray generating source, A, and a set of tube slits and/or a Göbel Mirror, or monochromator, B, collectively called the tube beam conditioning optics. The x-rays exit the conditioning optics either parallel if using a Göbel Mirror or divergent if using a monochromator and slits as shown in Figure 13.3.

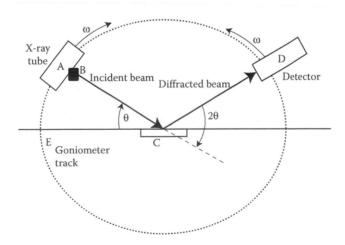

FIGURE 13.2 Abstract diagram of x-ray diffractometer.

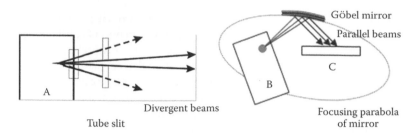

FIGURE 13.3 X-ray path from an x-ray generating tube using slits to minimize the divergence angle, A, and parallel x-rays emanating from a Gobel Mirror arrangement, B.

The x-rays then exit the beam conditioning optics and strike the sample labeled, C. The angle between the sample and the incident beam is θ as is the angle between the diffracted beam and the sample. The x-ray shown penetrating the sample equals an angle 2θ to the diffracting beam. The choice of this measurement as opposed to simply θ becomes clearer in the section of this chapter describing the phenomena of powder diffraction. The x-ray tube (source), A, and the detector, D, are mounted on a motorized goniometer, E. The goniometer provides very precise angular positioning of θ and is crucial in modern diffractometers for accurate measurement of the angular positioning. The angular speed at which the tube and detector move toward each other is equal and shown by ω. The geometry of the diffractometer described above is known as Bragg–Brentano θ–θ geometry since the tube and detector move in lockstep at an identical angular speed, ω. There is another variant described below in the section on beam optics in which the tube is positioned horizontally and the sample and detector move, and this is referred to as Bragg–Brentano θ–2θ geometry.

13.2.1.3 Beam Optics

The angle between the diffracting plane and the incident and diffracted beams is equal to 2θ. The angle between the diffracting plane and both the incident and diffracted beams therefore must be equal to half of that or, θ. This suggests that two geometries can be made for measuring the diffracted beams. In one setup the x-ray tube is mounted such that the incident beam lies in the horizontal plane. The sample is mounted on a platform that can rotate around an axis perpendicular to the plane of the drawing; the rotation angle is equal to θ and the angular rate is ω. The x-ray detector is mounted on a movable arm, which rotates around the same axis, but at twice the angular rate, 2ω, as shown in Figure 13.4 (De Graef, 2007).

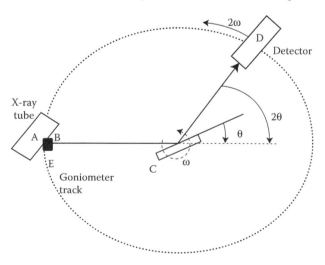

FIGURE 13.4 θ–2θ configuration for x-ray generator, sample, and detector.

This means that the angle between the detector axis and the incident beam direction is always equal to twice the angle between the plane of the specimen and the incident beam direction. This is known as Bragg–Brentano geometry, or, more commonly, the θ–2θ geometry. Another variant of the Bragg–Brentano geometry, known as θ–θ geometry has the sample stationary, and both the x-ray tube and detector move at angular rate ω toward each other, as shown in Figure 13.2. This latter type of geometry makes powder sample analysis easier as there is less chance of shifting sample due to rotation of the sample holder, especially for powders that do not pack well. In powder diffraction experiment the x-ray source exits the tube and the rays begin to diverge. A set of tube slits with widths from 0.2 to 1 mm are positioned between the beam and the sample. On striking the sample and diffracting, the beams then converge and if the system is properly aligned this convergence occurs directly at the detector face. One limitation of Bragg–Brentano geometry is that it is crucial for the sample height to be directly even with the focusing plane of the divergent beams. A sample displacement of 1 mm can shift the 2θ peak by as much at 0.5°. This can be problematic with samples that are not powder, have a rough texture, or are poorly prepared.

Göbel Mirrors (GM) have made sample analysis less sensitive to placement. The parallel beam geometry permits an accurate lattice parameter determination regardless of sample displacement or surface roughness, a measurement without preceding sample preparation and reduced time of measurements. GM convert the divergent x-ray beam coming from an x-ray tube into a parallel beam and additionally partially monochromatize it. A Göbel Mirror consists of multilayers with laterally graded thickness, deposited with accuracy below 1% onto a parabolic surface and typically has a divergence of 0.5 mrad (Michaelsen et al., 1998).

Due to its ability to make highly parallel x-ray beams, the parabolically shaped Göbel Mirror quickly became widely used to remedy this problem. In a Gobel Mirror the divergent beams of the source strike the parabolically curved mirror at different spots along the mirror that contain thin multilayered gradients of W/Si or Ni/C. The thickness of the films increases with length along the mirror. After striking the mirror the resultant beams are all parallel, as opposed to divergent. The parallel beams remove the traditional limitations of the Bragg–Brentano geometry for powder diffraction and thin film investigations. In contrast to conventional monochromator crystals, GM are synthetic crystals whose properties are defined by complex multilayers that are deposited on silicon or glass substrates under ultrahigh vacuum conditions. To shape the mirror, one of two methods is used. The most common is to carefully bend a silicon substrate, either before or after the multilayered structure is deposited, depending on the application. For highly demanding thin-film requirements, the mirror can be fabricated using a pre-shaped quartz substrate. Typically, the layer thicknesses of the mirrors are an order of magnitude larger than those of the commonly used monochromator crystals. Therefore, the "Bragg angles" for the diffracted beams are an order of magnitude smaller, typically near 1.5°. As a consequence, the reflectivity and the beam divergence of the beams conditioned by a Göbel Mirror are larger than the ones from a monochromator, which yields a monochromatized beam without sacrificing intensity. The ability to monochromatize the beam from the x-ray source is limited to suppressing unwanted Kβ-radiation and the high-energy Bremsspectrum. However, compared to instrumental setups with filters only, the background scattering is reduced, which results in better peak-to-background ratio and hence better data quality (Holz et al., 1998).

13.2.1.4 Grazing Incidence Angle

X-ray radiation has a large penetration depth into any matter. Due to this property XRD is not entirely surface sensitive. This can become problematic when measuring the particle size of a deconstructed MEA after operation. It is generally important to separate the particle size growth of the cathode from the anode. Grazing incidence diffraction (GID) is a technique to overcome this restriction. GID measurements are performed at very low incident angles to maximize the signal from the thin layers. The area illuminated by the incident beam follows a cos θ function. At low angles the beam is wider across the surface of the sample but as the incident beam angle increases the beam becomes narrower and the penetration depth increases. It is sometimes very difficult to analyze thin films due to their small diffracting volumes, which result in low diffracted intensities compared to the substrate and background. This combination of low-diffracted signal and high background make it very difficult to identify the phases present. The most common technique for analyzing films as thin as 10 nm is to use a grazing incidence angle arrangement combined with parallel beam geometry. By increasing the path length of the incident x-ray beam through the film, the intensity from the film can be increased, relative to the substrate material, so that conventional phase identification analysis can be run. For GID, the incident and diffracted beams are made nearly parallel by means of a narrow slit or GM on the incident beam and along Soller slits on the detector side. The stationary incident beam makes a very small angle with the sample surface (typically 0.3–3°), which increases the path length of the x-ray beam through the film. This helps to increase the diffracted intensity, while at the same time, reduces the diffracted intensity from the substrate. Overall, there is a dramatic increase in the film signal-to-background ratio. Since the path length increases when the grazing incidence angle is used, the diffracting volume increases proportionally. During the collection of the diffraction spectrum, only the detector rotates through the angular range, thus keeping the incident angle, the beam path length, and the irradiated area constant. A long Soller slit on the receiving side allows only those beams that are nearly parallel to arrive at the detector. The purpose of a Soller slit is to take a line source of radiation and slice it into smaller, parallel beams. This reduces axial divergence of the beam. This has an added advantage of reducing sensitivity to sample displacement from the rotation axis. A Soller slit is constructed by piling up a stack of parallel thin metal foils with an interposing spacer between them and is used to restrict vertical and/or horizontal divergence of x-rays. The metal foils of conventional Soller slit are formed from rolled stainless steel or brass (Cu–Zn) (Fujinawa and Umegaki, 2001).

13.2.1.5 Sample Holders

Modern diffraction equipment produces very accurate diffraction patterns from polycrystalline specimens. However, to get the most out of the data, it is still necessary to minimize all nondesirable scattering from reaching the detector during the experiment. The two largest contributors to the background are the air in the beam path and the specimen support. The air contribution may be minimized by evacuating the beam path or by using helium. The support contribution may be minimized by using specimen holders made from single crystals cut so no Bragg diffraction occurs. Quartz and silicon are the most favorable materials for this purpose.

The use of a cavity specimen mount does allow the use of small amounts of sample confined to the center of the support. However, the user must weigh the effect of compacting the specimen into a small area with depth to spreading the specimen over the full active area of the x-ray beam. If the sample is composed of a low atomic number material, then the beam will penetrate into the depth of the specimen. The penetration will show up in the diffractometer trace as a specimen displacement to lower diffraction angles. If the sample is composed of a high Z material, only the surface grains of the specimen are seen by the x-ray beam, and the more efficent use of the sample would be to spread it on the surface of a noncavity slide. There will be a minimal sample displacement error with the thin specimen, so that the measured angles will be more representative of the true values (Smith, 2010).

13.2.1.6 Detectors

In modern diffractometers, the diffracted beam is intercepted by a detector, either a charge-coupled device (CCD) or a scintillation counter, and the intensity of each reflection is recorded electronically. The detector for the diffractometer only senses the radiation scattered in the plane on the diffractometer circle and does not detect the other diffracted and scattered rays. For any position of the detector, the receiving slit F and the x-ray source, S, are always located on the diffractometer circle. The focusing circle is not of constant size but increases in radius as the angle 2θ decreases (Cullity and Stock, 2001). Use of a flat specimen causes some broadening of the diffracted beam at F and a small shift in line position toward smaller angles, particularly at 2θ angles less than 60°. These effects can be overcome by decreasing the divergence of the of the incident beam either through the use of Gobel mirrors and/or an appropriately sized beam slit. Sizing of the slit requires choosing the smallest slit that does not compromise the intensity of the intended diffraction peaks.

The quality of an XRD analysis is crucially affected by the full peak width at half-maximum height (FWHM), the peak-to-background ratio (PTBR), and the absolute count rate (ACR). Therefore, it is essential to select the most appropriate diffraction system setup in order to obtain the most detailed sample property measurements. FWHM is defined by the mechanical configuration and the x-ray optical components used, whereas the detector is the essential component for the appropriate PTBR and ACR. One of the main influences on the quality of day-to-day XRD experiments is the level of background obtained. The sample itself contributes to the overall background level particularly where the incident beam causes it to emit secondary x-rays (i.e., fluorescence). In addition the source produces a broad continuum (white radiation) that can also fulfill the Bragg condition and therefore contributes to the background.

Scintillation detectors contain a sodium iodide crystal activated with a trace amount of thallium that fluoresce light in the violet region of the spectrum when x-ray radiation passes through. Absorbed x-rays raise some of the electrons in the NaI from the valence to the conduction band. These electrons then transfer some of their energy to the Tl^+ ion. When the excited ion returns to its ground state, light is emitted. A flash of light (scintillation) is produced in the crystal for every x-ray quantum absorbed, and this light passes into the photomultiplier tube and ejects a number of electrons from a photosensitive material generally made of a cesium–antimony intermetallic compound. The photosensitive material is called a photocathode and it emits electrons when excited. The amount of light emitted is proportional to the x-ray intensity and can be measured by means of a photomultiplier tube.

Inside the photomultiplier are a series of dynodes, so named because they act as a middle point between the (photo) cathode and the anode, maintained at 100 V more positive than the dynode preceding it in the tube. As the incoming electron strikes the first dynode several electrons are ejected from the metal surface. This occurs multiple times increasing in a cascading manner at each dynode until reaching the final dynode attached to the measuring circuit. The dynodes act to multiply the signal by as much as 10^7 times. These types of detectors have been used for many years in the areas of XRD and are considered the gold standard for reliability. However, they suffer from the need to have long collection times in order to reduce the signal-to-noise ratio compared to more modern CCD detectors. Additionally, the pulses produced in a scintillation detector have sizes proportional to the energy of the x-ray quanta absorbed. This leads to a certain quantum of energy that is much less sharply defined and it is more difficult to discriminate between x-ray quanta of different wavelengths on the basis of pulse size. The result is that it is more difficult to minimize the contribution from the unwanted $K_{\alpha 2}$ and K_β incident radiation that was not filtered out in the optics. This leads to unwanted peak broadening. Another shortcoming of scintillation detectors is that they can be damaged by prolonged direct exposure, saturation, to the source beam at low angles such as those used in grazing incidence experiments.

More recently, charge-coupled device (CCD)-type detectors have replaced the older scintillation detectors and they can minimize the time required for data collection and decrease the width of the detected peak through better energy discrimination. Line detectors are available that measure a range of angles up to 12° 2θ are becoming more commonly used. They can be used in both snapshot mode and scanning mode. When compared with a point detector, they reduce measurement times drastically from over an hour for a scintillation detector to several minutes when scanning a range of 10–90° 2θ. The line detectors maintain angular resolution, with modern goniometers, on par with scintillation-type detectors. In scanning mode, these detectors can provide collection 100 times faster than scintillation detectors and thus drastically increase the throughput of an instrument (Bruker-AXS, 2010).

Solid-state detectors such as silicon CCDs (charge-coupled devices, similar to the CCDs in video cameras) consist of silicon (the standard computer chip material) doped with impurities to create sites where the conductivity is different. Other solid-state devices exist, using similar principles as for CCDs. Unlike optical CCDs, which measure light impacting on the surface of the chip, x-ray CCDs measure x-rays that penetrate into the middle of the CCD. There, the incoming x-rays pass through a Beryllium window (transparent to x-rays) and strike the Si(Li) lithium-doped silicon detector surface and create an electron–hole pair when it reacts with the silicon–impurities. The unit is held at a very high voltage and this causes the electron–hole pair to be swept (by voltages applied to the chip) across the chip and measured at the end as an electric pulse. The electrons and holes are highly mobile and therefore sweep the detector clear of charge carriers before the next incident photon creates new carriers. This decreased dead time allows the device to return to ground state for the next incoming quanta of x-rays much faster than in scintillation detectors. The charge measurement gives you a very accurate estimate of the energy of the original x-ray. Timing measurements are decent, since you have regular clock-like readouts of your CCD. Counting rates vary linearly with x-ray intensity up to rates of about 5–10 K counts per second. The majority of counting losses exist in the electronics system rather than the detector (Lochner, 2010).

13.2.2 Diffraction

13.2.2.1 Crystal Structure

The crystal structure of a material to a large extent determines the properties of a material. The pattern determined using XRD can be used as a fingerprint to identify the structure of crystalline and semicrystalline materials. It is important to note that identification of a material by XRD is not necessarily unique if two different materials share a similar crystal structure. Therefore, a pattern match is an evidence that an unknown material matches a given structure or the structure from a given database but for absolute identification other techniques such as x-ray-photoelectric spectroscopy (XPS) and inductively coupled plasma (ICP) should be used to confirm the electronic structure of the elements present and their atomic composition, respectively. The information given in an XRD pattern for use in assessing catalysts for PEM durability experiments consist of the following effects:

1. The size and shape of the unit cell (locations of the peaks)
2. Atomic positions within the unit cell (intensities of the peaks)
3. Crystallite size and lattice strain (peak broadening)
4. Quantitative phase amount (scaling factor)

The size and shape of the unit cells determine the relative positions of the diffraction peaks. The atomic positions partially determine the relative intensities of the diffraction peaks. Therefore, using this information from a collected diffraction pattern one can calculate the size and shape of the unit cell and, with a refinement program, determine the positions of the atoms in the unit cell. The intensities of the peaks do not allow direct determination due to other complicating factors that are described in the scattering section.

In the simplest case an infinite discrete set of points, called the basis, with arrangements that are identical from any other points, forms what is called the Bravais lattice. A Bravais lattice is a mathematical abstraction with application to the study of crystalline solids. A Bravais lattice is an infinite set of points in space with positions such that at every point the arrangement of the surrounding points looks exactly the same, and is separated by only a translation from the original point. In XRD the point or basis is comprised of atoms in the simplest case, or groups of atoms. For now consider that each lattice point refers to the placement of an atom. In one dimension this lattice forms a line with equally spaced points. In two dimensions there are five different bravais lattices namely: oblique, rectangular, centered rectangular, hexagonal, and square as shown in Figure 13.5.

When these two-dimensional lattices are then stacked appropriately into three-dimensional space with points (atoms) only occurring at the corners then a total of seven lattice systems are created. These lattice systems are as follows:

1. Cubic
2. Hexagonal

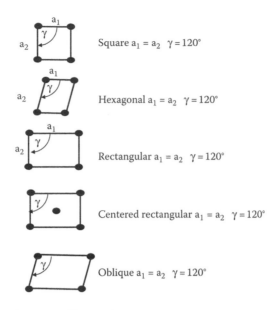

Square $a_1 = a_2$ $\gamma = 120°$

Hexagonal $a_1 = a_2$ $\gamma = 120°$

Rectangular $a_1 = a_2$ $\gamma = 120°$

Centered rectangular $a_1 = a_2$ $\gamma = 120°$

Oblique $a_1 = a_2$ $\gamma = 120°$

FIGURE 13.5   The five two-dimensional Bravais lattices.

3. Rhombohedral
4. Tetragonal
5. Orthorhombic
6. Monoclinic
7. Triclinic

By further defining the lattice centering for each lattice system a total of 14 Bravais lattices are possible in three dimensions, as shown in Figure 13.6. The total number is higher than 14 but by symmetry arguments they can be reduced to 14. The possible lattice centerings are as follows:

- Primitive centering (P): Lattice points on the cell corners only
- Body centered (I): One additional lattice point at the center of the cell
- Face centered (F): One additional lattice point at center of each of the faces of the cell
- Centered on a single face (*A*, *B*, or *C* centering): One additional lattice point at the center of one of the cell faces

The smallest identifiable unit of a crystal is a primitive unit cell whereby the crystal is constructed by taking many unit cells packed tightly together and extending them into space in one, two, or three dimensions. From the above description of the arrangements of atoms into three-dimensional lattices it should be evident that parallel planes containing various combinations of lattice points (atoms) can be imagined. These planes are important in understanding the phenomena of diffraction. The plane-to-plane spacing is different depending on which sets of atoms are chosen to draw the plane. The distance between these planes is known as d-spacing.

Miller indices are a common notation system in crystallography for describing these planes in crystal Bravais lattices. The Miller indices consist of a set of three integers (h, k, l) that denote either a single plane or a family of parallel equivalent planes. The integers are usually written in lowest terms, that is, their greatest common divisor should be 1. A three-dimensional crystal lattice can be described as a regular ordering of unit cells. Three vectors **a**, **b**, and **c**, emanating from a common origin, determine the size and shape of the unit cell by forming a parallelepiped. If **a**, **b**, and **c** are equal in length and the angles between the vectors are 90°, the shape is a special kind of parallelepiped: a cube, but this is not always

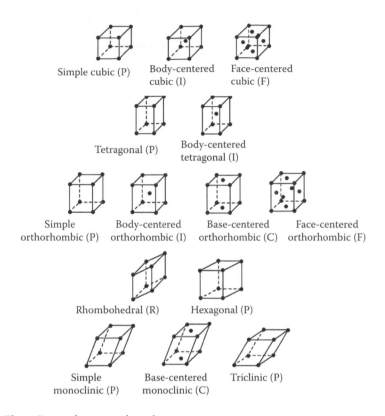

FIGURE 13.6 The 14 Bravais lattices in three dimensions.

the case, as lower symmetry unit cells are also possible. To find the Miller indices of a given plane, make sure that the plane does not pass through the origin. Then,

1. Determine the intercepts of the plane along the crystallographic axes in terms of the dimensions of the unit cell.
2. Take the reciprocal of the intercept values.
3. Multiply by the lowest common denominator.
4. Reduce to lowest terms.

Planes that run parallel to an axis (**a**, **b**, or **c**) have an intercept at infinity, and the Miller index is zero. A common analogy in describing the Miller indices is that they represent the floors and walls in a building. In this representation the Miller indices would be able to distinguish the floors, North/South walls, and East/West walls but would not describe the difference between the second and eighth floors. Planes with lower Miller indices have greater spacing between them and a higher density of lattice points (i.e., run through more atoms in the lattice per unit length). It is these properties of the crystallographic planes that allow us to characterize a crystal using XRD.

In three dimensions, the Miller indices can be determined as shown in Figure 13.7. The combination of each of the indices h, k, and l form the intercepts of a plane with the crystallographic axis, an example of which is shown in Figure 13.8. These planar constructs form the basis for understanding diffraction and will be further elaborated on below.

As mentioned above, the distance between crystallographic planes (formed by Miller indices) is called d-spacing. A crystal will have d-spacings that correspond to each set of Miller indices, and therefore the d value is usually denoted by a subscript to denote the plane being referred to, that is, the spacing between (100) planes in a simple cubic crystal is denoted d_{100}. A crystal will have more planes and thus more

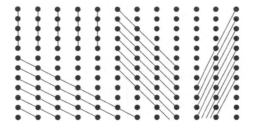

FIGURE 13.7 Illustration of different planes formed by connecting lattice points in a rectangular two-dimensional net.

d-spacings as the symmetry of the crystal decreases. The order of symmetry from higher to lower is cubic, tetragonal, orthorhombic, hexagonal, monoclinic, and triclinic.

13.2.2.2 Scattering by an Atom

When a monochromatic beam of x-rays strikes an atom, two scattering processes occur. Tightly bound electrons are set into oscillation and radiate x-rays of the same wavelength as that of the incident beam (coherent or unmodified scattering). More loosely bound electrons scatter part of the incident beam and slightly increase its wavelength in the process (incoherent or modified scattering), the amount dependent on the scattering angle. Each of these effects occurs simultaneously and in all directions. If the atom is part of a large group of atoms arranged in space in a regular periodic fashion as in a crystal then the coherently scattered radiation from all the atoms undergoes reinforcement in certain directions and cancellation in other directions, thus producing diffracted beams. Diffraction is therefore reinforced coherent scattering. The x-ray intensity of this reinforced coherent scattering forms the peaks evident in an XRD pattern also known as a diffractogram (Cullity and Stock, 2001).

Diffraction intensity varies at different angles as measured from the incident and diffracted beam. The variation in intensity is caused by phase relations between incident x-rays after interacting with the periodic structure of a material. The differences in phase are caused by differences in the path traveled between two or more incident beams. The introduction of phase differences produces a change in amplitude. The greater the path difference, the greater the difference in phase, and thus the path difference equals the phase difference. The wavelength of the incident radiation is close to the distances between the planes of atoms causing the diffraction so that the path difference is also measured in wavelengths. If the wavelengths were too large then there would be no interaction. As an analogy to XRD on the atomic scale consider a rock thrown into a pond. The waves that generate outward from the location of the rock's impact with the water will form concentric circles. These circles will not be perturbed or broken into smaller waves unless the size of an obstacle in the water is of the same order of magnitude as the distance between crests of the wave.

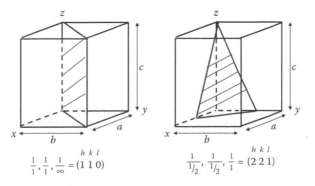

FIGURE 13.8 Determination of Miller indices and the plane formed by combination of *h*, *k*, and *l* indices.

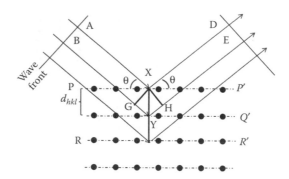

FIGURE 13.9 Origin of phase difference in diffracting x-rays as a consequence of the distance between planes P', Q', and R'.

Two rays are completely in phase whenever their path length difference is zero or some whole number of wavelengths. A diffracted beam may be defined as a beam composed of a large number of scattered rays mutually reinforcing each other. In this sense, diffraction is just a subset of the general phenomena of scattering. The atoms will scatter the incident x-rays in all directions but in some of these directions the scattered beams will be completely in phase and thus reinforce each other to form diffracted beams. If the scattering atoms are not arranged in a regular, periodic fashion but in some independent manner, then the rays scattered by them would have a random phase relationship to one another. In other words, there would be an equal probability of the phase differences between any two scattered rays having any value between zero and one wavelength. Neither constructive nor destructive interference takes place under these conditions, and the intensity of the beam scattered in a particular direction is simply the sum of the intensities of all the rays scattered in that direction.

The Bragg condition for the reflection of x-rays by a crystal is shown in Figure 13.8. The array of dots represents a section through a crystal plane and the lines joining the dots mark a set of parallel planes with Miller indices *hkl* and interplanar spacing d_{hkl}. A parallel beam of monochromatic x-rays enters along lines AX and BY which are incident to the planes at an angle θ. The ray A is scattered by atom X and the ray B is scattered by atom Y. For the scattered beams to emerge as a single beam with intensity above the background, they must reinforce, or arrive in phase with one another. This is known as constructive interference, and for constructive interference to occur, the path lengths of the interfering beams must differ by an integral number of wavelengths as shown in Figure 13.9. A visual representation of constructive and destructive interference is shown in Figure 13.10.

In other words the length represented by GYH, in Figure 13.9, must be nλ wavelengths for constructive interference to occur. If XG and XH are drawn at right angles to the beam, the difference in path length between the two beams is given by GY + YH. By trigonometry GY = YH = d_{hkl} sinθ. This number must equal an integral number, *n*, of wavelengths. If the wavelength of the x-rays is λ then,

$$n\lambda = d_{hkl}\ \sin\theta \text{ (Bragg equation)}. \tag{13.1}$$

To determine the possible directions in which a given crystal can diffract a monochromatic x-ray beam a relation is obtained by combining Bragg's law and the equations for plane spacing. The directions in which a beam of given wavelength is diffracted by a given set of lattice points is determined by the crystal system in which the crystal belongs and its lattice parameters. Thus, diffraction directions are determined solely by the shape and size of the unit cell. Therefore, it is important to realize the converse of this is that all that can be determined about an unknown crystal by measurements of the directions of the diffracted beams is the shape and size of its unit cell (Cullity and Stock, 2001).

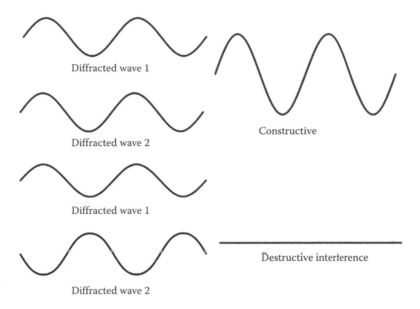

Diffracted wave 1

Diffracted wave 2

Constructive

Diffracted wave 1

Destructive interference

Diffracted wave 2

FIGURE 13.10   Diagram of constructive and destructive interference.

Consider now a primitive cubic unit cell. For a primitive cubic system $a = b = c$ and $\alpha = \beta = \gamma = 90°$. The distance between the planes (100), (010), and (001) planes, low-index planes, gives the largest separation. Any set of planes for a cubic system, and only a cubic system, can be determined by the following general equation:

$$d_{hkl} = \frac{a}{\sqrt{(h^2 + k^2 + l^2)}} \tag{13.2}$$

Combining this with the Bragg equation gives

$$\lambda = \frac{2a \sin \theta_{hkl}}{\sqrt{h^2 + k^2 + l^2}} \tag{13.3}$$

And rearranging gives

$$\sin^2 \theta_{hkl} = \frac{\lambda^2}{4a^2}(h^2 + k^2 + l^2) \tag{13.4}$$

For the primitive cubic class all integral values of the indices h, k, and l are possible. It becomes clear that for the case of a cubic cell the distances between the three nearest corner atoms can be calculated since in a cubic cell the distances are all the same. It should be noted that the (100), (101), and (001) planes would all give the same value as one would expect from such a symmetrical system. The number of identical planes is called the multiplicity. It can be further noted that by calculation of $h^2 + k^2 + l^2$ there will be certain missing integer combinations, also known as systematic absences, that are characteristic of the lattice type. In this way, indexing is a method of associating the observed peaks in an XRD pattern with the planes present in the material being examined. It is in essence a fingerprint of the lattice type but as mentioned earlier not a definitive identification of a particular chemical composition. In cubic systems, planes with higher Miller indices will diffract at higher values of θ. In summary, indexing occurs by the following steps:

1. Identify the peak locations.
2. Determine the value of $\sin^2 \theta$.
3. Calculate the ratio of $\sin^2 \theta/\sin^2 \theta_{min}$ and multiply by the appropriate integers.
4. Select the result from (3) that yields $h^2 + k^2 + l^2$ as an integer.
5. Compare results with the sequences of $h^2 + k^2 + l^2$ values to identify the Bravais lattice, by noting the systematic absences.
6. Calculate the lattice parameters.

In practice, most indexing is done by a computer program but it is helpful to have some understanding of the methodology used in assigning the peaks to the appropriate reflection planes. Table 13.1 shows the possible configurations and the selection rules for the given reflections.

Therefore, the peaks present or absent can be used to identify the crystal system. The differences arise because centering halfway between two planes leads to destructive interference for some reflections and these extra missing reflections are known as systematic absences. A closer inspection of the values of $h^2 + k^2 + l^2$ that favor diffraction reveals what are called the selection rules, for instance in a body-centered lattice the value of $h^2 + k^2 + l^2$ must be even otherwise no reflection occurs. Note that 0 is counted as an even number. To understand why certain lattices do not have a reflection, even though they would be predicted by the Bragg equation, the intensity of scattered radiation by a unit cell must be considered.

13.2.2.3 Scattering by a Unit Cell

The intensity of a diffracted beam is determined by the coherent scattering from all the atoms making up the crystal. The periodic arrangement of the atoms will limit this scattering to certain directions, determined by Bragg's law and will now be referred to collectively as a set of diffracted beams. However, due to the location of atoms in a three-dimensional structure, Bragg's law may be satisfied for a set of diffraction planes and yet no diffraction will be observed. The phase difference dependence of the waves scattered by the individual atoms of a unit cell is dependent on the atoms spatial relation and the presence of atoms at certain locations will cancel out a diffracted beam where one is predicted based on Bragg's law (Cullity and Stock, 2001).

TABLE 13.1   Example of Indexing and the Resultant Selection Rules for Crystal Structures of the Cubic Class

hkl	$(h^2 + k^2 + l^2)$	Cubic	Body-Centered Cubic	Face-Centered Cubic
100	1	✓	×	×
110	2	✓	✓	×
111	3	✓	×	✓
200	4	✓	✓	✓
210	5	✓	×	×
211	6	✓	✓	×
220	8	✓	✓	✓
300	9	✓	×	×
310	10	✓	✓	×
311	11	✓	×	✓
222	12	✓	✓	✓
320	13	✓	×	×
321	14	✓	✓	×
400	16	✓	✓	✓
Selection rules		All	Even	h,k,l all even or all odd

Consider two reflections, one from the planes formed by A atoms and one by B atoms aligned parallel to each other. The reflections will interfere depending on the path length between the two planes. The path difference, S, is given by

$$S = 2d' \sin\theta \qquad (13.5)$$

But remember that for Bragg's law to be satisfied

$$2d \sin\theta = n\lambda \; \rightarrow \; \sin\theta = \frac{n\lambda}{2d} \qquad (13.6)$$

so that

$$S = \frac{d'}{d(n\lambda)} \qquad (13.7)$$

Therefore, the path difference depends on the position of atom B in the unit cell and on which planes (Miller indices) the diffraction occurs. It can be shown that if B atoms are at (xa, yb, zc) with fractional coordinates (x, y, z) then

$$d'_{hkl} = (hx + ky + lz)d_{hkl} \qquad (13.8)$$

and substituting this into Equation 13.7 gives

$$d'_{hkl} = (hx + ky + lz) \qquad (13.9)$$

Since x-rays are waves of wavelength λ they go through 2π oscillations by going through a distance λ. The phase difference δ, in radians, of the reflected beam is therefore calculated as

$$\delta = S2\pi/\lambda = (hx + ky + lz)(2\pi n) \qquad (13.10)$$

13.2.2.4 Atomic Scattering Factor

The atomic scattering factor, f, is measure of the scattering power of an isolated atom. It is used to describe the efficiency of scattering of a given atom in a given direction. In the forward scattering direction this number is equal to the atomic number, Z. As θ increases, the waves scattered by individual electrons become more and more out of phase and f decreases. The actual calculation involves $\sin\theta$ so that the net effect is that f decreases as the quantity $(\sin\theta)/\lambda$ increases.

The scattered wave amplitude due to A atoms

$$f = \alpha f_A \qquad (13.11)$$

Then the scattered wave amplitude due to the B planes is given by

$$f = \alpha f_B e^{i2\pi(hx+ky+lz)} = \alpha f_B e^{i\delta} \qquad (13.12)$$

The total amplitude at the detector is then

$$F = \alpha(f_A + f_B e^{i\delta}) \tag{13.13}$$

where the relation $i\delta$ has the phase position of B in the unit cell encoded in it.

13.2.2.5 Structure Factor

The total amplitude at the detector, F, represents the structure factor and it is a complex number whereby the real part is the amplitude and the imaginary part represents the phase. The value δ has the position of B in the unit cell encoded in its value. Since the magnitude of the waves is measured and not the phase, there is a loss of information regarding the location of B in the lattice. X-ray sources are not coherent and therefore there are no definite phase relationships between different points in a cross section of the beam. Detection of the x-ray beam is not phase sensitive. The measured intensity is given by the square of the absolute value and is obtained by multiplying the complex number portion by its conjugate.

$$I_{hkl} = |F_{hkl}|^2 = F_{hkl}^* F_{hkl} \tag{13.14}$$

And since the complex conjugate is generally represented by $(a + ib)(a - ib)$

$$I_{hkl} = |F_{hkl}|^2 = \alpha^2 (f_A + f_B e^{i\delta})(f_A - f_B e^{i\delta}) \tag{13.15}$$

$$I_{hkl} = |F_{hkl}|^2 = \alpha^2 \left(f_A^2 + f_B^2 + f_A f_B [e^{2\pi i\delta} + e^{-2\pi i\delta}] \right) \tag{13.16}$$

Remembering that Euler's identity is

$$e^{i\theta} = \cos\theta + i\sin\theta \tag{13.17}$$

substituting into Equation 13.16 we have

$$I_{hkl} = |F_{hkl}|^2 = \alpha^2 \left(f_A^2 + f_B^2 + f_A f_B [\cos(2\pi\delta) + i\sin(2\pi\delta) + \cos(2\pi\delta) - i\sin(2\pi\delta)] \right) \tag{13.18}$$

$$I_{hkl} = |F_{hkl}|^2 = \alpha^2 \left(f_A^2 + f_B^2 + f_A f_B [2\cos(2\pi\delta)] \right) \tag{13.19}$$

Since $\delta_{hkl} = hx + ky + lz$, the vector representation of the d value in the unit cell then we can see that the intensity is modulated as follows:

$$I_{hkl} = \alpha^2 \left(f_A^2 + \alpha^2 f_B^2 + 2 f_A f_B [\cos(2\pi\delta)] \right) \tag{13.20}$$

(independent) (modulation depending on the position of B).

Then it follows that the relative intensity of the peaks can be used to help determine the position of B. In practice, there are many parameters that also affect the intensity of the peaks that must be accounted for in determining the structure of the material of interest. A procedure, known as Rietfeld refinement, is generally used to refine a structure in several iterations until the locations of all of the atoms are known.

13.2.2.6 Diffracted Beam Intensity

While the location, presence, and absence of diffraction peaks at given values of 2θ gives information about the arrangement of atoms in a unit cell, the intensity of the diffracted peaks contains convoluted information about total scattering from each plane in the crystal structure and is directly dependent on the distribution of particular atoms in the structure and the atomic number of the atoms responsible for the scattering. Intensities are therefore related to both the structure and the composition of the phase. The total diffraction intensity equation is summarized as follows (Connolly, 2010):

$$I_{(hkl)\alpha} = \frac{I_o\lambda^3}{64\pi r}\left(\frac{e^2}{m_ec^2}\right)^2 \frac{M_{(hkl)}}{V_\alpha^2}\left|F_{(hkl)\alpha}\right|^2\left(\frac{1+\cos^2(2\theta)\cos^2(2\theta_m)}{\sin^2\theta\cos\theta}\right)_{hkl}\frac{v_\alpha}{\mu_s} \tag{13.21}$$

- $I_{(hkl)\alpha}$ = Intensity of reflection of *hkl* in phase α
- I_o = incident beam intensity
- r = distance from specimen to detector
- λ = x-ray wavelength
- (e^2/m_ec^2) = square of classical electron radius^2
- μ_s = linear absorption coefficient of the specimen
- V_α = volume fraction of phase α
- $M_{(hkl)}$ = multiplicity of reflection *hkl* in phase α
- $\left((1+\cos^2(2\theta)\cos^2(2\theta_m))/(\sin^2\theta\cos\theta)\right)_{hkl}$ = Lorentz polarization and monochromator correction
- v_α = volume of the unit cell of phase α
- $2\theta_m$ = diffraction angle of the monochromator
- $F_{(hkl)\alpha}$ = structure factor for reflection of *hkl* of phase α

Many of the terms in the intensity calculation are a function of the experimental setup and can be combined into a constant, K_e called the experimental constant. For a given phase of material present, a second constant can be defined as $K_{(hkl)\alpha}$. This value is equal to the structure factor term for phase α. Substituting the weight fraction (X_α) for the volume fraction, the density of the phase (ρ_α) for the volume, and the mass absorption coefficient of the specimen (μ/ρ)$_s$ for the linear absorption coefficient yields the following equation:

$$I_{(hkl)\alpha} = \frac{K_e K_{(hkl)\alpha} X_\alpha}{\rho_\alpha(\mu/\rho)_s} \tag{13.22}$$

This equation describes the intensity for peak *hkl* in phase α. In most experiments the mass absorption coefficient for the sample, $(\mu/\rho)_s$ is not known. Since it is itself a function of the amount of the constituent phases, and that is usually the object of quantitative analysis, then peak intensity-related methods for doing quantitative analysis involve circumventing this problem to make the equation solvable. Table 13.2 lists the various factors which control the intensities in a powder diffraction pattern and the parameters that the factor is a function of.

Structure-sensitive factors are mostly included in the $K_{(hkl)\alpha}$ term of the intensity equation. Most of these factors are intrinsic properties of the phase producing the reflection, but their intensity can be modified by both the sample temperature and wavelength of the incident radiation.

TABLE 13.2   Factors and Parameters Affecting the Peak Intensity of Diffracted X-Rays

Factor	Parameter
Structure sensitive	Atomic scattering factor structure factor polarization multiplicity temperature
Instrument-sensitive	
(a) absolute intensities	Source intensity diffractometer efficiency volatge drift take off angle of tube receiving slit width axial divergence allowed
(b) relative intensities	Divergence slit aperture detector dead time
Sample sensitive	Microabsorption crystallite size degree of crystallinity residual stress degree of particle overlap particle orientation concentration
Measurement-sensitive	Method of peak area measurement degree of peak overlap method of background subtraction presence or absence of $K_{\alpha 2}$ stripping degree of data smoothing employed

13.3 Applications in Fuel Cells

13.3.1 Phase Determination

Determination of the phase present in a sample is usually performed by checking the obtained diffraction spectra, after background subtraction, to a database updated yearly by the International Center for Diffraction Data. They offer a variety of databases ranging from general databases to ones suited specifically toward organics, inorganics, or minerals. A search algorithm matches the given spectra to entries in the database and returns a list of possible matches. The researcher must then scan through the possible results or cards until a best match is determined. The possible matches are called cards because before the widespread adoption of electronic databases, the standards were recorded on index cards and filed for matching. Each material card contains information about relative intensities of each peak compared to the largest peak, a reference to the journal or group that collected the data, the lattice constants for that structure, the space group of the crystal system, and the indexing for each plane.

13.3.2 Particle Size Determination (Uvarov and Popov, 2007)

This has often been referred to as "particle size" in the literature. The size obtained with x-ray methods is the effective length, measured in the direction of the diffraction vector, along which diffraction is coherent. Thus, "crystallite size" is preferred, since individual particles may contain several crystallites, or domains having different orientations. The distinction between crystallite and particle is shown in Figure 13.11. The distinction is important if sizes obtained from diffraction broadening are compared with those given by other methods. XRD data, like transmission electron microscopy (TEM) are complimentary techniques that can both provide information on particle sizes in the nanometer range. The two techniques differ in that XRD analysis provides information on crystallite size rather than actual particle size. XRD provides an estimate of particle size from volume averaging across the sample whereas TEM gives localized information generated from counting resulting in number-averaged results. The analysis of profile parameters in x-ray patterns allows determination of crystallite sizes and/ or microstrain. Size analysis using profile widths gives the crystallite or domain size, which is not directly comparable with results from other techniques. Within a particle there may be crystals and/or grain domains; these are separated by domain walls, disturbing the coherent scattering of the x-rays. These domain walls are not commonly visible by TEM. The domain walls may be formed from tangled dislocations or low-angle boundaries. Usually, when particle size is determined with TEM or scanning electron microscopy (SEM), they do not give the domain size.

However, the domain walls are not always present. When optical techniques using visible light (or IR or UV) are used to determine particle size, a larger size may be seen if small particles form clusters (granulates) with one another or are attached to larger particles.

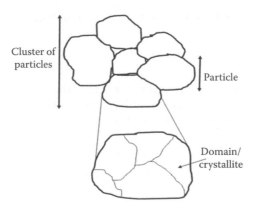

Cluster of particles

Particle

Domain/ crystallite

FIGURE 13.11 Distinction between crystallite, particle, and cluster.

As the crystallite size decreases, the width of the diffraction beam increases. To either side of the Bragg angle, the diffracted beam should destructively interfere and a sharp peak is expected. The destructive interference, to produce the sharp peak, is the result of the summation of all the diffracted beams so that close to the Bragg angle it takes diffraction from many planes to produce completed destructive interference. Due to the lack of many planes (in a small crystallite) to create the requisite destruction, a broadened peak is observed. A simple relation developed by Debye–Scherrer to estimate the thickness of a crystallite is commonly used in analysis of particle size growth due to catalyst growth resulting from certain fuel cell operating conditions.

In analysis of the structure factor it was shown that the destructive interference is just as much a consequence of the periodicity of atom arrangement as is constructive interference. If the path difference between x-ray photons scattered by the first two planes of atoms differs only slightly from integral number wavelengths, then the plane of atoms scattering x-rays exactly out of phase with the photons from the first plane will lie deep within the crystal. If the crystal is so small that this plane of atoms does not exist, then complete cancellation of all the scattered x-rays will not result. It follows that there is a connection between the amount of "out-of-phaseness" that can be tolerated and the size of the crystal. The result is that very small crystals cause broadening (a small angular divergence) of the diffracted beam, that is, diffraction (scattering) at angles near to but not equal to, the exact Bragg angle. Incident x-rays that make angles only slightly different from the Bragg angle, θ_B produce only incomplete destructive interference. For example, a beam that deviates only slightly from θ_B either larger or smaller, say θ_1 or θ_2, would normally be canceled out by planes located midway in the crystal populated by atoms scattering x-rays that are one-half wavelength out of phase with these angles. It follows that the diffracted intensities at angles near $2\theta_B$, but not greater than $2\theta_1$ or $2\theta_2$, are not zero but have values intermediate between zero and the maximum intensity of the beam diffracted at an angle $2\theta_B$.

The width of the diffraction curve increases as the thickness of the crystal decreases, because the angular range $(2\theta_1 - 2\theta_2)$ increases as the thickness of the crystal decreases. The width B of the diffraction peak from the small crystallite (in radians) is measured at an intensity equal to the width at half of the maximum peak intensity, termed full-width at half-maximum (FWHM). B is therefore roughly one-half the distance between the two extreme angles at which the intensity is zero, which amounts to assuming that the diffraction line is triangular in shape. Therefore,

$$B = \frac{1}{2}(2\theta_1 - 2\theta_2) = \theta_1 - \theta_2 \tag{13.23}$$

The path-difference equations for these two angles are similar to Bragg's law

$$n\lambda = 2d\sin\theta \qquad (13.24)$$

but related to the entire thickness of the crystal rather than to the distance between adjacent planes. If m represents the number of diffracting planes available in the crystallite (entire thickness of the crystal) rather than the distance between adjacent planes, Then

$$\begin{aligned} 2t\sin\theta_1 &= (m+1)\lambda, \\ 2t\sin\theta_2 &= (m-1)\lambda, \end{aligned} \qquad (13.25)$$

Subtracting these two equations gives

$$t(\sin\theta_1 - \sin\theta_2) = \lambda \qquad (13.26)$$

By the product to sum identities of trigonometric functions

$$\sin\theta_1 - \sin\theta_2 = 2\sin\left(\frac{\theta_1 - \theta_2}{2}\right)\cos\left(\frac{\theta_1 + \theta_2}{2}\right) \qquad (13.27)$$

$$2t\cos\left(\frac{\theta_1 + \theta_2}{2}\right)\sin\left(\frac{\theta_1 - \theta_2}{2}\right) = \lambda \qquad (13.28)$$

But θ_1 and θ_2 are close in value to θ_B so that

$$\theta_1 + \theta_2 = 2\theta_B\,(\text{approximately})$$

and given that the difference in the sin of two close angles is roughly equivalent to the difference between the two angles,

$$\sin\left(\frac{\theta_1 - \theta_2}{2}\right) = \left(\frac{\theta_1 - \theta_2}{2}\right)(\text{approximately}) \qquad (13.29)$$

Therefore,

$$2t\left(\frac{\theta_1 - \theta_2}{2}\right)\cos\theta_B = \lambda \qquad (13.30)$$

And substituting in $B = \theta_1 - \theta_2$ gives

$$t = \frac{\lambda}{B\cos\theta} \qquad (13.31)$$

This is known as the Scherrer equation named after Paul Scherrer (February 3, 1890–September 25, 1969). He was a Swiss physicist born in Herisau, Switzerland. He studied at Göttingen, Germany, before becoming a lecturer there. Later, Scherrer became head of the Department of Physics at ETH Zürich. A more formal use of the equation incorporates shape or structure factor, K.

$$t = \frac{K\lambda}{B\cos\theta} \tag{13.32}$$

where t is the crystallite size, λ is the wavelength of the x-rays, θ is the Bragg angle, and B is the full width at half-maximum (FWHM) of the peak in radians corrected for instrumental broadening (Smart and Moore, 2005). In practice, the correction for broadening due to the instrument is calculated by measuring a highly crystalline sample with a diffraction peak in a similar position to the sample. Commonly a sample of NIST 640b Si or LaB_6 is ground up with the sample of interest (taking care that there are no overlapping peaks) and measured at the same time. The broadening due to the standard B_M is subtracted from broadening due to the sample B_S as follows:

$$t = \frac{K\lambda}{\sqrt{B_M^2 - B_S^2}\cos\theta} \tag{13.33}$$

B_M and B_s are the FWHM of the sample and of a standard, respectively. A highly crystalline sample with a diffraction peak in a similar position to the sample is chosen and this gives the measure of the broadening due to instrumental effects. This method is particularly useful for plate-like crystals with distinctive shear planes (e.g., the 111) as measuring the peak width of this reflection gives the thickness of the crystallites perpendicular to these planes.

13.3.3 Examples in PEMFC Degradation Studies

13.3.3.1 Loss of Platinum Surface Area

In many instances it is useful to use grazing incidence XRD to examine a fresh and then cycled membrane electrode assembly MEA. This can reveal processes responsible for performance drop due to platinum surface area loss. Normal incidence XRD may not be useful since the x-rays can penetrate through the cathode, electrolyte, and anode and therefore the contributions of surface loss from one or both electrodes can be parsed out. For instance, work by Ferreira et al. (2005) describes a procedure whereby a Pt/Vulcan sample is analyzed by glancing angle mode XRD with 1° angle of incidence. The material-dependent depth of x-ray penetration in samples can be estimated by computing the x-ray intensity $I(x)$ at a depth x from the following relationship:

$$I(x) = I_0 e^{[-(\mu/\rho)_{electrode} * \rho_{electrode} * x]} \tag{13.34}$$

where (μ/ρ) is the mass attenuation coefficient of porous electrodes and $\rho_{electrode}$ is the density of the porous electrodes. Ferreira et al., used an identical fabrication procedure for both an anode and cathode (having 46 wt% Pt/Vulcan, a Nafion:C weight ratio of 0.8:1, and Nafion with an approximate net composition of CF_2 consist of 32 wt% Pt, 23 wt% F, and 45 wt% C), then estimated the mass attenuation coefficient as follows:

$$(\mu/\rho)_{electrode} = 0.32(\mu/\rho)_{Pt} + 0.23(\mu/\rho)_F + 0.45(\mu/\rho)_C = 48.9 \text{ cm}^2\,\text{g}^{-1} \tag{13.35}$$

By considering that the electrodes are 67% porous, the density of Pt/Vulcan porous structure and electrode densities were calculated from the following equations, respectively:

$$\rho_{Pt/C} = 0.33(\rho_{Pt}\rho_C)/(0.46\,\rho C + 0.54\,\rho_{Pt}) = 1.23 \text{ g cm}^{-3} \tag{13.36}$$

$$\rho_{electrode} = (\rho_{Pt/C}\,\rho_{CF2})/(0.35\,\rho_{Pt/C} + 0.65\,\rho_{CF2}) = 1.46\,g\,cm^{-3} \qquad (13.37)$$

The characteristic depth of x-ray penetration, $I(x) = I_o e^{-1}$, in the anode and cathode was calculated to be 140 μm. This value is deeper than the combined thickness of the MEA samples. This would give diffracted intensities from both the anode and cathode using normal incidence angles. The penetration depth of the x-ray bean follows a sin θ function and therefore has maximum penetration at 90°. Therefore, low-incidence angles of the incoming x-rays are required to collect only the anode or cathode diffraction intensities. Less than 0.5% of x-ray intensity passes through these electrodes at an incidence angle of 1° (13 μm/sin1° = 745 μm for x-ray beam path), and therefore when separating out the differences in degradation between the anode and cathode it is important to use a low-incident beam angle.

An x-ray powder diffraction pattern of the pristine Pt/Vulcan powder sample collected at normal incidence was compared with those of the anode and cathode in the cycled Pt/Vulcan MEA sample collected at 1° incidence in Figure 13.12.

The sharpening of the platinum diffraction peaks is evident for the cycled cathode, which indicated a considerable increase in the volume-averaged crystal size of platinum in the cycled MEA cathode. The average platinum crystal sizes (volume averaged) of pristine and cycled Pt/Vulcan samples were estimated using the Scherrer equation, by measuring the full-width half-maximum of the $\{111\}_{Pt}$, $\{200\}_{Pt}$, and the $\{220\}_{Pt}$ diffraction peaks. The pristine Pt/Vulcan sample was found to have a crystal size of 2.3 nm averaged over these three crystallographic planes. Upon cycling between 0.6 and 1.0 V, the volume-averaged platinum crystal size in the cycled MEA cathode increased significantly to 10.5 nm. Therefore, significant coarsening and surface loss of platinum crystals occurred in the MEA cathode during potential cycling, which is consistent with previous studies of potential-dependent platinum instability in PAFCs. In contrast, the average platinum crystal size in the cycled MEA anode grew only slightly to 3.9 nm, showing a considerable difference in loss on cycling of the cathode versus the anode.

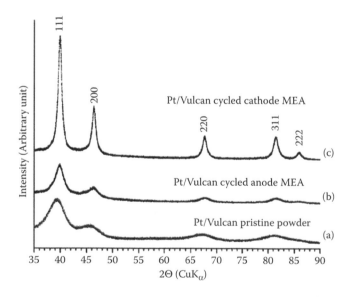

FIGURE 13.12 (a) X-ray powder diffraction pattern of the pristine Pt/Vulcan powder sample collected at normal incidence was compared to those of anode, (b) cathode, and (c) in the cycled Pt/Vulcan MEA sample collected at 1° incidence. (Reproduced from Ferreira, P. J. et al. 2005. *J. Electrochem. Soc.* 152(11): A2256–A2271. With permission.)

X-ray powder diffraction measures the volume-averaged diameter, d_v. It can be estimated from summarizing the products of individual platinum crystal diameters and their volume fractions, as shown in the following simplified expression:

$$d_{mean,vol} = \frac{\sum_i d_i^4}{\sum_i d_i^3}$$

(13.38)

where v_i is the volume fraction of platinum crystal d_i. In their study it was found that the volume-averaged diameter determined from x-ray powder diffraction was slightly different from the volume/area-averaged one from TEM measurement. It is believed that this difference results from the fact that platinum particles smaller than 1 nm were poorly represented in the TEM measurements, and therefore the volume/area-averaged diameter from TEM would be slightly larger than that of XRD. For the cycled MEA cathode, the volume-averaged diameter determined from x-ray powder diffraction is significantly different from the volume/area averaged one from TEM data collected from powder scrapped from the cycled MEA cathode surface. As x-ray powder diffraction determines the volume-averaged crystal size, it preferentially detects large crystal sizes to small crystals in powder samples. The large difference between the volume-averaged and volume/area-averaged diameters is attributed to a comparatively wide distribution of platinum crystal sizes in the cycled MEA cathode relative to that of the pristine sample.

13.3.3.2 Degradation Acceleration due to Temperature Effects

Active research is currently being pursued to push PEM operation temperatures above 80°C as increasing CO tolerance can be gained at temperatures of 120–150°C. It is expected that increased temperatures will accelerate many degradation processes. Bi and Fuller examined the degradation of carbon-supported Pt catalyst in common Nafion membrane electrode assemblies by using an accelerated protocol by square-wave potential cycling at temperatures of 40°C, 60°C, and 80°C (Bi and Fuller, 2008). To examine the effect on the platinum catalyst they used an x-ray diffractometer with Cu K_α radiation at a grazing

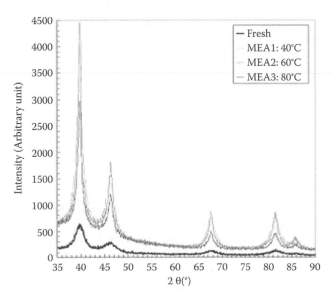

FIGURE 13.13 Accelerated agglomeration of platinum particles with increasing temperature. (Reproduced from Bi, W. and Fuller, T. F. 2008. *J. Electrochem. Soc.* 155(2): B215–B221. With permission.)

TABLE 13.3 Platinum Particle Size Estimated by the Scherrer Equation

MEAs	Pt{111}	Pt{200}	Pt{220}	Pt{311}	Average
Fresh	3.8	2.9	3.6	2.9	3.3
MEA 1	5.5	4.5	4.8	4.6	4.9
MEA 2	8	6.6	8.8	7	7.6
MEA 3	8.7	7.4	9.7	7	8.2

Source: Reproduced from Bi, W. and Fuller, T. F. 2008. *J. Electrochem. Soc.* 155(2): B215–B221. With permission.

incident angle of 1°. The patterns were collected by continuous scans of 1.8° per min from 35° to 90°, as shown in Figure 13.13. The particle sizes were calculated from the following crystal planes: Pt{111}, Pt{200}, Pt{220}, and Pt{311} using the Scherrer equation. The data were fitted by the Jade 7.5 software package (Material Data, Inc.). The observed diffraction patterns indicate a lattice parameter of 0.392 nm. The degraded cathodes showed a sharpening of the peaks compared to the fresh sample. The calculated Pt particle average size increased with an increase in temperature.

If a spherical shape to the Pt particles is assumed, then the catalyst surface area is given by the following formula:

$$S_{XRD}(m^2 g^{-1} Pt) = \frac{6 \times 1000}{D \times \rho}$$

(13.39)

where, D is the estimated particle size from the XRD data and ρ is the Pt density. The ripening of the particle size is shown in Table 13.3.

13.4 Conclusion

X-ray diffraction is a powerful tool for determination of the lattice parameters, crystallite size and phase identification of materials in the solid phase. It is particularly suited for analysis of platinum particle size increases often seen upon cycling of MEAs. The use of XRD is noninvasive and can usually be done in less than 10 min with newer line detectors. The field of XRD, and its many applications, is enormous and very well documented in the literature and textbooks. A more fundamental treatment of the topic can be found in Cullity's seminal work and extensive online resources such as Matter.org.uk, a nonprofit consortium of UK materials science departments, is available for further exploration of the topic complete with Java applets for better understanding of key concepts.

References

Bi, W. and Fuller, T. F. 2008. Temperature effects on PEM fuel cells Pt/C catalyst degradation. *J. Electrochem. Soc.* 155(2): B215–B221.

Bruker-AXS. 2010. Vantec-1 detector for super speed X-ray diffraction. [cited 11/09 2010]. Available from http://www.bruker-axs.de/uploads/tx_linkselector/VANTEC_Spec_Sheet_S88-E00017.pdf.

Connolly, J. R. 2010. Introduction to X-ray powder diffraction. [cited 11/12 2010]. Available from http://epswww.unm.edu/xrd/xrdclass/09-Quant-intro.pdf.

Cullity, B. D. and Stock, S. R. 2001. *Elements of X-Ray Diffraction*. Upper Saddle River, NJ: Prentice-Hall.

De Graef, M. 2007. *Structure of Materials* (1st edn.). Cambridge: Cambridge University Press.

Ferreira, P. J., la O', G. J., Shao-Horn, Y., Morgan, D., Makharia, R., Kocha, S., and Gasteiger, H. A. 2005. Instability of Pt/C electrocatalysts in proton exchange membrane fuel cells. *J. Electrochem. Soc.* 152(11): A2256–A2271.

Fujinawa, G. and Umegaki, S. 2001. Rigaku Corporation (Akishima, J. P.). Soller slit and manufacturing method of the same. US Patent 6,266,392. July 24, 2001.

Gene, I. and Pang, J. 2009. Tutorial on x-ray microLaue diffraction, *Mater. Charact.* 60(11): 1191–1201.

Harding, M. 1991. Laue diffraction studies of inorganic materials. *J. Phys. Chem. Solids.* 52(10): 1293–1298.

Holz, T., Dietsch, R., Mai, H., and Brügemann, L. 2000. *Application of Ni/C-Göbel Mirrors as Parallel Beam. X-ray Optics for cu ka and mo ka Radiation.* Proceedings of the Denver X-ray Conference, Advances in X-ray Analysis Vol. 43, Denver, CO.

Lochner, J. 2010. X-ray detectors in Goodard Space Flight Center [database online]. [cited 11/10 2010]. Available from http://imagine.gsfc.nasa.gov/docs/science/how_l1/xray_detectors.html.

Michaelsen, C., Ricardo, P., Anders, D., Schuster, M., Schilling, J., and Goebel, H. 1998. *Improved Graded Multilayer Mirrors for XRD Applications.* Proceedings of the Denver X-ray Conference, Advances in X-ray Analysis Vol. 42, Denver, CO.

Smart, L. E. and Moore, E. A. 2005. *Solid State Chemistry: An Introduction.* Boca Raton, FL: Taylor & Francis Group, LLC.

Smith, D. 2010. The gem dugout X-ray diffraction products. [cited 11/09 2010]. Available from http://www.thegemdugout.com/products.html.

Uvarov, V. and Popov, I. 2007. Metrological characterization of X-ray diffraction methods for determination of crystallite size in nano-scale materials. *Mater. Charact.* 58(10): 883–891.

14

Scanning Electron Microscopy

Robert Alink
Fraunhofer Institute for
Solar Energy Systems

Dietmar Gerteisen
Fraunhofer Institute for
Solar Energy Systems

14.1 Introduction

The three main functional parts of the PEM fuel cell are the polymer membrane, the catalyst layers, and the gas diffusion layers (GDL). They represent the core of a fuel cell and operate under harsh conditions with respect to potential, temperature, humidity, and pH. As a result, the materials used are subjected to strict durability requirements. One common way of learning more about the aging processes in PEM fuel cells and their mitigation is to examine the structural changes of these layers postmortem either after long-term operation or under accelerated stress tests. Suitable examination techniques have to be used to resolve the functional structure of PEM fuel cells in the electrodes or GDL, which are in the order of micrometers and nanometers. In most cases, this is beyond the range of conventional optical microscopes. SEMs, however, can measure well into this desired range. For this reason, SEMs are becoming increasingly interesting for the *ex situ* examination of degradation issues in PEM fuel cells. In addition, the decreasing prices along with the steadily improving capability and ease of handling foster its propagation and availability for fuel cell research.

Mostly, SEMs are combined with an energy-dispersive x-ray (EDX) unit that allows elemental analysis of the sample surface. This technique is very valuable in analyzing the change in the elemental composition of the investigated fuel cell layer due to, for example, particle migration, growth and washing-out of catalyst particles, or the contamination of fuel cell components due to fuel cell operation.

Visually examining an object means receiving information about the interaction between the object matter and the electromagnetic radiation within the visible band of the light. Human vision is sufficient to see structures in the dimensions common to our daily life, but if we want to analyze structures having dimensions within the range of visible light (400–700 nm) human vision is far beyond its limitations.

An SEM is an instrument that uses electrons to visualize structural as well as material properties of the object of investigation. Since the wavelength of electrons with high energy (10^{-3} nm) is five orders of

TABLE 14.1   Energies and Wavelengths of Particles Used for Sample Visualization

Type	Wavelength (nm)	Energy
Photons (visible)	400–700	2 eV
Photons (x-ray)	0.05–1.25	1–25 keV
Electrons	1e–3–3e–3	100 keV 1 MeV
Protons	1e–4	10 keV
Neutrons	0.1	0.025 eV

magnitude shorter than wavelengths of the visible spectrum (Table 14.1), resolutions of up to 1:1,000,000 with a very high focus depth can be obtained. The advantage of achieving a very high resolution is bought dearly by a few disadvantages associated with this technique. One disadvantage is that it is physically constricted to using one wavelength to analyze the object. Only the intensity of the detected signal can be transformed into a grayscale image. In contrast, the human eye can receive and process a specific bandwidth of electromagnetic waves in the form of colors.

Also in its spatial resolving power, the SEM shows disadvantages as compared to human vision. Two-dimensional resolution in the human eye is obtained by simultaneously receiving information on a two-dimensional detector, called the retina. In SEMs, only zero-dimensional information is received by a detector which is not spatially resolved. To generate two-dimensional pictures using the SEM, the electron beam is scanned in a certain pattern over the sample surface. At the same time, the interaction with the surface is analyzed by a zero-dimensional detector. Since the area of investigation needs to be scanned, a comparably low updating rate of the picture results. Further, the analyzed object may not be moved during the scanning process.

The lateral deflection of the electron beam can be controlled only to within a few micrometers. For larger sample navigation, the sample holder is mounted on a mechanical positioning unit. For many instruments, besides the navigation in the xy-plane (perpendicular to the direction of the electron beam), the sample can be moved in the z-direction, rotated around the z-axis, and tilted around the y-axis.

Since electrons strongly interact with any gaseous material located between the electron beam source, the sample, and the detector, the ray path from the electron gun to the sample, the sample itself and the detector are all placed in a vacuum chamber. This enforces a strong restriction on the specimen in terms of vacuum stability, which is not required for optical microscopes (Section 14.2.1). In a standard SEM, the chamber is evacuated to a minimum pressure, defined by the effectiveness of the vacuum system. In some special cases, it is beneficial to leave a certain amount of atmosphere in the chamber. This can be realized in an ESEM where the final pressure of the specimen chamber and the composition of the residual atmosphere can be adjusted to a defined value. With this technique it is not mandatory to have electrically conductive sample surfaces as needed for a standard SEM to avoid sample charging. Thus, the surface of the object under investigation can remain unchanged and working pressures over the triple point of water (Figure 14.1) can be adjusted which in turn allows the *ex situ* examination on the wettability of fuel cell components such as GDLs. However, ESEM is quite a recent technique and their distribution is not as advanced as for standard SEM instruments.

After briefly introducing the working principle of the SEM, the device configuration and the requirements for sample preparation are described. Further, an overview of recently published works in the field of fuel cell degradation using SEM analysis is given. Typical areas of application for using the techniques of SEM/ESEM/EDX to analyze fuel cell degradation are presented. This includes the analysis of the catalyst layer structure (carbon support structure, catalyst distribution, porosity, layer thickness, surface roughness, cracks, and pinhole formation), the polymer membrane (layer thickness, catalyst particle migration, wetting properties, swelling behavior), and the GDL (porosity, microcontact angle distribution, loss of hydrophobicity).

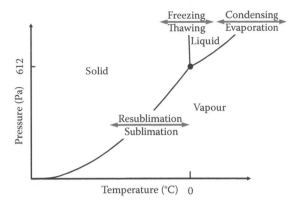

FIGURE 14.1 Phase diagram of water and phase transitions.

14.2 Principle

14.2.1 SEM

14.2.1.1 Physical Background

Since the working principle of an SEM is based on the interaction between the electron beam and the specimen, a nearly monochromatic electron beam with homogeneous energy distribution is necessary. Over the last few decades, a simple tungsten wire was used as an electron beam source, whereby a high temperature is applied to the wire to generate electrons with sufficient energy for emission. Today, a lanthanum hexaboride emitter (LaB6) is preferred as an electron beam emitter since this crystal produces a brighter beam with a smaller-sized electron source. This increases the resolution of the SEM and additionally allows the use of lower accelerating voltages to prevent beam induced damage to the sample. It reduces artifacts by sample charging as well.

After emission, the electrons are accelerated by an electrical field with a voltage ranging from several kilovolts to 40 kV. Then, the electrons are deflected by magnetic lenses and consequently focused by magnetic condenser lenses to a small spot on the specimen surface. Since the detector in SEM is zero dimensional, this spot is subsequently scanned over the desired visualization field of the sample in a specified two-dimensional pattern to obtain a two-dimensional image. A schematic of an SEM setup is shown in Figure 14.2.

When hitting the sample surface, the primary electron beam penetrates into the specimen while scattering with its atoms. The interaction with the sample matter results in an energy loss of the primary electrons, determining the penetration depth and the so-called interaction volume. The interaction volume has a pear-like shape and has dimensions of the order of 1 μm, depending on the acceleration voltage and the specimen material composition. Due to the conservation of energy, the energy loss of the primary electrons (PE) results in an energy transfer to the electrons in the sample material through a sequence of events. Thus, several types of electrons and radiation are emitted from the specimen: backscattered electrons (BSE) by elastic scattering, secondary electrons (SE) by inelastic scattering and electromagnetic radiation. Due to the elastic scattering of the primary electrons, the energy range of the BSE is in the same range of the PE (several keV) and is strongly dependent on the sample atom mass (Figure 14.3). Therefore, the signal of the BSE contains information about the material composition and the morphology of the sample. The origin of the BSE is the entire interaction volume of the electron beam with the sample matter.

The energy of the SE is only a few electron volts. The SEs originate from the atoms of the specimen, emitted when the primary electrons are decelerated in the sample matter by nonelastic collisions.

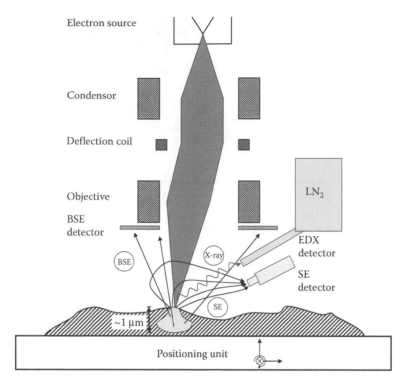

FIGURE 14.2 Schematic diagram of an SEM with EDX.

Due to their low energy, only the SEs generated near the surface can leave the sample and are detectable. Thus, the SE contains more information about the specimen surface morphology than the BSE. To increase the number of detectable secondary electrons, a small positive voltage is applied between the specimen and a lattice in front of the SE detector to accelerate the electrons toward the detector.

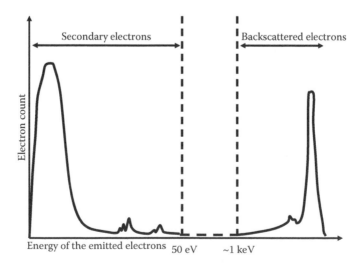

FIGURE 14.3 Energy distribution of the detectable backscattered, or secondary, electrons. (From Colliex, C. 2008. *Elektronenmikroskopie- eine anwendungsbezogene Einführung*, Stuttgart: Wissenschaftliche Verlagsgesellschaft mgH. Reprinted with permission from Presses Universitaires de France (PUF).)

Depending on the desired information, that is, the sample's surface topography (SE) or information about the material (BSE), different detectors with distinct energy ranges are used to detect the various signals. In Figure 14.4, the signals of the SE detector (top, left) and the BSE detector (top, right) are shown for the same cross section of a membrane electrode assembly (MEA) framed by GDL material. Obviously, the signal of the electrodes is more pronounced in the picture obtained with the BSE detector. Also, in the picture of the BSE, the cathode electrode on top is brighter than the lower one since the platinum loading within the cathode is higher than within the anode for this kind of MEA.

Not only electrons, but also photons with wavelengths in the range of x-rays up to the visible light are emitted when the electron beam is penetrating into the sample material. The emission of an electron from a core level of an atom leaves a vacancy that can be filled by an electron from a higher energy level. This process is accompanied by a release of energy in form of radiation. Since the height of the energy gaps between the electron levels is atom specific, these photons can be used for an analytical investigation of the composition of the volume penetrated by the electron beam. EDX uses a special type of detector to process the impinged photons and analyzes this information to obtain the composition of the sample surface (Section 14.2.2). The EDX method can be applied on a single spot of the sample or whole regions. For whole regions newer instruments can analyze the distribution of predefined elements beneath the average composition.

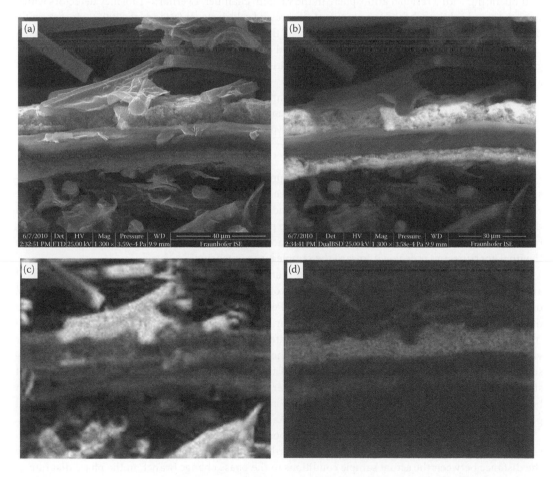

FIGURE 14.4 Cross section of an MEA sandwiched between two GDLs with lower catalyst loading on the anode side. (a) SE and (b) BSE images with the distribution of (c) carbon and (d) platinum, are determined by SEM and EDX analysis. The sample is clamped in a sample holder and cut with a fresh razor blade.

14.2.1.2 ESEM Principle

Owing to the high vacuum required in standard SEMs, limitations exist on the type of samples that can be analyzed. On the one hand, the analyzed sample has to endure the vacuum and on the other hand, the vacuum system has to be tolerant to any degassing of the sample. Furthermore, the analyzed sample surface has to be electrically connected to ground to avoid the charging of the sample. Charging would lead to local variations in electron emissions and deflecting of the primary electrons, resulting ultimately in artifacts. To obviate these restrictions, low-vacuum systems have been developed and commercialized in recent years. ElectroScan introduced the first widely known ESEM two decades ago (Rice and Knowles, 2005).

An ESEM is a special kind of SEM, where the type of residual gas and the absolute pressure in the sample vacuum chamber can be adjusted. For this purpose, the vacuum inside the beam-leading vacuum chamber and the vacuum in the sample chamber are separated. In the beam-leading chamber, high-vacuum conditions are still present to avoid beam adsorption, expanding or deflection while in the sample chamber absolute pressures higher than 10 Pa are adjustable (dependent on detector). A supplemental gas supply can be mounted to the ESEM to operate the sample chamber with a desired operating gas. When water vapor is used as chamber gas, it is evaporated from a liquid water reserve, which is mounted to the vacuum chamber.

If operating with a residual atmosphere in the vacuum chamber, Everhard–Thornley detectors which were the standard in early high-vacuum systems arc and self-destruct (Rice et al., 2005). More recently developed ESEM detectors locate the sample near the final aperture of the electron beam and use an additional amplifying voltage to attract the emitted electrons. The attracted electrons collide with the gas molecules in the atmosphere and generate additional gaseous secondary electrons that are accelerated toward the ESEM detector.

The most important advantage of the ESEM compared to standard SEMs is that less effort is necessary for sample preparation. Also, nonconductive samples can be analyzed without additional coating. Sample charging induced by the electron beam, is avoided by the ionized gas molecules that were generated during the formation of the gaseous secondary electrons. The negative charges in the sample are neutralized by these positive ions.

For the analytical EDX (Section 14.2.2) analysis, problems arise when operating with a residual atmosphere in the sample chamber. The primary beam electrons collide with atmospheric particles on the way from the final aperture to the sample surface and cause scattering. The greater the distance between the final aperture and the sample, the higher is the probability that the primary electron impinges far from the intended beam spot. To improve the EDX accuracy, gaseous analytical detectors (GAD) reduce the final aperture and redeploy it toward the sample by an extension.

Since the operating pressures of ESEM detectors are often near the triple point of water, conditions of all three phases can be adjusted by simultaneously controlling the sample temperature. External control of the sample temperature can be realized by using Peltier-based cryostatic tables which can be mounted onto the positioning unit within the sample vacuum chamber. Thus, condensation, evaporation, freezing, thawing, sublimation, and resublimation of water can be induced on the sample when using water vapor as chamber gas (Figure 14.1). In Figure 14.5, the formation of liquid water droplets during condensation, rearrangement, and subsequent freezing of the water in a GDL while going into the different states of aggregation around the triple point of water is shown. In fuel cell research, this procedure can be used to examine the interaction of the fuel cell components with water in the vapor, liquid, and solid phase, for example, on the freeze capability of MEAs (Section 14.4.5).

A further advantage of the ESEM is that the evaporation rate of substances in the sample can be controlled by changing the chamber pressure. In water-based samples, the phase change rate is defined by the distance between the actual sample conditions to the phase change branch in the phase diagram of water (Figure 14.1). When using other volatile materials, the evaporation rate can be significantly reduced by lowering the sample temperature and increasing the chamber pressure in general.

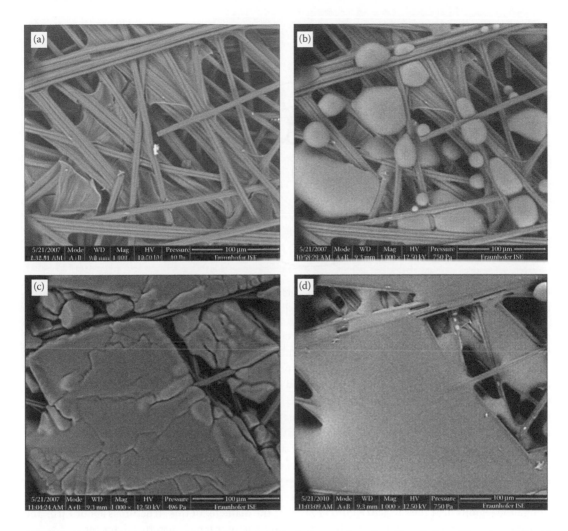

FIGURE 14.5 GDL in ESEM before condensation (a), during condensation (b and c), and freezing of liquid water (d).

14.2.2 Energy-Dispersive X-Ray

14.2.2.1 Physical Background

EDX is a nonquantitative analytical method to analyze the surface material composition. Similar to SEM, the measuring technique of the EDX also relies on the interaction of the electron beam with the matter in the sample. Therefore, EDX is often installed in combination with an SEM, sharing the electron gun. However, in contrast to SEM detectors, the EDX detector processes photons rather than electrons emitted from the sample.

Primary electrons at high-energy levels interact with ground-state electrons near the sample surface and eject them to an unbounded state. An electron from an outer shell fills the vacated space in the inner shell while emitting a photon which has the energy equivalent to the difference between these two atomic orbitals (Figure 14.6). The energy difference between the inner and the outer shell is characteristic for the atomic structure of the element and thus the emitted photons carry information about the element from which they originate. In general, only those elements can be detected in which the energy

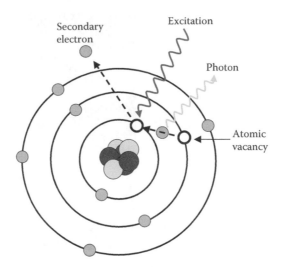

FIGURE 14.6 Atomic model and principle of events induced by electron collision. In SEMs, emitted secondary electrons are detected. In EDX, photons emitted when shell electrons change into a lower energy shell, are detected.

gap between the atomic shells is in the range of the energy of the primary electrons. On the other hand, hydrogen and helium have only one atomic shell and thus cannot be detected.

The photons are emitted in almost the whole penetration volume of the primary electron beam (approximately 1 μm in depth) and can be analyzed by an EDX detector, which is normally a semiconductor with a highly ohmic, x-ray-sensitive volume. To obtain a high resolution, the ohmic part has to be as low as possible and is therefore cooled to a maximum. There are mainly two main types of EDX detectors. The Li (Si) detector, consisting of a silicon crystal, has to be cooled by liquid nitrogen. The silicon drift detector is a newer type and is made out of silicon wafers. Cooling to −20°C by a Peltier element is sufficient for this detector type. With this kind of detector the efficiency of detecting high activation energies over 20 keV is significantly reduced.

When the electron beam hits the sample surface, the beam is scattered with a pear-shaped distribution in the regions near the surface (Figure 14.2). Since the diameter of this structure is about 1 μm, the maximum spatial resolution of the EDX is restricted to structures larger than this diameter.

14.2.2.2 Visualization Techniques

EDX analysis can be applied to single points or whole regions. If only one point is examined, the electron beam remains at that position and the detected counts for discrete activation energies can be plotted to obtain a spectrum (Figure 14.7). For analysis of a discrete area, the electron beam is scanned over that area and the counts for the discrete activation energies are integrated. An average spectrum for the whole region is then obtained which holds the information of the material composition near the surface.

Each element produces one or more spectral lines in the EDX-spectrum. The activation energies of the detectable elements are most often stored in the database of the evaluation software. This supplies a pretty good estimate automatically. However, the signals of certain elements can overlap or the counts of elements occurring less often are not sufficient to be recognized automatically. For these cases manual readjustment is needed.

When the spectrum is collected over a whole region, the spatial distribution of the selected elements can also be obtained. For this purpose, the elements of interest first have to be selected, for example, by collecting a spectrum over the whole region of interest. Then, the electron beam scans the whole area and accumulates the counts for the elements for each scanning point separately. The resulting picture gives a spatial distribution of the selected elements, mostly coded in color scale pictures. In Figure 14.4,

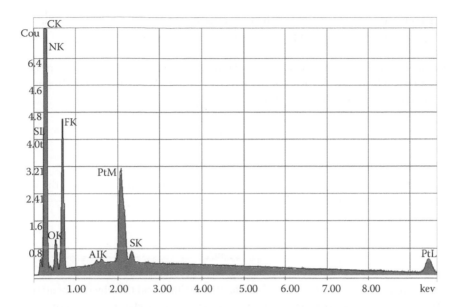

FIGURE 14.7 EDX-spectrum of Figure 14.4. Elements detected are sulphur, carbon, nitrogen, oxygen, fluorine, aluminum, and platinum by the activation energy of the K, L, and M shells. The signal of aluminum is most likely from sample holder or depositions of the razor blade. Nitrogen most likely originates from the atmosphere.

the distribution of carbon (bottom left) and platinum (bottom right) in the MEA and GDL cross section (shown in Figure 14.4 at the top) are displayed.

14.3 Instruments and Measurements

14.3.1 Different SEM Types

In contrast to the high temperatures of the electron emitter in standard SEMs, high voltages are applied to a wire to emit the electrons in field emission electron microscopes (FSEMs). This results in a more brilliant electron beam than for thermal cathodes, and the beam can be focused on a smaller spot on the sample surface while maintaining a high-emission current density. However, the higher resolution of such instruments is in contrast to the greater requirements on the vacuum system in the electron source chamber which are increasing to a maximum of 10^{-9} Pa (Colliex, 2008). Newer instruments overcome this discrepancy by combining a higher temperature and higher voltage to reduce the requirements on the vacuum system while maintaining the higher beam quality of the field emission cathodes.

In transmission electron microscopes (TEM), transmitted electrons, in contrast to the reflected electrons in standard SEMs, are processed. As a result, the requirement on the sample preparation is significantly increased, as well as the obtainable resolution. The two-dimensional visualization of the sample surface is obtained by a two-dimensional detector that also results in a higher temporal resolution of TEMs. However, scanning transmission electron microscopes (STEM), only have a zero-dimensional detector and the electron beam is scanned over the sample surface as in SEMs. The examination methodology of STEM and TEM, are described in detail in Chapter 15.

14.3.2 Restrictions/Obtainable Resolution

Besides the architecture of the SEM, the obtainable resolution is dependent on many factors. On the one hand, settings such as the acceleration voltage, spot size, and emission current influence the energy of the electrons in the beam and therefore the contrast of the obtained signal. On the other hand, the type of chamber gas and the gas pressure in the chamber as well as the working distance influence the interactions with gas molecules in the sample chamber and the ray path between electron source, sample, and detector.

As mentioned previously, the high resolution of SEMs compared to optical microscopes is mostly due to the lower wavelength of the electrons used for visualization. According to De Broglie, an electron with mass m and velocity v has the wavelength:

$$\lambda = \frac{h}{m \cdot v} \tag{14.1}$$

with h (6.62×10^{-34} J s) as the Planck quantum. When accelerated by the voltage U, the wavelength of each electron with charge e can be calculated to

$$\lambda = \frac{h}{\sqrt{2mUe}} \tag{14.2}$$

Obviously, the wavelength decreases with the acceleration voltage and thereby the theoretically obtainable resolution increases. However, the minimum spot size of the electron beam on the sample surface is some orders of magnitude higher than the wavelength of the electrons and therefore is more restricting in terms of resolution. Thus, the acceleration voltage has a much higher influence on the obtainable resolution due to the increased contrast resulting from the higher energy of the detected signal.

On the one hand, the spot size can be manually adjusted according to the distance between the scanning points. Otherwise either overlapping or free spaces between the scanning points will cause fuzziness or subsampling. On the other hand, similar to optical microscopes, there are also a couple of aberrations in the ray path that can significantly reduce the minimum spot size and the homogeneous energy distribution within the spot.

Thereby, the objective of the electron microscope plays the central role in determining the image quality. Spherical aberration and astigmatism within the objective influence the sharpness of the sample picture in optical microscopes as well as in electron microscopes. Even color aberration can occur within electron microscopes when the wavelengths of the electrons in the beam are not exactly the same. To reduce this aberration, field emission cathodes are often used which have less energy fluctuation in the emission current of the electron source and decrease the color aberration by a factor of 3–5. Much effort in correcting the spherical aberration has led to microscopes which have a resolution of up to 0.1 nm.

14.3.3 Sample Preparation

The most important requirements affecting sample preparation and the type of sample are that the sample will be subjected to a vacuum on the one hand and to a high-energy electron beam on the other hand.

When not operating in ESEM mode, the vacuum in the chamber will be lower than 10^{-2} Pa which causes many materials to go into the gaseous phase. Thus, the type of material that is brought into the vacuum chamber and the conditions in the vacuum chamber (vapor partial pressure and sample temperature) have to be considered carefully. The evaporation rate of materials depends on the difference between the vapor pressure of the material surrounding the sample and the temperature-dependent saturation pressure. When using water vapor as chamber gas, the evaporation of water within the samples can be controlled by setting sample temperature and pressure close to the phase transition line.

The evaporation of ingredients can also harm the vacuum system, for example, when preparing the sample by using nonspecific glue. To avoid this, most often carbon stickers are used for fixing the sample on the sample holder.

When the sample is subjected to the electron beam, the matter in the sample interacts with the electrons in the electron beam. Although the contrast of the sample picture increases with the electron energy in the beam, a highly energetic electron beam, in turn, causes a local temperature increase, charging of the sample, or even a structural change in the matter itself.

In Figure 14.8, the structural changes in the sample from Figure 14.4 are shown after being exposed to a 25 kV electron beam for half an hour. The most obvious changes are seen in the membrane material.

Charging will lead to local variations in the electron emissions and deflections of the primary beam electrons. Occurring hot spots can locally change or destroy the sample. Both charging and hot spots, however, can be avoided by thermally and electrically connecting the sample surface to a temperature and electron sink (ground). If the sample is not thermally and electrically conductive, sputtering with conductive material or operating in ESEM mode can help to analyze the sample.

Samples are most often mounted on specimen holders and are then assembled to multiple stub holders. In general, there are two ways of analyzing a sample: cross section or top view. Especially sample cross sections must be prepared with care when analyzing the consequences of degradation.

Cutting the sample in an adequate way is the easiest method to prepare cross sections. Scalpels often have specially formed blades which can cut askew, but even when using razor blades, the analyzed microstructure can be harmed during cutting. Damage by sample deformation can be avoided by casting in rosin, curing, and then cutting followed by polishing, as compared to cutting the bare sample. Care must be taken, however, since mechanical stress can occur during curing. Also, deformation of some softer materials can take place when cutting and polishing, creating cracks in the microstructures.

Another method to prepare cross sections is to freeze the sample by soaking it in liquid nitrogen and then subsequently breaking it. However, problems can arise due to the different thermal expansion coefficients of the composite materials or uneven temperature distribution within the sample during freezing. The resulting mechanical stress can also influence the inner sample structure.

When polishing or cutting the sample, a smearing of the material over the cross-sectional surface is unavoidable. As a result, the distribution of particles of interest for examining degradation no longer provides information on the degradation mechanisms. Electrode particles or high platinum concentration found by EDX and SEM analysis in the GDL or in the membrane, for example, can originate from the polishing or cutting process during cross-section preparation. Especially when using EDX analysis, the preparation of the sample can have an important influence on the distribution of the elements in the cross sections and again great care must be taken during preparation.

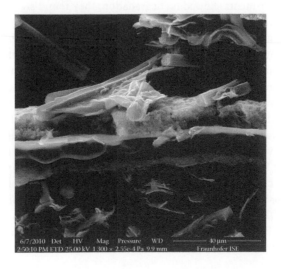

FIGURE 14.8 The same cross section as shown in Figure 14.4 but after being subjected to the 25 keV electron beam for approximately 20 min during EDX collecting.

14.4 Applications/Literature Review

To enhance long-term stability and durability, an improved understanding of the degradation processes that lead to irreversible performance loss or to complete component failure, in the worst case, is desired. Since several mechanisms are simultaneously at work during fuel cell aging, these effects must be separated out as much as possible by using different *in situ* and *ex situ* analysis techniques. Besides the *in situ* characterization techniques such as polarization curves, cyclovoltammetry, and impedance spectroscopy, postmortem analysis such as imaging techniques or mechanical testing are powerful tools for obtaining a clearer picture about the aging processes. Due to its high resolution, SEM is suitable for investigating changes in the functional structures of components which are often in the range of micrometers. In combination with an EDX unit, it is possible to perform a spatially resolved element analysis at the same time.

In the literature, an increasing number of recently published papers are focusing on durability tests of fuel cell components, single cells, and even fuel cell stacks, whereby SEM images are used to support their assumptions on the degradation mechanisms. In most cases, the SEM technique is used for investigating the structural change of the membrane (Section 14.4.1), the catalyst layer (Section 14.4.2), the diffusion media such as microporous layer and GDL (Section 14.4.3), and peripheral components (Section 14.4.4). The change in wetting properties of the porous media is also attributed to degradation and can be analyzed by means of liquid water visualization in an ESEM (Section 14.4.5).

14.4.1 Membrane Degradation

Membrane degradation can be classified into three different categories: mechanical, thermal, and chemical/electrochemical degradation (Collier et al., 2006; LaConti et al., 2003). An excessive mechanical and thermal stress often causes a complete breakdown due to perforation, cracks, or pinhole formation. Chemical, or electrochemical, stress leads to membrane thinning by radical attack on the polymer chains and therefore irreversible performance loss by decreasing proton conductivity or increasing gas permeability. Through the visualization of the physical deformations in top views and cross-sectional views of the membranes and MEAs, SEM can identify this kind of degradation. Fernandes and Ticianelli (2009) applied SEM imaging to investigate the chemical stability of Nafion® membranes (N212 and N112) by means of accelerated degradation procedures on Nafion membranes. Even if Fourier transform infrared spectroscopy experiments showed no degradation, they found significant formations of bubbles, tears, and bumps in the top view and cross sections of the membranes. The damage seemed to be more severe for the Nafion N212 membrane than for the Nafion N112 membrane. Kundu et al. (2008) also found severe degradation in the cross sections of Nafion membranes after applying two different styles of Fenton's test. In one method, the surface of the membrane actually had the appearance of foam, while the center of the membrane essentially remained as one continuous phase. Using the other method, they even found that the membrane had split in two halves.

During fuel cell operation, platinum dissolves (Darling and Meyers, 2003) and migrates from the cathode catalyst layer toward the anode side into the membrane (Sasaki et al., 2009). The solubility of platinum depends on cell voltage and is highest at open circuit voltage. The driving force underlying the crossover of the dissolved Pt^{2+} particles is the concentration gradient diffusion (Guilminot et al., 2007). In SEM images showing cross sections of previously operated MEAs, the platinum migration is evidenced by a distinct white band in the membrane (Bi et al., 2007; Ferreira et al., 2005; Péron et al., 2008; Yu et al., 2005b). Using EDX analysis, the reason for the band is verified to be a high platinum concentration (Péron et al., 2008). Bi et al. (2007) applied a simple model using the gradient of hydrogen and oxygen partial pressures within the membrane to predict the position of the band and validated their model using SEM imaging.

In the region of the Pt-band in form of membrane void, cracks, and delamination, localized membrane degradation has been observed with an SEM by Escobedo (2006). Damage due to freezing conditions in membranes was reported by Yan et al. (2006). They found an increased roughness, micro-cracking, and pinhole formation in membranes which were subjected to freeze/thaw cycling.

14.4.2 Catalyst Layer Degradation

According to experimental results, degradation of the catalyst layers during long-term operation can be evidenced by cracking or delamination of the layer, catalyst ripening, catalyst particle migration, catalyst washout, electrolyte dissolution, and carbon coarsening (Borup et al., 2007; Sasaki et al., 2009). Most of these mechanisms can be investigated postmortem by SEM or TEM imaging.

Young et al. characterized the structural degradation of PEMFC catalyst layers by SEM analysis (Young et al., 2009) among other techniques. A time measure of 30 h of accelerated stress tests by cycling the cathode potential from 0.1 to 1.5 V showed a clear thinning of the cathode catalyst layer and microporous layer. Over this corrosion period, there was no significant change in the anode or membrane thickness. Under dynamic load cycling, Lin et al. (2009) observed catalyst layer thinning leading to a rapid degradation of performance after 280 h. Platinum migration into the membrane was detected by EDX analysis. Cracks and gaps were found in the catalyst layer. As evidenced by SEM analysis of cross sections, increased contact resistance could be attributed to delamination of the electrode due to swelling and expansion of the electrolyte in the electrode (Liu et al., 2009).

Yu et al. investigated the degradation of MEAs using ESEM and EDX after 700 h operational time and 14% performance loss in a fuel cell (Yu et al., 2005a, 2006). Besides silicon in the MEA material that originated most likely from sealing material, the EDX analysis showed different fluorine contents for the anode and the cathode side. When water droplets were condensed on cross sections of the sample in the ESEM mode, the cathode side also showed some flat droplets with contact angles less than 60° indicating loss of hydrophobicity. The authors suppose that silicon in the sealing material and preprocessing could be the reason for the silicon content in the electrode. This resulted in an inhomogeneous loss of hydrophobicity during operation.

To investigate the degradation by freeze/thaw cycling with and without a preceding drying step as well as freeze start-ups from −20°C, Alink et al. (2008) used SEM analysis to examine cross sections and top views of MEAs after being operated in a 6-cell fuel cell stack. They found serious damage mainly in the cathode electrode, but only if the stack had been frozen without a preceding drying step. By analyzing 60 MEA cross-sectional images with an image processing algorithm, the loss of electrode cross-sectional area was calculated to be −13% for the anode side and −35.9% for the cathode side, respectively. When drying the stack with air before freezing, the loss was negligible on the anode side and −9.4% on the cathode side. The authors conclude that the expansion of water, present in the pores of the electrode structure during the phase transition from liquid to solid, leads to cracks and detachment of electrode particles. A drying step before freezing successfully prevents the most severe damage.

At Ballard Power Systems, Li et al. investigated the formation of ice in the catalyst layer a fuel cell stack at −25°C during start-up. They analyzed two different types of MEAs at three different check times during a passive freeze start: after purging without starting the stack, after a full freeze start-up attempt, and after having produced half the amount of water of the full freeze start-up attempt. After disassembling the stacks in a cold chamber at −35°C, the MEAs were stored and cut in liquid nitrogen to avoid melting of the ice which was formed within the porous structure of the electrodes during freeze start-up. In addition, they took care that no ice sublimation or condensation of water from the atmosphere took place on the sample. Then the MEA samples were analyzed at a temperature of −170°C in an FESEM on a cryo stage. The authors claim that at this temperature and at 1.33 Pa chamber pressure the sublimation rate of the water in the porous media is as low as 10^{-5} nm/s and thus sublimation can be neglected. The FESEM allowed using an electron beam acceleration voltage as low as 0.5–0.8 kV with sufficient resolution that prevented sample damage and ice melting. They found that even after purging the fuel cell for 5 min with dry gases, following 20 min with 65% RH gases and without subsequent operation, 70% of the pores were still filled with ice after cooling the cell to −25°C. In cross sections of the MEAs, there was a significant gradient of the ice density in the cathode electrode. The highest ice density was evidenced in the side of the electrode facing the GDL. Obviously this was due to the higher temperature of the side facing the membrane during fuel cell operation at −25°C and the water migrating toward the colder GDL-side. By calculating the amount of water produced during freeze start and comparing it with the water found in the porous structure, they found that 60% of the water has to be adsorbed by the membrane during subzero operation.

Degradation in the catalyst layer due to ice formation at subzero conditions was analyzed by several groups in recent years (Alink et al., 2008; Borup et al., 2007; Li et al., 2008; Lim et al., 2010; Luo et al., 2010; Oszcipok and Alink, 2009). Luo et al. (2010) investigated the cross-sectional morphologies of catalyst-coated membranes by SEM and found no delamination or segregation, no physical damage, or destruction. In contrast to these findings, delamination of the electrodes from the membrane is observed in cross sections in the SEM after freeze start or freeze/thaw cycling in many papers (Hou et al., 2006; Kim and Mench, 2007; Lim et al., 2010; Yan et al., 2006). Lim et al. (2010) tested the impact of various GDLs under the condition of 50 freeze/thaw cycles on the MEA surface by SEM (Lim et al., 2010). They found that the catalyst layer was severely damaged under the channel area where the MEA is not mechanically well supported by the carbon cloth and carbon paper GDL. Their examination showed that the surface morphology of MEA tested with carbon felt exhibited the lowest electrode damage. A mechanism for freeze/thaw durability was proposed in connection with the bending stiffness of GDLs (Lim et al., 2010).

Often, however, one cannot exclude that delamination found in SEM cross sections is caused by cutting, freezing in liquid nitrogen and breaking or casting of the samples after operation. Since delamination has a significant impact on the high-frequency resistance, the delamination in SEM cross sections should be at least correlated with *in situ* recording of the high-frequency resistance.

14.4.3 GDL Degradation

In PEM fuel cells, the GDL plays a crucial role in gas, electron, heat, and liquid water transport. In addition, the GDL has to mechanically support the MEA and homogeneously distribute the clamping pressure over the active area. Due to its multifunctionality, the GDL must be very carefully designed. Only small structural changes will most likely result in a loss of functionality.

With respect to mass transport losses, the GDL plays a key role in the high-current density region and as a result the maximum power density of PEM fuel cells. On the one hand, the GDL is responsible for the transport of reactants toward the catalyst layer but on the other hand it is also responsible for the water transport from the reaction zone to the gas channel. This combination makes the functional capability of this layer under wet operating conditions much more important since pore flooding reduces the diffusion pathway of the gases to the active zones.

To ensure a proper liquid water transport, the GDL normally contains a certain amount of PTFE to increase its hydrophobicity. However, it was found that the hydrophobicity of the GDL decreases with time, depending on operating conditions (Borup et al., 2007; Wood and Borup, 2009). The PTFE content and distribution within the GDL can be analyzed by EDX. With regard to changes in the wettability of the GDL microstructure, ESEM liquid water imaging is a new and powerful tool since the wetting of the microstructure can be visualized with high resolution (Yu et al., 2005a, 2006).

Wood and coworkers found microstructural changes of PTFE particles by SEM analysis after the GDL had been extensively operated (Wood et al., 2006).

Damage of the GDL by freeze/thaw cycling is discussed controversially in the literature. Yan et al. found some damage to the fibers and binder within the backing layer which was attributed to water expanding during phase transition from liquid to solid (Yan et al., 2006). However, Alink et al. did not find any difference of fresh samples compared to samples which were frozen in a wet status 62 times and after 9 freeze start-ups (Alink et al., 2008). They conclude that the fibers of the GDL have enough elasticity to sustain the mechanical stress imposed during freezing by the water expanding in the pores of the GDL during freezing. In the same paper, they also investigated the degradation by condensing and freezing water in the same GDL material in the ESEM and came to the same conclusion.

14.4.4 Degradation of Peripheral Components

Only a few articles are available on SEM investigations on the degradation phenomena in peripheral components such as bipolar plates, sealing gaskets, and tubing. Up to now, the main focus has been on the layers of the MEA since the structural changes here show the largest impact on fuel cell performance. However, changes in the wetting properties of the gas channels in the bipolar plates or a change in the contact resistance due to surface structure modification would be of interest. Guo et al. (2008) investigated the degradation of the current collector in a DMFC stack by SEM. They showed severe pit holes in the range of 10–250 μm on the surface of a Ni/Au-coated printed circuit board current collector. These results were consistent with *in situ* experiments and the authors suggested that pitting corrosion due to F⁻ attacking dominate during lifetime.

EDX in combination with SEM imaging was also used by Park et al. (2010) to investigate the degradation of titanium bipolar plates for direct methanol fuel cells. They found corroded parts and a high oxygen concentration on the lands of the flow field and conclude that oxygenation of the titanium material resulted in high ohmic losses during fuel cell operation. Also, Nikam and Reddy (2006) investigated the corrosion of copper alloy bipolar plates for PEM fuel cells. They investigated the surface of the material by means of SEM and EDX after chronoamperometry testing. They found pitting and significant signs of corrosion which nevertheless had no detrimental influence on the material conductivity.

In spite of not being subjected to much research activity, Bieringer et al. (2009) pointed out that the degradation of gaskets could also play an important role in fuel cell durability. They compared the low-temperature flexibility and high-temperature resistances of different elastomers used as a sealing material in fuel cells. They found that the class of silicones were prone to degradation under the harsh conditions of high humidity and high temperature and showed depositions of silicone fragments on the inner surfaces of the cell in SEM pictures.

14.4.5 Liquid Water Examination in ESEM

Water management in PEM fuel cells has a strong impact on durability in terms of degradation due to freezing water (Alink et al., 2008; Borup et al., 2007; Li et al., 2008; Lim et al., 2010; Luo et al., 2010; Oszcipok and Alink, 2009), high humidity, liquid water, and mechanical stress due to the swelling of the membrane. Also, degradation mechanisms can result in a change in water management by changing of the component wettability (Borup et al., 2007; Yu et al., 2005a).

Liquid water transport in PEM fuel cells is highly correlated to the wetting properties and liquid water transport behavior of the functional structures within fuel cells. ESEM are able to visualize liquid water formation and even dynamic processes induced by liquid water (Section 14.2.1.2). This capability can be very helpful in terms of investigating any change of material interaction with liquid water.

Nam and Kaviany (2003) were the first to investigate the wetting behavior of fuel cell materials in an ESEM. However, they only showed some pictures of condensed water in gas diffusion media to verify their model on the water transport in hydrophobic media.

Hebling's group investigated the liquid water transport and wetting properties of fuel cell materials in the ESEM (Yu et al., 2005a, 2006). They used the ESEM to investigate the contact angles on bare Nafion, a PTFE/Nafion and a SPEEK/PTFE composite membrane as well as on catalyst layers. The behavior was homogeneously hydrophilic both in cross section and top views of the Nafion membrane. The SPEEEK/PTFE composite membrane was also hydrophilic in the SPEEK part, but hydrophobic in the PTFE part. Locally high contact angles on the SPEEK part correlated with high fluorine contents

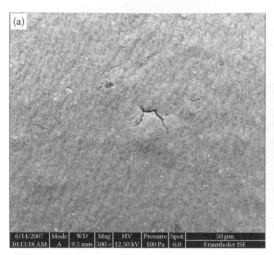
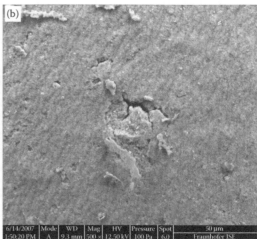

FIGURE 14.9 Top view of an MEA before (a) and after (b) 10 subsequent freeze/thaw cycles in ESEM. (Reprinted from Alink, R., Gerteisen, D., and Oszcipok, M. 2008. *J. Power Sources* 182: 175–187. With permission.)

derived by EDX measurements. By burning parts of the electrode with a laser, the contact angle of the bare membrane part and electrode part could be directly compared and a higher hydrophobicity in the catalyst part was found.

In a study of Hwang et al. (2009), they proclaimed that depending on ionomer water content there are four different pore water morphologies in the ionomer. They supported their assumption on the visualization of droplet formation and contact angle distribution during water condensing on the ionomer in an ESEM.

Alink et al. (2008) used liquid water visualization in the ESEM to examine the degradation effects of *ex situ* freeze/thaw cycling of MEAs and GDLs in the ESEM. For this purpose, water was first condensed, then frozen, and finally evaporated on GDL and MEA samples 10 times. For comparison, the liquid water was not frozen but directly evaporated after condensing on other samples. For comparing pre- and poststatus and for the visualization of liquid water and ice formation, different spots on the sample surface were used to avoid any influence by the electron beam. Only when including a freezing step by going into the solid phase of water, obvious degradation was observed in the top views of commercial CCM (Figure 14.9). However, the GDL material does not seem to have been harmed by the same procedure.

14.5 Concluding Remarks

SEMs play an important role in the *ex situ* investigation on the degradation mechanisms in PEM fuel cells. The visualization of structural changes can be taken with a very high resolution up to the nanometer range. The steadily improving ease of use and capabilities as well as the decreasing prices contribute to an increasing availability and utilization for fuel cell research.

Parallel to the SEM investigation, EDX analysis provides important information on the content and distribution of elements, with a resolution of a few micrometers. In the field of PEM fuel cells, these techniques are already applied to provide mostly nonquantitative information on the changes in gas diffusion media, electrodes, and membranes. Pinhole formation, electrode deformation, changes in the fibrous structure, and diffusion of elements can give important information on the mechanisms that lead to performance decay. In addition, the ESEM technique is a relatively new method which can be used to characterize the changes of the wetting properties at microscales which play a significant role in the durability of fuel cells. However, the high resolution of SEMs makes the related examination techniques prone to artifacts, resulting from inadequate sample handling or preparation. Also, the requirement of exposing the sample to a vacuum, charging of the sample by the electron beam, and the sensitivity of the materials to electron bombardment imposes additional restrictions on the examination technique. This is particularly important when excavating sample materials to analyze their inner structure (e.g., cross sections). Care is required for data analysis on samples that present microcracks or ablation damages.

References

Alink, R., Gerteisen, D., and Oszcipok, M. 2008. Degradation effects in polymer electrolyte membrane fuel cell stacks by sub-zero operation—An *in situ* and *ex situ* analysis. *J. Power Sources* 182: 175–187.

Bi, W., Gray, G. E., and Fuller, T. F. 2007. PEM fuel cell Pt/C dissolution and deposition in Nafion electrolyte. *Electrochem. Solid-State Lett.* 10: B101–B104.

Bieringer, R., Adler, M., Geiss, S., and Viol, M. 2009. Gaskets: Important durability issues. In *Polymer Electrolyte Fuel Cell Durability*, eds. F. N. Büchi, M. Inaba, and T. J. Schmidt, pp. 271–281. New York, NY: Springer.

Borup, R., Meyers, J., Bryan, P., et al. 2007. Scientific aspects of polymer electrolyte fuel cell durability and degradation. *Chem. Rev.* 107: 3904–3951.

Collier, A., Wang, H., Yuan, X. Z., Zhang, J., and Wilkinson, D. P. 2006. Degradation of polymer electrolyte membranes. *Int. J. Hydrogen Energy* 31: 1838–1854.

Colliex, C. 2008. *Elektronenmikroskopie- eine anwendungsbezogene Einführung*, Stuttgart: Wissenschaftliche Verlagsgesellschaft mgH.

Darling, R. M. and Meyers, J. P. 2003. Kinetic model of platinum dissolution in PEMFCs. *J. Electrochem. Soc.* 150: A1523–A1527.

Escobedo, G. 2006. Enabling commercial PEM fuel cells with breakthrough lifetime improvements. *Department of Energy Hydrogen Program Annual Merit Review Proceedings* V_b_3: 701–712.

Fernandes, A. C. and Ticianelli, E. A. 2009. A performance and degradation study of Nafion 212 membrane for proton exchange membrane fuel cells. *J Power Sources* 193: 547–554.

Ferreira, P. J., La O, G. J., Shao-Horn, Y., et al. 2005. Instability of Pt/C electrocatalysts in proton exchange membrane fuel cells–A mechanistic investigation. *J. Electrochem. Soc.* 152: A2256–A2271.

Guilminot, E., Corcella, A., Chatenet, M., et al. 2007. Membrane and active layer degradation upon PEMFC steady-state operation I. Platinum dissolution and redistribution within the MEA. *J. Electrochem. Soc.* 154: B1106–B1114.

Guo, J.-W., Xie, X.-F., Wang, J.-H., and Shang, Y.-M. 2008. Effect of current collector corrosion made from printed circuit board (PCB) on the degradation of self-breathing direct methanol fuel cell stack. *Electrochim. Acta* 53: 3056–3064.

Hou, J., Yu, H., and Zhang, S. et al. 2006. Analysis of PEMFC freeze degradation at −20°C after gas purging. *J. Power Sources* 162: 513–520.

Hwang, G. S., Kaviany, M., Nam, J. H., Kim, M. H., and Son, S. Y. 2009. Pore-water morphological transitions in polymer electrolyte of a fuel cell. *J. Electrochem. Soc.* 156: B1192–B1200.

Kim, S. and Mench, M. M. 2007. Physical degradation of membrane electrode assemblies undergoing freeze/thaw cycling: Micro-structure effects. *J. Power Sources* 174: 206–220.

Kundu, S., Simon, L. C., and Fowler, M. W. 2008. Comparison of two accelerated Nafion (TM) degradation experiments. *Polym. Degrad. Stab.* 93: 214–224.

Laconti, A. B., Hamdan, M., and Mcdonald, R. C. 2003. Fundamentals Technology and Applications. In *Handbook of Fuel Cells*, eds. W. Vielstich, A. Lamm, and H. A. Gasteiger, pp. 647–662. Chichester: John Wiley & Sons.

Li, J., Lee, S., and Roberts, J. 2008. Ice formation and distribution in the catalyst layer during freeze-start process—CRYO-SEM investigation. *Electrochim. Acta* 53: 5391–5396.

Lim, S.-J., Park, G.-G., Park, J.-S., et al. 2010. Investigation of freeze/thaw durability in polymer electrolyte fuel cells. *Int. J. Hydrogen Energy* 35: 13111–13117.

Lin, R., Li, B., Hou, Y. P., and Ma, J. M. 2009. Investigation of dynamic driving cycle effect on performance degradation and micro-structure change of PEM fuel cell. *Int. J. Hydrogen Energy* 34: 2369–2376.

Liu, P., Yin, G.-P., and Cai, K.-D. 2009. Investigation on cathode degradation of direct methanol fuel cell. *Electrochim. Acta* 54: 6178–6183.

Luo, M., Huang, C., Liu, W., Luo, Z., and Pan, M. 2010. Degradation behaviors of polymer electrolyte membrane fuel cell under freeze/thaw cycles. *Int. J. Hydrogen Energy* 35: 2986–2993.

Nam, J. H. and Kaviany, M. 2003. Effective diffusivity and water-saturation distribution in single- and two-layer PEMFC diffusion medium. *Int. J. Heat Mass Transfer* 46: 4595–4611.

Nikam, V. V. and Reddy, R. G. 2006. Copper alloy bipolar plates for polymer electrolyte membrane fuel cell. *Electrochim. Acta* 51: 6338–6345.

Oszcipok, M. and Alink, R. 2009. Fuel cells/protonexchange membrane fuel cells/freeze operational conditions. In *Encyclopedia of Electrochemical Power Sources*, ed J. Garche, pp. 931–940. Amsterdam: Elsevier.

Park, Y.-C., Lee, S.-H., Kim, S.-K., et al. 2010. Performance and long-term stability of Ti metal and stainless steels as a metal bipolar plate for a direct methanol fuel cell. *Int. J. Hydrogen Energy* 35: 4320–4328.

Péron, J., Nedellec, Y., Jones, D. J., and Rozière, J. 2008. The effect of dissolution, migration and precipitation of platinum in Nafion-based membrane electrode assemblies during fuel cell operation at high potential. *J. Power Sources* 185: 1209–1217.

Rice, T. and Knowles, R. 2005. Ultra high resolution SEM on insulators and contaminating samples. *Microscopy Today* 13: 40–42.

Sasaki, K., Shao, M., and Adzic, R. 2009. Dissolution and stabilization of Platinum in oxygen cathodes. In *Polymer Electrolyte Fuel Cell Durability*, eds. F. N. Büchi, M. Inaba, and T. J. Schmidt, pp. 7–27. New York, NY: Springer.

Wood, D. L. and Borup, R. L. 2009. Durability aspects of gas-diffusion and microporous layers. In *Polymer Electrolyte Fuel Cell Durability*, eds. F. N. Büchi, M. Inaba, and T. J. Schmidt, pp. 159–195. New York, NY: Springer.

Wood, D. L., Davey, J. R., Atanassov, P., and Borup, R. L. 2006. PEMFC component characterization and its relationship to mass-transport overpotentials during long-term testing. *ECS Trans.* 3: 753–763.

Yan, Q., Toghiani, H., Lee, Y.-W., Liang, K., and Causey, H. 2006. Effect of sub-freezing temperatures on a PEM fuel cell performance, startup and fuel cell components. *J. Power Sources* 160: 1242–1250.

Young, A. P., Stumper, J., and Gyenge, E. 2009. Characterizing the structural degradation in a PEMFC cathode catalyst layer: Carbon corrosion. *J. Electrochem. Soc.* 156: B913–B922.

Yu, P., Pemberton, M., and Plasse, P. 2005b. PtCo/C cathode catalyst for improved durability in PEMFCs. *J. Power Sources* 144: 11–20.

Yu, H. M., Schumacher, J. O., Zobel, M., and Hebling, C. 2005a. Analysis of membrane electrode assembly (MEA) by environmental scanning electron microscope (ESEM). *J. Power Sources* 145: 216–222.

Yu, H. M., Ziegler, C., Oszcipok, M., Zobel, M., and Hebling, C. 2006. Hydrophilicity and hydrophobicity study of catalyst layers in proton exchange membrane fuel cells. *Electrochim. Acta* 51: 1199–1207.

15

Transmission Electron Microscopy

Rui Lin
Tongji University

Junsheng Zheng
Tongji University

Jian-Xin Ma
Tongji University

15.1 Introduction

Transmission electron microscopy (TEM) is an important tool for characterization of material microstructures. A beam of electrons is transmitted through an ultra-thin specimen, interacting with the specimen as it passes through. An image is then formed from this interaction, and the image either can be magnified and focused onto an imaging device or can be detected by a sensor such as a charge-coupled device (CCD) camera (Williams and Carter, 2007).

TEM has revolutionized our understanding of materials by completing the processing–structure–properties link at the atomic level (Cecil, 1953). With nanotechnology becoming a major field in science and technology, TEM has become more and more important, potentially providing almost all structural, phase, and crystallographic data on materials.

15.1.1 History of TEM

Historically, TEM was developed in response to the limited image resolution of light microscopes. In traditional light microscopy, the wavelength of visible light limits resolution (Paul et al., 1993). In a TEM, electron energies are between 100 and 1000 keV, and the wavelengths of electrons corresponding to these energies range from 0.00370 to 0.00087 nm, which are very short in comparison to the wavelengths of visible light. In 1925, Louis de Broglie first theorized that the electron had wave-like characteristic, with a wavelength substantially less than visible light (Paul et al., 1993). Further, in 1927 Davisson and Germer, as well as Thompson and Reid, independently carried out classic electron diffraction experiments that demonstrated the wave nature of electrons (Dawn and Peter, 1990). After that, it did not take long for the idea of an electron microscope to be proposed. Knoll and Ruska brought the concept of electron lenses into reality, and in 1931 presented electron images taken on their instrument (Lambert and Mulvey, 1996; Williams and Carter, 2007). Only four years later, TEM was developed by commercial companies. After the advent of electron microscopes, many other electron applications were realized, most of which are to some extent manifested in a modern TEM. Figure 15.1 displays a modern TEM fabricated by JEOL Ltd.

15.1.2 Advantages and Disadvantages of TEM

TEM has three main important advantages as a material characterization technique. First, it can capture images of material structures with atomic-level resolution, so one can easily observe images of nanoscale material structures with a normal TEM. Second, it has a well-established theoretical foundation for analyses of observed images; these can be analyzed in detail, unambiguously, on the basis of highly developed theories of electron optics and electron diffraction (Lambert and Mulvey, 1996). Third, TEM is capable of microanalysis when combined with spectroscopic methods. For example, TEM can work with energy-dispersive x-ray spectroscopy (EDX) and thereby provide more information about materials.

The main disadvantage of TEM is the difficulty of preparing TEM samples. This preparation requires certain experimental skills, which is discussed later in this chapter. The other disadvantage is the sample degradation caused by high-energy electrons, which becomes a serious problem when observing organic materials (Williams and Carter, 2007).

15.2 Principles and Measurements of TEM

15.2.1 Principles

The principles of TEM are similar to those of an optical microscope, except that in a TEM, high-energy electrons are used instead of visible light. As mentioned above, the wavelength of an electron is shorter than that of light, so the resolution limits of conventional TEMs are nearly at the atomic level.

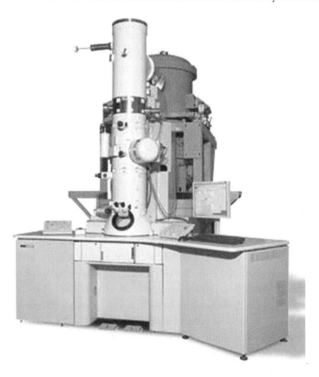

FIGURE 15.1   A modern TEM fabricated by JEOL Ltd. (Reproduced with permission from JEOL Ltd. China.)

Imaging methods in TEM utilize the information contained in the electron waves exiting from the sample to form an image. The projector lenses allow for the correct positioning of the electron wave distribution onto the viewing system. The observed intensity of the image, I, can be approximated as proportional to the time-average amplitude of the electron wave functions, where the wave that forms the exit beam is denoted by Ψ, t is the electron wave function time, and k is a parameter (John, 1995):

$$I(x) = \frac{k}{t_1 - t_0} \int_{t_0}^{t_1} \psi \psi^* \, dt \tag{15.1}$$

Different imaging methods therefore attempt to modify the electron waves exiting the sample to obtain information about the sample or about the beam itself. It can be deduced that the observed image depends not only on the beam's amplitude, but also on the phase of the electrons. It is also well known that contrast and diffraction patterns are the two most important factors for TEM imaging.

Complex imaging techniques, which utilize the unique ability to change lens strength or to deactivate a lens, allow for many operating modes that may be used to discern information of particular interest to the investigator. In TEM analysis, one's eyes cannot detect intensity changes of less than 5%; even <10% is difficult. So, unless the contrast of a specimen exceeds 5–10%, one cannot see anything on a screen or photograph. However, if the image is digitally recorded, low contrast can be enhanced electronically to levels at which the human eye can perceive the contrast.

A diffraction pattern is generated by adjusting the magnetic lenses rather than the imaging plane, which is placed on the imaging apparatus. For thin crystalline samples, this produces an image consisting of a pattern of dots for a single crystal, or a series of rings for a polycrystalline or amorphous solid material. In the case of a single crystal, the diffraction pattern is dependent on the specimen's orientation and the sample's structure (Earl, 1998).

Diffraction patterns can have a large, dynamic range, and for crystalline samples the patterns may have greater intensities than those recorded by a CCD. Hence, a TEM may still be equipped with film cartridges to obtain these images, as the film is a single-use detector (Ludwig, 1993).

Analysis of diffraction can be complex as the image is sensitive to a number of factors, such as thickness and orientation of the specimen, objective lens defocus, as well as spherical and chromatic aberration. Although quantitative interpretation of the contrast shown in lattice images is possible, it is inherently complicated and can require extensive computer simulation and analysis, such as electron multislice analysis (Cowley and Moodie, 1957).

Moreover, there is an important relationship between the diffraction pattern and a diffraction-contrast image. If the diffraction pattern is changed in any way, the contrast in the image will change as well. To relate the two, it is important to remember that one may have to calibrate the relationship between the image and the diffraction pattern; whenever magnification is changed, the image rotates but the diffraction pattern does not.

15.2.2 Hardware and System

Normally, a TEM is composed of several components, the most important being the illumination system, the objective lens and stage system, the imaging system, and the vacuum system.

The illumination system takes the electrons from the gun and transfers them to the specimen, giving either a broad or a focused beam.

The objective lens and stage system are the heart of a TEM. One can use the stage system to clamp the specimen holder in the correct position; then the objective lens can form images and diffraction patterns in a reproducible manner.

In TEMs the most common imaging system is the CCD. The system mounts on the diffraction camera ports located just above the viewing chamber on most TEMs. The image is formed on a detector, then transmitted to a camera through appropriate optics. The detector can be retracted at any time, restoring the image to the viewing screen for fine focusing or a macrograph. There is no shadowing or disturbance of the image when the detector is retracted. The imaging system should be calibrated when it is first installed and then periodically throughout its life, especially if one wishes to carry out accurate measurements from images or diffraction patterns.

In a standard TEM, the vacuum system evacuates the TEM to low pressures, typically in the order of 10^{-4} Pa, to increase the mean free path of the electron–gas interaction.

The TEM system also needs to have two alignments performed to ensure correct operation. The most important is the alignment of the objective lens center of rotation, and the second is the alignment of the diffraction pattern on the optic axis (Williams and Carter, 2007).

15.2.3 Sample Preparation

To be transparent to electrons, a specimen must be thin enough to transmit enough electrons that they fall with sufficient intensity on a screen or photographic film to yield an interpretable image in a reasonable time. At the same time, due to the strong interaction between electrons and sample materials, the latter must be quite thin to gain any information using the transmitted electrons. Thus, in TEM, the thinner the sample the better, and specimens below 100 nm thickness should be used if possible. To carry out high-resolution transmission electron microscopy (HRTEM), a specimen of thinness <50 nm is very important (Anderson, 1990; Williams and Carter, 2007).

Sample preparation in TEM is a complex procedure, and one of the most important processes for TEM analysis. The specimen being viewed by TEM or HRTEM may require extremely careful handling. To obtain a thin sample, one should first verify whether the sample can be self-supporting or should be supported on a grid or washer.

A self-supporting specimen means that the whole specimen consists of one material. To prepare a self-supporting disk for final thinning, one should first create a thin slice from the bulk sample, then thin the disk. The thinning methods for self-supporting samples can be divided into sample staining, mechanical milling, chemical etching, and ion etching (Porter, 1953; Phillips, 1961; Baram and Kaplan, 2008). The next step in preparing a self-supporting sample is to cut the sample into small discs.

The alternative to self-supporting disks is to make small electron-transparent portions of the specimen, or create particles and support them on a thin film on a grid or washer. Small particles should be deposited on amorphous carbon film, Cu film, or other films of uniform thickness (William, 1984); the particles may stick to the film or may have to be clamped between two grids.

TEM specimens must also be kept under optimal conditions. Usually this means keeping them dry (as water affects most materials), perhaps in an inert atmosphere (dry nitrogen works well, or a dry-pumped vacuum desiccator), and in an inert container (a Petri dish with filter paper).

15.2.4 High-Resolution TEM

HRTEM is an imaging mode of TEM that allows imaging of a sample at the atomic scale (Spence, 1981). The purpose of HRTEM is to maximize the useful detail in an image. The HRTEM picture is magnified by an electron-optical system and seen on a screen, typically at 10^6 magnification. Because of its high resolution, it is an invaluable tool for studying the nanoscale properties of crystalline materials such as semiconductors and metals. At present, the highest resolution realized is 0.08 nm with HRTEM (Buseck et al., 1988), and ongoing research and development will push HRTEM resolution to about 0.05 nm. At that scale, HRTEM is useful for direct atomic-level study of features such as interfaces, dislocations, defects, and so forth. Figure 15.2 shows a modem HRTEM fabricated by JEOL Ltd.

The diffraction pattern of HRTEM is the Fourier transform of the periodic potential for the electrons in two dimensions (Hirahara et al., 2001). In the objective lens all diffracted beams and the primary beam are brought together again; their interference provides a back-transformation and leads to an

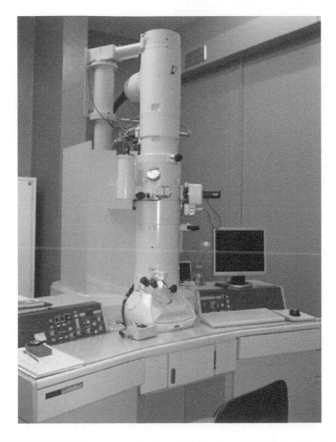

FIGURE 15.2 JEOL 2011 HRTEM. (Reproduced with permission from JEOL Ltd. China.)

enlarged picture of the periodic potential. But for TEM, the lens system is not perfect so the image is distorted and some data is lost (Abbe's theory) (Horiuchi, 1994). On the other hand, an HRTEM image is an interference pattern between the forward-scattered and diffracted electron waves from the specimen. HRTEM image formation relies on phase contrast. In phase-contrast imaging, contrast is not necessarily intuitively interpretable, as the image is influenced by strong aberrations in the imaging lenses of the microscope.

When conducting imaging, one must be prepared to use image simulation to assist in interpreting the images. For HRTEM to provide information about atom arrangements in a material, computer simulations of the image are usually required. If quantitative imaging is desired, simulation is an essential component of the process. Most materials scientists use one of the established software packages, such as Digital Micrograph (Williams and Carter, 2007).

15.3 Energy Dispersive X-Ray Spectroscopy

EDX is an analytical technique used for the elemental analysis or chemical characterization of materials (Goldstein and Newbury, 1995). To use the x-rays generated when the electron beam strikes a TEM specimen, one can detect them first and then identify them as coming from a particular element. It is well known that each element has a unique atomic structure allowing x-rays, and this is the fundamental principle by which EDX characterizes elements' atomic structures as distinct from one another. Different spectra indicate that each specimen must have a different elemental composition, and using EDX it is possible to obtain this information in a matter of minutes (Williams et al., 1995). EDX is the only x-ray spectrometry currently used in tandem with TEM. It is remarkably compact, efficient, and sensitive for material analysis.

Normally, there are four primary components of an EDX setup: the beam source, the x-ray detector, the pulse processor, and the analyzer. Although a number of free-standing EDX systems exist, EDX systems are most commonly found on scanning electron microscopes (SEM-EDX) or TEM. Electron microscopes are equipped with a cathode and magnetic lenses to create and focus a beam of electrons. A detector is used to convert x-ray energy into voltage signals; this information is sent to a pulse processor, which measures the signals and passes them onto an analyzer for data display and analysis.

The accuracy of EDX spectra is one of the most important topics for material analysis, and can be affected by many factors. First, EDX detectors cannot detect elements with an atomic number less than 4 (e.g., He and Li) (Goldstein and Newbury, 1995). Second, over-voltage settings in EDX may alter peak sizes. Third, the accuracy of the spectrum can also be affected by the nature of the sample, for many elements will have overlapping peaks (e.g., Ti K_β and V K_α, Mn K_β and Fe K_α). Fourth, the x-rays are emitted in all directions, so they may not all pass through the sample, which can result in reduced accuracy for heterogeneous and rough samples.

15.4 Applications of TEM (EDX) in PEM Fuel Cells

Proton exchange membrane fuel cells (PEMFCs) have been developed as power sources for portable and automotive equipment. Reliability and durability are the most important practical issues for such power sources. To improve PEMFC durability, researchers have been studying the factors that determine a PEMFC's lifetime (Ferreira et al., 2005; Borup, 2007).

Within a PEMFC, the individual components are exposed to an aggressive combination of strong oxidizing conditions, liquid water, strongly acidic conditions, high temperature, high electrochemical potentials, reactive intermediate reaction products, a chemically reducing atmosphere at the anode, high electrical current, and large potential gradients (Borup, 2007). PEMFC materials are not immune to corrosion and decomposition under such severe degradation conditions. To understand the roles of these various operating conditions in the degradation process, it is highly necessary to determine the causes and mechanisms of degradation under such conditions at both the micro- and the nanoscale. In order to establish microstructure–performance relationships and to elucidate MEA degradation and failure mechanisms, it is important to understand the structural and compositional changes that occur during long-term MEA aging (Ferreira et al., 2005; Akita et al., 2006).

In terms of PEMFC durability, deterioration in fuel cell performance is mostly due to degradation of the electrocatalyst and membrane. A number of mechanisms can contribute to catalyst degradation, including catalyst agglomeration and catalyst dissolution/migration (Ralph et al., 2002; Knights et al., 2002; Borup et al., 2006; Li et al., 2006; Ye et al., 2006).

TEM enables very detailed observation and investigation of particle size change during operational processes, and thus facilitates nanoscale examination of degradation mechanisms for Pt catalysts. With TEM alone, particle changes in Pt can clearly be observed. Combined with other technologies, like x-ray diffraction and electrochemical measurements, TEM enables us to determine particle sizes, catalyst agglomeration sizes, complete size distributions, and the mechanisms of fuel cell degradation (Akita et al., 2006).

In what follows, a literature-based analysis is carried out in an attempt to achieve an understanding of degradation mechanisms for cells under different operating conditions. The effects of humidity, fuel starvation, driving cycles, and operating temperatures is discussed in detail below.

15.4.1 Effect of High Humidity

Water management is one of the critical issues to be solved in the design and operation of PEMFCs. Water is produced at the cathode, and if it is not removed effectively it accumulates there, causing electrode flooding. The consequence is oxygen starvation, which increases the cathode's concentration overpotential. In the worst scenarios, a proton (H^+) reduction reaction, instead of the oxygen reduction reaction (ORR), might occur at the cathode. Not only will this cause a cathode potential drop, but the output voltage of a single cell will likely be reversed due to oxygen starvation (Tüber, 2003; Wei, 2006).

A detailed study of the morphological changes in the anode and cathode catalyst layers during high-humidity PEMFC durability testing has been carried out using TEM techniques. Xie (2005a,b) reported on the long-term durability of hydrogen-air PEMFCs and microstructural changes in membrane electrode assemblies during PEMFC durability testing at high humidity (Xie, 2005a,b). The chromium in a Pt_3Cr binary alloy catalyst was found to migrate from cathode to anode during the course of life testing when operating within an oversaturated, or high-humidity, gas feeding regime. Comparison of TEM images of an MEA after testing for 2200 h and TEM images of a fresh MEA show significant agglomeration/growth of metal catalyst clusters over the course of the life test. Catalyst agglomeration and/or growth, and/or dissolution (ripening) of metal clusters from the catalyst layer may be a major cause of MEA degradation during middle-term life tests under high-humidity conditions.

To establish a baseline for evaluating morphological changes in catalyst layers as a function of MEA testing time, fresh MEAs were initially examined through TEM. A TEM image of a Pt_3Cr/C cathode catalyst is shown in Figure 15.3a and b. In both parts of the figure, some catalyst particles are clearly observable within the recast ionomer/epoxy regions rather than attached to the carbon-support surfaces, suggesting that the Pt_3Cr nanoparticles are not initially well bonded to the carbon. Weak bonding between the catalyst particles and carbon surface may also explain why catalyst metal agglomeration occurred so readily over the course of life testing. Individual Pt_3Cr particles grew from diameters of 4–12 nm (Figure 15.3a diameters of 6–>20 nm; Figure 15.3c after approximately 500 h of operation).

Catalyst particle growth is much faster during the first 500 h of testing than during the second 500 h (Figure 15.4a through c), for 0, 500, and 1000 h of testing, respectively. The average Pt_3Cr particle size for the fresh MEA increased from 4–12 nm to 6 to >20 nm during the initial ~500 h (Figure 15.4b), but increased much less with an additional 500 h of testing of the second MEA (Figure 15.4c). The TEM and electrochemically active surface area (ECSA) data confirm that the bulk of catalyst metal agglomeration occurred during the first 500 operation hours. This might be due to the growth/agglomeration process, the inherently small catalyst particle size, the high surface tension of the nanoparticles, and the weak bonding of catalyst particles to carbon surfaces.

15.4.2 Effect of Fuel Starvation

Corrosion of the catalyst carbon support is another important issue pertaining to electrocatalyst and catalyst layer durability that has lately attracted considerable attention in academic and industry research. In PEM fuel cells and stacks, two modes are believed to induce carbon corrosion: (1) transitioning between startup and shutdown cycles, and (2) fuel starvation due to the blockage of H_2 from a portion of the anode under steady-state conditions.

Taniguchi reported on degradation caused by fuel starvation (Taniguchi, 2004). Degradation phenomena in an electrocatalyst, caused by cell reversal during operation under fuel starvation conditions, are characterized using TEM, EDX, and electrochemical methods, detecting the dissolution of Ru from PtRu/C electrocatalyst at the anode, and the growth of particles at anode and cathode.

The time-dependent changes in anode and cathode potential during this cell voltage reversal experiment were observed to rapidly drop to a negative voltage, and the MEA polarity changed due to cell reversal as soon as the experiment started (Taniguchi, 2004).

On the basis of the relative changes in the Pt:Ru ratio, the platinum ratio of the anode catalyst layer clearly increased after cell reversal. Evidently, ruthenium dissolution from individual catalyst particles occurred in the anode catalyst layer.

Platinum ratios of individual catalyst particles in the fuel outlet region before and after a 10 min degradation experiment were evaluated from on-particle EDX spectra. The tendency toward Pt-rich composition in smaller particles was observable in the catalyst before the experiment, and the Ru

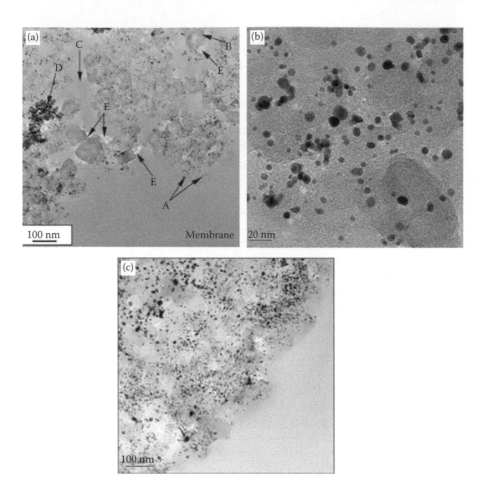

FIGURE 15.3 (a) TEM cross-sectional micrograph of fresh cathode catalyst layer and Nafion membrane (0.20 mg Pt/cm^2, 20% Pt$_3$Cr/C, and 28 wt% Nafion). Arrow demarcation as follows: A, Pt$_3$Cr/C catalyst clusters, 4–12 nm in diameter; B, carbon aggregate, 80–320 nm in diameter; C, recast ionomer region epoxy; D, agglomerated Pt$_3$Cr catalyst particles; and E, large (secondary) pores ~40–80 nm in diameter. (b) Higher magnification of (a); detailed distribution of catalyst particles inside cathode catalyst; (c) TEM cross-sectional micrograph of the Pt$_3$Cr/C cathode catalyst layer and membrane after ~500 h of life testing (0.20 mg Pt/cm^2, 20% Pt$_3$Cr/C, 28 wt% Nafion). (Reproduced from Xie, J. et al. 2005b. *J. Electrochem. Soc.* 152: A104–A113. With permission from the Electrochemical Society.)

concentration of individual particles decreased during cell reversal. Ruthenium was detected neither in the membrane nor on the cathode side. It was also evident from HRTEM images (Figure 15.5a and b) that the particle size distribution of the catalyst particles was broader after the experiment, with the average particle size increasing from 2.64 to 4.95 nm. Pt was also considered to have been dissolved and reprecipitated during cell reversal.

15.4.3 Effect of Potential Cycling and Driving Cycle

Potential cycling has also been used as an accelerated testing method to examine operating conditions that lead to loss of electrocatalyst surface. Pt particle size increase was accelerated by potential cycling, with the size and morphologies of the Pt nanoparticles varying greatly from the cathode surface to the cathode–membrane interface after potential cycling (Ye, 2006a,b; Borup, 2007).

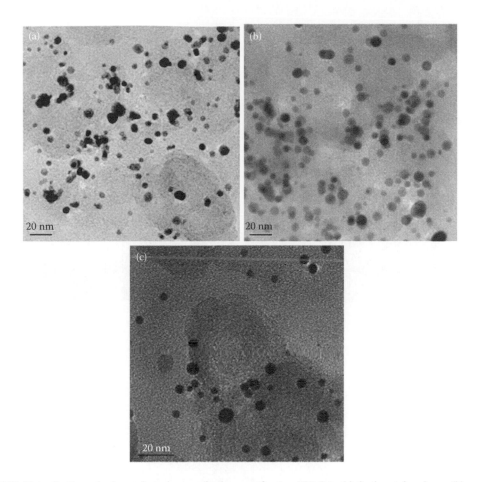

FIGURE 15.4 Pt₃Cr cathode catalyst clusters (high-magnification TEM) in (a) fresh catalyst layer, (b) catalyst layer after ~500 h of life testing, and (c) catalyst layer after 1000 h of life testing (0.20 mg Pt/cm², 20% Pt₃Cr/C, 28 wt% Nafion). (Reproduced from Xie, J. et al. 2005b. *J. Electrochem. Soc.* 152: A104–A113. With permission from the Electrochemical Society.)

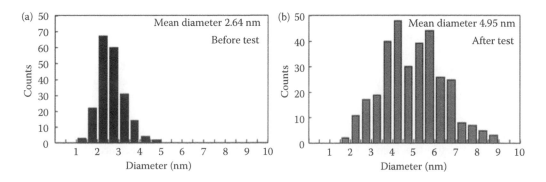

FIGURE 15.5 Particle size distribution for an anode catalyst: (a) before, and (b) after a degradation experiment for 10 min, determined from high-resolution TEM images. (Reprinted from *J. Power Sources*, 130, Taniguchi, A. et al. Analysis of electrocatalyst degradation in PEMFC caused by cell reversal during fuel starvation. 42–49. Copyright (2004), with permission from Elsevier.)

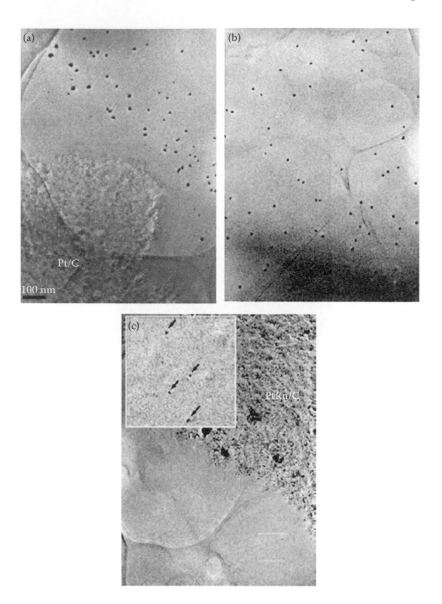

FIGURE 15.6 Cross-sectional TEM images of an MEA after 1.0 V was applied for 87 h. (a) TEM image near the interface between the cathode catalyst layer and the PEM, (b) TEM image of the PEM 10 nm from the cathode layer, and (c) TEM image near the interface between the anode catalyst layer and the PEM. (Reprinted from *J. Power Sources*, 159, Akita, T. et al. Analytical TEM study of Pt particle deposition in the proton-exchange membrane of a membrane-electrode-assembly. 461–467. Copyright (2006), with permission from Elsevier.)

PEMFC performance loss under steady-state and cycling conditions has been attributed in part to a loss of ECSA for the high-surface-area, carbon-supported platinum electrocatalyst. Platinum dissolution plays a major role in the ECSA loss, especially in the cathode catalyst, where high potentials are encountered. The dissolved platinum can either deposit on existing platinum particles to form large particles, or diffuse into electrochemically inaccessible portions of the MEA. It has been speculated that platinum dissolution occurs both as a result of potential cycling caused by varying loads on the PEMFC stack and under constant potential high voltage typical of fuel cell "idling" conditions (Borup, 2006).

Analytical TEM has been used to investigate structural changes in the PtRu/C anode and Pt/C cathode catalysts and the PEM of a single cell in a PEMFC after an acceleration test in which constant high potential was applied to the cathode; Pt dissolved and diffused, and Pt crystals grew (Akita et al., 2006). The structure of the MEA after the application of a potential was successfully observed by analytical TEM at the nano and atomic scales.

Observation of both a thin area of the electrode catalyst layer and the PEM indicated that large Pt particles of about 10–100 nm were formed near the Pt/C electrocatalyst layer (Figure 15.6a) after an accelerated test in which a constant potential was applied, while particles in the electrocatalyst were 2–5 nm. The large Pt particles are also observable in Figure 15.6b, which was taken far from the cathode layer in the PEM. The Pt particles gradually decrease in size, depending on their distance from the cathode layer. Small Pt particles are observable even in the PEM near the anode side, as shown in Figure 15.6c. An enlarged TEM image is shown in the white rectangle in Figure 15.6c. The Pt atoms seem to have diffused from the cathode to the anode layer during the application of a constant potential.

Figure 15.7 shows the EDX spectrum obtained from a Pt particle in the PEM. The particle mainly consists of Pt. A peak for Cu appears as a background signal from the Cu-grid supporting the sample. Carbon and fluorine are detectable from the PEM in the measurement area. Small peaks of potassium and iron are also detectable as impurities. HRTEM images and electron diffraction patterns after the test showed some small Pt particles aggregated into a large Pt particle, and the corresponding diffraction pattern indicated that the Pt particle consisted almost entirely of a single crystal. Aggregates of Pt particles were often observed in the electrocatalyst layer, even before the MEA underwent degradation testing. The distribution of Pt particles in the PEM depends on the kind of gas supplied for the cathode. These TEM images were taken after accelerated tests in which the gas supplied for the cathode was varied. Figure 15.8a through c shows TEM images after 1.0 V is applied for 87 h with N_2 supplied for the cathode; in the cathode catalyst layer large particles exist near the membrane and few Pt particles appear far from the PEM. The distribution of Pt particles also depends on the thickness of the PEM. When air is supplied for the cathode for 87 h, the Pt particles almost reach the anode layer through the PEM.

The platinum surface area measurements in the cathode cycled between 0.6 and 1.0 V were performed under H_2–N_2 at 80°C and 100% RH. It was found the electrochemical surface area of platinum dropped from 63 $m^2\,g^{-1}$ Pt (before the test) to 23 $m^2\,g^{-1}$ Pt after 10,000 cycles. TEM images at four different magnifications collected from the fresh Pt/C sample are shown in Figure 15.9a through d. High-resolution

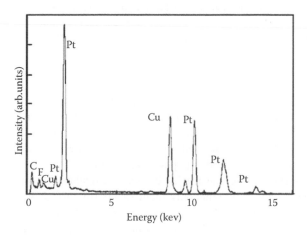

FIGURE 15.7 EDX spectrum obtained from a Pt aggregate in a PEM. (Reprinted from *J. Power Sources*, 159, Akita, T. et al. Analytical TEM study of Pt particle deposition in the proton-exchange membrane of a membrane-electrode-assembly. 461–467. Copyright (2006), with permission from Elsevier.)

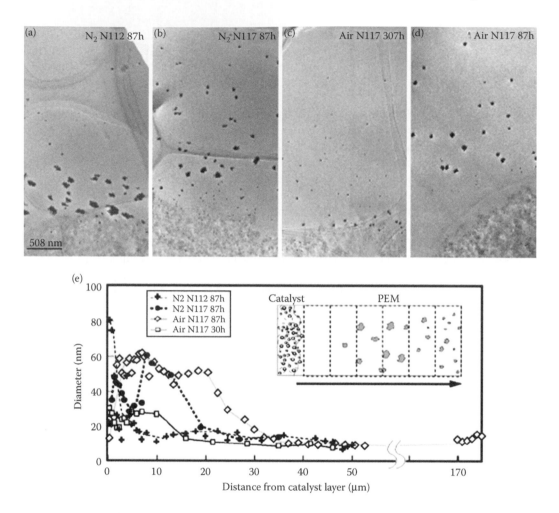

FIGURE 15.8 TEM images of the interface between the cathode and PEM after acceleration tests and the size distribution of Pt particles in the PEM. 1.0 V versus RHE for the cathode with N_2 gas supplied for 87 h (a) using Nafion 112, (b) using Nafion 117, and with air supplied (c) for 30 h using Nafion 117, and (d) for 87 h, and (e) Size distribution of Pt particles in the PEM. (Reprinted from *J. Power Sources*, 159, Akita, T. et al. Analytical TEM study of Pt particle deposition in the proton-exchange membrane of a membrane-electrode-assembly. 461–467. Copyright (2006), with permission from Elsevier.)

lattice imaging revealed that most spherically shaped platinum particles were single crystals. Closer observation identified that most of these single particles had a size of 2–4 nm. Figure 15.9c and d shows agglomeration of individual platinum crystals (Ferreira et al., 2005); these clusters are not electrochemically active, which reduces the electrochemical surface area. Considerable coarsening of the spherically shaped platinum nanoparticles was found after potential cycling.

The electrochemical surface area of the pristine Pt catalyst calculated through the TEM pictures was much larger than the ECSA, indicating that only about 70% of the surface was chemically active. TEM cross-section studies of the cycled MEA cathode were performed to study variation in the Pt particle size distribution as a function of cathode thickness. From Figure 15.10 it is observable that the locations inside the cathode heavily affected the agglomeration of the Pt catalyst. Near the diffusion media (DM)/cathode interface most platinum particles remain spherical and have an almost uniform size distribution. Moving away from the DM/cathode surface toward the cathode/membrane interface, nonspherical

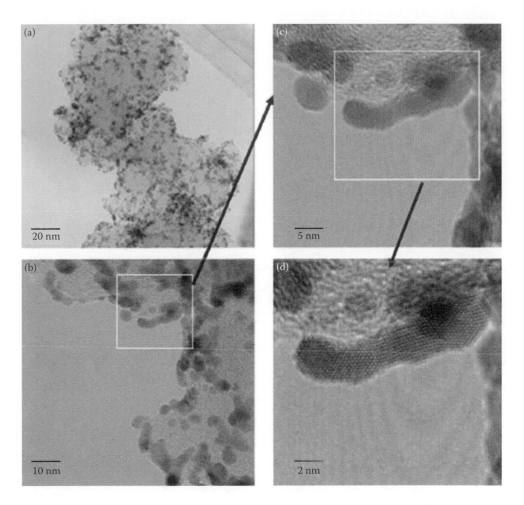

FIGURE 15.9 (a–d) Sequence of TEM images from a fresh Pt/Vulcan sample, taken at different magnifications, showing the distribution of platinum nanoparticles on the carbon support. As shown in image d, particle agglomeration is notable in some cases. (Reproduced from Ferreira, P. J. et al. 2005. *J. Electrochem. Soc.* 152: A2256–A2271. With permission from the Electrochemical Society.)

platinum agglomerates on carbon and nonspherical platinum particles of carbon increases considerably at the expense of spherical platinum particles on carbon.

During real road operation of a vehicle, a fuel cell goes through dynamic automotive working conditions. It is generally acknowledged that the performance of a fuel cell is impacted by a dynamic load cycle much more strongly than by constant load conditions, because under a dynamic driving cycle the variety of operating condition is far more extreme. This may result in oxidant starvation, local "hotspots," and physical degradation, with the fuel cell's performance being much more seriously degraded. Considerable effort has been put into investigating the performance of PEMFCs over dynamic driving cycles, to improve the design and manufacturing of a robust and durable air/hydrogen fuel cell for transportation applications.

Dynamic driving cycles were designed to simulate real vehicle operation for accelerated testing (Lin, 2009). A moderate degradation rate in the voltage could be detected after about 280 h of operation, followed by a much more rapid decrease in performance. Figure 15.11 illustrates TEM images of the initial and degraded catalyst. In the initial MEA, Pt particles were uniformly dispersed on the carbon support. After 370 h of driving cycles, catalysts obviously grew up. Driving cycles clearly had a dramatic effect on

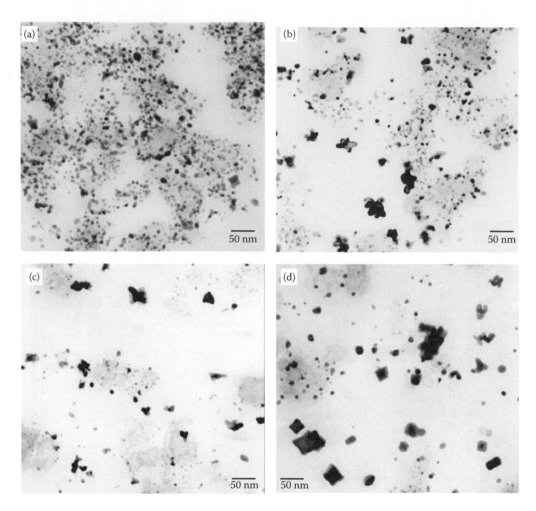

FIGURE 15.10 TEM micrographs obtained from locations a, b, c, and d in a cross-sectional MEA cathode. (Reproduced from Ferreira, P. J. et al. 2005. *J. Electrochem. Soc.* 152: A2256–A2271. With permission from the Electrochemical Society.)

the catalyst layers. As the fuel cell was operated under 100% humidification, Pt growth possibly occurred because Pt mobility (dissolution or reprecipitation) was enhanced by water.

15.4.4 Effect of Operating Temperature

At high temperatures, Pt nanocatalysts aggregate and grow, which decreases the catalyst's specific surface area and mass activity. Zhai (2007) investigated the stability of phosphoric acid-doped PBI (H_3PO_4/ PBI) in a PEMFC operated at high temperature, using a series of intermittent life tests (100, 300, and 520 h) for a single PEMFC with H_3PO_4/PBI membranes. During the first 300 h, ECSA decreased rapidly, then experienced only a small decrease in the following 210 h. The size of Pt particles increased after the life tests. Some coalescence of particles was apparent after only 100 h, but numerous small particles were still evident. After 300 h, the majority of small crystallites appeared to be absent. Figure 15.12 shows the histogram of Pt/C particle size distribution from TEM photographs; evidently, the mean particle size increased from 4.0 to 9.0 nm during 520 h of life testing, the distribution of Pt particle sizes became

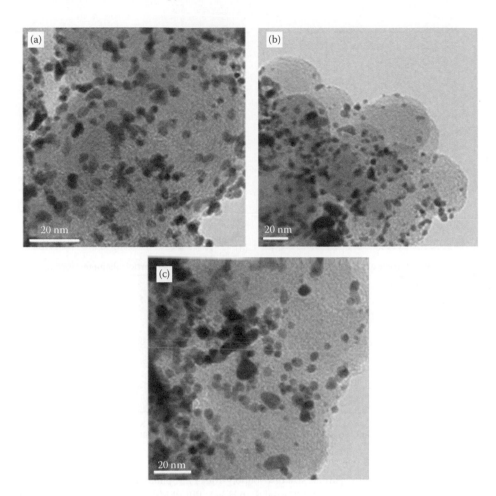

FIGURE 15.11 TEM micrographs of a catalyst: (a) initial catalyst, (b) degraded catalyst from the anode side, and (c) degraded catalyst from the cathode side. (Reprinted from Inter. *J. Hydrogen Energy*, 34, Lin, R. et al. Investigation of dynamic driving cycle effect on performance degradation and micro-structure change of PEM fuel cell. 2369–2376. Copyright (2009), with permission from Elsevier.)

gradually broader over time, and the number of Pt particles with small diameters (4.0 nm) decreased while larger Pt particles (>10.0 nm) appeared.

The tested MEA was also analyzed by EDX to investigate Pt deposition in the H_3PO_4/PBI membrane, but no Pt was found there. The study's results indicated that Pt particle agglomeration is likely responsible for ECSA loss in Pt/C electrocatalysts. Further analyses of the study's TEM data suggested that Pt particle agglomeration occurred on carbon through a coalescence mechanism at the nanometer scale.

In conclusion, during H_3PO_4/PBI high-temperature PEMFC single-cell life tests, considerable performance degradation occurred due to significant ECSA loss in the Pt/C cathode electrocatalyst, which resulted from serious agglomeration of Pt particles on the carbon support during the first ~300 h. In more prolonged life tests, the growth of Pt particles in the cathode was not obvious and the ECSA of the Pt/C cathode electrocatalyst decreased only slightly. On the other hand, the absence of Pt in the tested H_3PO_4/PBI membrane suggested that Pt dissolution probably had not occurred.

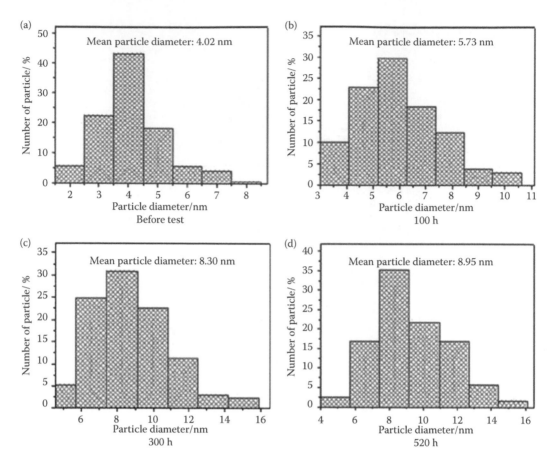

FIGURE 15.12 Histogram of Pt/C particle size distribution from TEM photographs. (Reprinted from *J. Power Sources*, 164, Zhai, Y. F. et al. The stability of Pt/C catalyst in H_3PO_4/PBI PEMFC during high temperature life test. 126–133. Copyright (2007), with permission from Elsevier.)

15.5 Outlook

From particle imaging using TEM techniques, we can detect changes in catalyst particle size, distribution, and morphology following electrochemical aging. Particle data measured directly from TEM images can be used to elucidate the predominant mechanisms of particle coursing that contribute to the reduction of a catalyst's ECSA and the concomitant fuel cell performance degradation.

Nevertheless, TEM is destructive in the study of electrocatalysts, since the catalyst has to be removed from the working electrode after treatment in the electrolyte. With developments in TEM technology, it is now possible to obtain an improved understanding of the degradation mechanism of fuel cell catalysts without destroying the sample. A novel, nondestructive method based on TEM has been developed that enables the investigation of identical locations on a catalyst before and after electrochemical treatments, and can reveal particle growth, spatial distribution of components, and the composition of a local area at the nano-to-atomic scale (Akita et al., 2004).

In summary, TEM can yield very detailed observations for examining the degradation mechanisms of Pt catalysts. Further advances in TEM technology itself offer the promise of even more remarkable progress.

Acknowledgments

The authors gratefully acknowledge the financial support by the 111 Project (No. B08019), the National Natural Science Foundation (No. 20703031 and No. 21006073), the Shanghai Rising-Star Program (11QA1407200), Young Teacher Research Fund of the Doctoral Program of Higher Education (No. 20070247055), and Program for Young Excellent Talents in Tongji University (No. 2006KJ022).

References

Akita, T., Taniguchi, A., Maekawa, J., Siroma, Z., Tanaka, K., and Kohyama, M. 2006. Analytical TEM study of Pt particle deposition in the proton-exchange membrane of a membrane-electrode-assembly. *J. Power Sources* 159: 461–467.

Anderson, R. M. 1990. Specimen preparation for transmission electron microscopy of materials. II. *Materials Research Society Symposium Proceedings* 199. Garland Science, San Francisco, California, USA.

Borup, R. L., Davey, J. R., Garzon, F. H., Wood, D. L., and Inbody M. A. 2006. PEM fuel cell electrocatalyst durability measurements. *J. Power Sources* 163: 76–81.

Borup, R., Meyers, J., Pivovar, B. et al. 2007. Scientific aspects of polymer electrolyte fuel cell durability and degradation. *Chem. Rev.* 107: 3904–3951.

Buseck, P., Cowley, J., and Eyring, L. 1988. *High-Resolution Transmission Electron Microscopy and Associated Techniques*. New York: Oxford University Press.

Cecil, H. E. 1953. *Introduce to Electron Microscopy*. New York: McGraw-Hill.

Cowley, J. M. and Moodie, A. F. 1957. The scattering of electrons by atoms and crystals. I. A new theoretical approach. *Acta Crystallographica* 199: 609–619.

Dawn, C. and Peter, J. G., 1990. The operation of transmission and scanning electron microscopes, *Royal Microscopically Society Handbook*. New York: Oxford University Press.

Earl, J. K. 1998. *Advanced Computing in Electron Microscopy*. New York: Springer.

Goldstein, J. I., Newbury, D. E., Echlin, P. et al. 1995. *Scanning Electron Microscopy and X-ray Microanalysis*. New York: Plenum Press.

Ferreira, P. J., la, O. G. J., Shao-Horn, Y. et al. 2005. Instability of Pt/C electrocatalysts in proton exchange membrane fuel cells, a mechanistic investigation. *J. Electrochem. Soc.* 152: A2256–A2271.

Hirahara, K., Bandow, S., Suenaga, K. et al. 2001. Electron diffraction study of one-dimensional crystals of fullerenes. *Phys. Rev.* B 64: 115420–115421.

Horiuchi, S. 1994. *Fundamentals of High-Resolution Transmission Electron Microscopy*. The Netherlands: North-Holland.

John, M. C. 1995. *Diffraction Physics*. New York: Elsevier Science B. V.

Lambert, L. and Mulvey, T. 1996. Ernst Ruska (1906–1988) designer extraordinaire of the electron microscope: A memoir. *Adv. Imag. Electron Phys.* 95: 2–62.

Knights, S. D., Colbow, K. M., St-Pierre, J., and Wilkinson, D. P. 2002. Aging mechanisms and lifetime of PEFC and DMFC. *J. Power Sources* 127: 127–134.

Li, J., He, P., and Wang, K. 2006. Characterization of catalyst layer structural changes in PEMFC as a function of durability testing. *ECS Trans.* 3: 743–751.

Lin, R., Li, B., Hou, Y. P., and Ma, J. X., 2009. Investigation of dynamic driving cycle effect on performance degradation and micro-structure change of PEM fuel cell. *Inter. J. Hydrogen Energy* 34:2369–2376.

Ludwig, R. 1993. *Transmission Electron Microscopy: Physics of Image Formation and Microanalysis* (3rd ed.). New York: Springer-Verlag.

MRS. Baram, M. and Kaplan, W. D. 2008. Quantitative HRTEM analysis of FIB prepared specimens. *J. Microsc.* 232: 395–405.

Paul, M. F., Stephen, G., and Stephen, T. T. 1993. *Physics for Scientists and Engineers*. Englewood Cliffs, NJ: Prentice-Hall.

Phillips, R. 1961. Diamond knife ultra microtomy of metals and the structure of microtomed sections. *Br. J. Appl. Phys.* 12: 554–558.

Porter, K. and Blum, J. 1953. A study in microtomy for electron microscopy. *Anat Record* 117: 685–712.

Ralph, T. R. and Hogarth M. P. 2002. Catalysis for low-temperature fuel cells, Part II: The anode challenges. *Plat. Metal. Rev.* 46: 117–135.

Spence, J. C. H. 1981. *Experimental High-Resolution Electron Microscopy.* Oxford, UK: Oxford University Press.

Taniguchi, A., Akita, T., Yasuda, K. and Miyazaki, Y. 2004. Analysis of electrocatalyst degradation in PEMFC caused by cell reversal during fuel starvation. *J. Power Sources* 130: 42–49.

Tüber, K, Pócza, D., and Hebling, C. 2003. Visualization of water buildup in the cathode of a transparent PEM fuel cell. *J. Power Sources* 124: 403–414.

Wei, Z. D., Ji, M. B., Hong, Y., Sun, C. S., Chan, S. H., and Shen, P. K. 2006. MnO_2–Pt/C composite electrodes for preventing voltage reversal effects with polymer electrolyte membrane fuel cells. *J. Power Sources* 160: 246–251.

William, D. B. 1984. *Practical Analytical Electron Microscopy in Materials Science* (2nd ed.). Mahwah, NJ: Philips Electron Optics Publishing Group.

Williams, D. B. and Carter, B. C. 2007. *Transmission Electron Microscopy: A Textbook for Materials Science (I–IV).* London, UK: Springer-Verlag Media Inc.

Williams, D. B., Goldstein, J. I., and Newbury, D. E. 1995. *X-ray Spectrometry in Electron Beam Instruments.* New York: Plenum Press.

Xie, J., Wood III, D. L., More, K. L., Atanassov, P., and Borup, R. L. 2005a. Microstructural changes of membrane electrode assemblies during PEFC durability testing at high humidity conditions. *J. Electrochem. Soc.* 152(5): A1011–A1020.

Xie, J., Wood III, D. L., Wayne, D. M., Zawodzinski, T. A., Atanassov, P., and Borup, R. L. 2005b. Durability of PEFCs at high humidity conditions. *J. Electrochem. Soc.* 152: A104–A113.

Ye, S., Hall, M., Cao, H., and He, P. 2006b. Degradation resistant cathodes in polymer electrolyte membrane fuel cells. *ECS Trans.* 3: 657–666.

Zhai, Y. F., Zhang, H. M., Xing, D. M., and Shao, Z. G. 2007. The stability of Pt/C catalyst in H_3PO_4/PBI PEMFC during high temperature life test. *J. Power Sources* 164: 126–133.

16

Infrared Imaging

Xiao Zi Yuan
*Institute for Fuel Cell
Innovation*

Shengsheng Zhang
*Institute for Fuel Cell
Innovation*

Jinfeng Wu
*Institute for Fuel Cell
Innovation*

Haijiang Wang
*Institute for Fuel Cell
Innovation*

16.1 Introduction

16.1.1 Infrared Radiation

The electromagnetic spectrum consists of radiation that is differentiated by wavelength: gamma rays (<1 pm), x-rays (0.001–10 nm), ultraviolet (10–400 nm), a thin region of visible light (400–700 nm), infrared (IR) (0.7 μm–1 mm), microwaves (1 mm–1 m), and radio waves (>1 m).

All objects emit a certain amount of IR radiation, also known as blackbody radiation, as a function of their temperatures. Generally speaking, the hotter the object is, the more radiation it emits.

16.1.2 IR Camera

An IR camera is a device that creates images from IR radiation. Also known as a thermographic camera, an IR camera uses wavelengths longer than visible light to create images even in total darkness, similar to a regular camera that forms an image using visible light. The two main types of IR cameras are those with (1) cooled or (2) uncooled IR detectors.

Instead of collecting radiation and converting it to an image on its own, a cooled IR detector, typically contained in a vacuum-sealed case, employs a cooling technique known as the Joule–Thomson effect to produce the image. Cooling the semiconductor materials with cryogenic coolers—for example, in the range of 60–100 K—allows the radiation detected in an object to be singled out to produce a much clearer image than with a traditional IR camera having uncooled detectors. Although cooled IR cameras provide superior image quality, they are expensive to produce and to run. The cameras also may need several minutes to cool down before they can begin working, and special training and certification is required to operate them.

An uncooled thermal camera, which is a less costly alternative, simply functions at or close to the ambient temperature by sensing the heat emitted from an object and converting it into an image using changes in resistance, voltage, or current that occur when the object is heated by IR radiation. Uncooled IR sensors, mostly based on pyroelectric and ferroelectric materials or microbolometer technology to form pixels with highly temperature-dependent properties, do not require bulky, expensive cryogenic coolers, making these IR cameras smaller and less costly. While ferroelectric detectors operate close to the phase transition temperature of the sensor material, with a noise equivalent temperature difference (NETD) of 70–80 mK (with f/1 optics and 320×240 sensors), silicon microbolometers can reach a NETD as low as 20 mK. However, their resolution and image quality all tend to be lower than those of cooled detectors (Wikimedia Foundation, Inc., 2010).

16.1.3 Applications of IR Thermography

Thermographic cameras were originally developed for military use but have slowly migrated into many other fields. Lower prices, advanced optics, and sophisticated software interfaces have greatly enhanced the versatility of IR cameras. Applications of IR cameras include

- Astronomical devices such as the Spitzer Space Telescope
- Night vision equipment
- Firefighting operations
- Military and police target detection and acquisition
- Search and rescue operations
- Predictive maintenance (early failure warning) on mechanical and electrical equipment
- Process monitoring
- Moisture detection in walls and roofs
- Chemical imaging
- Medical testing for diagnosis
- Nondestructive testing
- Quality control in production environments
- Research and development of new products
- Quarantine monitoring of visitors to a country
- Flame detection

16.1.4 Importance of Perforation Detection in PEM Fuel Cells

IR imaging has been used to nondestructively detect and locate cracks within material samples (Sun and Deemer, 2003), and to determine the depth and lateral position of subsurface flaws using transient thermography to display the flaws (Ringermacher et al., 1998). Attempts have also been made to demonstrate the usefulness of thermal wave imaging nondestructive evaluation (NDE) in detecting various types of damage, such as scratches, folds, and pinholes in membrane materials.

Nafion®, manufactured by Dupont, is a widely used membrane material for PEM fuel cells. In the fabrication of membranes and MEAs for solid polymer fuel cells, the detection of perforations in the membrane is an important aspect of quality control because of the need to maintain fluid isolation of the fuel and oxidant streams during fuel cell operation. Since the PEM is fragile, it is handled and processed carefully to minimize physical tears or thinning. However, during operation the PEM is subjected to relatively high-stress conditions, including high temperature and pressure, and alternating humidification. Overall, the components and assembly associated with the MEA often lead to imperfections or defects (Murphy and Litteer, 2004).

A perforation or leak in the membrane of a fuel cell can cause the fuel and oxidant streams to fluidly communicate and chemically react, thereby degrading the electrochemical potential of the fuel cell.

Serious membrane degradation can also occur if the fuel combusts in the presence of the catalyst and oxygen. It is therefore highly beneficial to detect and locate perforations in the membrane after its fabrication and/or after fabrication of the MEA. Further, the ability to detect and locate perforations in the membrane of an MEA after it has been used in an operating fuel cell can assist in diagnosing degradation and failure mechanisms. Thus, a method of detecting such imperfections or defects is certainly needed.

In the following sections, we will introduce "nondestructive testing" that uses IR cameras to produce thermographic images, particularly for the investigation of PEM fuel cells and their failure modes and degradation mechanisms.

16.2 Principles and Measurements

16.2.1 Principles

The method of using IR imaging to detect and locate perforations in MEAs, particularly their membranes, of PEM fuel cells was first patented by Ballard Power Systems in 1998 (Lamont and Wilkinson, 1998). As described in the patent, one side of the membrane is exposed to a reactant fluid, preferably a gaseous mixture containing hydrogen, while the other surface is exposed to another reactant fluid, preferably ambient air containing oxygen. When no perforations are present in the membrane, the two fluids remain isolated from each other by the MEA. Upon contact, the fluids exothermically react in the presence of a catalyst:

$$H_2 + \tfrac{1}{2}O_2 \rightarrow H_2O(l) + 286 kJ/mol \tag{16.1}$$

The generated heat is then detected using an IR thermal detector or thermal imaging device. The temperature distribution obtained from, for example, an IR thermal detector creates a vivid image of the thermal variation across the MEA surface, demonstrating the level of reactant crossover through the PEM. In a durability study, the same test can also be conducted on a fresh MEA for comparison (Zhang et al., 2010).

16.2.2 Equipment

In our study, IR imaging was performed using an IR camera (InfraTech GmbH) and a specially designed cell (50 cm^2) with an open cathode. Figure 16.1 shows pictures of the VarioCAM® head and the cell.

The VarioCAM head is a robust IR camera for precise real-time thermography based on an uncooled microbolometer array. The detector is thermally stabilized with high precision using a Peltier element to make it independent of the ambient temperature. Having an uncooled detector means short camera ramp-up times and long operating life. The camera lens captures the IR radiation emitted by the object into the field of view (FOV) and projects it onto the detector array. The FOV and resolution (IFOV) are determined by the focal length of the installed lens. The standard lens ($f = 25$ mm) has an FOV of 32×25 and an IFOV of 1.8 mrad (1.8 mm at 1 m object focus range). The object range can be set between 0.28 m and infinity (Infrared Solutions Inc., 2005). The camera enables quick, contactless recording of an object's surface temperature, and its radiometric measuring function resolves temperature differences <0.1 K with a spectral range of 7.5–14 μm. The IR lens of the camera displays the object scenery on a microbolometer array with a resolution of 320×240 pixels. Detailed parameters of the VarioCAM head are listed in Table 16.1.

Another preferred IR camera is the ThermaCAM® PM 695 manufactured by FLIR Systems™. The detection system of this device monitors IR surface emissions within the spectral wavelength range of 7.5×10^{-6}–13×10^{-6} m. This camera measures temperatures from about −40°C to ~1500°C, with the preferred detection range being about −40°C to ~120°C. Measurement modes include spot, area,

(a) 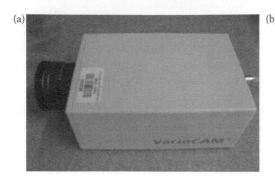   (b)

FIGURE 16.1 (a) VarioCAM head (InfraTech GmbH); (b) specially designed fuel cell with an open cathode and an active area of 50 cm^2.

isotherms, line profiles, and temperature gradients. The accuracy of the measurements is typically ±2°C or ±2% (Murphy and Litteer, 2004).

16.2.3 Measurements

16.2.3.1 Procedures and Software

The cathode of a self-designed cell is open, providing sufficient air for the reaction described in Equation 16.1. The anode requires hydrogen or a diluted hydrogen supply. The entire experimental setup, as illustrated in Figure 16.2, includes hydrogen cylinder, hydrogen regulator, flow meter, dummy cell, IR camera, and computer. A fresh or degraded MEA is installed in the dummy cell with pure or diluted hydrogen passing through the anode. The IR camera is placed towards the open cathode to observe the temperature distribution.

Various commercially available software programs are compatible with thermal IR cameras to enhance data acquisition and analysis. These programs can, for example, perform temperature analysis, static image analysis of stored images, and plotting of temperature over time. Further, such software enhances pattern recognition for defect occurrences, in addition to enabling more sophisticated analysis of IR thermographic images. In our study, we used IRBIS online 2.4 to observe a live picture of the temperature distribution or obtain a snapshot. To view the saved image or adjust its parameters, we used VarioAnalyze.

TABLE 16.1 Technical Data for the VarioCAM Head

Spectral Range	7.5–14 μm
Temperature measurement range	−40–1200°C, optional >2000°C
Temperature resolution at 30°C	better than 0.1 K
Measurement accuracy	± 2K or ±2%
Detector	Uncooled microbolometer array
IR imaging frequency	50/60 Hz
Interface	RS232, IEEE1394 (FireWire)
Working temperature range	−15–50°C
Standard lens, FOV, IFOV, min. focus	25 mm, 32 × 25, 1.8 mrad, 0.28 m
Emissivity	Adjustable from 0.1 to 1.0 in steps of 0.01

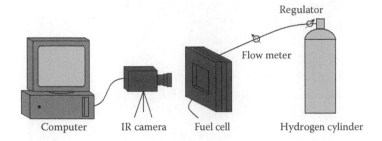

FIGURE 16.2 The experimental scheme of an *ex situ* IR imaging camera for detecting defects in an MEA.

16.2.3.2 Effect of Hydrogen Concentration

The effect of hydrogen concentration on the heat intensity of thermal images was investigated using samples before and after a relative humidity (RH) cycling test (Yuan et al., 2010). Three levels of hydrogen concentration were tested: 100% (from the building's hydrogen supply pipeline), 20% H_2 in N_2 (from Linde), and 5% H_2 in N_2 (from Linde). The fresh and degraded samples, provided by the University of Waterloo, were assembled with a catalyzed Gore 57 catalyst-coated membrane (CCM) and SGL 30BC gas diffusion layers (GDLs). The degraded Gore 57 MEA sample was run under RH cycling for about 400 h. At a flow rate of 30 sccm min^{-1}, room temperature, and a pressure of 5 psi, IR images of the Gore 57 MEA samples before and after degradation under RH cycling were taken using the setup shown in Figure 16.2. The results for 5% H_2 in N_2, 20% H_2 in N_2, and 100% H_2 are compared in Figures 16.3, 16.4, and 16.5, respectively.

In Figures 16.3 through 16.5 the same hot spot is observable, regardless of different hydrogen concentration levels and with no significant membrane thinning for the rest of the area. However, as the hydrogen concentration increases, the maximum temperature of the hot spot for the degraded sample also increases because more hydrogen has crossed over and more heat has been generated. Note that in Figure 16.4b, the maximum temperature of the hot spot seems higher than in Figure 16.5b. This is because the room temperature when the test occurred using 5% H_2 was about 2°C higher.

16.3 Applications in PEM Fuel Cells

16.3.1 Hydrogen Crossover/Membrane Thickness

Membrane thickness and hydrogen crossover through membranes can be compared or determined qualitatively using IR images. Figures 16.6a–d compare four fresh MEAs with Nafion membranes of different thicknesses (N117:183 μm; N115:127 μm; NR212:50.8 μm; and NR211:25 μm). These MEAs were assembled using Ion Power customized CCMs with a Pt catalyst loading of 0.3 mg cm^{-2}, on both sides, and SGL GDLs.

As can be seen from the above four images, temperature is evenly distributed throughout the MEA for all the fresh samples with membranes of different thicknesses. No obvious hot spots can be observed, even for the thinnest membrane. However, from the color of the IR images or the average temperature for each sample (N117: 23.09°C; N115: 23.43°C; NR212: 23.72°C; NR211: 24.06°C), we can conclude that the thinner the membrane, the more hydrogen that crosses through it, which agrees very well with the results of hydrogen crossover current measured using linear scan voltammetry (LSV).

16.3.2 Membrane Degradation

Degradation testing of the aforementioned four MEAs with membranes of different thicknesses was conducted for 1000 h using a four-cell stack with an active area of 50 cm^2 under idle conditions. During the degradation test, the four-cell stack was operated at a constant current of 0.5 A (10 mA cm^{-2}). The

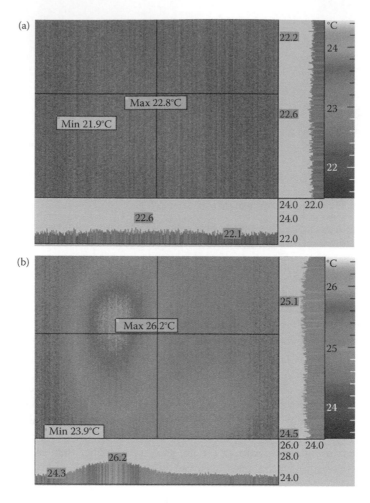

FIGURE 16.3 Comparison of IR images for Gore 57 MEA samples (from University of Waterloo) before and after degradation under RH cycling using 5% H$_2$ in N$_2$. (a) Fresh sample; (b) degraded sample.

fuel cell stack temperature was kept at 70°C, and air and hydrogen were fully humidified prior to delivery into the fuel cell. The flow rates for air and hydrogen were set at 1.0 and 2.0 standard liter per minute (slpm), respectively.

IR images for MEA samples with membranes of different thicknesses before and after degradation under idle conditions were taken using 20% H$_2$ in N$_2$. The results are compared in Figures 16.7 through 16.10. As can be seen, for thicker membranes (N117 and N115) there is a very slight increase in the average temperature, shown by the color of the images. However, this temperature difference before and after degradation, caused by the change in hydrogen crossover, is not significant, and agrees with the slight performance decay observed from the polarization curve.

As for thinner membranes (NR212 and NR211), hydrogen crosses over significantly through weak areas or pinholes after 1000 h of operation under idle conditions. This result agrees very well with the LSV curves, and explains the significant open-circuit voltage (OCV) drop as well as the performance degradation measured by the polarization curves. We further speculate that for thinner membranes the pinhole formed is the major reason that accounts for the membrane degradation rather than the membrane thinning indicated by a tiny, slight increase in the background temperature.

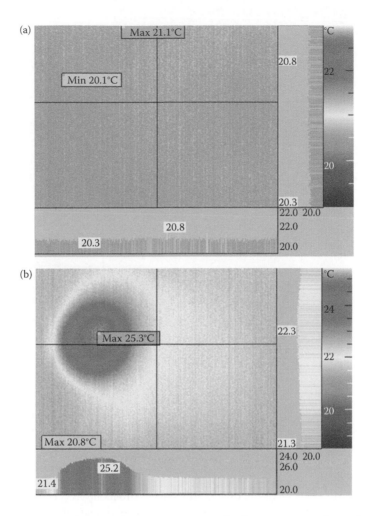

FIGURE 16.4 Comparison of IR images for Gore 57 MEA samples (from University of Waterloo) before and after degradation under RH cycling using 20% H_2. (a) Fresh sample; (b) degraded sample.

16.3.3 *In situ* Temperature Mapping

IR imaging has been used for temperature mapping of an operating fuel cell with an IR transmissive window. Using an IR detector array, the IR camera captures a thermographic profile of the fuel cell while simultaneously measuring the electrochemical output of the cell (Roscoe et al., 2006). The following are some examples of IR imaging for temperature mapping.

Wang et al. (2003, 2004, 2006) developed a fuel cell with a window made out of barium fluoride (transparent to IR light) at the anode side. After producing the IR images, they measured the temperature at certain points along the flow channel, thereby developing plots of temperature as a function of position in the flow field. Similarly, Hakenjos et al. (2004) designed a cell for combined measurement of current and temperature distribution. For temperature distribution an IR transparent window made of zinc selenide was located at the cathode side of the fuel cell. This allowed IR thermography of water droplets and flow-field flooding. Similar studies were presented by Shimoi et al. (2004), who used a narrow, straight channel flow-field and sapphire as the IR transparent window, and Kondo et al. (2003), who studied different flow-field designs and used a calcium fluoride crystal as the IR transparent window. NASA's Jet Propulsion Laboratory (NASA, 2001) has also employed IR thermography to record the

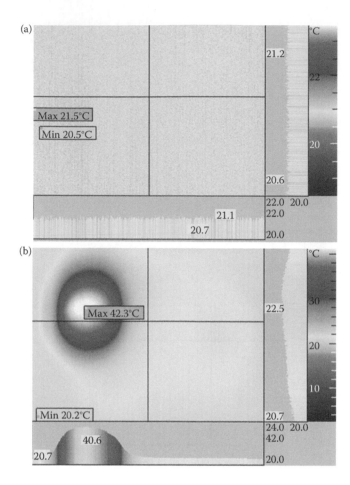

FIGURE 16.5 Comparison of IR images for Gore 57 MEA samples (from University of Waterloo) before and after degradation under RH cycling using 100% H_2. (a) Fresh sample; (b) degraded sample.

temperature distribution of fuel cells and stacks, and thereby further optimize their designs using an IR camera equipped with quantum-well infrared photodetectors (QWIPs), which can detect temperature differences as small as 0.005 K. A series of studies on temperature mapping has also been done by Ishikawa et al. (2005, 2006, 2007, 2008) to clarify the phenomenon of water freezing below its usual freezing point. In order to achieve this, they used visible and IR images simultaneously to observe the water generated on the surface of the catalyst layer and found that surprisingly, water generated below freezing point was in a liquid state at the time of generation and maintained a super-cooled state.

16.3.4 Material Screening

Spatially resolved IR imaging has been demonstrated by Olk (2005) for screening hydrogen storage candidates. Using a combination of pulsed laser deposition and magnetron sputtering, multicompositional samples were prepared for hydrogen sorption studies. A sample consisting of 16 separate Mg–Ni–Fe ternary pads and 32 Mg–Ni or Mg–Fe binary pads was analyzed. The results showed that depending on the sample composition, hydrogen-sorption-related emissivity changes indicated a substantial decrease in hydriding temperatures.

Yamada et al. (2004) used IR thermography to rapidly evaluate the CO tolerance of anode catalysts for PEM fuel cells. A series of Pt/C modified metal oxide catalysts were tested for the combustion of

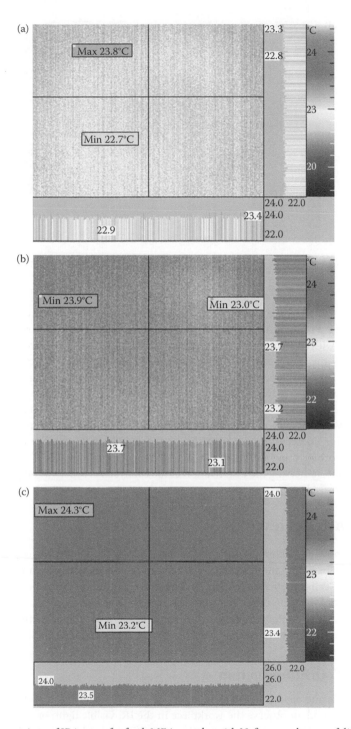

FIGURE 16.6 Comparison of IR images for fresh MEA samples with Nafion membranes of different thicknesses, using 5% H_2 in N_2. (a) N117; (b) N115; (c) NR212; (d) NR211.

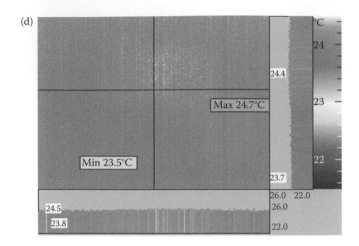

FIGURE 16.6   *(Continued)*

diluted CO, H_2, and CO + H_2 in air. Nine metal oxides were found to improve the CO tolerance of Pt/C in gas-phase reactions. Pt/C modified separately with niobium oxide and with tantalum oxide was electrochemically tested with conventional cyclic voltammetry using a rotating electrode. With the addition of these metal oxides, the CV measurements also indicated the possibility of improved CO tolerance.

Son et al. (2003) have developed an improved Pt catalyst on alumina using a water-pretreatment method. However, at temperatures below 100°C, water from the hydrogen-oxidation side reaction accumulates on the catalyst, causing it to deactivate. To better understand this deactivation, the water-pretreated catalyst was characterized at room temperature (28°C) and at 50°C under various reaction conditions (O_2 and H_2O concentrations) by analyzing the activity of the catalyst and by measuring the reactor temperature profile using an IR camera. The IR images clearly showed a decrease in the reactor temperature over time, indicating the deactivation of the catalyst. As deactivation progressed, the reactor temperature decreased.

16.4 Literature Review

As discussed in the previous section, IR imaging has been widely used in PEM fuel cell research, such as perforation detection in membrane degradation, temperature/current mapping, and material screening. To ensure that a fuel cell can operate reliably, it is necessary to test for MEA perforations. Given this handbook's substantial focus on durability and degradation, a review of methods to detect pinholes/perforations in MEAs is of particular interest. This section describes various methods, based on transillumination and other techniques that can be used to detect perforations in an MEA.

16.4.1 Varied Transillumination

Existing test methods and apparatus for recognizing flaws or damage based on transillumination use X-radiation or ultrasound, or observe the workpiece in the IR, visible light, or UV regions. Particular difficulties arise when testing workpieces consisting of multiple joined layers of different materials. The materials properties of the MEA do not permit inspection by means of X-radiation or ultrasound. It is therefore desirable to carry out tests and quality inspections on MEAs using observations and measurements in the IR region, by means of IR cameras (Boehmisch and Haas, 2006a). Alternatively, exothermically generated heat is detected and located using a layer of thermally sensitive film (including, e.g., thermally sensitive paper) positioned in proximity to the membrane (Lamont and Wilkinson, 1998).

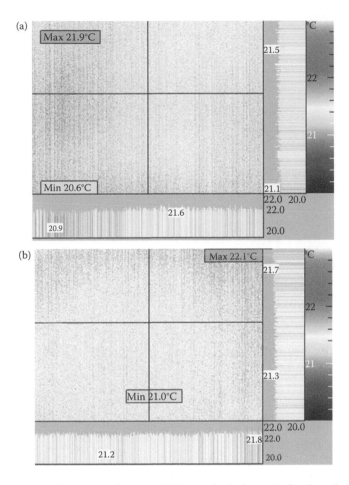

FIGURE 16.7 Comparison of IR images for N117 MEA samples before and after degradation under idle conditions using 20% H_2 in N_2. (a) Before degradation; (b) after degradation.

16.4.1.1 Heat Flow

Similar to the IR imaging technique described earlier in this chapter, another patented apparatus (Boehmisch and Haas, 2006a) includes a heat source, a sensor to determine temperature distribution (e.g., IR camera), and an evaluation apparatus. By applying a heat flow to the surface of the workpiece and continuously measuring the temperature distribution, defects in the workpiece can be recognized by inhomogeneous temperature distribution, on the basis of deviation between the temperatures measured at two or more adjacent locations.

16.4.1.2 Gas Source

Another test apparatus for examining sheet-like components for perforations has been patented (Boehmisch and Haas, 2006b). This test apparatus includes a gas source, an electromagnetic radiation source/laser light source, and a detector (transmitted-light sensor apparatus and/or scattered-light sensor apparatus) to detect spontaneous emission of electromagnetic radiation by the test gas molecules (CO_2). The test apparatus features two chambers (a pressure chamber connected to the test gas and a reduced-pressure chamber where the radiation source and detector are located) separated by the MEA. The electromagnetic radiation source emits electromagnetic radiation in a wavelength region adapted to the test gas (CO_2). The test molecules are excited by absorbing part of the electromagnetic radiation and

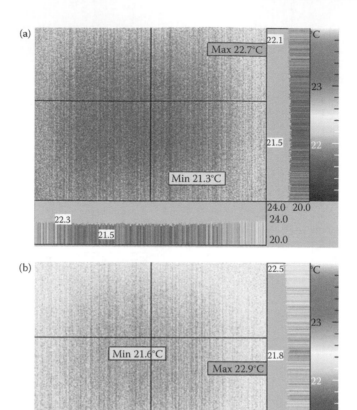

FIGURE 16.8 Comparison of IR images for N115 MEA samples before and after degradation under idle conditions using 20% H_2 in N_2. (a) Before degradation; (b) after degradation.

escape from the pressure chamber through perforations. The detector records electromagnetic radiation that has been absorbed by the test gas in the wavelength region. This method can be used not only for MEAs, but also for bipolar plates or their subcomponents, to detect perforations or pinholes during the production process or goods inspection.

16.4.1.3 Detecting Electrical Defects by Applying a Voltage

Another invention for detecting defects within a PEM is to apply a voltage across the MEA through electrically conductive substrates that contact the two surfaces of the MEA. The location of a defect is determined using thermographic images taken by IR camera to monitor variations in the intensity level of IR radiation emitted from the surface of the MEA. In the absence of a defect, the potential difference is maintained across the membrane. In the presence of a defect, the potential difference decreases as current flows between the two electrodes, generating hot spots.

Using this method, it is often advantageous to analyze the MEA and GDLs prior to final assembly. The PEM, when operating properly without defects, should behave as a dielectric with nearly infinite resistance. Any current passing through the MEA therefore reveals electrical defects. If current does pass through the MEA, the material surrounding the short circuit heats up due to its resistant properties, creating a hot spot. The intensity of the heat generated at the hot spot establishes the severity of the

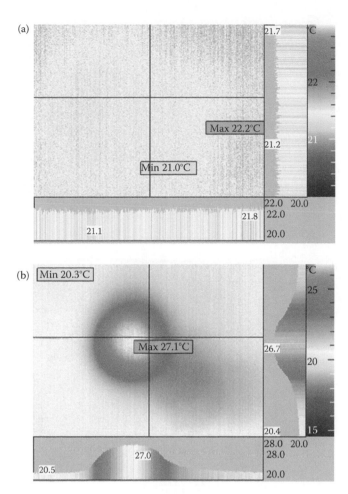

FIGURE 16.9 Comparison of IR images for NR212 MEA samples before and after degradation under idle conditions using 20% H_2 in N_2. (a) Before degradation; (b) after degradation.

defect. This method may also be used to perform a postmortem analysis of a disassembled fuel cell to identify the source of MEA failure (Murphy and Litteer, 2004).

16.4.2 Liquid Penetration

16.4.2.1 Water Bubbling

Using the same setup as depicted in Figure 16.2 but without the IR camera, pinholes can also be observed by putting a certain amount of deionized water on top of the open cathode, with diluted hydrogen passing through the anode channel of the dummy cell.

To prevent the damaged edges from affecting the outcome, the degraded samples were reframed using Kapton films. Diluted hydrogen was used at a pressure of 5 psi and a flow rate of 30 sccm/min. Distinct bubbles can be seen on the NR212 sample after 1000 h of degradation under idle conditions (Figure 16.11b), whereas no bubbles were visible on the fresh NR212 sample before degradation (Figure 16.11a). This further confirms that pinholes were generated during the degradation test. Unfortunately, the bubble locations do not exactly match the locations of the hot spots observed from IR images, shown in Figure 16.9, due to the GDL's porous nature.

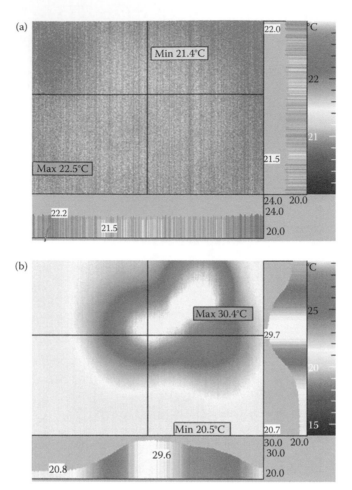

FIGURE 16.10 Comparison of IR images for NR211 MEA samples before and after degradation under idle conditions using 20% H_2 in N_2. (a) Before degradation; (b) after degradation.

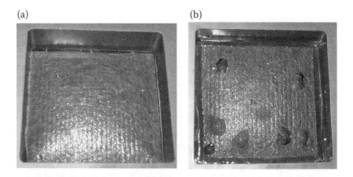

FIGURE 16.11 Pictures of NR212 MEA samples before and after 1000 h of degradation under idle conditions using 5% H_2 in N_2. (a) Before degradation; (b) after degradation.

16.4.2.2 Other Liquids

By introducing CCl_4 into either the cathodic or the anodic chamber of an electrolytic cell and air into the opposite chamber, leaks in an ion-exchange membrane can be detected, under a sealed condition partitioned with the ion-exchange membrane. Air is then withdrawn from the opposite chamber, and the presence of phosgene is indicated using a halogen detector. Thus, the existence of pinholes with diameters above 0.1 mm can be detected (Takenaka et al., 1983).

Easy detection of a breakage in an ion-exchange membrane has also been reported using an NaCl or NaOH solution. An electrolytic cell fitted with an ion-exchange membrane is filled with an electrolytic solution of NaCl or NaOH. An electric current, generally small, is supplied to the cell, and then the change in the electrolytic cell voltage is measured. When a breakage such as a pinhole is present in the ion exchange membrane, the voltage is lower than normal by ≥10%, so a breakage is inferred when a cell's electrolytic voltage drops this much below normal (Takenaka et al., 1986).

16.4.3 Gas Permeability

Holding either of the polar chambers of a cathodic or an anodic chamber in a reduced pressure state and measuring the change in pressure of the other polar chamber enables the easy, safe detection of abnormalities such as pinholes in an ion-exchange membrane (Takagi et al., 1989). Detailed descriptions can be found in Chapter 21.

16.4.4 Other Methods

Other methods for detecting pinholes in MEAs include the plasma technique and the use of an integral sensor. The plasma technique can accurately detect a pinhole defect even in a film whose pinhole is difficult to discover using a conventional detection method such as a transparent or translucent film formed on an opaque film. By supplying an inert gas to one surface of a polymer film for a fuel cell and supplying power to a plasma head, the plasma head is scanned on the other surface of the polymer film, and light generated from this surface is received. The existence of a pinhole in the polymer film for the fuel cell is detected based on the quantity of received light (Mase et al., 2005).

The integral sensor approach involves a sensor disposed on the surface of the PEM. The sensor monitors the physical, thermal, chemical, or electrical state of the MEA. Information obtained from the sensor is used to identify a defective MEA, and the operation of the fuel cell is altered based on this identification (Kelley et al., 2000).

16.5 Advantages and Limitations

16.5.1 Advantages

IR imaging provides a fast, efficient, nonintrusive method of detecting defects, pinholes, or perforations in an MEA, one that is readily adaptable to continuous manufacturing settings and quality control analysis. Other forms of thermal-monitoring equipment and sensors, such as heat-sensitive paper, can also be used interchangeably to determine temperature variations on the surface of an MEA. However, IR cameras have proven to be the most accurate and reliable thermal-imaging equipment currently commercially available.

Methods of leak detection, such as measurement of the leak rate of a fluid through the membrane with an applied pressure differential across the membrane, or measurement of the gas composition on one side of the membrane when a reference gas is applied to the other side, do not typically provide information on the location of perforations. Bubble observation methods, requiring liquid contact with one face of the membrane or MEA, also do not precisely locate perforations. Current methods of quality control rely on leak detection by physical infiltration of gases through perforations in the membrane of an MEA; however, these methods do not provide the information required to identify the intensity and/ or location of defects.

The IR imaging described in this chapter identifies perforations in membranes by detecting heat generated by the exothermic reaction of a pair of reactants that are substantially isolated on opposite sides of the membrane. With the use of an IR camera, the location of the membrane perforations may also be determined.

The method and apparatus of IR imaging are also suitable to detect and locate perforations in (1) ion-exchange membranes and separators used in other types of electrochemical cells, such as electrolytic cells, chloroalkali cells, and batteries; (2) electrowinning; and (3) membranes employed in the plate-and-frame humidification sections of solid polymer electrochemical fuel cell stacks (Lamont and Wilkinson, 1998).

16.5.2 Limitations

One drawback of this method is the potential danger of simultaneously handling gases that react with one another, and the attendant risk of spontaneous combustion of the MEA. This danger may also occur if the test apparatus is not handled correctly—for example, if an excessively high pressure is set on the anode side and the MEA consequently bursts. If only small quantities of gas penetrate through the MEA at the perforations, only small amounts of heat will be generated, leading to errors arising from slight temperature differences (e.g., caused by the body heat of the people who are present).

Although IR imaging is easy to conduct, the obtained images can only qualitatively determine thinning or pinhole generation in the membrane. The existence of pinholes still has to be confirmed by other techniques, such as water bubble observation. In addition, other questions—for example, How thin is the membrane after degradation? How many pinholes are generated? How big are the pinholes and how do they distribute?—cannot be answered using IR imaging alone.

16.6 Concluding Remarks

To summarize, IR imaging is a simple, easy technique to diagnose PEM fuel cell degradation, especially in the membrane. However, attention must be paid to the use of hydrogen. In order not to burn the sample or damage/degrade it with the heat caused by high hydrogen crossover, diluted hydrogen is preferable to pure hydrogen. Despite the risks involved in handling and the fact that it yields only qualitative results, IR imaging is still a useful method to diagnose membrane degradation. Thermal imaging is expected to be widely used in fuel cell research, development, and manufacturing.

References

Boehmisch, M. and Haas, C. 2006a. Test apparatus and test method for the non-destructive testing in particular of membrane electrode assembly for use in fuel cells, which can be integrated in production. US2006/0029121 A1. Stuttgart Germany: C. DaimlerCrysler AG.

Boehmisch, M. and Haas, C. 2006b. Test apparatus and method for examining sheet-like components for perforations. US2006/0028640 A1. Stuttgart Germany: C. DaimlerCrysler AG.

Hakenjos, A., Muenter, H., Wittstadt, U., and Hebling, C. 2004. A PEM fuel cell for combined measurement of current and temperature distribution, and flow field flooding. *J. Power Sources* 131: 213–216.

Infrared Solutions Inc. 2005. *VarioCAM® Head User Manual*. Plymouth, MN.

Ishikawa, Y., Hamada, H., Uehara, M., and Shiozawa, M. 2008. Super-cooled water behavior inside polymer electrolyte fuel cell cross-section below freezing temperature. *J. Power Sources* 179: 547–552.

Ishikawa, Y., Morita, T., Nakata, K., Yoshida, K., and Shiozawa, M. 2005. Verification of generated water in a super-cooled state below freezing point in PEFC. *ECS Trans.* 1: 359–364.

Ishikawa, Y., Morita, T., Nakata, K., Yoshida, K., and Shiozawa, M. 2007. Behavior of water below the freezing point in PEFCs. *J. Power Sources* 163: 708–712.

Ishikawa, Y., Morita, T., and Shiozawa, M. 2006. Behavior of water below the freezing point in PEFCs. *ECS Trans.* 3: 889–895.

Kelley, R. J., Mulligan, R. J., Pratt S. D. et al. 2000. Integral sensors for monitoring a fuel cell membrane and methods of monitoring. WO200045450. Schaumburg IL: Motorola Inc.

Kondo, Y., Daiguji, H., and Hihara, E., Kobe/Japan. 2003. Optimization of the flow-field for PEMFCs and prospect from visualization of temperature distribution. *Proc. Int. Conf. Power Eng.* 2: 463–468.

Lamont, G. J. and Wilkinson, D. P. 1998. Method and apparatus for detecting and locating perforations in membranes employed in electrochemical cells. US 005763765A. Burnaby Canada: Ballard Power Systems Inc.

Mase, K., Nagasaki, T., Okumura, T., and Yoshinaga, M. 2005. Pinhole detection method and pinhole detection device. JP2005134218. Tokyo Japan: Matsushita Electric Ind. Co. ltd. (now Panasonic Corporation).

Murphy, M. W. and Litteer B. A. 2004. Method for detecting electrical defects in membrane electrode assemblies. US 2004/0048113 A1. Bloomfield Hills, MI: Harness Dickey & Piece, P.L.C.

NASA's Jet Propulsion Laboratory. Pasadena/California. Nov. 2001. Thermal imaging for diagnosing fuel cells. http://findarticles.com/p/articles/mi_qa3957/is_200111/ai_n9008284. Accessed May 2010.

Olk, C. H. 2005. Combinatorial approach to material synthesis and screening of hydrogen storage alloys. *Meas. Sci. Technol.* 16: 14–20.

Ringermacher, H. L., Archacki, Jr., R. J., and Veronesi, W. A. 1998. Nondestructive testing: Transient depth thermography. US005711603A. United Technologies Corporation, Hartford Conn.

Roscoe, S. B., Rakow, N. A., Atanasoski, R., Jackson, E. R., Thomas III, J. H., McIntosh III, L. H., Fuel cell. 2006. US2006/0127729 A1. ST. Paul, MN: 3M Innovative Properties Company.

Shimoi, R., Masuda, M., Fushinobu, K., Kozawa, Y., and Okazaki, K. 2004. Visualization of the membrane temperature field of a polymer electrolyte fuel cell. *J. Energy Res. Technol.* 126: 258–261.

Son, I. H., Lane, A. M., and Johnson, D. T. 2003. The study of the deactivation of water-pretreated Pt/γ-Al$_2$O$_3$ for low-temperature selective CO oxidation in hydrogen. *J. Power Sources* 124: 415–419.

Sun, J. and Deemer, C. 2003. Thermal imaging measurement of lateral diffusivity and non-invasive material defect detection. US 6517238 B2. Washington DC: The United States of America as represented by the United States Department of Energy.

Takagi, T., Watanabe, K., and Sugimoto, S. 1989. Method for detecting breakage of ion exchange membrane. JP01255682. Tokyo Japan: Mitsui Toatsu Chem. Inc.

Takenaka, M., Maeda, K., and Aoki, K. 1986. Inspection of electrolytic cell incorporating ion-exchange membrane in it. JP58193383. Tokyo Japan: Asahi Glass Co Ltd.

Takenaka, M., Maeda, K., and Aoki, K. 1986. Method for detecting breakage in ion-exchange membrane. JP61153295. Tokyo Japan: Tokuyama Soda KK (originally Tokuyana Soda Co. Ltd.).

Wang, M., Guo, H., and Ma, C. 2006. Temperature distribution on the MEA surface of a PEMFC with serpentine channel flow bed. *J. Power Sources* 157: 181–187.

Wang, M. H., Guo, H., Ma, C. F., Liu, X., Yu, J., and Wang, C. Y. 2004. Measurement of surface temperature distribution in a proton exchange membrane fuel cell. *Chin. J. Power Sources* 28: 764–766.

Wang, M. H., Guo, H., Ma, C. F. et al. 2003. Temperature measurement technologies and their application in the research of fuel cells. *Proceedings of the 1st International Fuel Cell Science, Engineering and Technology Conference*, April 21–23, Rochester, New York, NY, pp. 95–100.

Wikimedia Foundation, Inc., Wikipedia®. Thermographic camera. http://en.wikipedia.org/wiki/Thermographic_camera Accessed May 2010.

Yamada, Y., Ueda, A., Shioyama, H., and Kobayashi, T. 2004. High-throughput screening of PEMFC anode catalysts by IR thermography. *Appl. Surf. Sci.* 223: 220–223.

Yuan, X. Z., Zhang, S., Wu, J., Sun, C., and Wang, H. 2010. Post analysis on MEA degradation: IR imaging. 2010 controlled technical report #102 of Institute for Fuel Cell Innovation (IFCI)'s PEMFC group (IFCI-PEMFC-CTR-102).

Zhang, S., Yuan, X. Z., Ng Cheng Hin, J. et al. 2010. Effects of open circuit operation on membrane and catalyst layer degradation in PEM fuel cells. *J. Power Sources* 195: 1142–1148.

17

Fourier Transform Infrared Spectroscopy

Andrea Haug
Institute of Technical Thermodynamics

Renate Hiesgen
Institute of Technical Thermodynamics

Mathias Schulze
Institute of Technical Thermodynamics

Günter Schiller
Institute of Technical Thermodynamics

K. Andreas Friedrich
Institute of Technical Thermodynamics

17.1 Introduction

The major importance of infrared (IR) spectroscopy is based on the considerable information that can be obtained from a spectrum, as well as the diverse possibilities for sample preparation and measuring modes to identify compounds and investigate sample composition. IR spectroscopy has therefore been developed in the past decades into one of the most important methods for analytical investigation of materials and material changes, in numerous fields of technological development.

IR exploits the fact that molecules absorb specific frequencies of light that are characteristic for their structure. These absorptions are resonant frequencies—that is, the frequency of the absorbed radiation matches the frequency of the bond or group that vibrates, which can then be related to a particular bond type. An essential requirement for a molecule to be IR active is to be associated with changes in the dipole moment. Examination of the transmitted light of an IR spectrum reveals how much energy is absorbed from the molecule at each wavelength. The IR portion of the electromagnetic spectrum is usually divided into three regions: the near-, mid-, and far-IR, named for their relation to the visible spectrum. The far-IR, approximately 400–10 cm^{-1} (λ = 1000–30 µm), has low energy and may be used for rotational spectroscopy. The mid-IR, approximately 4000–400 cm^{-1} (30–2.5 µm), may be used to study the fundamental vibrations and associated rotational–vibrational structure. The higher energy near-IR, approximately 14,000–4000 cm^{-1} (2.5–0.8 µm), can excite overtones or harmonic vibrations.

An IR spectrum can be recorded by means of a monochromatic beam that changes in wavelength over time, or in the present day typically by using a Fourier transform instrument to measure the wavelengths in a wide range all at once. The FTIR method, whereby the interferogram measured in a time domain has to be retransferred to a spectrum by a mathematical operation (Fourier transformation), requires high computational effort and has therefore prevailed on the market since powerful computers became available. The position and intensity of the absorption bands of a sample are exceptionally specific for a substance and can be used as a highly characteristic "fingerprint" for identification. By applying computer-aided search/match programs for a huge number of reference spectra, each component of a multicomponent sample can be determined quantitatively. Furthermore, concentration ranges have been enormously improved, enabling detection of very low concentrations of substances. A derivative method is IR microscopy, which allows very small areas to be analyzed.

With quantitative IR spectroscopic measurements it is possible to study time-dependent processes such as chemical reactions and degradation processes. IR spectroscopy is widely used in both research and industry as a simple and reliable technique for routine measurements, quality control, and dynamic measurement. By measuring at a specific frequency over time, changes in the structure or quantity of a particular bond can be measured. This is especially useful, for instance, in polymer manufacture to measure the degree of polymerization. But IR spectroscopy has also been highly successful for applications in both organic and inorganic chemistry, such as in the field of semiconductor microelectronics.

In recent years, FTIR spectroscopy has also been increasingly applied in PEM fuel cell research and development as a diagnostic tool for investigating PEM fuel cell durability and the relevant degradation processes taking place in the fuel cell. This chapter provides an overview of the FTIR method and the possibilities for studying degradation in PEM fuel cells. It covers instrumentation as well as experimental techniques for different measuring modes, such as transmission, attenuated total reflectance, and diffuse reflectance. Furthermore, it gives examples of the interpretation of spectra for various applications in PEM fuel cell development, and discusses the method's advantages and limitations.

17.2 Instrumentation and Experimental Techniques

This section offers a short description of the methods available in the German Aerospace Center, Institute for Technical Thermodynamics. More detailed information about the different FTIR techniques is available in the literature (Alpert, 1973; Günzler and Gremlich, 2002; Wartewig, 2003; Stuart, 2004).

17.2.1 Transmission Method

The transmission technique is the simplest method for measuring samples with IR spectroscopy. The theoretical foundation for this method is the Lambert–Beer law, which relates the absorption of light to the properties of the material through which the light passes. The investigated sample can be solid, liquid, or gas, and is generally embedded in a cell or a KBr pellet.

$$E = \varepsilon \cdot c \cdot d \tag{17.1}$$

Equation 17.1 expressed the Lambert–Beer law, where E is the absorbance, ε is the molar decadic absorption coefficient, and d is the cell thickness.

Figure 17.1 depicts a schematic diagram of the transmission technique: IR light with an intensity I_0 is going from the source through the sample. Light of various wavelengths will be absorbed from the sample and an attenuated intensity, I, will be measured by the detector (Haug, 2009).

17.2.2 Reflectance Methods

17.2.2.1 Attenuated Total Reflectance

Attenuated total reflectance (ATR), also referred to as internal reflection spectroscopy (IRS), is a versatile method for analyzing samples without causing them any damage. ATR provides an IR spectrum from the surface of the sample, and thus is also applicable for samples that are either too thick or the absorption is too high for the standard transmission technique.

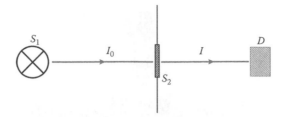

FIGURE 17.1 Schematic diagram of an IR beam going through a sample, using the transmission technique: S_1: source; S_2: sample; D: detector; I_0: incident beam; I: emergent beam.

The ATR technique is based on the work of Newton (1952), who investigated light reflectance phenomena at the interface of two media with different refraction indices. He determined that the total reflected light in the medium with the higher refraction index penetrated a few wavelengths into the medium with the lower refraction index. IRS has been developed since 1959 (Harrick, 1979), after Fahrenfort (1959) published that optical absorption spectra can be obtained easily from the interaction of total reflected light from the medium with the higher refraction index and that one with the lower refraction index (Fahrenfort, 1962). Figure 17.2 (Günzler and Gremlich, 2003) shows schematic examples of IRS (a) with one reflection, and (b) with multiple reflections.

In the media with the low optical density an evanescence wave is formed under the conditions of total reflection. The evanescence field interacts with the dipole-active molecules; ATR spectroscopy uses this theoretical background (Goos and Lindberg-Hänchen, 1949; Goos and Hänchen, 1947; Drexhage, 1970; Kramer, 1997; Haug, 2004).

Typical materials for ATR/IRS crystals will have a high refraction index; they include, for example, zinc selenide, (ZnSe), silicon (Si), germanium (Ge), diamond, and KRS-5 (Figure 17.3).

17.2.2.2 Diffuse Reflectance

The surface of powder samples or inner surfaces of porous samples like GDL, MPL, or reaction layers are not directly accessible for light as needed for the ATR technique. In order to investigate inner surfaces diffuse reflections must be used. The diffuse reflectance technique has been employed as versatile sample analysis tool for ultraviolet, visible, and IR spectroscopic analysis (Wendland and Hecht, 1966;

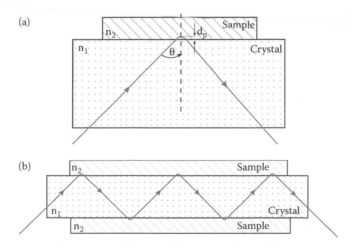

FIGURE 17.2 Schematic diagram of an attenuated total reflection going through an ATR crystal: (a) single reflection, (b) multiple reflections.

FIGURE 17.3   ATR unit (MIRacle ATR, PIKE Technologies, Madison, WI, USA).

Harbeck, 2005). The use is restricted to the UV/vis or near-IR regions of the spectrum caused by the low quantum efficiency of multiple diffuse reflections and restrictions in the hardware.

With the development of FTIR spectrometers and improvements in diffuse reflectance accessory design, DRIFT became more impact as serious method for FTIR spectroscopy. Today the DRIFT spectroscopy is one of the most suitable methods for an examination of rough and/or opaque samples, as well as strong scattering and/or absorbing samples. The technique is especially suitable for the detection of surface species in low concentration. Additionally, diffuse reflectance measurements can be conducted on heated samples.

Besides the possibility to investigate powder, rough, or porous samples, one of the main advantages of the DRIFT spectroscopy is that no special sample preparation is needed—meaning that the sample can be measured directly. Further advantages are the high sensitivity, a high versatility, and the capability of performing the measurements under real conditions like *in situ* measurements. Disadvantages are the separation of the diffuse reflectance from the specular reflectance, the adjustment of the system, and the sample positioning (Harbeck, 2005). A schematic diagram of a DRIFT unit is depicted in Figure 17.4.

17.2.3 FTIR Microscopy

For FTIR microscopy the above-mentioned methods (transmission, reflectance, and ATR) are also available. Samples can be examined under optical and IR microscopy. FTIR microscopy offers the advantages of defined local resolution and the possibility of recording images with the information derived from the IR spectra. This allows one to study inhomogeneous distributions of chemical components and properties. In general, this method is very important in the field of surface science (Salzer and Siesler, 2009).

17.2.4 Spectrometers

Two types of IR spectrometers are known—the dispersive and FTIR spectrometers. Recently, the FTIR spectrometer has come into common use because of its reduced measurement time. Two FTIR spectrometers currently used at the German Aerospace Center, Institute of Technical Thermodynamics, are shown in Figures 17.5 and 17.6. The Vertex 80v (Bruker Optik GmbH, Ettlingen, Germany, Figure 17.5) with coupled microscope (Hyperion3000, Bruker Optik GmbH, Ettlingen, Germany) possesses an evacuated optics bench to eliminate moisture absorption and CO_2. This high-end spectrometer is used for surface science analysis, especially in the field of fuel cell components and battery research (gas analytics,

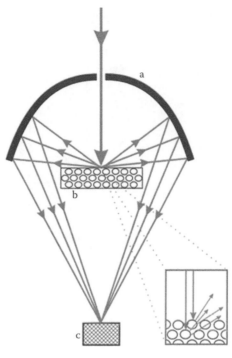

a: mirrors
b: sample
c: detector

FIGURE 17.4 Schematic diagram of a DRIFT unit.

solid-state spectroscopy, surface analytics, etc.). The low-cost Alpha spectrometer (Bruker Optik GmbH, Ettlingen, Germany, Figure 17.6) is used for initial tests on the work bench. For this small spectrometer the above-mentioned techniques and mirror units are also available in a smaller version.

An IR spectrum can be recorded by means of a monochromatic beam that changes in wavelength over time or, as is more usual these days, by using a Fourier transform instrument to measure

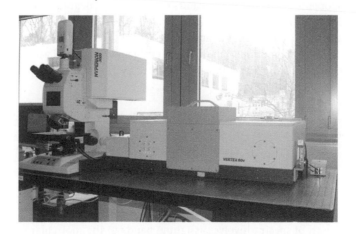

FIGURE 17.5 FTIR spectrometer Vertex 80v with IR microscope Hyperion3000 (Bruker Optik GmbH, Ettlingen, Germany).

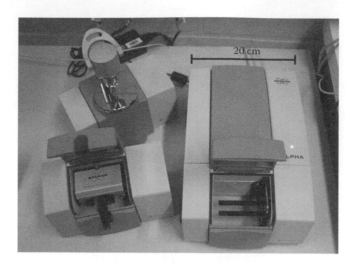

FIGURE 17.6 FTIR spectrometer Alpha (Bruker Optik GmbH, Ettlingen, Germany).

simultaneously all wavelengths in a wide range. This FTIR method, whereby the interferogram measured in a time domain has to be retransferred to a spectrum by a mathematical operation (Fourier transformation), requires high computational effort and has therefore prevailed on the market since powerful computers became available.

17.3 Applications

FTIR spectroscopy plays a significant role in a wide range of research and quality control (e.g., the pharmaceutical, chemical, and automotive industries, and academic research). It is used to identify molecules, in particular organic molecules. The technique therefore has a broad range of applications in, for example, materials science, quality control, and process control.

In the field of fuel cell research this spectroscopy is important for the characterization and development of polymeric components like membranes or hydrophobic agents. This kind of spectroscopy allows one to investigate alterations in polymers induced by the degradation processes of fuel cell components. Changes in the binding states as well as in the inter- and intramolecular interactions of the polymers used in fuel cell components can be useful indicators for identifying degradation mechanisms.

Furthermore, FTIR spectroscopy can be used to investigate the chemical reactions occurring on catalyst surfaces and in the liquid and gaseous components inside fuel cells. This technique can principally be employed under operating conditions, but the main problem is the liquid water phase, which causes a broad absorption band that cannot be eliminated and thus overlays the other spectra.

17.4 Interpretation of Spectra

The IR spectra include information about binding states, inter- and intramolecular interactions, and the concentration of IR-active groups. The various methods of evaluating these spectra fall into two categories: qualitative and quantitative interpretation.

17.4.1 Qualitative Interpretation

Qualitative interpretation of spectra involves assigning bands to the molecular components. Various approaches can be taken—for example, using spectra databases and libraries, IR handbooks, and literature data, which contain either the full spectra for a multitude of molecules or the IR-active groups for

TABLE 17.1 Absorption Bands of Nafion

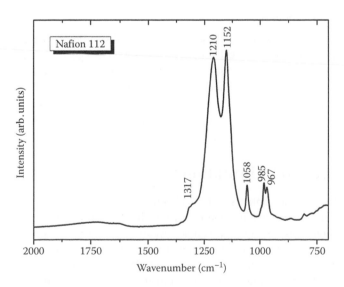

Wavenumber (cm^{-1})	Assignment
1317	v_{as} R–SO$_2$–OH
1210	v_{as} CF$_2$
1152	v_s CF$_2$
1058	v_s S–O
985	C–F stretching (–CF$_2$–CF(CF$_3$)-group)
697	C–O–C stretching, symmetric

comparison and assignment. Another possible method of assigning and confirming experimental data is to simulate the molecular vibrations with established methods like density functional theory (DFT) and Hartree–Fock calculations (Haug et al., 2008).

An accurate determination of the energy position of the absorption band is necessary to evaluate the peak shift compared to reference values and/or further recorded spectra from the same material under different experimental conditions. Alteration of the peak position can yield information about different molecular structures, different molecular phases (crystalline, amorphous, etc.), as well as inter- and intramolecular interactions (e.g., hydrogen bonds or hydrated molecules).

An example of the qualitative interpretation of a fuel cell component (Nafion membrane) is given in Figure 17.7 and the assignment of vibrational bands in Table 17.1 (Liang et al., 2004).

FIGURE 17.7 FTIR spectrum of a Nafion membrane. Absorption bands are assigned in correlation to their energy position (cm^{-1}).

17.4.2 Quantitative Interpretation

A quantitative analysis of the IR spectra is necessary to achieve more detailed information about the intensity of different absorption bands that have been qualitatively assigned. The intensity of the different molecules/functional groups can be determined by peak fit analysis, whereby the peak area is related to the quantity of molecules/functional groups. Detailed band analysis allows a differentiated assignment of the absorption bands to the specific molecules/functional groups and their specific inter- and intramolecular interactions. Analysis of the intensity of the vibration bands allows one to compare the absorption bands and gives further information about their interactions. The ratio of different species, determined by peak analysis, can be used as a characteristic in the investigation of reaction mechanisms or other molecular changes caused by, for example, degradation processes in the fuel cell components. Time-resolved measurements or experiments under different conditions (*ex situ and in situ*) are helpful for this kind of investigation. FTIR spectroscopic measurements before and after fuel cell operation are commonly used to study degradation mechanisms.

A broadening of vibrational bands can also be determined by peak fit analysis. This determination is necessary to observe changes in molecular structure and in inter- and intramolecular interactions.

17.4.3 Examples of FTIR Spectroscopy in Fuel Cell Research

Two examples of the application of FTIR spectroscopy in fuel cell degradation research are briefly described in the following two sections.

17.4.3.1 Mapping with FTIR Microscopy (ATR)

FTIR-ATR microscopy imaging of a used anode after 300 h of operation is shown in Figure 17.8. The left picture shows an optical microscope image of the used anode with the impressed flow field diagonally oriented. The crosses mark the measuring points (49 points, $150 \times 150 \ \mu m^2$). A typical single spectrum in the range of $2500-600 \ cm^{-1}$ is shown in the right-hand image. The intensities of the different C–F vibration modes (C–F, C–F$_2$) ($1263-1078 \ cm^{-1}$, v_{C-F}) have been determined by means of peak fit analysis for each of the 49 spectra. The results are depicted in the middle image as a two-dimensional chart. The colors are correlated to the intensities of the C–F absorption bands and indicate the distribution of the anode's C–F components.

17.4.3.2 Degradation Processes

Figure 17.9 illustrates the result of a peak fit analysis of C–F absorption bands in the region of $1263-1078 \ cm^{-1}$ (v_{C-F}). The spectra are recorded with a Vertex 80v FTIR spectrometer using the ATR technique in the range of $4000-600 \ cm^{-1}$. Figure 17.9 compares 16 segmented cells (Gülzow et al., 2003; Schulze et al., 2007) on the anode and cathode sides of a microporous layer, which degraded after 250 h of operation under harsh conditions (Hiesgen et al., 2011). Both anode and cathode sides display a different behavior in each segment. The inhomogeneous alteration in C–F intensity correlates with the inhomogeneous loss of performance during the fuel cell experiment.

17.5 Advantages and Limitations

FTIR spectroscopy has many advantages in the field of fuel cell research. One of the main benefits is that IR spectroscopy is a nondestructive method because IR light is a low-energy radiation and therefore does not dissociate molecules. In addition, sample preparation is very easy and does not influence the sample. Since samples are not destroyed by IR light or preparation, the same samples can then be investigated with other methods (e.g., Raman spectroscopy, XPS). FTIR spectroscopy is also a very fast method, with the time for recording spectra determined by the chosen resolution and the number of repetitions. Depending on the selected method, the sensitivity can be very high, allowing small amounts of IR-active groups to be detected.

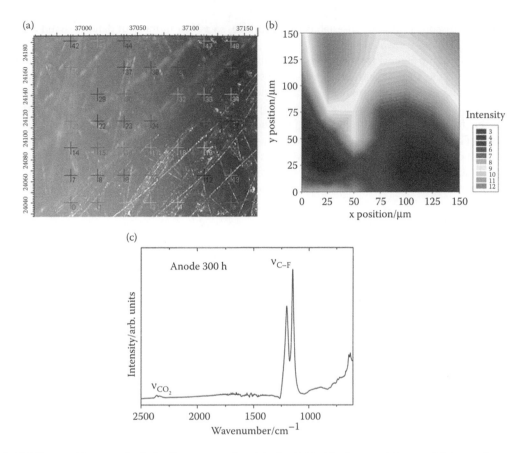

FIGURE 17.8 Mapping of a fuel cell component (anode) after 300 h: (a) microscopic image with inserted measuring points (150 ×150 μm; 49 points); (b) two-dimensional diagram of the C–F vibrations of all 49 measuring points; (c) single spectrum of one measuring point.

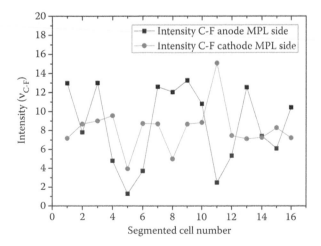

FIGURE 17.9 Result of a peak fit analysis of C–F vibrational modes of segmented cell (16 segments) after an operating time of 250 h (80°C, 100% relative humidity on both sides). The intensities of the C–F absorption bands of the anode and cathode side of the MPL are compared and show different degradation.

FTIR spectroscopy is also a well-established technique, so a high number of databases and a great deal of literature data are available.

The modular setup of most spectrometers allows the insertion of several mirror units for the different methods (e.g., ATR, DRIFT) with only one basic system. The use of an FTIR microscope enables high local resolution as well as measurements of images derived by mapping (solid) samples. *In situ* analysis (inside the cells) is possible but expensive because a special type of fiber optics is needed. FTIR methods are suitable for the investigation of solid, liquid, and gaseous samples, provided they fulfill the basic condition of being IR-active, meaning they possess a changing dipole moment.

The main disadvantage/limitation is that when water is present in samples, its very broad absorption bands (vibrational bands for solid/liquid water and rotational bands for gaseous water) overlap with the other vibrational bands. This is a very strong limitation for the *in situ* characterization of fuel cells with IR spectroscopy.

17.6 Conclusions and Outlook

FTIR spectroscopy is a powerful tool for the investigation of fuel cell components containing organic compounds, such as the polymer electrolyte membrane, the reaction layer, and the gas diffusion layer (including the microporous layer). The method offers many techniques to deal with various challenges—that is, the ATR technique for surface analysis, the DRIFT technique for powders and rough surfaces, and the transmission technique for bulk material analysis. Selection of the right method depends mainly on the optical properties of the material to be studied. For investigating degradation processes in the carbon-containing components of fuel cells (GDL, reaction layer), ATR is the method of choice because the IR light will be absorbed in the carbon structure and consequently no transmittance will occur. The main advantages of IR spectroscopy—nondestructiveness, easy sample preparation, high sensitivity, and short recording times—make it a standard tool for fuel cell analysis, especially because a high number of experiments can be performed in a short time. Additionally, time-resolved and/or locally resolved experiments can be carried out with a suitable experimental setup.

The development of *in situ* FTIR measurements should continue, despite water being a problematic factor. Another option is to couple the above-mentioned methods with time-resolved measurements.

References

Alpert, N. L. 1973. Theory and practice of Infrared spectroscopy (2nd edn.), New York, NY: Plenum Pub. Co.

Drexhage, K. H. 1970. Monomolecular layers and light. *Sci. Am.* 222: 108–118.

Fahrenfort, J. 1959/1962. *Adv. Mol. Spectrosc. Proc. Int. Meet. 4th 1959* 2: 701.

Goos, H. and Hänchen, H. 1947. Ein neuer und fundamentaler versuch zur totalreflexion. *Ann. Phys.* 1: 333–346.

Goos, H. and Lindberg-Hänchen, H. 1949. Neumessung des strahlversetzungseffektes bei totalreflexion. *Ann. Phys.* 5: 251–252.

Gülzow, E., Weißhaar, S., Reißner, R. et al. 2003. Fully automatic test facilities for the characterisation of DMFC and PEFC MEAs. *J. Power Sources* 118: 405–410.

Günzler, H. and Gremlich, H.-U. 2002. *IR Spectroscopy–An Introduction*. Weinheim: Wiley-VCH.

Günzler, H. and Gremlich, H.-U. 2003. *IR-Spektroskopie* (4th edn.). Weinheim: Wiley-VCH.

Harbeck, S. 2005. Characterisation and Functionality of SnO_2 Gas Sensor Using Vibrational Spectroscopy. PhD Thesis, University of Tübingen, Germany.

Harrick, N. J. 1979. *Internal Reflection Spectroscopy*. Ossining, NY: Harrick Scientific Corporation.

Haug, A. 2004. Charakterisierung von alkyl-substituierten phthalocyanin-dünnschichtsystemen mittels FTIR-ATR-spektroskopie. Diploma Thesis, University of Tübingen, Germany.

Haug, A. 2009. Oberflächenanalytische untersuchungen von dünnen cytosin-, thymin-, uracil- und hydrogen-silsesquioxan-filmen. PhD Thesis, University of Tübingen, Germany.

Haug, A., Schweizer, S., Latteyer, F. et al. 2008. Thin-film properties of DNA and RNA bases: A combined experimental and theoretical study. *Chem. Phys. Chem.* 9: 740–747.

Hiesgen, R., Wehl, I., Helmly, S. et al. 2011. AFM and IR analysis of aging processes of fuel cell components. *J. Electroanal. Chem.*, submitted.

Kramer, M. H.-M. 1997. FT-IR-ATR-spektroskopische untersuchung zur morphologie dünner organischer schichten und ihrer wechselwirkung mit analyten. PhD thesis, University of Tübingen, Germany.

Liang, Z., Chen, W., Liu, J. et al. 2004. FT-IR study of the microstructure of Nafion® membrane. *J. Membr. Sci.* 233: 39–44.

Newton, I. 1952. *Opticks.* New York, NY: Dover.

Salzer, R. and Siesler, H. W. 2009. *Infrared and Raman Spectroscopic Imaging.* Weinheim: Wiley-VCH.

Schulze, M., Gülzow, E., Schönbauer, St. et al. 2007. Segmented cells as tool for development of fuel cells and error prevention/prediagnostic in fuel cell stacks. *J. Power Sources* 173: 19–27.

Stuart, B. H. 2004. *Infrared Spectroscopy: Fundamentals and Applications (Analytical Techniques in the Sciences).* New York, NY: John Wiley & Sons.

Wartewig, S. 2003. *IR and Raman Spectroscopy: Fundamental Processing (Spectroscopic Techniques: An Interactive Course).* Weinheim: Wiley-VCH.

Wendland, W. W. and Hecht, H. G. 1966. *Reflectance Spectroscopy.* New York, NY: John Wiley & Sons.

18

X-Ray Photoelectron Spectroscopy

Mathias Schulze
Institute of Technical Thermodynamics

Andrea Haug
Institute of Technical Thermodynamics

K. Andreas Friedrich
Institute of Technical Thermodynamics

18.1 Introduction

Low-temperature fuel cells consist of various components with different functions, determined by their bulk and surface properties. Important bulk properties are, for example, the porosity, hardness, and conductivity of the GDL material, which is important for transport processes in the cell, the ionic conductivity of the membrane material, the leakage rate of the membrane material, the conductivity and tightness of the bipolar plates, and the elasticity of the sealing materials. Surface properties are also very important. The chemical composition of a few atomic layers determines the catalytic properties of catalysts and the surface composition of the GDL material—for example, the ratio of PTFE in the covered and uncovered areas of the carbon structure—determines the GDL's hydrophobic/hydrophilic behavior. This means that the ratio of carbon to PTFE in the GDL in the bulk is not relevant. A further example of an important surface property is the surface composition of the bipolar plates, which has a significant influence on the degradation or corrosion of this component and on the contact resistance.

To investigate fuel cell components, characterization methods for the different properties are needed—methods for investigating surfaces and for characterizing bulk or volume properties. Especially in research on degradation phenomena, study of alterations in surface compositions is needed to understand changes in catalytic behavior, changes in water management induced by a decrease in the hydrophobicity of the GDL and reaction layer, to investigate poisoning of the catalyst, or other active surfaces by decomposition products or external impurities. Many surface analytical methods are available, each with different advantages and disadvantages. X-ray photoelectron spectroscopy (XPS) is a surface science method suitable for many different investigations of fuel cell components.

18.2 Methodology

To determine the chemical composition of a surface, three principle approaches can be used. One is to investigate the atoms on the surface by using the inner electrons to determine the surface elements. The second method is to remove atoms from the surface and investigate them, for example, by mass spectrometry. The third approach is to investigate the interactions between the atoms on the surface; for that type of investigation, different methods can be used, such as vibration spectroscopies like infra-red (IR) (see Chapter 17) or Raman spectroscopy, measurement of the band structure as in ultraviolet photoelectron spectroscopy, and scanning tunneling spectroscopy.

18.2.1 Principle of XPS

For the direct determination of surface elements, measurement of the interactions between the excitation radiation and the inner electrons is a common method. Four different methods use the inner electrons for spectroscopy in the following way. Induced by an excitation radiation, an electron will be removed from an inner energy level and the radiation leaving the surface will be measured.

For XPS a characteristic monochromatic x-ray radiation is used. The complete energy of the x-ray quant is transferred to an electron in an inner energy level. This electron leaves the surface, and the kinetic energy of this photoelectron is the energy of the x-ray quant, $h\nu$, minus the energy required to leave the surface.

$$E_{kin} = h\nu - E_{bind} \tag{18.1}$$

The energy needed to leave the surface is mainly determined by the energy level in the atom from which the electron is removed. An additional energy shift is given by the chemical environment of the atom, meaning its binding state; this additional shift is called chemical shift.

The unoccupied inner level induced by the photoelectron or induced by a higher energy electron beam can be filled by an electron from a higher energy level. During this intra-atomic electron transfer the energy difference between the two electron configurations will be emitted by an x-ray quant or an electron (typically from the same higher energy level from which the electron in the inner energy level will be transferred). This kind of electron—called an Auger electron, as well as x-ray radiation—when leaving the surface also has a characteristic energy, which allows one to determine the element.

If x-ray radiation is used for excitation as well as for spectroscopy the method is called x-ray fluorescence spectroscopy (XRF). Due to the large penetration depth of x-ray radiation the surface sensitivity of this method is not very high.

In scanning electron microscopes, energy-dispersive x-ray spectroscopy (EDX) is frequently employed to determine surface composition; here, electrons are used for excitation and x-ray radiation for spectroscopy. The information depth is determined by the penetration depth of the electrons and therefore is related to the electron energy. Because of its high-energy requirement, this method is not as surface sensitive as Auger electron spectroscopy or XPS but is significantly more sensitive than XRF.

In Auger electron spectroscopy, an electron beam is used for excitation and the emitted electrons are analyzed. In this method electrons are used for both excitation and spectroscopy, the penetration depth of the electrons is not high and consequently the surface sensitivity is very high comparable to it of XPS, with an information depth of only a few nanometers.

The choice of the method determines not only the information depth but also the kind of information obtained. All of the methods yield information about element concentrations but only those methods using electron spectroscopy allow one to derive information about chemical state, because the energy resolution with detectors commonly used in electron spectroscopy is a few magnitudes better than the resolution obtainable using x-ray radiation analyzers.

XPS yields more information about chemical states than Auger electron spectroscopy does because evaluation of the XP spectra is easier. The detected energy of the electrons is the difference between the states before and after electron emission; in the case of Auger electron spectroscopy, the evaluation is more complicated because the initial state is an excited (ionic) one. An advantage of Auger electron spectroscopy is that very high local resolution is possible, limited only by the focus of the electron source.

XPS yields information not only about surface elements and their binding configuration but also about the elements' concentrations, by analyzing the peak intensities (peak area). With known sensitivity factors the element concentrations can easily be calculated, making a quantitative analysis of the surface composition possible.

In fuel cell research, especially the investigation of degradation processes, high surface sensitivity and information about chemical states is vital, so XPS is an important method for such research.

Significantly more detail about these methods can be found in any surface analytic textbook, such as (Ibach, 1977; Ertl and Küppers, 1985; Briggs and Seah, 1990; Henzler and Göpel, 1991; Hüfner, 1996).

18.2.2 Instrumentation

An XPS system consists of an x-ray source and an electron analyzer, as well as numerous supporting systems and components.

18.2.2.1 X-ray Sources

In the simplest x-ray source, electrons emitted from a cathode will accelerate onto an anode; x-ray radiation is emitted on this target. Secondary electrons will be rejected by a thin metallic foil of a light metal (typically aluminum). The emitted characteristic x-ray radiation depends on what material is used for the anode, magnesium and aluminum being very common. These kinds of x-ray sources emit their radiation in a wide angle range; therefore, the x-ray flux on the surface is normally not very high. To increase the local x-ray intensity the x-ray source can be combined with a monochromator consisting of x-ray mirrors (grids with a suitable lattice constant). The x-ray radiation will be monochromatized from a typical energy full width at half maximum (FWHM) of 0.8–0.2 eV, and this radiation can also be focused on the surface.

Electron synchrotrons can also be used as x-ray sources for XPS experiments; however, while the radiation is very intense and monochromatic, the experimental effort is exceedingly high. These XPS systems must be fitted to a synchrotron and thus the experiments can only be performed a few days per year.

One problem with x-ray radiation is that it can disintegrate organic molecules, such as polymers.

18.2.2.2 Electron Energy Analyzer

The second important component of an XPS system is the electron energy analyzer. Various types of electron energy analyzers can be used (Ibach, 1977; Ertl and Küppers, 1985; Henzler and Göpel, 1991). For XPS a hemispherical analyzer is the most common type, but cylindrical mirror analyzers are also used. The advantage of the hemispherical analyzer is that the distance between the optimal position and the real position does not influence the measured energy of the photoelectrons. In contrast, if a cylindrical mirror analyzer is used the sample must be accurately positioned. Using electron optics, the local position for analysis as well as the local resolution can be changed. Depending on the electron optics and the analyzer, a local resolution of a few micrometers is possible. By scanning the surface maps of element distribution and the chemical binding states of the elements can be measured. With some analyzers it is possible to image the surface energy selectively on the electron detector, allowing the direct measurement of element and binding states.

The other analyzer types are not commonly used in XPS but can be integrated into more exotic XPS systems.

Some of the analyzers (e.g., hemispherical) have an electron optics in front of the analyzer entrance. With the electron optics and additional mechanical apertures the local resolution can be changed from large spot analysis (in which the diameter of the measured area is a few mm) to small spot analysis (a diameter of a few μm).

A method of studying inhomogeneities by XPS is to scan the surface or to measure XPS image, which is possible with some modern XPS systems. For the imaging technique, the surface image will be created on an detector like in an electron microscope (not an scanning electron microscope) with electrons of a selected energy.

18.2.2.3 Charge Compensation

During XPS spectroscopy measurements photoelectrons as well as Auger electrons are leaving the surface. In electrically isolated samples this loss of electrons leads to a charging of the surface. As a result of this electrical charging of the sample, the spectra will be shifted. To eliminate the charging effect the sample surfaces must be electrically neutralized. For this purpose the sample surface can be flooded with low-energy electrons (only a few eV). If the samples become charged, electrons from the low-energy electron gun in front of the surface are used to compensate for the photoelectron current.

In some situations this type of charge compensation is insufficient, in which case an additional source of positively charged particles needs to be operated together with the electron gun. An ion source that creates noble gas ions (mostly argon) with low energy is used as the positive charge source.

18.2.2.4 Ion Gun

Frequently, XPS systems are equipped with ion sources that are used to etch the surface. With an ion beam the atoms from the first layer can be remove, and the second layer becomes the surface. By sequential performance of XPS measurements and ion etching, surface depth profiles can be recorded. However, ion etching has two main problems: the ion beam can disintegrate the surface composition—for example, polymers can be decomposed, and the oxidization state of oxides can be changed; the second problem is that not all elements will be removed from the surface with the same efficiency, so ion etching can change the surface composition.

18.2.2.5 Vacuum System

Owing to the use of photoelectrons, the low penetration depth of electrons and high surface sensitivity, XPS measurements must be performed under vacuum conditions. Normally, XPS systems are ultrahigh vacuum systems (UHV) operated in a pressure range between 10^{-10} and 10^{-13} Pa. Under UHV conditions the time for the sample surface to become covered by adsorbates can increase—this time is approximately 1 h at a pressure of 10–12 Pa. During the investigation of fuel cell components, the surface is normally covered by an adsorbate layer, which must be taken into account in the evaluation of the spectra.

The need for vacuum conditions necessitates several supporting components. A vacuum system will have the same pumps being used for different pressure ranges—for example, rotating pumps for the fine vacuum range (10^{-5}–10^5 Pa), and turbo pumps or ion-getter pumps for the high or ultrahigh vacuum range. Another component in the UHV system is a method for measuring the vacuum pressure.

Normally, for sample management a sample lock and a manipulator are integrated into the UHV system. The sample lock is used to insert the samples without breaking the vacuum. The samples are then transferred to the manipulator, which allows them to be precisely positioned.

18.2.3 Advantages and Limitations

The various surface analytical methods have individual advantages and disadvantages.

One of its disadvantages is the considerable experimental effort required because of the need for UHV conditions, which means an XPS system consists of many components that are costly and require time-consuming maintenance. In addition, measurements take longer than for methods not requiring UHV conditions, because sample insertion takes time, and the measurement procedure itself is longer than for other methods.

In many cases XPS is a destructive method, especially if the investigated samples contain polymers or organic compounds. X-ray radiation can disintegrate organic molecules, and during depth-profile measurements the sample surface will change for each type of sample. Alteration/destruction of the sample is one of the main disadvantages of XPS.

XPS has a very high surface sensitivity, meaning concentration in the submonolayer range (a few percent of a monolayer) can be detected, so preparation of the sample surface must be performed very carefully to avoid surface contamination. As well, in the investigation of fuel cell components the separation of the individual components from an operated MEA is difficult.

However, high surface sensitivity is the main advantage of XPS in fuel cell research. This sensitivity allows the gathering of relevant data about the surface compositions of the various fuel cell components, which is relevant for understanding many properties that influence fuel cell behavior.

A quantitative analysis of XPS spectra is very easy. The excellent energy resolution of electron energy analyzers allows the measurement of the chemical shift and the determination of the binding state of elements, so XPS yields very exact information on surface composition.

Depending on the XPS system, spectroscopy can be performed with different local resolutions, or mappings of the surface composition can be carried out. This allows the gathering of data on different local scales, which can be used to gain information about inhomogeneities in surface composition.

For rough samples like GDLs quantitative analysis can be problematic, because geometric effects (shadows) can distort the results. In combination with ion etching for depth profiling measurements, shadow effects distort the quantitative analysis, only a qualitative evaluation of these measurements should be applied.

18.3 Applications in Fuel Cell Research

This section includes examples of XPS analysis being used in fuel cell research at the DLR, in studies of the reaction layer and investigations of catalysts and of hydrophobic agents in the GDL.

18.3.1 Characterization of the Reaction Layer

The first example (Gülzow et al., 2000) presents depth profiles measured for two different electrodes. In both cases the reaction layers are prepared on the GDL. The depth profiles are measured by alternately recording XP spectra and ion etching the surface (Ar$^+$ ions, 2.5 keV). The ion-etching time between the measurement of two XP spectra is increased every five cycles by a factor of 3 to record the chemical composition up to a depth of a few 10s of nanometers.

At the top of Figure 18.1 the depth-profile of a commercial electrode is shown, in which the reaction layer is prepared from a suspension containing the catalyst, hydrophobic agents, and additives. The high concentration of fluorine and carbon on the surface and the low concentration of platinum clearly show that the surface is covered by a polymer film (PTFE or Nafion). This polymer film is partially removed and decomposed during the depth-profile measurement and as a consequence the Pt signal increases.

At the bottom of Figure 18.1 is shown the depth profile of an electrode prepared by a dry preparation technique at the DLR (Bevers et al., 1998; Gülzow et al., 1999, 2000, 2002, 2003; Kaz and Gülzow, 2002; Schulze et al., 2002). A mixture consisting of carbon-supported Pt catalyst, PTFE powder, and Nafion powder is sprayed with a nitrogen flow onto the GDL surface. In the depth profile of this electrode the concentrations of the elements do not alternate as significantly as in the depth profile of the commercial electrode. The platinum concentration remains nearly constant, indicating that this electrode does not have a polymer film covering the surface. The increase in the carbon concentration and the decrease in the fluorine signal are induced by decomposition of the polymer and by the different ion etching efficiencies for the two elements; fluorine is significantly better removed from the surface than carbon.

Owing to the different structures of these two electrodes, their behavior also differs. To investigate the activity of both electrodes in the initial state, they are analyzed using cyclic voltammetry (CV) in a half-cell configuration with 0.5 M sulfuric acid as the electrolyte at room temperature (Rheaume et al., 1998). The CV measurements are performed between 0 and 1.5 V with a sweep rate of 100 m V s^{-1}.

Figure 18.2 shows the peak intensities of the oxidation and reduction peaks during the CV measurements; on top the commercial electrode measurements and on the bottom the dry prepared electrode measurements is shown. In the former the values of both currents become larger throughout the cycling process in a nearly linear fashion, indicating that the transferred charge steadily increases and

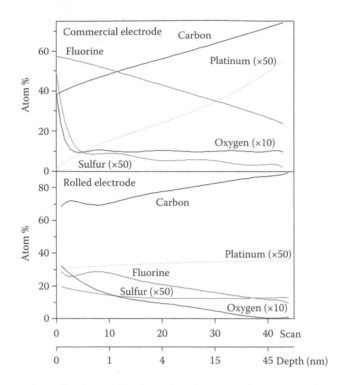

FIGURE 18.1 Top: Depth profile of an E-TEK electrode with Pt/C catalyst as received. Bottom: Depth profile of an electrode prepared at the DLR with Pt/C catalyst. (Reprinted from *J. Power Sources*, 86, Gülzow, E. et al., Dry layer preparation and characterisation of polymer electrolyte fuel cell components, 352–362, Copyright (2000), with permission from Elsevier.)

correspondingly the active catalytic surface area grows. In contrast, the dry prepared electrode peak heights do not change significantly, meaning the electrochemical active surface area does not change during the CV experiment.

Due to the fact that the polymer film covers the surface of the commercial electrode the platinum catalyst is not active at the beginning of the CV experiment and consequently the measured peak heights are very small—1% of the intensity measured for the dry-prepared electrode. The increases in the peaks indicated that the electrochemical active surface area had grown and thus the platinum surface became increasingly accessible. The surface of the commercial electrode is investigated by XPS before and after the CV experiment. Figure 18.3 shows both XP spectra, with on top before the experiment and on the bottom after the experiment. Comparing the spectra one can see that the platinum signal has increased significantly, meaning the platinum concentration on the surface increased.

18.3.2 Investigation of Catalysts

XPS is used to investigate degradation processes in low-temperature fuel cells since the 1980s. The following example presents alteration in the chemical composition of the catalyst layer in an alkaline fuel cell electrode (anode). Figure 18.4 reveals the depth profiles of an activated electrode and an electrode that is operated for 1344 h (Schulze and Gülzow, 2004). The concentration of the nickel catalyst decreases, as does that of fluorine, indicating a decrease in the concentration of PTFE, which is used in these electrodes as an organic binder and to adjust their hydrophobicity. In addition to the decrease in nickel concentration the nickel catalyst's oxidation state changes, which is depicted in Figure 18.5. After activation of the electrode the nickel is mainly in the metallic state, whereas after fuel cell operation it is in an oxidized state.

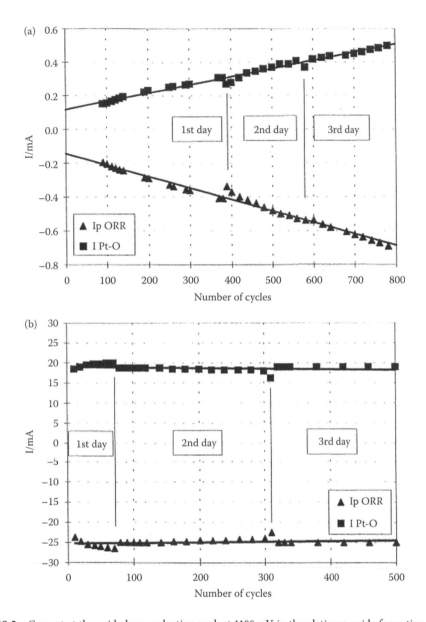

FIGURE 18.2 Current at the oxide-layer reduction peak at 1180 mV in the platinum oxide formation area versus number of cycles for Pt/C electrodes subject to cyclic voltammetry continuously over a three-day period, between 0 and 1.5 V at 100 mV s^{-1} in 0.5 M H$_2$SO$_4$ with ongoing Ar bubbling and at room temperature. Top: E-TEK 20 wt% Pt/C electrode. Bottom: DLR rolled Pt/C electrode. (Reprinted from *J. Power Sources*, 76, Rheaume, J. M., Müller, B., and Schulze, M., XPS analysis of carbon-supported platinum electrodes and characterization of CO oxidation on PEM fuel cell anodes by electrochemical half cell methods, 60–68, Copyright (1998), with permission from Elsevier.)

PtRu is used as an anode catalyst in direct methanol fuel cells (DMFCs) or PEMFCs that will be operated with hydrogen-rich fuels containing some CO.

During fuel cell operation the composition of the anode catalyst changes (Schulze et al., 2007a). In the XP spectra of the new and the operated DMFC anode, in the range of the C 1s and Ru 3d spectra (Figure 18.6), a decrease in the Ru signal can be observed whereas the Pt signal is unchanged (not shown here). The change in the catalyst surface results in a change in the catalytic behavior, which was investigated in a temperature-programmed reduction (TPR) experiment (shown in Figure 18.7). In

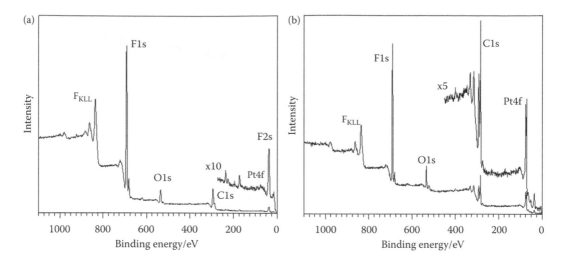

FIGURE 18.3 XPS measurement of intensity versus binding energy of E-TEK electrodes. (a) As received. (b) After 800 cycles from 0 to 1.5 V at 100 mV s^{-1} in 0.5 M H$_2$SO$_4$ with ongoing Ar bubbling. (Reprinted from *J. Power Sources*, 76, Rheaume, J. M., Müller, B., and Schulze, M., XPS analysis of carbon-supported platinum electrodes and characterization of CO oxidation on PEM fuel cell anodes by electrochemical half cell methods, 60–68, Copyright (1998), with permission from Elsevier.)

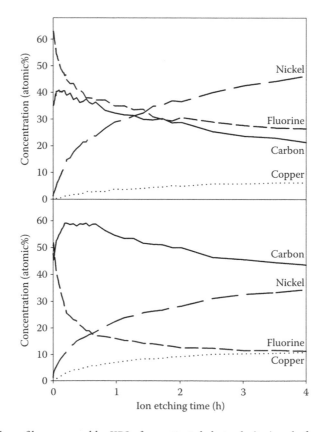

FIGURE 18.4 Depth profiles measured by XPS of an activated electrode (top) and of an electrode operated for 1344 h at a current density of 100 mA cm^{-2} (bottom). (Reprinted from *J. Power Sources*, 127, Schulze, M. and Gülzow, E., Degradation of nickel anodes in alkaline fuel cells, 252–263, Copyright (2004), with permission from Elsevier.)

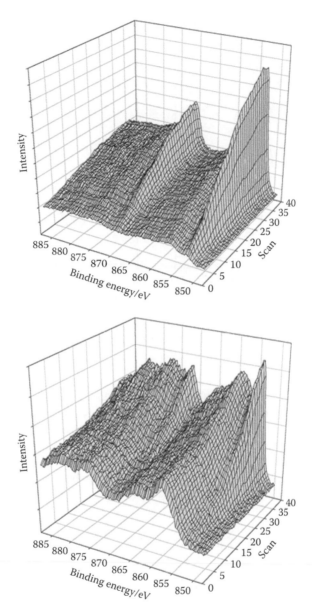

FIGURE 18.5 Ni 2p spectra recorded during XPS depth profile measurements of an activated electrode (top) and after long-term operation (approximately 5000 h) at 50 mA cm^{-2} (bottom). (Reprinted from *J. Power Sources*, 127, Schulze, M. and Gülzow, E., Degradation of nickel anodes in alkaline fuel cells, 252–263, Copyright (2004), with permission from Elsevier.)

the TPR experiment the characteristics of the used DMFC MEA are similar to those of an MEA consisting of electrodes with a platinum catalyst, but are significantly different from those of the unused DMFC MEA.

18.3.3 Investigation of Polymers

The concentration of the hydrophobic agents on a surface determines its hydrophobic/hydrophilic character. PTFE is typically used as a hydrophobic agent in low-temperature fuel cells. Using the dry

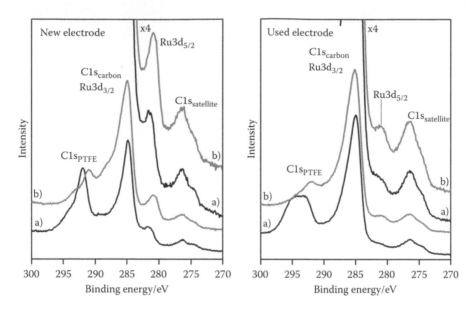

FIGURE 18.6 XP spectra of a new and a used DMFC anode, (a) before ion etching, and (b) after ion etching. (Reproduced from Schulze, M., Reissner, R., and Christenn, C. 2007a. *ECS Trans.* 5: 95–106. With permission from The Electrochemical Society.)

preparation technique developed at the DLR (Gülzow et al., 2000; Schulze et al., 2002), GDLs with various mixtures of PTFE and carbon black are prepared, and their hydrophobic/hydrophilic behavior and surface composition are investigated. To change their surface composition the surfaces of the GDLs are subjected to mechanical stress by touching the surface with a lint-free paper cloth. The hydrophobic character is determined by wetting experiments and the surface composition by XPS (Figure 18.8). With a concentration of carbon black over 80 wt% (PTFE concentration below 20 wt%) the surface is hydrophilic, but with a surface concentration of carbon black below 80 wt% (PTFE concentration over 20 wt%)

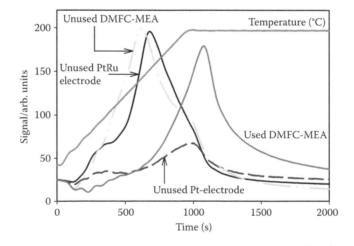

FIGURE 18.7 Temperature-programmed reduction measurement of a PtRu electrode, a Pt electrode, an unused MEA for a DMFC, and a MEA operated 85 h in a DMFC. (Reproduced from Schulze, M., Reissner, R., and Christenn, C. 2007a. *ECS Trans.* 5: 95–106. With permission from The Electrochemical Society.)

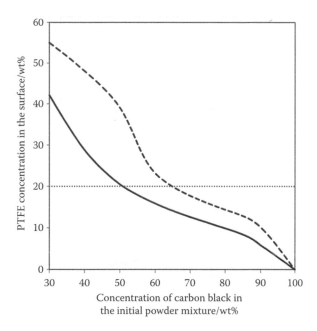

FIGURE 18.8 PTFE concentration in dry prepared electrodes before (solid line) and after (dashed line) mechanical stressing of the surface.

the surface is hydrophobic. This investigation shows that XPS is a suitable method for studying the hydrophilic/hydrophobic character of a GDL surface.

Investigation of degradation processes has also been carried out at the DLR using a combination of various methods—XPS, scanning electron microscopy, and electrochemical impedance spectroscopy—leading to the identification of "reversible" and "irreversible" degradation effects (Schulze et al., 2007b). "Reversible" degradation means performance loss that can be compensated for by adapting the operating conditions, whereas "irreversible" degradation cannot be compensated for. "Reversible" degradation is related to an alteration in water management, induced by a loss of hydrophobicity. Figures 18.9 and 18.10 show details of the C 1s spectra of new and used reaction layers. The spectra in Figure 18.9 are obtained from the reaction layers on a GDL (catalyst-coated backing = CCB, E-TEK electrode) and the spectra in Figure 18.10 from the reaction layers of a catalyst-coated membrane (CCM, IonPower). For both types of electrodes a significant reduction in the PTFE-related C 1s signal is observable on the anode side and, less significantly, on the cathode side. From evaluation of the electrochemical impedance spectra described in Schulze et al. (2007b) it is known that "reversible" degradation occurs mainly on the anode side, which is consistent with the XPS results.

18.3.4 Poisoning of the Surface by Sealing Decomposition

The last example in this chapter is the investigation of sealing degradation. Silicone sealings are frequently used in low-temperature fuel cells. In the spectra of a used electrode, silicon from the silicone sealing was detectede by XPS (Schulze et al., 2004), as displayed in Figure 18.11. Decomposition products can thus lead to poisoning of the catalyst surface.

18.4 Summary

XPS is a very suitable method for investigating the surface compositions of fuel cell components. Detection of surface composition using XPS allows the characterization of surface properties such as hydrophobicity and consequently also enables some predictions of surface behavior in fuel cells.

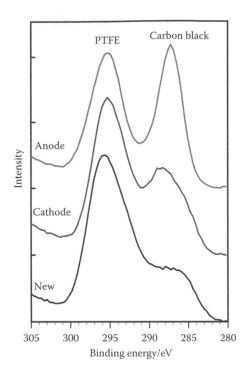

FIGURE 18.9 C 1s spectra of E-TEK electrodes. From bottom to top: as received, used as cathode, and used as anode. (Reprinted from *J. Power Sources*, 86, Gülzow, E. et al., Dry layer preparation and characterisation of polymer electrolyte fuel cell components, 352–362, Copyright (2000), with permission from Elsevier.)

XPS investigation of fuel cell component degradation allows one to detect changes in the surface composition and thereby determine some degradation mechanisms. In many cases, alterations in the surface composition can be observed by analysis of the fuel cell components before degradation has affected the fuel cell's performance and behavior. This is a considerable advantage because lifetime tests can be accelerated using such analytic *ex situ* methods.

Quantitative analysis of the surface composition of fuel cell components is sometimes distorted, but qualitative analysis is a powerful tool for investigating degradation processes. XPS does not allow

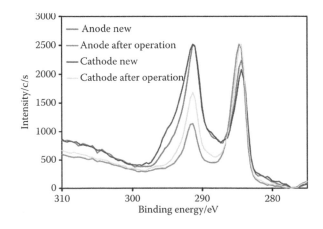

FIGURE 18.10   C 1s spectra of reaction layers of an IonPower CCM, as received and after fuel cell operation.

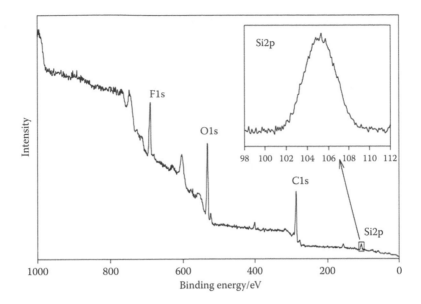

FIGURE 18.11   XP spectrum of a used electrode (the Si 2p region is enlarged). (Reproduced from *J. Power Sources*, 127, Schulze, M., Knöri, T., and Schneider, A., Degradation of sealings for PEFC test cells during fuel cell operation, 222–229, Copyright (2004), with permission from Elsevier.)

determination of the quantitative effects of degradation processes on fuel cell performance, but can explain and predict changes in fuel cell behavior that are related to surface properties.

References

Bevers, D., Wagner, N., and von Bradke, M. 1998. Innovative production procedure for low cost PEFC electrodes and electrode/membrane structures. *Int. J. Hydrogen Energy* 23: 57–63.

Briggs, D. and Seah, M. P. 1990. *Practical Surface Analysis—Auger and X-ray Photoelectron Spectroscopy* (Vol. 1). Chichester: John Wiley & Sons.

Ertl, G. and Küppers, J. 1985. *Low Energy Electrons and Surface Chemistry.* Weinheim: VCH.

Gülzow, E., Kaz, T., Reißner, R. et al. 2002. Study of membrane-electrode assemblies for direct-methanol fuel cells. *J. Power Sources* 105: 261 266.

Gülzow, E., Reisser, R., Weisshaar, S. et al. 2003. Progress in DMFC development using the dry spraying preparation technique. *Fuel Cells* 3: 48–51.

Gülzow, E., Schulze, M., Wagner, N. et al. 1999. New dry preparation technique of membrane-electrode-assemblies for polymer electrolyte fuel cells. *Fuel Cells Bull.* 15: 8–12.

Gülzow, E., Schulze, M., Wagner, N. et al. 2000. Dry layer preparation and characterisation of polymer electrolyte fuel cell components. *J. Power Sources* 86: 352–362.

Henzler, M. and Göpel, W. 1991. *Oberflächenphysik des Festkörpers.* Stuttgart: Teubner.

Hüfner, S. 1996. *Photoelectron Spectroscopy, Principles and Applications. Springer Series in Solid-State Sciences 82.* Berlin: Springer.

Ibach, H. 1977. *Electron Spectroscopy for Surface Analysis, Topics in Current Physics 4.* Berlin: Springer.

Kaz, T. and Gülzow, E. 2002. New results of PEFC electrodes produced by the DLR dry preparation technique. *J. Power Sources* 106: 122–125.

Rheaume, J. M., Müller, B., and Schulze, M. 1998. XPS analysis of carbon-supported platinum electrodes and characterization of CO oxidation on PEM fuel cell anodes by electrochemical half cell methods. *J. Power Sources* 76: 60–68.

Schulze, M. and Gülzow, E. 2004. Degradation of nickel anodes in alkaline fuel cells. *J. Power Sources* 127: 252–263.

Schulze, M., Kaz, T., and Lorenz, M. 2002. XPS study of electrodes formed from a mixture of carbon black and PTFE powder. *Surf. Interface Anal.* 34: 646–651.

Schulze, M., Knöri, T., and Schneider, A. 2004. Degradation of sealings for PEFC test cells during fuel cell operation. *J. Power Sources* 127: 222–229.

Schulze, M., Reissner, R., and Christenn, C. 2007a. Surface science study on the stability of various catalyst materials for DMFC. *ECS Trans.* 5: 95–106.

Schulze, M., Wagner, N., and Kaz, T. 2007b. Combined electrochemical and surface analysis investigation of degradation processes in polymer electrolyte membrane fuel cells. *Electrochim. Acta* 52: 2328–2336.

19

Atomic Force Microscopy

Renate Hiesgen
*University of Applied
Sciences Esslingen*

**K. Andreas
Friedrich**
*Institute of Technical
Thermodynamics*

19.1 Introduction

In a proton exchange membrane (PEM) fuel cell the function of the different components is strongly based on the nanoscale properties of the materials:

- Electrodes with a complex porous mixture of ionomer, Pt nanoparticles, and graphite
- Electrolyte membranes with a phase separation into hydrophilic and hydrophobic parts in the nanometer range providing an ionic conducting network for the current
- Corrosion on metallic bipolar plates
- Highly porous gas diffusion layers (GDLs) with functional hydrophilic/hydrophobic pores providing gas-, water-, and current transport

Although atomic force microscopy (AFM) is an *ex situ* technique, the measuring conditions can be chosen rather close to operating conditions in a PEM fuel cell. The flexibility in the analyzed areas allows an overview of the sample as well as a highly resolved measurement on the nanometer scale. The signals recorded by AFM are on the one hand based on force interaction of the AFM tip with the sample surface, thereby delivering, that is, local friction, adhesion, stiffness, and energy loss. On the other hand, electrical properties such as electronic conductivity, ionic conductivity as well as surface potential, electrostatic force or reactivity can be retrieved. The AFM works in ambient air and in a humid or liquid environment, including an electrochemical cell. Hydrophilic and hydrophobic surface properties are especially important for water management in a PEM fuel cell and degradation processes can be followed by a comparison of samples in a fresh state and after operation.

In this chapter, the basic principles and hardware related to AFM are described, as well as the different advanced AFM techniques and the available material properties. Examples of measurements on different components of a PEM fuel cell are discussed. A literature review of related work is also provided.

19.1.1 Advantages of AFM

- No restriction of materials
- Measurements possible in air, liquid, or vacuum
- Resolution better than 1 nm in all dimensions
- Three-dimensional measurement of surface profiles and roughness
- Analysis of additional surface properties, that is, friction forces, adhesion forces, hardness, dissipation energy, conductivity, surface potential, electric fields, magnetic field, chemical composition

19.1.2 Disadvantages of AFM

- The scanned area typically does not exceed 200 μm × 200 μm, the maximum height is in the order of micrometers, statistical analysis of several different areas is often required.
- Image quality strongly depends on probe geometry (as in all probe microscopy techniques).
- The scanning speed is typically slow compared to scanning electron microscopy (SEM).

19.2 AFM Imaging Principle

19.2.1 Experimental AFM Setup

The main part of an AFM setup is the sharp AFM tip mounted onto a flexible cantilever. This system is comparable in its properties to a "mass on a spring system" and is sensitive to small forces acting on the tip-cantilever-system. In "contact-mode" a resulting bending of the cantilever is sensed during the mechanical scanning of the surface by the tip. When the tip is brought into proximity of a sample surface, the forces between the tip and sample lead to a deflection of the cantilever according to Hooke's law. In "noncontact-mode" a forced oscillation of the tip–cantilever system is damped resulting in an amplitude and phase shift of the oscillation. Depending on the situation, forces that are measured in AFM include mechanical contact forces, van der Waals forces, capillary forces, chemical bonding, electrostatic forces, magnetic forces, Casimir forces, solvation forces, and so on. Using a conductive AFM tip, the system can be used as a nanometric electrical sensor for current and voltage; advanced systems use lock-in techniques for getting more information of the sample properties.

Typically, the cantilever deflection is measured optically by the deflection of a laser diode beam from a metallic mirror coating on the back side of the AFM cantilever. The position change of the reflected beam is sensed by a split (two- or fourfold) photo detector diode. The movement of the reflected laser beam on a split photodiode is dependent on the bending of the cantilever. By deflection from the cantilever, the reflected laser beam changes its position on the photo diode array proportional to the deflection, but with a magnification dependent on the geometric proportions. The position of the laser spot is sensed by the voltage difference between the split photodiode parts; with a fourfold diode array the horizontal as well as the vertical movement can be detected (Figure 19.1). Traditionally, the sample is mounted on a piezoelectric tube that can move the sample in the z direction to maintain a constant force or distance, and the x and y directions to scan the sample. Alternatively, a "tripod" configuration of three piezo crystals may be employed, with each responsible for scanning in the x, y, and z direction. The resulting map of the area represents, for example, the topography or another property of the sample.

For measurements in liquid or electrolyte solution, a fluid cell containing the sample is needed which may be equipped with electrodes for potential control. The whole system is often positioned in a

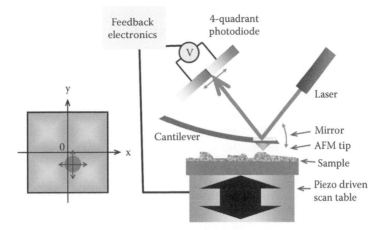

FIGURE 19.1 Optical measuring system with four-quadrant diode (left), laser diode (right), and piezo trans-former for *z*-movement of sample.

chamber for shielding the instrument from noise and other annoying influences and, depending on the analysis, for control of the ambient conditions such as surrounding gas atmosphere, temperature, or humidity. The measured data are stored in a computer system which also controls the function of the instrument. By the software, the display of the measured data as image or curve, and the further evalu-ation of the data, is possible. A drawing of a typical setup is given in Figure 19.2. Often it is necessary to isolate the instrument from oscillation of the surrounding building by, for example, positioning the instrument on a large mass or onto an air-damped table or by hanging it on springs.

Some instruments operate in ultra high vacuum (UHV) or in a glove box at ambient pressure but without oxygen and water, or in liquids, that is, in an electrochemical cell. The majority operates in ambient atmosphere or a gas chamber with controlled humidity and temperature. An important part of the measuring system is the computer/software system. All functions of the AFM are controlled by the software; the measured data are digitalized and stored as files in the computer, either as images with the measured signal (height, current) corresponding to the *x*, *y*- coordinates or as spectroscopic data, that is, for example, force values with separation data. Since the image data are digitally stored, a bundle of image-processing tools is normally available to measure, evaluate, or filter the images. For example, the

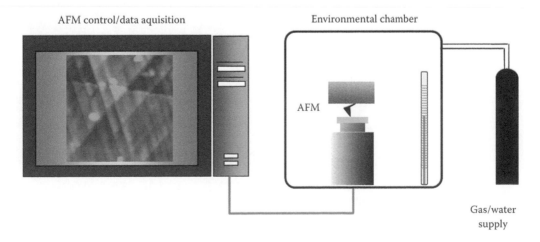

FIGURE 19.2 Typical setup of an AFM system.

measurements of profiles height along a pathway, and so on or the statistical evaluation of the distribution of sample heights or magnitude of current for every pixel delivers valuable information.

19.2.2 Interactions Relevant for AFM Imaging

19.2.2.1 Force–Distance Curve

At a fixed sample position a "force–distance curve" displays the different forces acting as a function of tip–sample distance. These include forces between tip and sample as well as properties of the sample with the tip putting pressure onto the sample.

During the classical procedure of recording a force–distance curve, the probe tip is brought close to a sample surface, indented into it, and afterwards pulled away again. Experimentally, this is done by applying a triangle-wave voltage pattern to the electrodes of the *z*-axis scanner. This causes the scanner to expand and then contract in the vertical direction, generating relative motion between cantilever and sample. The deflection of the free end of the cantilever, and subsequently the force acting on the tip, is measured while the vertical axis scanner extends the cantilever toward the surface and then retracts it again. A typical curve is given in Figure 19.3. This technique can be used to measure the long range attractive or repulsive forces between the probe tip and the sample surface, elucidating local chemical and mechanical properties of the sample surface like adhesion, give information of the subsurface material like elasticity (storage modulus) and, due to material deformation dissipation energy, even measure thickness of adsorbed molecular layers or chemical bond rupture lengths. Another fast approach to measure a force–distance curve is the reconstruction from the higher harmonic oscillations of the AFM tip after applying a pulse using a Fourier transformation. Depending on the application, a specially designed tip with especially strong higher harmonics is needed. With a fast computer system this procedure is possible for every image point and delivers images of the distribution of the different forces/ signals obtained from the curve.

19.2.2.2 Anatomy of a Force–Distance Curve

In the following, the different stages of a force–distance curve are described in more detail and are visualized in Figure 19.4.

a. The cantilever starts in a certain distance over the surface not touching the sample. If there are long-range attractive (or repulsive) forces acting on the cantilever, it will deflect downwards (or upwards) before making contact with the surface.

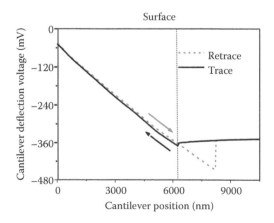

FIGURE 19.3 Force–distance curve on Pt.

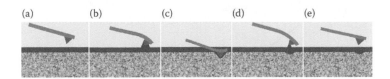

FIGURE 19.4 Bending of AFM cantilever at different stages of extension and retraction: (a) no forces, (b) tip jumps to surface, (c) tip is forced into the surface, (d) tip is bended by adhesion forces during retracting, and (e) no forces.

b. As the probe tip is brought very close to the surface, it may jump into contact if there are sufficient short-range attractive forces.

c. Once the tip is in contact with the surface (contact regime), cantilever deflection will increase as the fixed end of the cantilever is brought closer to the sample. If the cantilever is sufficiently stiff, the probe tip may indent into the surface at this point. In this case, the slope or shape of the contact part of the force curve can provide information about the elasticity of the sample surface. The indentation of the tip may be elastic or give rise to plastic deformation of the sample.

d. After pushing the cantilever down to reach a desired force value, the process is reversed. As the cantilever is withdrawn, adhesion or bonds formed during contact with the surface may cause the cantilever to adhere to the sample some distance past the initial contact point on the approach curve (b).

e. A key measurement of the AFM force curve is the point at which the adhesion is broken and the cantilever comes free from the surface. This can be used to measure the rupture force required to break chemical bonds or the magnitude of adhesion forces.

The force–distance curves can deliver additional information on the sample properties. In Figure 19.3, the case of an ideal elastic sample is drawn. The horizontal line represents the starting line before the tip is in contact with the surface. The slope is linear, the force is increasing proportional to the traveling way of the tip, and there is no difference in the path of the curve during lowering and retracting the tip, besides the final jump caused by an adhesion force. At the end, the tip again reaches the initial (zero) position. The different stages of tip movement, according to Figure 19.4, are also marked in Figure 19.5.

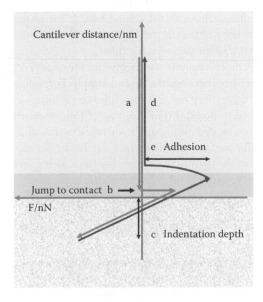

FIGURE 19.5 Force–distance curves with stages according to Figure 19.4.

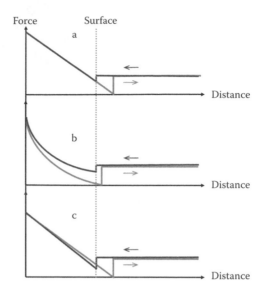

FIGURE 19.6 Main types of force–distance curves: (a) ideal case of elastic material, (b) typical case of visco-elastic material, and (c) typical artifacts due to nonlinear piezo elements.

In Figure 19.6, three main types of force–distance curves are drawn. In Figure 19.6a the ideal case of a totally elastic tip indentation is illustrated. In Figure 19.6b a force–distance curve often measured on different types of samples is shown. Here in the contact regime, the curve is not longer linearly, but has a smaller slope in the beginning and ends with a steeper slope. This is caused by elasto-plastic properties of the material or the rising influence of a harder layer beneath a softer surface, for example, on very thin samples. From the hysteresis between the two curves, the work (dissipation energy) on the sample can be calculated. The peak force recorded at a certain traveling distance gives information on the hardness of the sample at this point, and by comparison of force–distance curves of different sample sites, the hardness of small subsurface volume of the sample is recorded.

In Figure 19.6c, the most often seen artifact is demonstrated. The retracting curve is situated above the curve measured during lowering of the tip, indicating a smaller force during retraction. This artifact is caused by a nonlinear behavior of the piezoelectric tubes.

One of the first uses of force measurements was to improve the quality of AFM images by monitoring and minimizing the attractive forces between the tip and sample. Force measurements were also used to demonstrate similarly reduced capillary forces for samples in vacuum, in reduced humidity environments, or completely immersed in liquid. From mapping of force measurements on a certain sample area, images on the different forces, depending on the surface distance and conditions, that is, distribution of adhesion forces or stiffness of the sample surface, can be displayed.

During elastic indentation of the tip into the sample the force acting on the AFM probe can be measured from the slope of the curve using Hookés law for elastic materials:

$$F = -k\,\Delta x \tag{19.1}$$

where F is the force acting on the tip, k the spring constant, typically given by the deliverer, and Δx the traveling distance of the tip.

From a calibration measurement the sensitivity of the specific cantilever has to be determined in order to convert voltage into distance.

19.2.2.3 AFM Probes

The standard tip–cantilever assembly typically is micro-fabricated from Si or Si_3N_4. The radius of curvature of these tips is about 5–10 nm. For high-resolution imaging tips with a very high aspect ratio are necessary. There are several techniques to micro-fabricate sharper tips. One possibility is sharpening the prefabricated tip with an ion beam or the growth of nanotips by irradiation of the Si tip with an electron beam generating electron polymerization layers from the gas molecules in the electron beam chamber.

For specialized applications, specially fabricated tips are needed. AFM tips can be prepared from functional materials like, that is, for magnetic measurements where tips are made from magnetic material like CoNi alloys. Conductivity measurements need metal (i.e., platinum, gold, or aluminum) coated tips or tips prepared from highly doped silicon or diamond. It is also possible to coat a tip with chemical or biological functional molecules to change the chemical bonding and thereby the adhesion forces between tip and sample or enhance a chemical or biological reaction. A special design is, for example, needed in the so-called "HarmoniX-Mode" (Bruker Corp.) where the very tip is positioned asymmetrically onto the tip holder to give a stronger signal in higher harmonic vibrations. Scanning electron microscope (SEM) images of a standard and a specialized "HarmoniX-Mode"-tip are shown in Figure 19.7.

19.2.2.4 Artifacts

The evaluation and interpretation of the measured data and images needs further analysis since there is a variety of disturbance which may influence the reliability of the data. Artifacts are caused either by the system or by the environment.

19.2.2.4.1 System Inherent Artifacts

A basic problem with all nanoprobe methods using a real probe with certain dimensions is that the measured data do not show the real surface, but the convolution of the geometry of the tip with structures on the sample. This problem becomes serious in some cases where the dimensions of probe and sample structure have comparable size or the latter are even smaller. In this case, the tip is even imaged by the surface features and it is impossible to get a true image of the surface. Due to the convolution with the probe, for example, the diameter of small particles on the surfaces can easily be measured as being too large. This problem becomes even worse with higher surface roughness. Therefore, the tip has to have a high aspect ratio (quotient of width and length), which means it has to be very narrow and long (Figure 19.8). Limits are only set by stability aspects. Height measurements are insensitive to the tip geometry.

AFM images can be affected by hysteresis of the piezoelectric material and cross-talk between the (x, y, z) axes that may require software enhancement and filtering. Such filtering could "flatten" out real topographical features. However, newer AFM use real-time correction software or closed-loop scanners

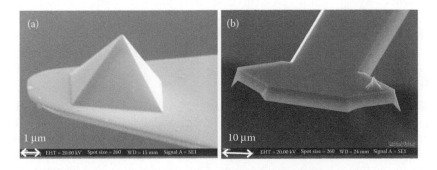

FIGURE 19.7 (a) Standard AFM probe and (b) asymmetrical HarmoniX probe.

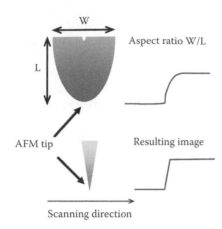

FIGURE 19.8 Geometric dimensions of AFM tip and resulting image profile.

which practically eliminate these problems. Some AFMs also use separated orthogonal scanners (as opposed to a single tube) which also serve to eliminate cross-talk problems.

19.2.2.4.2 Environmental Disturbances

19.2.2.4.2.1 Thermal Drift   Owing to thermal expansion of sample and cantilever, a displacement of the tip position during scanning may lead to a deformation of the image. This problem becomes more serious for high-resolution imaging. The relatively slow rate of scanning during AFM imaging often leads to influences of thermal drift in the image, making the AFM microscope less suited for measuring accurate distances between structures on the image. However, several fast-acting designs were suggested to increase microscope scanning productivity including what is being termed video-AFM (reasonable quality images are being obtained with video-AFM at video rate—faster than the average SEM).

19.2.2.4.2.2 Vibrations   Vibrations are caused by the building or by foot fall sound. For isolation purposes, the AFM system can be placed on air-damped tables or actively, by piezoelectric-elements, damped tables. In addition, the acoustic sound may disturb the measurements, especially if its frequency comes close to the resonance frequency of the cantilever. To avoid this problem, the AFM can be placed in noise protection boxes. Another possibility may be, under certain conditions, working in vacuum.

19.2.2.4.2.3 Interference   In the case of a highly reflecting sample surface, the reflected laser beam may cause inference stripes with the light reflected from the photodiode. These interference stripes are visible in the image additionally to the height profile and have a direction vertical to the actual scanning direction.

19.2.2.4.2.4 Static Charging   This is a problem observed on nonmetallic samples and a nonconducting AFM probe, especially during magnetic force measurements. Charges collected during scanning by the tip cannot flow away and cause electrostatic forces. In this case, the use of a conducting tip is necessary.

19.2.3 Imaging Techniques

There is a variety of imaging techniques based on combinations of basic modes with novel elements and an ongoing development of novel techniques based on the possibilities of fast computer technique. The AFM imaging techniques can be roughly broken down into three different categories. Imaging modes can be divided into direct imaging with static modes, direct imaging with dynamic modes using an oscillating tip and an indirect method with a point by point evaluation of locally measured properties leading to an image of a special sample property.

The two main modes of AFM tip operation are the contact-mode (static) and the noncontact mode (dynamic) technique. Also, the dynamic intermittent contact mode or tapping-mode technique uses an oscillating tip but touches the surface with varying amounts.

19.2.3.1 Contact Mode AFM

By contact of AFM, tip and surface repulsive forces dominate the tip-sample interaction and the overall force is repulsive. Consequently, this technique is called contact mode. Due to the disadvantage of contact mode AFM concerning resolution, it is nowadays only seldom used for topography measurement, but often used for DC current measurements. The repulsive force on the tip has a mean value of 10^{-9} N. It is set by pushing the cantilever against the sample surface with a piezoelectric positioning element. In contact mode AFM, the deflection of the cantilever is sensed and compared in a DC feedback amplifier to some desired value of deflection. Maintaining a constant deflection, the force between the tip and the surface is kept constant during scanning. The voltage measuring the feedback amplifier is a measure of the height differences and displayed as a function of the lateral position of the sample; it represents the topography.

A large class of samples, including semiconductors and insulators, can trap electrostatic charge (partially dissipated and screened in liquid). This charge can contribute to additional substantial attractive forces between the probe and sample. All of these forces combine to define a minimum normal force that can be applied in a defined way by the probe to the sample. This normal force creates a substantial frictional force as the probe scans over the sample. In practice, it appears that these frictional forces are far more destructive than the normal force and can damage the sample, dull the cantilever probe, and distort the resulting data.

These effects can be reduced by minimizing tracking force of the probe on the sample, but there are practical limits to the magnitude of the force that can be controlled by the user during operation in ambient environments. Under ambient conditions, sample surfaces are covered by a layer of adsorbed gases consisting primarily of water vapor and nitrogen which is 10–30 monolayers thick. When the probe touches this contaminant layer, a meniscus forms and the cantilever is pulled by surface tension toward the sample surface. The magnitude of this adhesion force depends on the details of the probe geometry, but is typically on the order of 100 nN. This meniscus force and other attractive forces may be neutralized by operating with the probe and with part or the entire sample totally immersed in liquid. There are many advantages to operate AFM with the sample and cantilever immersed in liquid. These advantages include the elimination of capillary forces, the reduction of van der Waals' forces and the ability to study technologically or biologically important processes at liquid–solid interfaces. However, there are also some disadvantages involved in working in liquids. These range from nuisances, such as leaks, to more fundamental problems such as sample damage on hydrated and vulnerable biological samples. Also, many samples such as semiconductor wafers cannot practically be immersed in liquid.

The resolution is limited by the contact area and at surface protrusions by the geometry of the tip. Mechanically stable samples like metal, glass, and ceramics can be measured in this mode since the effort is comparably small. Sensitive surfaces may be disturbed or damaged by the tip contact. In particular, the imaging of particles is quite difficult since they are prone to be pushed away.

19.2.3.2 Noncontact Mode AFM

19.2.3.2.1 Dynamic Mode

Although it is possible to measure within the attractive force regime with a small tip–sample separation, this mode has no importance in practice. Most of the measurements are performed in the dynamic mode where the cantilever is externally oscillated at or close to its resonance frequency (10–100 kHz, applied by the z-Piezo).

Noncontact AFM is performed such that the tip–sample interaction is in the attractive or van der Waals regime. In order to perform measurements in this attractive force region, the cantilever is oscillated with a low amplitude (<5 nm) near its resonant frequency by a small piezoelectric element.

For noncontact AFM the force is measured by comparing the frequency and/or amplitude of the cantilever oscillation relative to the driving signal. The tip performs a forced oscillation. The oscillation amplitude, phase and resonance frequency of the tip are modified by tip–sample interaction forces; these changes, during oscillation with respect to the external reference oscillation, provide information about the sample's characteristics. The resonance frequency is dependent on the mass and spring constant of the cantilever. The equation for the resonant frequency of a spring free of damping is given by

$$f_{res} = \frac{1}{2\pi}\sqrt{\frac{k}{m}} \qquad (19.2)$$

where f_{res} is the resonant frequency, k the spring constant of cantilever, and m the mass of the cantilever.

AFM cantilevers generally have spring constants of about 0.1 N m⁻¹. Forces between sample and tip cause an energy loss and thereby a damping of the tip oscillation. Depending on the amount of damping, the frequency of the forced tip oscillation is changed and a phase shift between generating and forced oscillation appears.

Schemes for dynamic mode operation include frequency modulation and the more common amplitude modulation. In frequency modulation, changes in the oscillation frequency provide information about tip–sample interactions. Frequency can be measured with very high sensitivity and thus the frequency modulation mode allows for the use of very stiff cantilevers. Stiff cantilevers provide stability very close to the surface and, as a result, this technique was the first AFM technique to provide true atomic resolution in ultra-high vacuum conditions.

19.2.3.2.2 *Tapping Mode*

Under ambient conditions, most samples develop a liquid-meniscus layer. It is difficult to keep the probe tip close enough to the sample to detect short-range forces while preventing the tip from sticking to the surface. The so-called "tapping mode" was developed to bypass this problem. It is also done by oscillating the cantilever near its resonant frequency, but with a significantly higher amplitude (>20 nm). Now, the tip is touching the surface once during each oscillation such that enough restoring force is provided by the cantilever to detach the tip from the sample. During tapping-mode operation, the cantilever oscillation amplitude is kept constant by a feedback loop. Thereby, only vertically directed forces are present; critical horizontal forces which may disturb sensitive samples like, that is, living cells, polymers, liquids, or particles are minimized. Choice of the amplitude regulates the maximal force. In amplitude modulation, changes in the oscillation amplitude or phase provide the feedback signal for imaging. Changes in the phase of oscillation can be used to discriminate between different types of materials on the surface. Amplitude modulation can be operated either in the noncontact or in the intermittent contact regime depending on the imaging parameters. Tapping mode is the most common mode for imaging surfaces. It allows high-resolution topographic imaging of sample surfaces, especially of those that are easily damaged, loosely hold to their substrate, or difficult to image by other AFM techniques. Tapping mode overcomes to some extent problems associated with friction, adhesion, electrostatic forces, and other difficulties in conventional AFM scanning methods and is a key advance in AFM.

19.3 AFM Measurement

19.3.1 Sample Properties Relevant for Fuel Cell Application Accessible by AFM

In the following chapter the most important imaging techniques for the measurement of a variety of sample properties relevant for PEM fuel cells are described, some with examples of AFM measurements of fuel cell components. An overview of the most important image modes is given in Table 19.1.

In Figure 19.9a, the dependence of van der Waals and chemical force, as most important forces acting on the tip as well as their sum, are drawn. A schematic overview on additional forces acting at different distances is drawn in Figure 19.9b.

TABLE 19.1 Tip–Sample Interaction Used for Imaging

Interaction	Property	Application in PEMFCs
Force		
Van der Waals force	Topography	All components: Membranes, catalyst, electrodes, GDL, bipolar plates
Friction force: Different material	Discerning of different surface layers	All components: Membranes, catalyst, electrodes, GDL, bipolar plates
Magnetic force: Magnetic domains	Magnetic force: Magnetic domains	Magnetic catalyst particles
Coating of AFM tip with special molecules	Special chemically/biologically active sites	Catalyst particles, biological fuel cells
Electrostatic force	Electric fields- electrically active regions	Membranes, catalyst, electrodes, GDL, bipolar plates
Adhesion force	Discerning of different surface layers	Membranes, catalyst, electrodes, GDL, bipolar plates
Maximum (peak force)	Discerning of subsurface material with different hardness	Membranes, catalyst, electrodes, GDL, bipolar plates
Stiffness	Discerning materials with different elasticity	Membranes, catalyst, electrodes, GDL, bipolar plates
Electric Measurements		
Electronic current	Electronic conductivity of sample material	Membranes, catalyst, GDL, bipolar plates
Ionic current	Ionically conductive domains	Membranes, electrodes
Potential	Discerning different materials by surface potential	Membranes, catalyst, electrodes
Electrochemical AFM	Surface changes due to electrochemical reactions	Catalyst, electrodes
Reactivity	Identification of reactive regions/ material- comparison of magnitude of reactivity	Catalyst, electrodes
Voltammograms	Identification of reactions	Catalyst, electrodes, membranes
Indirect methods		
Phase shift	Discerning of different surface materials	All components: membranes, catalyst, electrodes, GDL, bipolar plates
evaluation of force curve		
Adhesion force	Discerning of different surface layers	All components: membranes, catalyst, electrodes, GDL, bipolar plates
Maximum (peak) force	Discerning of subsurface material with different hardness	All components: membranes, catalyst, electrodes, GDL, bipolar plates
Stiffness	Discerning materials with different elasticity	All components: membranes, catalyst, electrodes, GDL, bipolar plates
Energy dissipation	Identification of deformable material in small subsurface volume	All components: Membranes, catalyst, electrodes, GDL, bipolar plates
Evaluation of current–voltage curve	Resistance	Membranes, electrodes, catalyst
Voltage steps	Dynamic behavior of conductive material-relaxation times	Membranes, electrodes, catalyst
Alternating current/-voltage	Impedance	Membranes, electrodes

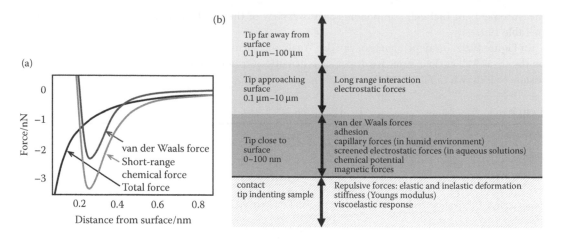

FIGURE 19.9   (a) Force-separation dependence of van der Waals-, chemical force, and total force and (b) interaction forces between AFM tip and surface at different distances.

19.3.1.1 Directly Accessible Properties

By using the AFM as an analytic tool, a variety of physical and chemically relevant properties can be directly measured by using a certain material or design of the tip, applying a voltage to the tip, or working in a special environment.

19.3.1.1.1 Topography

The topography of the sample surface is the main purpose of using an AFM. Topography is normally always recorded in parallel if another signal is measured. It can be measured with contact as well as with noncontact mode techniques. Since in contact mode a comparably high force is acting on the surface, the resolution is typically lower than in noncontact mode. In addition, the lateral forces applied to the tip can lead to a replacement of surface material and a considerable change or even damage of the surface. Some applications like DC current measurements are typically performed in contact mode. In the topography image, the stored data combine the lateral coordinates with the measured property, that is, the height. Figure 19.10a displays the topography image of a metallic bipolar plate measured by tapping-mode.

As a standard analysis, the topography of the sample surface can be displayed as 2D—or a quasi 3D-image. These images represent the topography of the samples and morphological details, that is, steps or particles can be discerned and their lateral and vertical dimensions can be measured with very

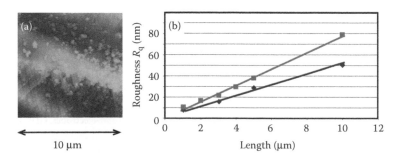

FIGURE 19.10   (a) Tapping mode image of a bipolar plate before operation, image length 10 μm and (b) increase of RMS roughness on a bipolar plate after fuel cell operation measured by tapping mode AFM.

high resolution. The height profiles at any line, providing with very precise information on height and width of structures on the surface, can be drawn and measured. Compared with a classical profilometer with a tip curvature of about 1 μm, the resolution of the measurement is much higher, typically in the range of (laterally) some nanometer; the resolution in *z*-direction is even higher. In addition, the roughness data from a chosen area are calculated by the software. This has become a standard measurement for the analysis of surface quality, for example, corrosion of the surface of bipolar plates before use and after operation.

The following data can typically be derived from a profile line or on an area:

R_a	arithmetic mean roughness
R_p	profile height compared to mean value
R_t	maximum height difference
R_q (RMS)	standard deviation of the mean roughness value

$$\text{RMS} = \sqrt{\frac{\sum_{i=1}^{N}(Z_i - Z_{ave})^2}{N}} \tag{19.3}$$

where Z_{ave} is the average and Z_i the current number of points within the analyzed area. The RMS roughness of the bipolar plate in Figure 19.10a, imaged by tapping mode AFM before and after fuel cell operation, is displayed in Figure 19.10b. A histogram of adhesion forces is also found in Figure 19.21.

19.3.1.1.2 Electronic Conductivity

Using a conducting AFM tip, a measurement of the local electronic conductivity of the sample in addition to the topographical data is possible. Typically, only a thin metallic coating of the tip is necessary, since the total current flow is small. For that measurement the sample is mounted electrically isolated from the AFM housing and typically a voltage is applied to the back side of the sample. The AFM tip is used as a counter electrode and also needs to be mounted isolated from a metallic tip holder or mounted onto a nonconductive one to measure the current. Since the current values are small and eager to be disturbed by electronic noise, a close-by current–voltage amplifier transforms the current into a voltage which is fed to the AFM system as additional input signal. A schematic drawing is given in Figure 19.11.

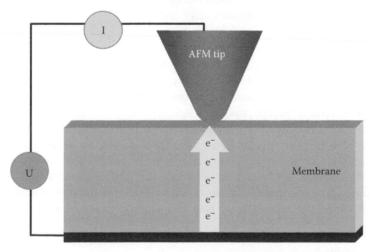

FIGURE 19.11 Setup of current measurement.

FIGURE 19.12   Current image of an electrode composed of 70% and Pt/C and 30% Nafion with a conductive area of 91% measured by AFM.

The measurement of direct currents is typically performed in contact mode, but also possible with tapping mode AFM. The measured current, and also the resolution in the current image, depend on the contact area (and the contact force) of the tip. On rough surfaces this area may change due to the surface topography. Apparently, large currents are often caused by edges and steep holes of the sample where the edges of the tip are in contact with a larger surface area. In addition, due to the mechanical stress on a hard sample surface, the conductive coating may break off from the end of the tip, and as an artifact, a current flow is only observed at edges. In this case, a completely conducting AFM tip can be chosen. A quantitative evaluation of the current has to take these artifacts into account including the occurrence of transient currents when the tip scans from isolating to conductive areas. Figure 19.12 displays the current image measured on an electrode composed of Pt/C and Nafion where different components differ in terms of their characteristic shape and current: Pt particles with high conductivity are bright spots and the carbon particles are shaped like a flower.

19.3.1.1.3 Current–Voltage Curve

At a defined sample site, without scanning the tip, a current–voltage curve can be measured by applying a voltage ramp to the tip while measuring the current. This technique is valuable for all conductive samples. Depending on the material, the rate of voltage change delivers different information. A small rate of voltage change probes the material under steady-state conditions. Here, the slope is a measure for the system resistance. A fast rate of change, like a voltage step, probes the dynamic behavior of the material to return to a steady state with a different voltage applied.

19.3.1.1.4 Electrochemical AFM

The electrochemical AFM (EC-AFM) has a high importance for fuel cell research and can be used with the sample fully immersed in liquid electrolyte in an electrochemical cell under potentiostatic or galvanostatic control as well as in a humid environment with a two-electrode arrangement for ionic conductivity measurements of solid electrolytes.

A big advantage of measurement in a liquid electrolyte is, to some extent, the independence of the measurement from the environment. Using the AFM in an electrolyte or in electrochemical cell can provide information on the structure of surfaces at the solid–liquid interface under different potential and on the change of surfaces during an electrochemical reaction. With EC-AFM, the structure of the interface can be measured *in situ* and has provided immense information on the structure of overlayers, that is, the CO adsorption on platinum single-crystal surfaces relevant for platinum catalyst particles.

19.3.1.1.5 Ionic Conductivity

A special application of electrochemical AFM is the measurement in a humid environment. Using a solid state electrolyte, the measurement of its proton conductivity is possible if suitable electrodes for

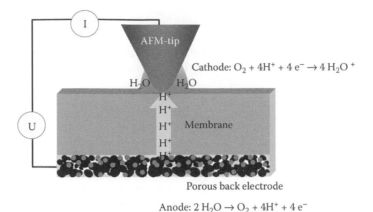

FIGURE 19.13 Measurement of ionic current.

electrochemical reactions are available at both sides of the electrolyte. The electrochemical reactions need to take place on both sides of the sample as, for example, with a polymer electrolyte membrane. The bottom side of the membrane is in connection to or coated with an extended conductive porous electrode containing catalyst (i.e., platinum particles). The platinum coated AFM tip, which serves as a point-like counter electrode, is biased negatively with respect to the back electrode; there is no further electrode at this side of the membrane. By applying a positive voltage, with respect to the AFM tip, to the back electrode, an electrochemical reaction (water splitting) provides protons which can pass through the ion-conductive membrane channels. The protons are consumed at the platinum-coated AFM tip by another electrochemical reaction, thereby closing the current circle as shown in Figure 19.13 (Aleksandrova et al., 2007a,b). The visualization of the ionic active areas on the surface of Nafion membranes, as used in PEM-fuel cells under controlled humidity, has been demonstrated with a resolution down to 2 nm (Hiesgen et al., 2009a, 2010). A typical measurement of current distribution on Nafion 112 membrane as a 3D- and 2D image, as well as the current profile along the marked line, show the inhomogeneous conductivity of Nafion membranes in Figure 19.14 (Friedrich et al., 2009).

19.3.1.1.6 Phase Imaging

In phase imaging, the phase lag of the cantilever oscillation, relative to the signal sent to the cantilever's piezo driver, is simultaneously monitored and recorded by the controller. A detailed study of the origin

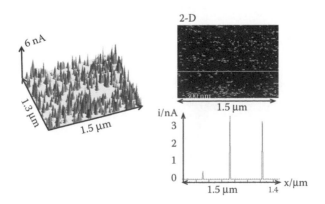

FIGURE 19.14 Current on Nafion 112 as 3D and 2D image with current profile at marked line measured at 70% relative humidity and room temperature.

of contrast in phase images can be found in the paper of James et al. (2001). The phase lag is caused by damping (energy loss) of the tip oscillation. There are two main reasons for energy loss: one is the presence of friction due to adhesion forces between surface and tip, for example in the presence of a water layer, the other is a visco-elastic behavior of the surface material. The value of the phase has a complex dependence on both components:

$$\sin \psi = \left(\frac{\omega \cdot A}{\omega_0 \cdot A_0} \right) + \frac{Q \cdot E_{Dis}}{\pi \cdot k \cdot A \cdot A_0} \qquad (19.4)$$

with ψ phase angle ω and ω_0 working and resonance frequencies, respectively; A set point amplitude; A_0 free amplitude; Q quality factor; k cantilever spring constant. The phase lag ψ is very sensitive to variations of these two components, and thereby the change of the surface layer, and can be used to obtain a material contrast, although it is difficult to interpret in detail reported by Cleveland et al. (1998). For receiving an unambiguous image of only the adhesion force or the change of visco-elastic properties, an evaluation of the force–distance curve is necessary to retrieve both components (see Section 19.2.2.1).

19.3.1.2 Indirectly Accessible Signals by AFM

19.3.1.2.1 Electrostatic Force

On materials with a mixed conductivity like electrodes the measurement of local electric fields indicates the position and distribution of conductive particles. One rather precise measuring technique of electrostatic forces relies on the change in resonant frequency of the cantilever due to vertical electrostatic force gradients from the sample. The samples should have a smooth surface to detect electric fields. The complete measurement needs two steps: first, the height profile of the image line is measured and stored; second, the tip follows this height profile thereby keeping a constant distance to the surface ($d < 100$ nm) to minimize the influence of surface topography. Depending on the local field strength, the conductive tip is bound toward or away from the surface (Figure 19.15). The cantilever resonant frequency changes in response to any additional force gradient and also the electrostatic force. Attractive forces reduce the cantilever resonant frequency; repulsive forces increase the resonant frequency (Figure 19.16). The change in resonant frequency of the cantilever is sensed by phase, frequency, or amplitude detection.

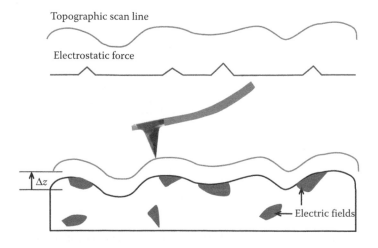

FIGURE 19.15   Detection of electric fields. (After manual of Multimode 5, Bruker Corp.)

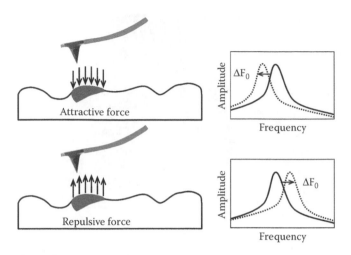

FIGURE 19.16 Principle of force detection for measurement of electric fields and surface potential. (After manual of Multimode 5, Bruker Corp.)

19.3.1.2.2 Surface Potential

The knowledge of the local surface potential may be valid for the investigation of electrodes as complex material with differently conductive components. Surface potential detection is a two-step measurement. The surface topography is obtained by tapping mode in the first scan. In the second scan, the tip follows the stored height profile to minimize the influence of topography and the surface potential is measured. The tip is kept in close proximity to the surface. On the first scan line, the sample topography is measured by standard tapping mode where the cantilever is mechanically vibrated near its resonant frequency. On the second scan line, the drive piezo that normally vibrates the cantilever is turned off. An oscillating voltage is applied directly to the cantilever tip to measure the surface potential. If there is a DC voltage difference between the tip and sample, then there will be an oscillating electric force on the cantilever at the frequency. This causes the cantilever to vibrate, and the oscillation amplitude can be detected. If the tip and sample are at the same DC voltage, there is no resulting force on the cantilever. Samples with regions of different materials also show contrast due to contact potential differences. Within a single image, quantitative voltage measurements can be made. The potential distribution on a microporous layer (MPL) of a GDL is imaged in Figure 19.17.

19.3.1.2.3 Force Spectroscopy

From the force–distance curve the local elastic properties of a sample, that is, Young's modulus, can be derived and calculated. A detailed study of the evaluation of force–distance curves, including theoretical background and experiments, can be found in the paper of Franceschini and Corti (2009).

The measurement of a force curve delivers information on the mechanical, and to some part chemical, properties of the material at this location. Measurement and interpretation are described in detail in Section 19.2.2.1. From the different parts of the curve the following properties can be retrieved simultaneously: the adhesion force (minimal force) of the surface, the peak force (maximal force at certain travelling path) of the affected sample volume, the DMT modulus (stiffness, slope of force curve during mechanical contact) as a measure of the elasticity, the dissipation energy (work, enclosed area of curve at tip approach and tip retract) as a measure of deformability of the effected sample volume, and the phase shift of the oscillation relative to the drive frequency as a measure of the total energy loss (damping) including adhesive and dissipative forces. An example of a measurement of an MPL of a GDL is shown in Figure 19.18. The different components, carbon and PTFE, are clearly distinguishable from their different properties. Besides the direct measurement, the force–distance

FIGURE 19.17 (a) Topography ($\Delta z = 1\ \mu m$), (b) phase shift (material contrast), and (c) surface potential ($\Delta V = 8\ mV$) on the same area of MPL/GDL with image length of $3\ \mu m$.

curve can be reconstructed from the resulting higher harmonic vibrations of the tip after a mechanical pulse to the sample. A specially formed cantilever/tip can be used where the tip is put asymmetrically onto the end of the cantilever (Figure 19.7b) which is designed to enhance the higher harmonics for a better reconstruction of the force. The reconstruction of the whole force curve, and subsequent evaluation at every image points, delivers the distribution of all different properties of the sample mentioned above as separate images. This mode (HarmoniX-Mode by Bruker Corp.) is especially useful for the investigation of compound material as GDLs or MPLs and electrodes, but can also be used for the investigation of solid electrolyte to investigate the phase separation of ionic conductive and PTFE phases.

19.3.2 Which Mode for What Measurement?

In Figure 19.19, four different images of the same area of $3\ \mu m$ length on a solid Teflon rod measured by Tapping AFM are displayed. The sample has a uniform composition of PTFE polymer and a

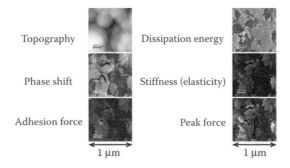

FIGURE 19.18 Example of different properties measured by HarmoniX mode on an MPL of a GDL.

FIGURE 19.19 (a) 2D-Topography, (b) 3D-topography, (c) amplitude error, and (d) phase image of solid PTFE (image size 1 μm).

comparable flat surface. The topography is displayed in Figure 19.19a where the height is encoded with color from white (high) to black (low), but, nevertheless, it is not easy to get the right idea of the surface morphology. A better imagination of the topography is given in the 3D-view in Figure 19.19b. Finer structures of the polymer fibers are more visible from the "Amplitude error" image in Figure 19.19c, where only changes in height are recorded, which serves as a high frequency filter enhancing topographic details. The visualization of the drilled polymer strands is best in the phase image (Figure 19.19d), where different damping of the cantilever oscillation differs between groves or holes and hills (compare the PTFE strands in Figure 19.22, AFM images of electrodes). Although there are no differences in composition on a macroscopic scale, in a high-resolution image differences form crystalline polymer strands and an amorphous polymer part is expected. These different morphological parts have different properties concerning hardness, adhesion, elasticity and deformability. Therefore, even if the topography would be very flat without any details, a contrast is expected in images displaying these properties, which are accessible for example with HarmoniX Mode (Bruker Corp.) (*cf.* Section 19.3.1.2.3), or another comparable technique. In Figure 19.20, images from the

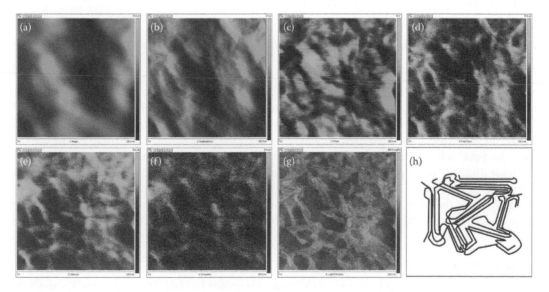

FIGURE 19.20 (a) Topography, (b) amplitude error, (c) phase image, (d) peak force, (e) adhesion, (f) dissipation energy, (g) DMT modulus of solid PTFE, size 300 nm, and (h) schematic drawing of PTFE strands forming crystalline areas in amorphous matrix.

same sample, but with a much higher resolution and an image length of 300 nm, are recorded:Figure 19.19a,b:

Topography: Flat elliptical features.

Figure 19.20c: Amplitude error: Sharp 2 nm broad lines along the elliptical protrusions
Figure 19.20d: Phase: Strong contrast between elongated features (dark) associated with dark lines in (c) and surrounding (bright) area
Figure 19.20e: Peak force image: Large peak force at regions where dark lines are in (d)
Figure 19.20f: Adhesion force: higher adhesion at elongated features with high peak force
Figure 19.20g: Dissipation energy distribution: higher energy dissipation parallel to high adhesion and high peak force
Figure 19.20h: DMT modulus (elasticity): sharply defined areas with different elasticity parallel to elongated elliptical features in a, small elongated surface structures in c, high peak force, adhesion, and dissipation in d,e,f, respectively

Taking the different surface and material information together, it can be concluded that the image shows elongated crystalline polymer regions from 2 to about 50 nm widths formed by parallel arranged PTFE strands embedded in an amorphous matrix with a smaller elasticity. The crystalline area measures about 80% of the surface in accordance with values found for a typical PTFE polymer (Ehrenstein, 1999).

19.4 Literature Review

The use of AFM for the analysis of fuel cell components is mostly restricted to the analysis of the PEM electrolyte membranes, its morphology and the estimation of ionic clusters from phase imaging. Some studies have measured catalysts particles or the morphology of electrodes. Only recently the other analytical options of AFM come more into focus. Only few authors report on the evaluation of mechanical properties (Section 19.3.1.2.3) or electrical properties like ionic conductivity (Section 19.3.1.1.5) or surface potential (Section 19.3.2.1.2).

In the following, a few publications are shortly introduced.

19.4.1 Application for Fuel Cell Components

19.4.1.1 Degradation of GDL

A statistical evaluation of adhesion forces and dissipation energy on different areas with the same size before and after fuel cell operation for 650 h on anode and cathode, respectively, has been performed using the HarmoniX mode (Section 19.3.1.2.3). As reference for PTFE properties, the sample shown in Figures 19.19 and 19.20 has been used. As a measure for the adhesion (and dissipation) on the imaged area, the peak value of the histogram of adhesion forces was taken (Figure 19.21). The diagrams in Figure 19.22a,b show an enhanced decrease of adhesion forces as well as dissipation at cathode compared to the anode. Compared to a fresh sample measured at dry conditions it indicates an increased loss of PTFE at the cathode in this experiment. In a wet environment, the adhesion at the cathode is the largest due to a thicker water layer at PTFE-free surface areas seen in Figure 19.22c (Hiesgen et al., 2010).

19.4.1.2 Electrodes

In Figure 19.23, three differently composed electrodes are images by HarmoniX-mode and the topography, the distribution of energy dissipation, and stiffness (DMT modulus or elasticity). A high contrast between carbon, platinum and PTFE is found in the stiffness images. On the surface, PTFE strands are seen as being comparable to those on the solid PTFE in Figures 19.19 and 19.20. A calculation of the

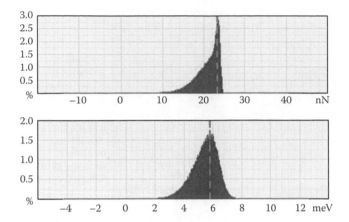

FIGURE 19.21 Histogram of adhesion force (top) and dissipation energy (bottom) of imaged area.

surface percentages of each component is in good accordance with the nominal compositions (Friedrich et al., 2009).

For applications to membranes, GDL/MPL, and bipolar plate, see Figures 19.14, 19.17, and 19.10, respectively.

19.4.2 Recent Advances

The use of AFM for the measurement of local ionic conductivity was first published by Hiesgen's group, University of Applied Sciences Esslingen, in Aleksandrova et al. (2007a) for Nafion membranes, but has also been reported with different aspects by other authors and applications. These techniques allow the direct visualization and analysis of the active ionic regions on the surface of a solid electrolyte with a resolution down to single active ionic channels (Hiesgen et al., 2009a,b).

Franceschini et al. (2009) provides the determination of Young's modulus of different membrane materials, including Nafion, from force–distance curves and correlated it to polymer properties. The authors discuss in detail the theory, application, and limitations of an evaluation of mechanical properties through force–distance curves.

A detailed study of morphology and local conductivity of different multiblock copolymer membranes is described in the work of Takimoto et al. (2009). A correlation of block length measured by AFM with fuel cell performance is described.

In the work of Hiesgen et al. (2010), an advanced AFM technique (HarmoniX from Bruker Corp.) is used for an analysis of the degradation of the MPL at anode and cathode side after fuel cell operation.

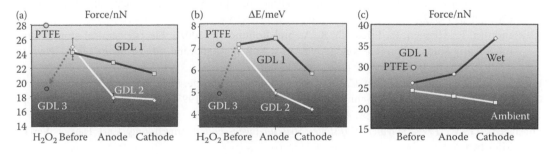

FIGURE 19.22 Statistical evaluation of change of adhesion (a,c) and dissipation (b) in dry, and (c) wet environment on an MPL after 650 h of fuel cell operation on cathode and anode, respectively.

Topography   Energy dissipation   Stiffness

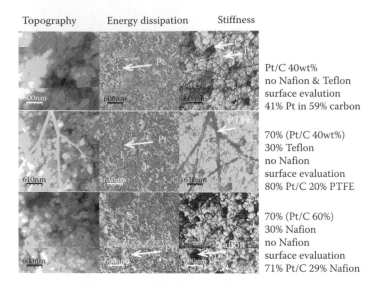

Pt/C 40wt%
no Nafion & Teflon
surface evalution
41% Pt in 59% carbon

70% (Pt/C 40wt%)
30% Teflon
no Nafion
surface evaluation
80% Pt/C 20% PTFE

70% (Pt/C 60%)
30% Nafion
no Nafion
surface evaluation
71% Pt/C 29% Nafion

FIGURE 19.23 Topography, energy dissipation, and stiffness of differently composed electrodes measured by AFM HarmoniX-Mode with comparison of measured surface composition with nominal values.

Here, the complete force–distance curve is reconstructed at every image point and, besides topography and phase images, provides local data on adhesion force, peak force, dissipation energy, stiffness (elastic modulus), and DMT modulus.

19.5 Outlook

Binnig, Quate, and Gerber invented the first atomic force microscope in 1986 (Binnig et al., 1986). Since that time, great improvements have been achieved. Despite its confinement to the surface or a surface-near analysis volume in most cases, AFM has presented a considerable potential to be a useful analysis tool for a variety of surface properties of all components in PEMFC examination, and also for solid oxide fuel cells (SOFCs). It is reasonable to believe it will have a promising future considering the development of fast personal computers and the proceeding application of complex techniques, where, that is, at every image point the evaluation of a whole force curve by Fourier transform techniques already became possible.

References

Aleksandrova, E., Hiesgen, R., Eberhard D., Friedrich, K. A., Kaz, T., and Roduner, E. 2007a. Nanometer scale visualization of ionic channels at the surface of a proton exchange membrane. *Chem. Phys. Chem.* 8: 519–522.

Aleksandrova, E., Hiesgen, R., Friedrich, K. A., and Roduner, E. 2007b. Electrochemical atomic force microscopy study of proton conductivity in a Nafion Membrane. *Phys. Chem. Chem. Phys.* 9: 2735–2743.

Binnig, G., Quate, C. F., and Gerber, C. 1986. Atomic force microscope. *Phys. Rev. Lett.* 56: 930–933.

Cleveland, J. P., Anczykowski, B., Schmidt, A. E., and Ehlings, V. E. 1998. Energy dissipation in tapping-mode atomic force microscopy. *Appl. Phys. Lett.* 72: 2613– 2615.

Ehrenstein, G. W. 1999. *Polymer-Werkstoffe: Struktur-Eigenschaften-Anwendung*, Carl Hanser Verlag. München.

Franceschini, E. A. and Corti, H. R. 2009. Elastic properties of Nafion, polybenzimidazole and poly [2,5-benzimidazole] membranes determined by AFM tip nano-indentation. *J. Power Sources* 188: 379–386.

Friedrich, A., Schulze, M., Bauder, A., Hiesgen, R., Wehl, I., Yuan, X. Z. et al. 2009. Nanoscale investigation of Nafion membranes after artificial degradation. *ECS Trans.* 25: 395–403.

Hiesgen, R., Aleksandrova, E., Meichsner, G., Roduner, E., and Friedrich K. A. 2009. High-resolution imaging of ion conductivity of Nafion® membranes with electrochemical atomic force microscopy. *Electrochim. Acta* 55: 423–429.

Hiesgen, R., Wehl, I., Friedrich, A., Schulze, M., Haug, A., Bauder, A. et al. 2010a. Atomic force microscopy investigation of polymer fuel cell gas diffusion layers before and after operation. *ECS Trans.* 28: 79–84.

James, P. J., Antognozzi, M., Tamayo, J., McMaster, T. J., Newton, J. M., and Miles, M. J. 2001. Interpretation of contrast in tapping mode AFM and shear force microscopy. A study of Nafion. *Langmuir* 17: 349–360.

Takimoto, N., Takamuku, S., Abe, M., Ohira, A., Lee, H. S., and McGrath, J. E. 2009. Conductive area ratio of multiblock copolymer electrolyte membranes evaluated by e-AFM and its impact to fuel cell performance. *J. Power Sources* 194: 662–667.

Appendix A: Classified Literature of AFM Application to Fuel Cells

A.1 Proton Exchange Membrane

A.1.1 Morphology of Surfaces

A.1.1.1 General

James, P. J., Antognozzi, M., Tamayo, J., McMaster, T. J., Newton, J. M., and Miles, M. J. 2001. Interpretation of contrast in tapping mode AFM and shear force microscopy. A study of Nafion. *Langmuir* 17: 349–360.

A.1.1.2 Nafion Membranes

Bass, M., Berman, A., Singh, A., Konovalov, O., and Freger, V. 2006. Surface structure of Nafion in vapor and liquid. *J. Phys. Chem. B* 114: 3784–3790.

Bertoncello, P., Notargiacomo, A., and Nicolini, C. 2005. Langmuir–Schaefer films of Nafion with incorporated TiO$_2$ nanoparticles. *Langmuir* 21: 172–177.

Bertoncello, P., Peruffo, M., Li, F., and Unwin, P. R. 2008. Functional electrochemically-active ultra-thin Nafion films. *Colloids Surf. A* 321: 222–226.

Bertoncello, P., Wilson, N. R., and Unwin, P. R. 2007. One-step formation of ultra-thin chemically functionalized redox-active Langmuir–Schaefer Nafion films. *Soft Matter* 3: 1300–1307.

Gargas, D. J., Bussian, D. A., and Buratto, S. K. 2005. Investigation of the connectivity of hydrophilic domains in nafion using electrochemical pore-directed nanolithography. *Nano Lett.* 5: 2184–2187.

Maeda, Y., Gao, Y., Nagai, M., Nakayama, Y., Ichinose, T., Kuroda, R., and Umemura, K. 2008. Study of the nanoscopic deformation of an annealed Nafion film by using atomic force microscopy and a patterned substrate. *Ultramicroscopy* 108: 529–535.

Ramdutt, D., Charles, C., Hudspeth, J., Ladewig, B., Gengenbach, T., Boswell, R., Dicks, A., and Brault, P. 2007. Low energy plasma treatment of Nafion (R) membranes for PEM fuel cells. *J. Power Sources* 165: 41–48.

Tazi, B. and Savadogo, O. 1997. New cation exchange membranes based on Nafion, silicotungstic acid and thiophene. Proceedings of the Second International Symposium on New Materials for Fuel-Cell and Modern Battery Systems II. 864–871.

Wei, H. Y., Kim, S. N., Marcus, H. L., and Papadimitrakopoulos, F. 2006. Preferential forest assembly of single-wall carbon nanotubes on low-energy electron-beam patterned Nafion films. *Chem. Mater.* 18: 1100–1106.

A.1.1.3 Other Membrane Materials

Gomes, D., Marschall, R., Nunes, S. P., and Wark, M. 2008. Development of polyoxadiazole nano-composites for high temperature polymer electrolyte membrane fuel cells. *J. Membr. Sci.* 322: 406–415.

Gromadzki, D., Cernoch, P., Janata, M., Kudela, V., Nallet, F., Diat, O., and Stepanek, P. 2006. Morphological studies and ionic transport properties of partially sulfonated diblock copolymers. *Eur. Polym. J.* 42: 2486–2496.

A.1.1.4 Catalyst/Electrodes

Bessarabov, D. and Sanderson, R. 2004. Solid polyelectrolyte (SPE) membranes with textured surface. *J. Membr. Sci.* 244: 69–76.

Kowal, A., Olszewski, P., Tripkovic, D., and Stevanovic, R. 2006. Nanoscale topography of GC/Pt-C and GC/Pt-Ru-C electrodes studied by means of STM, AFM and XRD methods. *Recent Developments in Advanced Materials and Processes* 518: 271–275.

Markovic, N. M. and Ross, P. N. 2002. Surface science studies of model fuel cell electrocatalysts. *Surf. Sci. Rep.* 45: 121–229.

Schmidt, T. J., Noeske, M., Gasteiger, H. A., Behm, R. J., Britz, P., and Bonnemann, H. 1998a. PtRu alloy colloids as precursors for fuel cell catalysts.—A combined XPS, AFM, HRTEM, and RDE study. *J. Electrochem. Soc.* 145: 925–931.

Schmidt, T. J., Noeske, M., Gasteiger, H. A., Behm, R. J., Britz, P., and Bonnemann, H. 1998b. PtRu alloy colloids as precursors for fuel cell catalysts. A combined XPS, AFM, HRTEM, and RDE study. *J. Electrochem. Soc.* 145: 3697–3697.

Siroma, Z., Ishii, K., Yasuda, K., Inaba, M., and Tasaka, A. 2007. Stability of platinum particles on a carbon substrate investigated by atomic force microscopy and scanning electron microscopy. *J. Power Sources* 171: 524–529.

A.1.2 Tapping-Mode Phase Imaging

A.1.2.1 Nafion Membranes

Affoune, A. M., Yamada, A., and Umeda, M. 2004. Surface observation of solvent-impregnated Nafion membrane with atomic force microscopy. *Langmuir* 20: 6965–6968.

Affoune, A. M., Yamada, A, and Umeda, M. 2005. Conductivity and surface morphology of Nafion membrane in water and alcohol environments. *J. Power Sources* 148: 9–17.

James, P. J., Elliott, J. A., McMaster, T. J., Newton, J. M., Elliott, A. M. S., Hanna, S., and Miles, M. J. 2000. Hydration of Nafion (R) studied by AFM and x-ray scattering. *J. Mater. Sci.* 35: 5111–5119.

James, P. J., McMaster, T. J., Newton, J. M., and Miles, M. J. 2000. *In situ* rehydration of perfluorosulpho-nate ion-exchange membrane studied by AFM. *Polymer* 41: 4223–4231.

Umeda, M. and Uchida, I. 2006. Electric-field oriented polymer blend film for proton conduction. *Langmuir* 22: 4476–4479.

Umemura, K., Kuroda, R., Gao, Y. F., Nagai, M., and Maeda, Y. 2008. Direct observation of deformation of Nafion surfaces induced by methanol treatment by using atomic force microscopy. *Appl. Surf. Sci.* 254: 7980–7984.

A.1.2.2 Other Membrane Materials

Bai, Z. W., Price, G. E., Yoonessi, M., Juhl, S. B., Durstock, M. F., and Dang, T. D. 2007. Proton exchange membranes based on sulfonated polyarylenethioethersulfone and sulfonated polybenzimidazole for fuel cell applications. *J. Membr. Sci.* 305: 69–76.

Cho, C. G., Kim, S. H., Park, Y. C., Kim, H., and Park, J. W. 2008. Fuel cell membranes based on blends of PPO with poly(styrene-b-vinylbenzylphosphonic acid) copolymers. *J. Membr. Sci.* 308: 96–106.

A.1.3 Conductivity

A.1.3.1 Nafion Membranes

Aleksandrova, E., Hiesgen, R., Eberhard D., Friedrich, K. A., Kaz, T., and Roduner, E. 2007a. Nanometer Scale Visualization of ionic channels at the surface of a proton exchange membrane. *Chem. Phys. Chem.* 8: 519–522.

Aleksandrova, E., Hiesgen, R., Friedrich, K. A., and Roduner, E. 2007b. Electrochemical atomic force microscopy study of proton conductivity in a Nafion Membrane. *Phys. Chem. Chem. Phys.* 9: 2735–2743.

Aleksandrova, E., Hiesgen, R., and Roduner, E. 2008. Proton conductivity and micro-morphology of Nafion fuel cell membranes determined by electrochemical atomic force microscopy in electrochemical scanning probe microscopy: From theory to real-world. *ECS Trans.* 11: 1–9.

Aleksandrova, E., Hiesgen, R., and Roduner, E. 2008. Visualisierung Ionen leitender Kanäle auf Polymer-Elektrolyt-Membranen mit dem elektrochemischen Rasterkraftmikroskop. *Horizonte* 31: 3–6.

Bussian, D. A., O'Dea, J. R., Metiu, H., and Buratto, S. K. 2007. Nanoscale current imaging of the conducting channels in proton exchange membrane fuel cells. *Nano Lett.* 7: 227–232.

Friedrich, A., Schulze, M., Bauder, A. Hiesgen, R., Wehl, I., Yuan, X., and Wang, H. 2009. Nanoscale investigation of Nafion membranes after artificial degradation. *ECS Trans.* 25: 395–403.

Hiesgen, R., Aleksandrova, E., Meichsner, G., Roduner, E., and Friedrich K. A. 2009. High-resolution imaging of ion conductivity of Nafion membranes with electrochemical atomic force microscopy. *Electrochim. Acta* 55: 423–429.

Hiesgen, R., Wehl, I., Aleksandrova, E., Roduner, E., Friedrich, A., and Bauder, A. 2010b. Nanoscale properties of polymer fuel cell materials—A selected review. *Int. J. Energy Res.* 34: 1223–1238.

Roduner, E. and Hiesgen, R. 2009. Membranes: Spatially resolved measurements. In: eds. G. Juergen, D. Chris, M. Patrick, O. Zempachi, R. David, and S. Bruno . *Encyclopedia of Electrochemical Power Sources* 2, 775–786. Amsterdam: Elsevier.

Sanchez, D. l. G., Friedrich, K. A., Hiesgen, R., and Wehl, I. 2010. Oscillations of polymer electrolyte fuel cells at low cathode humidification. *J. Electroanal. Chem.* 649: 219–231.

Takimoto, N., Ohira, A., Takeoka, Y., and Rikukawa, M. 2008. Surface morphology and proton conduction imaging of Nafion membrane. *Chem. Lett.* 37: 164–165.

A.1.3.2 Other Membrane Materials

Takimoto, N., Takamuku, S., Abe, M., Ohira, A., Lee, H. S., and McGrath, J. E. 2009. Conductive area ratio of multiblock copolymer electrolyte membranes evaluated by e-AFM and its impact to fuel cell performance. *J. Power Sources* 194: 662–667.

A.1.4 Surface Potential

A.1.4.1 Nafion Membranes

Kanamura, K., Morikawa, H., and Umegaki, T. 2003. Observation of interface between Pt electrode and Nafion membrane. *J. Electrochem. Soc.* 150: A193–A198.

A.1.5 Force Spectroscopy: Mechanical Properties

A.1.5.1 Nafion

Franceschini, E. A. and Corti, H. R. 2009. Elastic properties of Nafion, polybenzimidazole and poly [2.5-benzimidazole] membranes determined by AFM tip nano-indentation. *J. Power Sources* 188: 379–386.

Hiesgen, R., Wehl, I., Friedrich, A., Schulze, Haug, A., M., Bauder, Carreras, A., A. Yuan, X., and Wang, H. 2010a. Atomic force microscopy investigation of polymer fuel cell gas diffusion layers before and after operation. *ECS Trans.* 28: 79–84.

Umemura, K., Wang, T., Hara, M., Kuroda, R., Uchida, O., and Nagai, M. 2006. Nanocharacterization and nanofabrication of a Nafion thin film in liquids by atomic force microscopy. *Langmuir* 22: 3306–3312.

A.1.5.2 *Other Membrane Materials*

Franceschini, E. A., and Corti, H. R. 2009. Elastic properties of Nafion, polybenzimidazole and poly [2,5-benzimidazole] membranes determined by AFM tip nano-indentation. *J. Power Sources* 188: 379–386.

A.1.6 Impedance

A.1.6.1 *Nafion Membranes*

O'Hayre, R., Lee, M., and Prinz, F. B. 2004. Ionic and electronic impedance imaging using atomic force microscopy. *J. Appl. Phys.* 95: 8382–8392.

A.1.7 Electrochemistry

Kucernak, A. R., Chowdhury, P. B., Wilde, C. P., Kelsall, G. H., Zhu, Y. Y., and Williams, D. E. 2000. Scanning electrochemical microscopy of a fuel-cell electrocatalyst deposited onto highly oriented pyrolytic graphite. *Electrochim. Acta* 45: 4483–4491.

A.2 Solid Oxide Fuel Cells

A.2.1 Morphology of Surfaces

Baker, R. T., Salar, R., Potter, A. R., Metcalfe, I. S., and Sahibzada, M. 2009. Influence of morphology on the behaviour of electrodes in a proton-conducting solid oxide fuel cell. *J. Power Sources* 191: 448–455.

Eom, T. W., Yang, H. K., Kim, K. H., Yoon, H. H., Kim, J. S., and Park, S. J. 2008. Effect of interlayer on structure and performance of anode-supported SOFC single cells. *Ultramicroscopy* 108: 1283–1287.

Appendix B: Literature on AFM Technique

B.1 Books

Bowen, W. R. and Hilal, N. 2009. *Atomic Force Microscopy in Process Engineering. An Introduction to AFM for Improved Processes and Products.* Burlington, MA: Butterworth-Heinemann (Imprint of Elsevier Oxford, UK).

Colton, R. J., Engel, A., Frommer, J. E., Gaub, H. E., Gewirth, A. A., Guckenberger, R., Rabe, J., Heckl, W. M., Parkinson, B. 1998. Procedures in Scanning Probe Microscopies. Chichester: John Wiley & Sons Ltd.

Eaton, P. and West, P. 2010. *Atomic Force Microscopy.* New York: Oxford University Press, Inc.

Hiesgen R. and Haiber J. 2009. *Structural Properties: Atomic Force Microscopy.* In: J. Garche, C. Dyer, P. Moseley, Z. Ogumi, D. Rand, and B. Scrosati, eds *Encyclopedia of Electrochemical Power Sources.* 3. Amsterdam. Elsevier: 696–717.

Lindsay, S. 2009. *Introduction to Nanoscience.* New York: Oxford University Press, Inc. 978-0-19-954421-9.

Morita, S., Wiesendanger, R., and Meyer, E. (eds.) 2002. *Noncontact Atomic Force Microscopy in Springer: Series: NanoScience and Technology.* Berlin: Springer, ISBN: 978-3-540-43117-6.

Sarid, D. 1994. *Scanning Force Microscopy in: Oxford Series in Optical and Imaging Sciences.* New York, NY: Oxford University Press, Inc., ISBN 0-19-509204-X.

Wiesendanger, R. 1994. *Scanning Probe Microscopy and Spectroscopy—Methods and Applications.* Cambridge: Cambridge University Press.

Wiesendanger, R. 1998. *Scanning Probe Microscopy in Analytical Methods (Nanoscience and Technology).* Berlin: Springer.

Wiesendanger, R. and Guntherodt, H. J. (ed.) 1995a. *Scanning Tunneling Microscopy* II (2nd edn). Berlin: Springer.

Wiesendanger, R. and Güntherodt, H. J. (ed.) 1995b. *Scanning Tunneling Microscopy III: Theory of STM and Related Scanning Probe Method in Springer Series in Surface Sciences.* Berlin: Springer.

Wiesendanger, R., Guntherodt, H. J., and Workman, P. 1994. *Scanning Tunneling Microscopy I: General Principles and Applications to Clean and Adsorbate-Covered Surfaces in: Springer Series in Surface Sciences.* Berlin: Springer.

B.2 Web Links

en.wikibooks.org/wiki/Nanotechnology/AFM
en.wikipedia.org/wiki/Atomic_force_microscopy
www.afmuniversity.org/Cover.html
www.nanoscience.de/group_r/afm/introduction/Agilent
www.afmuniversity.org/index.cgi?CONTENT_ID = 1
www.veeco.com
www.ParkAFM.com
www.nanosurf.com/*atomicforce.de*
www.angstrom-advanced.com/index.asp
www.asmicro.com
www.olympus.com
www.spectroscopynow.com
www.nano.unr.edu/images.asp
www.pacificnanotech.com
www.quesant.com
www.jpk.com
www.nano.geo.uni-muenchen.de
www.thch.uni-bonn.de/pc/bargon/sensorik/Piezoelektrizitaet.html
www.micro.ecs.soton.ac.uk/activities/mcms/spm/
www.mcdb.colorado.edu/courses/3280/lectures/class04.html
www.weizmann.ac.il/surflab/peter/afmworks
www.witec.de/en/home/

20

Binary Gas Diffusion

Jun Shen
*Institute for Fuel Cell
Innovation*

20.1 Introduction

In a proton exchange membrane (PEM) fuel cell, oxygen is transported from airflow channels, through a gas diffusion layer (GDL) and then into a cathode catalyst layer (CCL), where an oxygen reduction reaction occurs. A CCL is a thin coating with a porosity of 30–60% and pore size distribution from several nanometers to 100 nm, permitting the diffusion of reactant oxygen into the CCL and the transport of product water vapor out of the CCL. The oxygen reduction reaction then can take place through the whole depth of the CCL. The rate of oxygen diffusion significantly affects the uniformity of oxygen reduction reaction through the whole CCL, the CCL lifetime, and the power output of the PEM fuel cell (Shen et al., 2011). In this context, the knowledge of the effective gas diffusion coefficients of the GDL and CCL are crucial for accurately predicting the performance and optimizing the design of PEM fuel cells.

Usually, the effective diffusion coefficient (EDC) of a porous material is correlated to the bulk diffusion coefficient (i.e., the corresponding binary gas diffusion coefficient D_{21} through empty space), and some statistical parameters of porous microstructure, such as porosity ε. On the subject of binary gas bulk diffusion, over a century, considerable efforts were made to develop a dependable and general equation to predict the binary gas diffusion coefficients D_{21} for various binary systems. On the basis of Stefan–Maxwell's model, Arnold (1930), Gilliland (1934), Chapman and Cowling (1952), Hirschfelder et al. (1954), Chen and Othmer (1962), Fuller et al. (1966), and Huang et al. (1972) developed empirical equations of the kinetic theory of gases. Among them the Fuller, Schettler and Giddings (FSG) equation

$$D_{21} = \frac{1.00 \times 10^{-3} T^{1.75} \left(1/M_1 + 1/M_2\right)^{1/2}}{P\left[\left(\Sigma \upsilon\right)_1^{1/3} + \left(\Sigma \upsilon\right)_2^{1/3}\right]^2} \tag{20.1}$$

provides the best practical combination of simplicity and accuracy (Karaiskakis and Gavril, 2004). In this equation, T is temperature (K), and M_i is the mole mass of gas i (g mol^{-1}). P is the absolute pressure (atm.) with the atomic and structural diffusion volume increment υ to be summed over the atoms, group of atoms, and structural features of each diffusion species. The υ-values were determined by a nonlinear least-squares analysis of over 500 experimental data; one may find the υ-values elsewhere (Fuller et al., 1969). This equation can be employed to calculate the bulk binary gas diffusion coefficient, which is required for the prediction of the EDC of porous materials.

The binary gas diffusion in porous materials can be dominated by bulk diffusion, Knudson diffusion, or a combination of them (Zhang et al., 2004). Bulk diffusion takes place when the mean free path of the gas molecules is much smaller than the pore diameter of the porous material, and therefore collisions between gas molecules occur more frequently than that between the molecules and pore walls. The Knudsen diffusion would be more dominant when the pore size is smaller compared with the mean free path, and the collisions between the molecules and the pore walls become more often (Kolaczkowski, 2003). Some theoretical models for binary gas diffusion in porous materials are available (Park et al., 1996; Kolaczkowski, 2003; Zamel et al., 2009). Among them, Bruggeman model (Bruggeman, 1935) is very frequently used to calculate the EDC. The Bruggeman approximation was derived for electrical conductivity and the dielectric constant of a medium composed of uniformly distributed spheres (Zamel et al., 2009). Bruggeman model has been commonly written as

$$D_{\text{eff}} = D_{21}\varepsilon^{m}, \qquad (20.2)$$

where m is the Bruggeman exponent, and its widely used value is 1.5. Here D_{eff} is the EDC of a porous material. The Bruggeman model does not consider the Knudsen diffusion with the diffusion coefficient D_{k}, which can be mathematically expressed as (Zhang et al., 2004)

$$D_{k} = 48.5 d_{p}\sqrt{T/M}, \qquad (20.3)$$

where d_{p} is the pore size (diameter), and M is the mole mass. When a pore size distribution of a porous material falls in the region of less than 1 μm, the Knudsen effect must be considered (Mu et al., 2008). The diffusion coefficient D_{p} in a cylindrical capillary of a diameter equal to a pore size d_{p} is given by the reciprocal additivity relation of Bosanquet formula (Pollard et al., 1948)

$$\frac{1}{D_{p}} = \frac{1}{D_{b}} + \frac{1}{D_{k}}. \qquad (20.4)$$

For a porous material, such as porous catalyst, the EDC sometimes can be expressed by the general relationship between the EDC and the straight pore diffusion coefficient D_{p} (Zhang et al., 2004)

$$D_{\text{eff}} = \frac{\varepsilon}{\varsigma} D_{p} \qquad (20.5)$$

with the tortuosity, denoted as ς, to take account of the deviation of the pore from straight capillaries.

Although people have made abundant efforts in theoretical study, satisfactory theoretical models are not easy to develop due to the complexity of the diffusion in porous materials of various pore networks (Park et al., 1996; Zamel et al., 2009). On the other hand, experimental measurement techniques, for example, gas chromatography (Karaiskakis and Gavril, 2004), nuclear magnetic resonance (Majors et al., 1991; Callaghan, 1994), electrochemical impedance spectroscopy (Flückiger et al., 2008; Kramer et al., 2008), and diffusion cell methods (Park et al., 1996; Zhang et al., 2004), have been developed to determine the EDCs of porous materials. One of the most reliable methods to measure binary diffusion coefficients of gases is a closed-tube method with a Loschmidt diffusion cell (Branski et al., 2003). Bulk binary diffusion coefficients were precisely measured (Rohling et al., 2007) with a Loschmidt diffusion cell. Moreover the EDC of a porous sample of a stainless-steel film with simple straight pores was measured and found to be consistent with the result of numerical computation of three-dimensional mass diffusion through the sample (Astrath et al., 2010), demonstrating that a Loschmidt diffusion cell can be used to measure EDC of a porous medium.

This chapter presents binary gas diffusion theory in a Loschmidt diffusion cell, experimental apparatus, and the applications of the diffusion cell in the determination of EDCs of porous materials of PEM fuel cells.

20.2 Theory of Binary Gas Diffusion

Diffusion is the process by which matter is transported from one part of a system to another (Crank, 1975). On the subject of binary gas diffusion in a Loschmidt diffusion cell, there are two gases, gas 1 and gas 2 (e.g., N_2 and O_2), in the diffusion cell. For the purpose of determining the EDC of a porous material using a Loschmidt diffusion cell, this section will theoretically describe the behavior of the binary gas diffusion in the diffusion cell.

Figure 20.1 schematically shows a Loschmidt diffusion cell. At the beginning ($t = 0$), the concentrations of gas 2 in the top and bottom chambers are C_2^t and C_2^b, respectively, and the initial condition can be written as

$$C_2 = C_2^b \quad (-L/2 \leq z < 0, t = 0), \tag{20.6}$$

$$C_2 = C_2^t \quad (0 < z \leq +L/2), t = 0. \tag{20.7}$$

Here t is time. When the diffusion coefficient D is a constant, Fick's second law can describe the one-dimensional diffusion in the Loschmidt diffusion cell (Crank, 1975)

$$\frac{\partial C}{\partial t} = D \frac{\partial^2 C}{\partial z^2}. \tag{20.8}$$

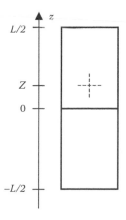

FIGURE 20.1 A schematic diagram of a Loschmidt diffusion cell. A gas sensor is located at $z = Z$ to detect the concentration of gas 2. (With kind permission from Springer Science+Business Media: Determination of binary diffusion coefficients of gases using photothermal deflection technique. *Appl. Phys. B* 87, 2007, 355–362, Rohling, J. R. et al.)

The impermeable boundary condition at $z = \pm L/2$, which means zero flow across the boundaries, is

$$\left(\frac{\partial C_2}{\partial z}\right)_{z=\pm L/2} = 0, \quad t > 0 \tag{20.9}$$

The general solution of the diffusion Equation 20.8 could be

$$C_2(z,t) = C + \sum_{n=1}^{\infty} \left[A_n \sin(\lambda_n z) + B_n \cos(\lambda_n z) \right] \exp\left(-\lambda_n^2 D t\right). \tag{20.10}$$

Here C, A_n, B_n, and λ_n are the constants to be determined by the initial and boundary conditions. The final solution is (Baranski et al., 2003)

$$C_2(z,t) = \frac{1}{2}\left(C_2^b + C_2^t\right) - \frac{2\left(C_2^b - C_2^t\right)}{\pi} \sum_{m=0}^{\infty} \frac{\exp\left(-(2m+1)^2 \pi t/\tau\right)}{(2m+1)} \sin\left[\frac{(2m+1)\pi z}{L}\right]. \tag{20.11}$$

Here

$$\tau = \frac{L^2}{\pi D} \tag{20.12}$$

is the characteristic diffusion time. Rohling et al. (2007) exhibited that if the pure gas 1 (e.g., N_2) and pure gas 2 (e.g., O_2) are filled in top and bottom chambers of the diffusion cell, respectively, at the beginning of the diffusion, the diffusion takes about 2τ to reach the steady state. When $t = \tau$, the concentration of gas 2 is about 95% of the concentration of the steady state at the ceiling of the top chamber, $z = (1/2)L$. When $t = 0.1\tau$, the concentration of gas 2 at $z = (1/2)L$ is less than 10% of that at the steady state. Therefore, one may consider that when $t < 0.1\tau$, gas 2 has not diffused to the ceiling of the top chamber, and the diffusion cell can be thought as an infinite space for gas 2 to diffuse when $t \leq 0.1\tau$ (Rohling et al., 2007).

Equation 20.11 is the trigonometrical-series type of solution and fully describes the gas diffusion process. No difference was found between the numerical simulations using the first 100 terms and that using the first 300 terms. It may be considered enough to calculate the equation with the summation of the first 100 terms. However, it is ideal, in reality, to use fewer terms to process experimental data. As presented in Figure 20.2, the trigonometrical series converge rapidly for moderate and large times ($t > 0.2\tau$), and the main difference is at the beginning of the diffusion if using fewer terms to approximate Equation 20.11.

In practice, the measurement is usually for a short time and requires a solution for the diffusion Equation 20.8, which converges for a short time. To find the solution, consider a long cylinder with a diffusing substance, gas 2, confined in the region $-h < x < 0$, shown in Figure 20.3. The initial conditions can be mathematically expressed as

$$C_2 = C_2^b \quad (-h < z < 0, \, t = 0) \tag{20.13}$$

and

$$C_2 = 0 \quad (0 < z, \, z < -h, \, t = 0). \tag{20.14}$$

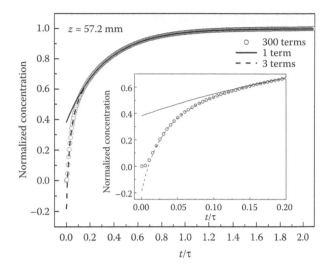

FIGURE 20.2 Numerical simulations of Equation 20.11. Circles: the sum of the 300 terms ($m = 0$ to 299), a solid line: the first-term ($m = 0$) approximation, and a dash line: the sum of the first-3-term ($m = 0$ to 2) approximation. The scale of the concentration is normalized by the concentration at the steady state. $L = 355$ mm; $z = 57.2$ mm. $C_2^b = 1$ and $C_2^i = 0$ The inset contains details from $t = 0$ to 0.20τ. (Adapted from Rohling, J. R. et al. 2007. *Appl. Phys. B* 87: 355–362.)

The substance diffuses in both directions of positive and negative z when $t > 0$. Following the calculation of Crank (1975, pp. 14, 15), the solution of Equation 20.8 is

$$C_2(z,t) = \frac{C_2^b}{2\sqrt{\pi Dt}} \int_z^{z+h} \exp\left(-\xi^2/\sqrt{4Dt}\right) d\xi. \tag{20.15}$$

The result of the integration of Equation 20.15 is

$$C_2(z,t) = \frac{C_2^b}{2}\left[\operatorname{erf}\left(\frac{z+h}{2\sqrt{Dt}}\right) - \operatorname{erf}\left(\frac{z}{2\sqrt{Dt}}\right)\right]. \tag{20.16}$$

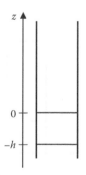

FIGURE 20.3 A schematic illustration of a long cylinder with a diffusing substance confined in the region $C_2^b = 1$.

In Equation 20.16 erf(x) is the error function defined by

$$\text{erf}(x) = \frac{2}{\sqrt{\pi}} \int_0^x \exp\left(-\xi^2\right) d\xi. \tag{20.17}$$

The error function has the properties

$$\text{erf}(-x) = -\text{erf}(x), \quad \text{erf}(0) = 1, \quad \text{erf}(\infty) = 1. \tag{20.18}$$

For an infinite space, $h \to \infty$, the solution of the diffusion Equation 20.8 is

$$C_2(z,t) = \frac{C_2^b}{2} \text{erfc}\left(\frac{z}{2\sqrt{Dt}}\right). \tag{20.19}$$

Here,

$$\text{erfc}(x) = 1 - \text{erf}(x) \tag{20.20}$$

is the complementary error function, which is also called the error function complement. As expected, Equation 20.19 is the same as Equation 2.14 of Crank (1975, p. 14), which describes the diffusion for a long cylinder with the diffusing substance in the area of $z < 0$ ($t = 0$).

In a finite cylinder with impermeable boundaries, as presented in Figure 20.4, the flux of the diffusant is considered reflected at $z = l$, and then at $z = -h$, and so on. The result of each successive reflection is superposed on the original solution Equation 20.16. The complete expression for the concentration in the finite cylinder is an infinite series of error functions, that is,

$$C_2(z,t) = \frac{C_2^b}{2} \sum_{m=-\infty}^{\infty} \left\{ \text{erf}\left[\frac{2m(l+h)-z}{2\sqrt{Dt}}\right] + \text{erf}\left[\frac{2h-2m(l+h)+z}{2\sqrt{Dt}}\right] \right\}. \tag{20.21}$$

If gas 2 is not confined in the area $z < 0$ at the beginning of the diffusion, the concentration of gas 2 in the area $z > 0$ is not zero, and the initial conditions are

$$C_2 = C_2^b \quad (-h < z < 0, t = 0) \tag{20.22}$$

FIGURE 20.4   A schematic diagram of a finite cylinder with impermeable boundaries at $z = l$ and $z = -h$.

and

$$C_2 = C_2^t \quad (0 < z < l, t = 0).$$ (20.23)

with the initial conditions, Equations 20.22 and 20.23, Equation 20.19 turns out to be

$$C_2(z,t) = \frac{1}{2}\left[C_2^b + C_2^t - \left(C_2^b - C_2^t\right)\mathrm{erf}\left(\frac{z}{2\sqrt{Dt}}\right)\right]$$ (20.24)

for the diffusion in an infinite space, and Equation 20.21 develops to

$$C_2(z,t) = \frac{1}{2}\left(C_2^b + C_2^t - \left(C_2^b - C_2^t\right)\left\{ 1 - \sum_{m=-\infty}^{\infty}\left[\mathrm{erf}\left(\frac{2m(l+h)-z}{2\sqrt{Dt}}\right) + \mathrm{erf}\left(\frac{2h-2m(l+h)+z}{2\sqrt{Dt}}\right)\right]\right\}\right).$$ (20.25)

In the case of a Loschmidt diffusion cell, Equation 20.25 becomes

$$C_2 = \frac{1}{2}\left(C_2^b + C_2^t - \left(C_2^b - C_2^t\right)\left\{ 1 - \sum_{m=-\infty}^{\infty}\left[\mathrm{erf}\left(\frac{2mL-z}{2\sqrt{Dt}}\right) + \mathrm{erf}\left(\frac{(1-2m)L+z}{2\sqrt{Dt}}\right)\right]\right\}\right).$$ (20.26)

Figure 20.5 presents a comparison of the solutions of the one-dimensional diffusion equation (Equation 20.8) in a Loschmidt diffusion cell in a trigonometrical series (Equation 20.11) and in an error-function series (Equation 20.26). The first 300 terms ($m = 0$ to 299) of Equation 20.11 and 3 terms ($m = -1$ to 1) of Equation 20.26 are employed to calculate the curves in Figure 20.5 for the trigonometrical-series-type and error-function-series-type solutions, respectively. At the beginning of the diffusion, the two curves are consistent, and the relative difference between them is about 1% at $t = 2\tau$, exhibiting that the error-function-series-type solutions converge rapidly for short and moderate times. The difference between the solutions for a Loschmidt diffusion cell and for an infinite space (Equation 20.24) is also shown in Figure 20.5. At the beginning, two solutions are the same, and the differences between them are less than 1% and about 2.5% at $t = 0.1\tau$ and $t = 0.2\tau$, respectively. Recall that the top chamber can be thought as an infinite space for gas 2 to diffuse when $t < 0.1\tau$ (Rohling et al., 2007); the solution, Equation 20.24 well describes the diffusion behavior in a Loschmidt diffusion cell when $t < 0.1\tau$ with the advantage of only one error-function term required for processing experimental data. Henceforth, in this chapter, Equation 20.24 will be employed for diffusion data processing.

Without any porous material in the diffusion cell, Equation 20.24 can be used to measure the bulk gas diffusion when D represents D_{21} in the equation. With the concept of equivalent diffusion coefficient D_{eq}, Equation 20.24 can also be employed to determine the EDC when a porous sample is placed in the Loschmidt diffusion cell, shown in Figure 20.1 at a position $Z > z > 0$. In this case, an equivalent resistance R_{eq} to binary gas diffusion can be defined as (Zhang et al., 2004)

$$R_{eq} = \frac{Z}{D_{eq}A}.$$ (20.27)

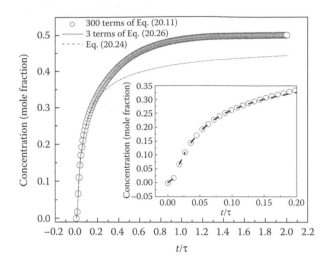

FIGURE 20.5 Numerical simulations of the solutions of the one-dimensional diffusion equation, Equations 20.11, 20.26, and 20.24. Circles: the sum of 300 terms of the trigonometrical series (Equation 20.11), a solid line: 3-term sum of the error-function series (Equation 20.26), and a dash line: the infinite space solution (Equation 20.24). $L = 355$ mm; $z = 57.2$ mm. $C_2^b = 1$, and C_2^t. The inset is the details of the comparison between Equations 20.11 and 20.24 from $t = 0$ to 0.20τ.

In Equation 20.27, D_{eq} is the equivalent diffusion coefficient of the media from $z = 0$ to $z = Z$, and A is the cross-sectional area available to diffusion. The concept of equivalent resistance originates from electrical conduction, and sometimes heat conduction analysis adapts the concept. In this method, the object is considered to be comprised of a number of individual resistances, which are connected to form an overall circuit (Zhang et al., 2004; Shen et al., 2011). When a single-layer porous sample is placed at a position of $Z > z > 0$, the bulk diffusion resistance and the porous sample resistance connect in series, and the equivalent resistance R_{eq} can be expressed as (Zamel et al., 2010)

$$R_{eq} = \frac{Z}{D_{eq}A} = \frac{Z-l}{D_{21}A} + \frac{l}{D_{eff}A}. \tag{20.28}$$

Here l and D_{eff} are the thickness and EDC of the porous sample, respectively. The gas sensor at $z = Z$ measures the concentration evolution of gas 2 in the top chamber, and the equivalent diffusion coefficient D_{eq} can be deduced using Equation 20.24 if D in Equation 20.24 is replaced by D_{eq}. With the measured D_{21} and D_{eq}, the EDC of the porous sample can be calculated with the flowing formula (Zamel et al., 2010)

$$D_{eff} = \frac{l}{Z/D_{eq} - (Z-l)/D_{21}} \tag{20.29}$$

In the case that the porous sample is of double layers, such as a catalyst layer deposited on a substrate, the diffusion resistances of the bulk diffusion, the substrate, and the catalyst layer connect in series, forming an equivalent diffusion resistance R_{ceq}. Mathematically, the R_{ceq} can be written as (Shen et al., 2011)

$$R_{ceq} = \frac{Z}{D_{ceq}A} = \frac{Z-l_s-l_c}{DA} + \frac{l_s}{D_{seff}A} + \frac{l_c}{D_{ceff}A}. \tag{20.30}$$

In Equation 20.30), l_s and l_c are the thicknesses of the substrate and catalyst layer, respectively. D_{seff} and D_{ceff} are the EDCs of the substrate and catalyst layer, respectively. D_{ceq} is the equivalent diffusion coefficient in the presence of the catalyst layer and the substrate. The EDC of the catalyst layer D_{ceff} can be found as (Shen et al., 2011)

$$D_{ceff} = \frac{l_c}{Z/D_{ceq} - (Z - l_s - l_c)/D_{21} - l_s/D_{seff}}. \tag{20.31}$$

In summary, it is convenient to use Equation 20.24 to process experimental data for a short-time ($t \leq 0.1\tau$) diffusion experiment. In this short period, the Loschmidt diffusion cell can be considered an infinite space for the diffusion. The diffusion coefficient D in Equation 20.24 represents D_{21} or D_{eq}, when one measures bulk diffusion coefficient or the EDC of a porous sample. With the measured D_{21} and D_{eq}, the EDC can be deduced using Equation 20.29 or 20.31 for a single-layer or double-layer porous sample.

20.3 Experimental Apparatus and Measurement

For the research and manufacture of PEM fuel cells, the knowledge of EDC under different temperature and relative humidity (RH) conditions is of great significance. The experimental apparatus for measuring the EDC consequently should have the capability to determine the EDC under various experimental conditions. The Loschmidt diffusion cell is the heart of the apparatus. With the diffusion cell, the experimental system that provides gases, temperature and RH control, and proper experimental procedure, the EDC can be precisely determined.

20.3.1 A Loschmidt Diffusion Cell

Rohling et al. (2007), Astrath et al. (2009), and Zamel et al. (2010) demonstrated an in-house made Loschmidt diffusion cell, consisting of two chambers, similar to the one shown in Figure 20.6. As a part of the bottom chamber, a flat sliding gate 5 connects the top and bottom chambers. The upper side of the gate is at the middle of the Loschmidt diffusion cell ($z = 0$). To measure the oxygen concentration evolution in the top chamber, an oxygen sensor 6 is placed in the top chamber and close to the gate (e.g., $Z = 1.68$ cm in Figure 20.1). In order to maintain one-dimensional diffusion, the size of the oxygen sensor should be much smaller than the lateral dimension of the Loschmidt diffusion cell. Consequently, an optic fiber oxygen sensor is a good candidate, such as Ocean Optics FOXY-AL300 sensor with a diameter of 0.3 mm, which is small enough for a Loschmidt diffusion cell of a diameter of 20.6 mm (Rohling et al., 2007). Between the gate and oxygen sensor, a sample holder 9 is placed. Without or with a porous sample in the sample holder, the bulk or equivalent diffusion coefficient (D_{21} or D_{eq}) can be measured. Two humidity sensors, 7 and 8, are placed in the top and bottom chambers, respectively, to monitor the RH inside of the diffusion cell. These humidity sensors can also be employed to measure water vapor diffusion when the RH values in the two chambers are not identical.

20.3.2 Experimental System

An experimental system for the EDC measurement is shown in Figure 20.7. The system consists of four major parts: a diffusion cell, temperature control, a dry and humidified gas supply, and system automation.

The temperature control consists of temperature controllers, resistance heating wires, and type T thermocouples to maintain the desired temperature (e.g., from room temperature to 80°C), accurate temperature control, and homogeneous temperature distribution in the diffusion cell.

To supply humidified gas to the diffusion cell, a bubble humidifier can be used. After passing through the humidifier to get desired RH, two gases (such as, O_2 and N_2) fill into the bottom and top chambers

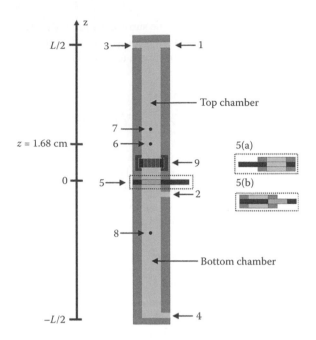

FIGURE 20.6 A schematic diagram of a cylindrical Loschmidt diffusion cell 1: inlet 1 for gas 1 (e.g., N_2); 2: Inlet 2 for gas 2 (e.g., O_2); 3 and 4: outlets; 5: a flat sliding gate; 5(a): gate opened; 5(b): gate closed; 6: oxygen sensor; 7 and 8: humidity sensors; 9: sample holder.

of the diffusion cell through the inlets 2 and 1 in Figure 20.6, respectively. Two mass flow meters are employed to control the flow rates of the gases. To avoid water condensation in the hoses connecting the humidifier and the diffusion cell, custom-made hot hoses with thermocouples integrated are used. A temperature controller is employed to control the temperature of the hoses.

Together with the temperature control of the diffusion cell up to 80°C, the integration of the humidifier into the diffusion cell enables the effective-diffusion-coefficient measurement of a porous sample, such as a catalyst layer, performed with different temperatures and humidity content, which is similar to the environment inside a PEM fuel cell. The EDC thus measured has practical significance for fuel cell research on both catalyst layer development and water management.

The diffusion measurement system can be fully automated, except changing porous material samples. As shown in Figure 20.7, a computer is used to control the experimental apparatus with a computer program, for example, a LabView program, including changing gas filling into the diffusion cell as well as oxygen concentration and relative humidity measurements. The system can be controlled remotely through Internet. The automaton of the apparatus provides an experimental foundation for highly repeatable measurements. Furthermore, it saves researcher's time, accelerating the progress of the measurement.

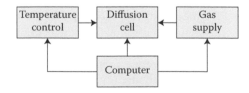

FIGURE 20.7 A diagram of an experimental system for the measurement of the effective binary gas diffusion of porous materials.

20.3.3 Experimental Procedure

The procedure of the EDC measurement includes filling gases into the diffusion cell chambers, measuring the diffusion gas concentration evolution in the top chamber, and processing the gas concentration evolution signal. The procedure can be found in literature (Rohling et al., 2007; Astrath et al., 2009; Zamel et al., 2010).

To fill the gases into the diffusion cell, there are four steps. Referring to Figure 20.6, as the first step, with the inlet 2 and outlet 3 closed as well as the gate opened (5a), both chambers (top and bottom) are filled with gas 1 (e.g., N_2) through inlet 1 for 15 min at a flow rate of 800 mL per minute. Meanwhile, outlet 4 is open to expel the originally existed gas. After both inlet 1 and outlet 4 are closed, the gate is closed (5b), and the bottom chamber is filled with gas 2 (e.g., O_2) through inlet 2 for 12 min at a flow rate of 800 mL/min as the second step; at the same time, outlet 4 is open to drive out gas 1. As the third step, both inlet 2 and outlet 4 are closed. Finally, both outlet 3 and outlet 4 are opened for around 15 s and then closed as the fourth step. Consequently, the pressure inside of the diffusion cell is kept as the ambient pressure (1 atm).

After the gases are filled into the diffusion cell, the O_2 concentration evolution in the top chamber is measured using the oxygen sensor. Before the diffusion starts, signal from the oxygen sensor is monitored for some time (e.g., 120 s) to confirm the system is stable. This signal is also served as a baseline of the oxygen concentration evolution. When the flat sliding gate is opened, the gas diffusion starts and the concentration of O_2 in the top chamber begins to increase. The time-resolved signal from the O_2 sensor is recorded. Least-squares curve-fitting method can be used to fit the signal to Equation 20.24 to find the bulk diffusion coefficient D_{21} or the D_{eq} for a porous sample. With the D_{21} and D_{eq}, the EDC D_{eff} of the porous sample can be calculated.

20.4 Applications

In this section, application examples of the diffusion theory and apparatus will be presented, including the measurements of the bulk binary gas diffusion and the EDCs of porous materials of PEM fuel cells.

20.4.1 The Effects of Temperature and Humidity on Bulk Binary Gas Diffusion

The EDC of a porous material is correlated to bulk gas diffusion coefficient, and PEM fuel cells run at around 80°C with high humidity. Realizing the importance of the bulk diffusion changing with temperature and relative humidity, Astrath et al. studied the effects of temperature and humidity on bulk diffusion (Astrath et al., 2009). Figure 20.8 exhibits the bulk diffusion coefficients of dry O_2–N_2 gas mixture at different temperatures, which are consistent with the Fuller, Schettler, and Giddings (FSG) equation (Fuller et al., 1966). The bulk diffusion coefficient increases with the increase of temperature, due to that the mean thermal speed (V) of the gas mixture rises at a higher temperature.

Different from the dry gas diffusion, the bulk O_2–N_2 gas diffusion in the presence of the humidity involves oxygen, nitrogen, and water vapor and could be ternary diffusion. Under the condition that the relative humidity values remained the same in both chambers of a Loschmidt diffusion cell, using Stefan–Maxwell's model, Astrath et al. (2009) proved that no effective water vapor diffusion would occur in the diffusion cell, and the O_2–N_2 diffusion in the presence of water vapor could be considered binary diffusion. Using Equation 20.24, Astrath et al. measured the bulk O_2–N_2 gas diffusion coefficients vs. various RH values. Their experimental results are presented in Figure 20.9.

In the experiment, the total pressure in the diffusion cell was remained at 1 atm. According to Dalton's law of partial pressures, the presence of water vapor reduces the partial pressure or the mass density of oxygen and nitrogen in the mixture. In comparison with the dry O_2–N_2 mixture, the mole mass of the O_2–N_2–water vapor mixture is smaller. Therefore, the presence of water vapor increases the mean

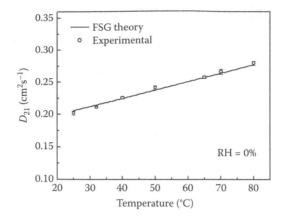

FIGURE 20.8 Temperature dependence (25–79°C) of the diffusion coefficient of binary gas mixture of O_2–N_2 measured under dry condition Open circles, experimental data; solid line, the theoretical prediction using the FSG equation. The following parameters were used in the theoretical calculations: $(\Sigma \upsilon)_{N_2} = 18.5$; $(\Sigma \upsilon)_{O_2} = 16.3$ $P = 1$ atm; $M_{N_2} = 28$ g mol^{-1} and $M_{O_2} = 32$ g mol^{-1}. (Adapted from Astrath, N. G. C. et al. 2009. *J. Phys. Chem. B* 113: 8369–8374.)

thermal speed (V), which is directly proportional to the temperature T and inversely proportional to the mole mass of the gas mixture. Besides, water vapor has smaller atomic and structural diffusion volume increment υ compared with oxygen and nitrogen (Fuller et al., 1966), which means that the presence of water vapor molecules reduces the effective cross section. Consequently, the presence of water vapor increases the mean free path (λ), which is inversely proportional to the product molecule number per unit volume and the effective cross-section area for collision in the gas mixture. According to the hard-sphere approximation, diffusion coefficient is directly proportional to the product of the mean thermal speed (V) and mean free path (λ): $D_{21} \propto V\lambda$. In the presence of water vapor content measured by RH, the behavior of the diffusion coefficient is governed by the increase of the mean thermal speed V and mean

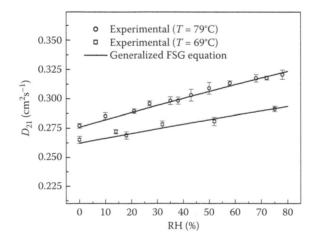

FIGURE 20.9 Humidity dependence of the diffusion coefficient of a binary gas mixture of O_2–N_2 at 69°C (open squares) and at 79°C (open circles). The continuous lines represent the theoretical prediction using the generalized FSG equation, Equation 20.32. The following parameters were used in the theoretical calculations: $\upsilon_{H_2O} = 13.1$, $\upsilon_{N_2} = 18.5$, $\upsilon_{O_2} = 16.3$, $M_{H_2O} = 18$ g mol^{-1}, $M_{N_2} = 28$ g mol^{-1}, $M_{O_2} = 32$ g mol^{-1}, $P = 1$ atm. (Reprinted from Astrath, N. G. C. et al. 2009. *J. Phys. Chem. B* 113: 8369–8374. With permission.)

free path λ, due to the molecular barriers created by the vapor induce less resistance to the mass transport (Astrath et al., 2009).

The FSG equation probably could be generalized to accommodate the influence of RH on the bulk diffusion coefficient and can be mathematically expressed as (Astrath et al., 2009)

$$D_{21}^{RH} = \frac{1.00 \times 10^{-3} T^{1.75} \left(1/(\eta_1^{RH} M_1) + 1/(\eta_2^{RH} M_2) \right)^{1/2}}{P \left[\left(\eta_1^{RH} \upsilon_1 + \eta_{H_2O}^{RH} \upsilon_{H_2O} \right)^{1/3} + \left(\eta_2^{RH} \upsilon_2 + \eta_{H_2O}^{RH} \upsilon_{H_2O} \right)^{1/3} \right]^2}. \tag{20.32}$$

Here η_1^{RH}, η_2^{RH}, and $\eta_{H_2O}^{RH}$ are the mole fractions of gas 1, gas 2, and water vapor, respectively. υ_{H_2O} is the atomic and structural diffusion volume increment of water vapor. The prediction of the generalized FSG equation is consistent with the experimental results as shown in Figure 20.9.

20.4.2 Binary Gas Diffusion in GDL

PEM fuel cells are increasingly designed to operate at high-current densities (2 A cm^{-2} or even higher). However, at these high-current densities, the loss owing to mass transport is significant (Zamel et al., 2010). As a result, the knowledge of the mass transport limitation in porous materials (such as, GDLs and catalyst layers) of PEM fuel cells is critical for the design. For this reason, Zamel et al. measured the EDCs of GDLs made of carbon paper using a Loschmidt diffusion cell and compared their experimental results with the predictions by existing theoretical models (Zamel et al., 2010).

Figure 20.10 shows that the EDCs of a GDL (TGP-H-120) increase with the rise of temperature. However, the slope of the increasing EDCs is less than that of the bulk diffusion coefficient. The diffusibility, the ratio of the EDC to the bulk one, $Q = D_{eff}/D_{bulk}$, changing with temperature is shown in Figure 20.11. The least-squares linear fitting to the experimental data reveals the relation between the diffusibility (Q) and the temperature (T in °C) is

$$Q = 0.251 + 3.10 \times 10^{-4} T. \tag{20.33}$$

The slope 3.10×10^{-4} is very small, exhibiting that the temperature has little effect on the overall diffusibility (Zamel et al., 2010).

Zamel et al. (2010) pointed out that the existing models available in literature over predicted the EDC significantly, by as much as a factor of 2.5. For example, using a porosity of 75.5% the Bruggeman model predicts that $Q = 0.650$, which is much larger than the measured one, probably due to that the model was developed based on spherical particles that form the porous materials, and the carbon paper is obviously not spherical-particle formed. The structure of the carbon paper is then the determinant of the magnitude of diffusibility (Zamel et al., 2010). Using a Loschmidt diffusion cell, Zamel et al. (2010) also experimentally examined the effects of Teflon treatment (wt%) and porosity of the GDL on the diffusibility.

20.4.3 Binary Gas Diffusion of Catalyst Layers

The Loschmidt diffusion cell has also been employed to measure the EDCs of CCLs of PEM fuel cells (Shen et al., 2011). Each CCL sample contained 30 wt% Nafion ionomer and 46 wt% Pt/graphitized carbon; the sample was deposited onto the Al_2O_3 membrane substrates, Anopore Inorganic Membranes (Anodisc 25, Whatman) using an automated spray coater. The thickness of eight CCL samples varied from 6 to 29 μm, corresponding to the Pt loadings of 0.2–0.8 mg cm^{-2}. A Hg porosimetry was employed to measure porosity and pore size distribution of the catalyst layers. In general, the pore size distribution

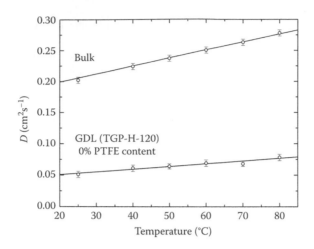

FIGURE 20.10 The temperature dependence of bulk and EDCs of a GDL (TGP-H-120) of dry O_2–N_2 mixture at 1 atm measured using a Loschmidt diffusion cell. (Reprinted from *Chem. Eng. Sc.*, 65, Zamel, N. et al. Experimental measurements of effective diffusion coefficient of oxygen–nitrogen mixture in PEM fuel cell electrode. 931–937, Copyright (2010), with permission from Elsevier.)

of these CCLs ranged from 10 to 200 nm, including meso (2–50 nm diameter) and macro (>50 nm) pores. The peak pore size for these CCLs was about 50 nm. The porosities of the CCLs with different thicknesses were subsequently calculated to be 30–40% (Shen et al., 2011).

The assembly of a CCL on the top of a substrate is a two-layer porous medium. In order to determine the EDC of the CCL using Equations 20.30 and 20.31, the knowledge of the substrate EDC is required. The average thickness of the substrate is 60 μm, and the EDC of the substrate can be measured as a single-layer porous sample using a Loschmidt diffusion cell as well as Equations 20.24, 20.28, and 20.29. Figure 20.12 presents the typical evolution curves of oxygen concentration in the top chamber of the diffusion cell with an Anodisc 25 substrate. The average equivalent binary gas diffusion coefficient D_{seq} of three Anodisc was found to be $D_{seq} = (1.90 \pm 0.03) \times 10^{-5} \, \text{m}^2 \, \text{s}^{-1}$. Using Equation 20.29 and bulk O_2–N_2

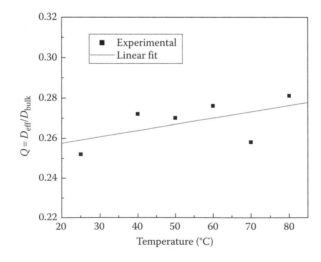

FIGURE 20.11 Linear fitting to the diffusibility Q of a GDL (TGP-H-120) changing with temperature.

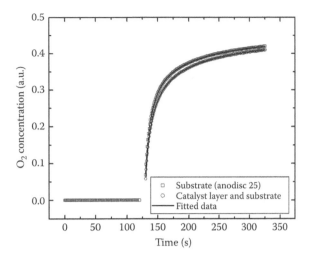

FIGURE 20.12 The evolution curves of oxygen concentration in the top chamber of the Loschmidt diffusion cell with an Anodisc 25 substrate and a 29 μm thick CCL on the top of the substrate, respectively. (Reprinted from *J. Power Sources*, 196, Shen, J. et al. Measurement of effective gas diffusion coefficients of catalyst layers of PEM fuel cells with a Loschmidt diffusion cell. 674–678, Copyright (2011), with permission from Elsevier.)

diffusion coefficient $D_{21} = 2.02 \times 10^{-5} \, \text{m}^2 \, \text{s}^{-1}$ (Astrath et al., 2009), the EDC of the substrate was calculated, and the arithmetic average of the EGCs of the three substrates was found to be $D_{\text{seff}} = (1.3 \pm 0.3) \times 10^{-6} \, \text{m}^2 \, \text{s}^{-1}$ (Shen et al., 2011).

With the value of D_{seff}, the EDC of a CCL can be deduced using Equation 20.31 after measuring the equivalent diffusion coefficient D_{ceq}. With the fact that the CCLs were fabricated with the same composition and the same deposition conditions, one may consider the microstructures of these CCLs to be similar in a statistical sense. Consequently, regression method may be used to deduce the D_{ceff} of the CCLs. Figure 20.13 shows the values of D_{ceq} of each assembly of a CCL on the top of a substrate. Rearranging Equation 20.31, one may have

$$D_{\text{ceq}} = \frac{Z}{(Z - l_s - l_c)/D_{21} + l_s/D_{\text{seff}} + l_c/D_{\text{ceff}}} \tag{20.34}$$

The averaged D_{ceff} value of the eight CCLs was attained by fitting Equation 20.34 to the experimental data and found to be $D_{\text{ceff}} = (1.47 \pm 0.05) \times 10^{-7} \, \text{m}^2 \, \text{s}^{-1}$.

In comparison with the prediction of Bruggeman model (Equation 20.2, $m = 1.5$), this EDC is much smaller. The porosity of the CCL samples is 30–40%, the EDC calculated with Bruggeman model would be around $D_{\text{ceff}} = 2.02 \times 10^{-5} (0.35)^{1.5} = 4.18 \times 10^{-6} \, \text{m}^2 \, \text{s}^{-1}$, more than one order larger than the measured one. This significant difference probably comes from the Knudsen diffusion (Equation 20.3). When pore size is as small as one micrometer, the Knudsen effect is not negligible (Mu et al., 2008). When pore size is less than 100 nm, the Knudsen effect becomes dominant, and Equations 20.4 and 20.5 should be used to calculate the EDC. Mu et al. (2008) exhibited that if the mean pore size was 50 nm and porosity was 40% of a catalyst layer, the EDC of the catalyst layer after taking into account the Knudsen effect would be about 15% of that without considering the Knudsen effect. The measured EDC was basically consistent with the prediction by Mu et al. The Bruggeman model overestimates the EDC mainly because this model does not consider the Knudsen effect.

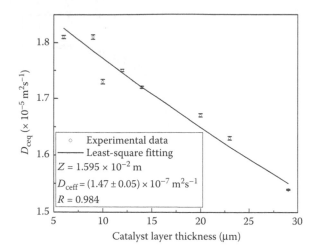

FIGURE 20.13 Experimental data of the equivalent diffusion coefficients of eight CCLs of different thicknesses and the least-squares curve-fitting Equation 20.34 to the experimental data: $Z = 1.595 \times 10^{-2}$ m, D_{seff} $=(1.3 \pm 0.3) \times 10^{-6}$ m^2 s^{-1}, and $l_s = 6.0 \times 10^{-5}$ m were used in the fitting. The correlation coefficient between the fitting and the experimental data $R = 0.984$ was found. (Reprinted from *J. Power Sources*, 196, Shen, J. et al. Measurement of effective gas diffusion coefficients of catalyst layers of PEM fuel cells with a Loschmidt diffusion cell. 674–678, Copyright (2011), with permission from Elsevier.)

In Equation 20.34, the possible diffusion resistance of the interface between the catalyst layer and the substrate was not considered. For the purpose to find the possible diffusion resistance, the Equation 20.30 can be rewritten as

$$R_{ceq}A = \frac{Z}{D_{ceq}} = \frac{Z - l_s - l_c}{D_{21}} + \frac{l_s}{D_{seff}} + \frac{l_c}{D_{ceff}}$$
$$= \left(\frac{1}{D_{ceff}} - \frac{1}{D_{21}}\right)l_c + \left(\frac{Z - l_s}{D_{21}} + \frac{l_s}{D_{seff}}\right).$$

(20.35)

with Equation 20.28, Equation 20.35 can be obtained

$$R_{ceq}A = \left(\frac{1}{D_{ceff}} - \frac{1}{D_{21}}\right)l_c + \frac{Z}{D_{seq}},$$

(20.36)

which assumes that there is no diffusion resistance in the interface. Equation 20.36 reveals that the product of the equivalent diffusion resistance and the diffusion area is a linear function of the thickness l_c of the catalyst layer, and the intercept is Z/D_{seq} if there is no diffusion resistance in the interface. To find the value of the intercept, a least-squares linear fitting of Equation 20.36 to experimental data was performed, as shown in Figure 20.14. The intercept thus found was 8.33×10^3 m^{-1} s. With $D_{seq} = (1.90 \pm 0.03) \times 10^{-5}$ m^2 s^{-1}, the calculated Z/D_{seq} was in the range of $8.26-8.53 \times 10^3$ m^{-1} s. The intercept 8.33×10^3 m^{-1} s was in the range, and one could conclude that the interface resistance was negligible.

20.5 Outlook

The Loschmidt diffusion cell has been used to precisely measure bulk binary gas diffusion coefficients and the through plane EDCs of porous materials (e.g., GDLs and CCLs) of PEM fuel cells. The binary

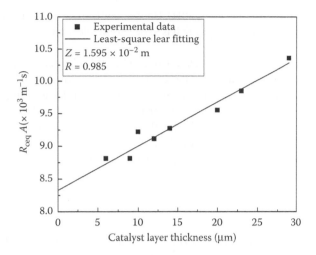

FIGURE 20.14 A best linear fitting to the experimental data of $R_{ceq}A$ (i.e., Z/D_{ceq}). (Reprinted from *J. Power Sources*, 196, Shen, J. et al. Measurement of effective gas diffusion coefficients of catalyst layers of PEM fuel cells with a Loschmidt diffusion cell. 674–678, Copyright (2011), with permission from Elsevier.)

gas diffusion theory presented in this chapter is based on Fick's second law of diffusion and the concept of equivalent diffusion resistance.

For a double-layer porous material, such as a catalyst layer deposited on the top of a substrate, Equation 20.31 can be used to deduce the EDC of the catalyst layer. Any fluctuation (ΔD_{ceq}) in the measurement of D_{ceq} can lead to a variation (ΔD_{ceff}) of D_{ceff}, as shown in

$$\Delta D_{ceff} = \frac{dD_{ceff}}{dD_{ceq}} \Delta D_{ceq}. \tag{20.37}$$

dD_{ceff}/dD_{ceq} can be calculated from Equation 20.31. Figure 20.15 reveals that dD_{ceff}/dD_{ceq} increases when the catalyst layer thickness decreases, and when the thickness of the catalyst layer is thinner than 10 μm,

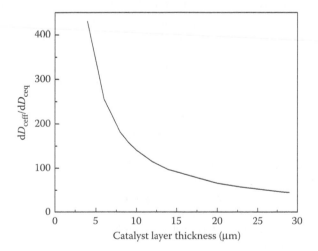

FIGURE 20.15 Numerical simulation of dD_{cef}/dD_{ceq} of Equation 20.31. (Reprinted from *J. Power Sources*, 196, Shen, J. et al. Measurement of effective gas diffusion coefficients of catalyst layers of PEM fuel cells with a Loschmidt diffusion cell. 674–678, Copyright (2011), with permission from Elsevier.)

the dD_{ceff}/dD_{ceq} increases rapidly (Shen et al., 2011). Consequently in order to measure the EDC of a thin catalyst layer precisely, the experimental system has to be very stable, that is, less fluctuation in the measurement of D_{ceq}.

The applications of Loschmidt diffusion cell introduced in this chapter are under the experimental condition of RH = 0. The fundamental understanding of the effects of the composition (e.g., ionomer content) and microstructure of a catalyst layer on the EDC are of scientific and engineering significance. The behavior of the EDCs of GDLs and CCLs with various temperatures and humidity content of gases will be an important research topic, especially for the optimization of GDLs and CCLs. The theory, experimental apparatus, and experimental procedure of the Loschmidt diffusion cell method presented in this chapter have laid a solid foundation for binary gas diffusion measurement and will play a significant role in the research on PEM fuel cells.

References

Arnold, J. H. 1930. Studies in diffusion. I-estimation of diffusivities in gaseous systems. *Ind. Eng. Chem.* 22: 1091–1095.

Astrath, N. G. C., Shen, J., Astrath, F. B. G. et al. 2010. Determination of effective gas diffusion coefficients of stainless films with differently shaped holes using a Loschmidt diffusion cell. *Rev. Sci. Instrum.* 81: 046104-1-3.

Astrath, N. G. C., Shen, J., Song, D. et al. 2009. The effect of relative humidity on binary gas diffusion. *J. Phys. Chem. B* 113: 8369–8374.

Baranski, J., Bith, E., and Lehmann, J. K. 2003. Determination of binary diffusion coefficients of gases using holographic interfereometry in a Loschmid's cell. *Int. J. Thermophys.* 24:1207–1220.

Bruggeman, D. A. G. 1935. Berechnung verschiedener physicalischer knostaten von heterogenen substanzen. *Ann. Phys.* 24: 636–664.

Callaghan, P. T. 1994. *Principles of Nuclear Magnetic Resonance Microscopy*. New York, NY: Oxford University Press.

Chapman, S. and Cowling, T. G. 1952. *The Mathematical Theory of Non-Uniform Gases*. London: Cambridge University Press.

Chen, N. H. and Othmer, D. P. 1962. New generalized equation for gas diffusion coefficient. *J. Chem. Eng. Data* 7: 37–41.

Crank, J. 1975. *The Mathematics of Diffusion* (2nd edn.). New York, NY: Oxford University Press.

Gilliland, E. R. 1934. Diffusion coefficients in gaseous systems. *Ind. Eng. Chem.* 26: 681–685.

Flückiger, R., Freunberger, S. A., Kramer, D. et al. 2008. Anisotropic, effective diffusivity of porous gas diffusion layer materials for PEFC. *Electrochim. Acta* 54: 551–559.

Fuller, E. N., Ensley, K., and Giddings, J. C. 1969. Diffusion of halogenated hydrocarbons in helium. The effect of structure on collision cross sections. *J. Phys. Chem.* 73: 3679–3685.

Fuller, E. N. Schettler, P. D., and Giddings, J. C. 1966. A new method for prediction of binary gas-phase diffusion coefficients. *Ind. Eng. Chem.* 58: 19–27.

Hirschfelder, J. O., Curtiss, C. F., and Bird, R. B. 1954. *Molecular Theory of Gases and Liquids* (2nd edn.). New York, NY: Wiley.

Huang, T.-C., Yang, F. J. F., Huang, C.-J., and Kuo, C. H. 1972. Measurements of diffusion coefficients by the method of gas chromatography. *J. Chromatogr.* 70: 13–24.

Kramer, D., Freunberger, S. A., Flückiger, R. et al. 2008. Electrochemical diffusimetry of fuel cell gas diffusion layers. *J. Electroanal. Chem.* 612: 63–77.

Karaiskakis, G. and Gavril, D. 2004. Determination of diffusion coefficients by gas chromatography. *J. Chromatogr. A* 70: 147–189.

Kolaczkowski, S. T. 2003. Measurement of effective diffusivity in catalyst-coated monoliths. *Catalysis Today* 83: 85–95.

Majors, P. D., Smith, D. M., and Davis, P. J. 1991. Effective diffusivity measurement in porous media via NMR radial imaging. *Chem. Eng. Sci.* 46: 3037–3043.

Mu, D., Liu, Z-S., and Huang, C. 2008. Determination of the effective diffusion coefficient in porous media including Knudsen effects. *Microfluid Nanofluid* 4: 257–260.

Park, I.-S. and Do, D. D. 1996. Measurement of the effective diffusivity in porous media by the diffusion cell method. *Catal. Rev. Sci. Eng.* 38: 189–247.

Pollard, W. G. and Present, R. D. 1948. On gaseous self-diffusion in long capillary tubes. *Phys. Rev.* 73: 762–774.

Rohling, J. R., Shen, J., Wang, C. et al. 2007. Determination of binary diffusion coefficients of gases using photothermal deflection technique. *Appl. Phys. B* 87: 355–362.

Shen, J., Zhou, J., Astrath, N. G. C. et al. 2011. Measurement of effective gas diffusion coefficients of catalyst layers of PEM fuel cells with a Loschmidt diffusion cell. *J. Power Sources* 196:674–678.

Zamel, N., Li, X., Astrath, N. G. et al. 2010. Experimental measurements of effective diffusion coefficient of oxygen–nitrogen mixture in PEM fuel cell electrode. *Chem. Eng. Sci.* 65: 931–937.

Zamel, N., Li, X., and Shen, J. 2009. Correlation for the effective gas diffusion coefficient in carbon paper diffusion media. *Energy Fuels* 23: 6070–6078.

Zhang, F., Hayes, R. E., and Kolaczkowski, S. T. 2004. A new technique to measure the effective diffusivity in a catalyst monolith washcoat. *Trans IChemE*, Part A, *Chem. Eng. Res. Des.* 82(A4): 481–489.

21

Gas Permeability of Proton-Exchange Membranes

Dmitri Bessarabov
North-West University

21.1 Introduction and Background

21.1.1 Gas Permeability in the MEA Components

Gas permeation in the membrane electrode assembly (MEA) components such as proton-exchange membranes (PEMs), anode and cathode catalyst layers (CLs), gas-diffusion layers (GDLs), and seals includes a number of complex mass-transport phenomena, the importance of which is well recognized, but detailed mechanisms of the processes are not fully understood in some cases.

Figure 21.1 provides schematic of gas and vapor transport phenomena in a typical catalyst-coated membrane (CCM) (Bessarabov and Hitchcock, 2009) in a hydrogen fuel cell. Figure 21.1 does not discuss either actual electrochemical processes of electrode kinetics or gas transport in GDLs. In Figure 21.1, process 1 denotes back diffusion of water in a PEM from the cathode to the anode. Process 2 denotes hydrogen permeation from the anode to the cathode without electrochemical oxidation.

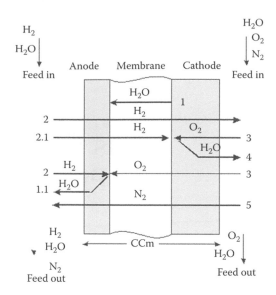

FIGURE 21.1 Schematic diagram of gas transport phenomena occurring in the CCM of a fuel cell.

Process 2.1 denotes hydrogen permeation from the anode to the cathode with electrochemical oxidation coupled with oxygen permeation (process 3) and related water generation (process 4). Similarly, oxygen can permeate the PEM (process 3) and react with hydrogen producing water (process 1.1). Finally, nitrogen permeation is denoted by process 5. Other gases such as He, CO, CO_2, NH_3, and H_2S may also be present in the air and/or fuel feed. Water and water vapor permeation phenomena in PEMs were covered, for example, by Majsztrik et al. (2007) and are not discussed in this chapter.

Gas permeability term is also often applied to such hydrogen fuel cell stack components as bipolar plates, GDLs and seals. For example, Blunk et al. (2006) investigated hydrogen permeation in association with using thin, highly filled composite plates. The factors affecting gas permeation, such as plate temperature, thickness, graphite loading, and aging were discussed. Hydrogen permeation through the anode plate into the liquid coolant may result in the build-up of the hydrogen bubbles to the unsafe concentration levels. According to Blunk et al. (2006), these bubbles may also result in poor heat rejection, as well as formation of hot spots and premature thermally induced MEA failures.

Gas-diffusion media or GDLs of the fuel cell MEA provide oxygen and hydrogen with access from flow-field channels to CLs. In this case, gas permeability is referred to the property of GDL. More specifically, in-plane and through-plane gas permeability processes are considered. Gas transport models as applied to diffusion in micro-porous media are typically used to discuss mechanism of permeation in GDLs (Mathias et al., 2003; Pharoah, 2005; Astrath et al., 2010; Yu and Carter, 2010) and are not discussed in this chapter.

Gas permeability of CLs of the fuel cell MEA was discussed by Hiramitsu et al. (2010) in terms of efficiency of oxygen permeation. It was shown that the nature of an ionomer in the CL may influence cold start-up of PEM fuel cells. Some related measurement details and hardware description were presented by Shen et al. (2011).

The effect of gas permeability of seals in large fuel cell stacks on durability and fuel efficiency is not fully understood yet and is a subject of ongoing research.

This chapter will only focus on a detailed discussion of gas permeation in fuel cell membranes.

21.1.2 Membrane Gas Permeability

An advanced hydrogen-based fuel cell power train system for the automotive industry must meet a number of demanding requirements to become commercially attractive. These include requirements of

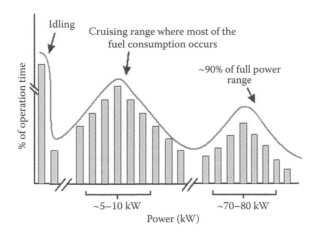

FIGURE 21.2 Driving profile of a FCV: operation time (%) vs. power consumption (kW). (From Lee, S., Bessarabov, D., and Vohra, R. 2009. *Int. J. Green Energy* 6: 594–606. With permission.)

fuel cell vehicles (FCVs), such as improvements in fuel economy, power efficiency, and durability to provide suitable vehicle range and costs.

Key challenges that are directly associated with MEA components include relatively high MEA costs, relatively low mass activity of the cathode catalyst under current catalyst loadings, relatively low MEA durability, poor ionic conductivity of solid polyelectrolyte (SPE) membranes at low relative humidity and high temperature, and excessive gas permeability of the perfluorosulfonic acid (PFSA) SPE membranes (Lee et al., 2009). For example, nitrogen gas permeability of humidified membranes in automotive fuel cell stacks is required to be less than 4 Barrer permeability units (Kundu et al., 2009). Other membrane requirements for automotive application are also listed on the DOE website (Web ref (a)).

An approximate representation of the time spent by an FCV at certain power levels during a drive cycle is shown in Figure 21.2. For example, the low-power (5–10 kW) "cruising regime" is where most of the fuel is consumed. To meet this requirement, a value of 0.84 V at 0.1 A cm^{-2} was estimated as a baseline reference and tentative requirement point (Lee et al., 2009).

Initial PEM fuel cell development was based on using relatively thick PEMs (~0.175 mm). One of the ways to increase power efficiency of modern PEM-based fuel cells includes fabrication and use of thinner PEMs. As membranes in automotive fuel cells become thinner (~25 μm range and less), gas crossover in a membrane is becoming one of the critical variables that limits membrane thickness and directly affects fuel economy.

Increased gas permeation rate in fuel cell membranes not only results in inefficient fuel utilization, but also leads to mixed electrochemical potentials and other fuel-cell-specific operational problems (Bessarabov and Kozak, 2007a). Hydrogen and oxygen gases that permeate through the membrane are consumed producing heat and water leading to fuel inefficiency. The cell voltage obtained at low current densities is strongly affected (reduced) by gas crossover.

Extensive diffusion of oxygen from the cathode to the anode may lead to the formation of hydrogen peroxide, which can significantly shorten the lifespan of the membrane under certain conditions in fuel cell stacks. On the other hand, an increase in nitrogen concentration in a fuel tank due to the permeation process might cause a significant change in overall fuel cell performance if hydrogen fuel is recycled during a fuel cell stack operation and the diffusion rate of nitrogen gas in membranes is high.

New polymeric materials for automotive fuel cell applications (perfluorinated and hydrocarbon-based) are also being developed and, therefore, further research in membrane gas permeation properties is required.

21.2 Effect of Membrane Gas Crossover on Fuel Cell Operations

21.2.1 Membrane Durability

Research and development in the area of hydrogen fuel cells composed of polymeric PEMs has led to an understanding of an important effect of gas permeation, which takes place in automotive fuel cell stacks, and how it affects their overall performance. The effects of hydrogen and oxygen permeation on cell polarization, as well as the effects of the crossover on membrane degradation, are well known (Zhang et al., 2006; Mittelsteadt and Liu, 2009).

As membranes in automotive fuel cells become thinner, the risk of membrane failure becomes considerably higher because of various stressors. These stressors, which affect membrane durability, could be related to

1. Membrane chemical degradation due to high gas permeation rate
2. Cell design parameters (such as compression at specific locations)
3. MEA properties (such as specific membrane mechanical and/or chemical properties)
4. Operating conditions (such as wet–dry cycles and temperature)

The diffusion of oxygen from the cathode to the anode in a PEM automotive fuel cell (process 3 in Figure 21.1) can result in the formation of aqueous hydrogen peroxide. Decomposition of hydrogen peroxide in the presence of various cationic impurities (e.g., Fe^{2+}) could further produce active radicals (hydroxy- and peroxy-) such as OH• and OOH•.

Hydroxy- and peroxy- radicals are believed to attack H-containing terminal bonds formed during the polymer and membrane manufacturing processes and initiate decomposition (Curtin et al., 2004). The specific H-containing end groups such as carboxylic acid, perfluorovinyl, and difluoromethyl were discussed by Pianca et al. (1999). Recent research results indicate that sulfonic acid groups of membranes are possible sites for radical attack (Qiao et al., 2006). Electrochemical and chemical processes along with other membrane stressors, such as changes in membrane water content, temperature, and mechanical stress, would accelerate membrane degradation, resulting in membrane thinning, tensile strength loss, and pinhole formation. Other stress factors that could accelerate membrane degradation by means of a radical attack mechanism include elevated temperature (above 90°C) and low relative humidity.

An additional contribution to membrane degradation, though possibly a minor one, includes localized heat generation during direct combustion of hydrogen and oxygen on the anode side of the fuel cell, leading to fuel inefficiency (Inaba et al., 2006).

According to Mittal et al. (2007), a catalytic mechanism of oxygen "activation" under reductive conditions may also result in the formation of active radicals. It is further proposed that the reaction mechanism of oxygen activation in fuel cells may be independent of potential and is chemical in nature (Mittal et al., 2007). The reaction of active oxygen species with a membrane could result in membrane degradation on the cathode side of a fuel cell.

Electrochemical mechanisms leading to radical formation in a PEM were discussed earlier in terms of "reductively activated oxygen" by Yamanaka and Otsuka (1991) is shown in Figure 21.3.

Figure 21.4a,b show dynamics of pinhole formation in a PEM subjected to various stressors. Formation of pinholes results in an increase in gas-crossover (Weber, 2008). This chapter does not discuss gas-crossover via pinholes and focuses on the fundamentals of gas permeation in beginning-of-life (BOL) membranes. For detailed report on the pinhole formation in fuel cell membranes caused by such factors as membrane aging, excessive local heat generation, fuel cell configuration (design and components), and operating conditions see report by Stanic and Hoberecht (2004).

21.2.2 Fuel Efficiency

The range of a typical FCV is significantly limited by the amount of hydrogen that can be stored in such tanks because compressed hydrogen tanks store fuel at lower energy density compared to liquid

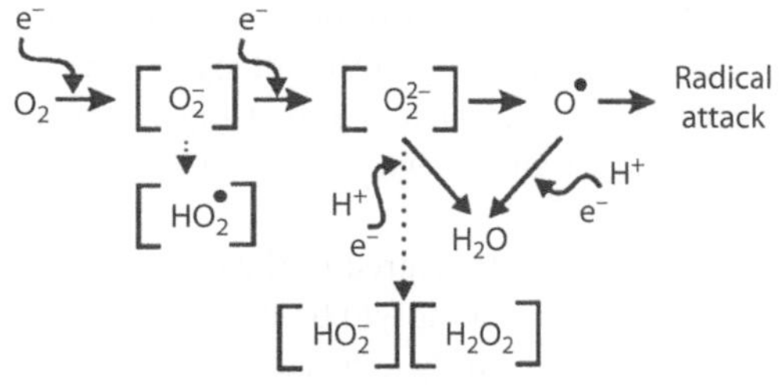

$$O_2 \xrightarrow{e^-} \left[O_2^- \right] \xrightarrow{e^-} \left[O_2^{2-} \right] \rightarrow O^\bullet \rightarrow \begin{array}{c} \text{Radical} \\ \text{attack} \end{array}$$

$$\left[HO_2^\bullet \right] \qquad H^+ \begin{Bmatrix} \\ e^- \end{Bmatrix} \quad H_2O \xleftarrow{\raisebox{1ex}{H^+}}_{e^-}$$

$$\left[HO_2^- \right]\left[H_2O_2 \right]$$

FIGURE 21.3 "Reductively activated oxygen" mechanism of active radical formation enhanced by the presence of "Fenton ions" in a PEM system that may play an important role in membrane degradation on the cathode side of a fuel cell. Fenton-like cationic impurities accelerate ionomer degradation. (From Yamanaka, I. and Otsuka, K. 1991. *J. Electrochem. Soc.* 138: 1033–1038. With permission.).

fuels. Modern hydrogen tanks on board of FCVs provide compressed hydrogen storage of up to 70 MPa so far.

Recent research activities have been mainly addressing fundamentals of hydrogen and oxygen diffusion in PEMs. In hydrogen/air fuel cells diffusion of nitrogen to the fuel circulation loop determines the minimum stoichiometry that can be used for operation. Much less information is available in the area of nitrogen permeation in automotive thin PEMs and how nitrogen permeation affects fuel cell performance (Kocha et al., 2006).

Nitrogen permeating from the cathode side does not have a large effect on the polarization of the cell; however, it will dilute the hydrogen. According to Kocha et al. (2006), localized accumulation of nitrogen in fuel cells of a stack may contribute to a localized fuel starvation and may lead to the cathode carbon corrosion described by Zhang et al. (2009) and Liang et al. (2009).

As nitrogen accumulates in the anode loop, it may cause other losses. In FCVs fuel is typically recirculated by a pump (Kocha et al., 2006; Ahluwalia and Wang, 2007). The energy used by the recirculation pump is proportional to the gas volume recirculated. With an increase in nitrogen concentration

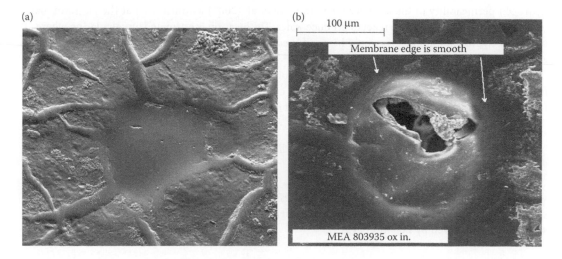

FIGURE 21.4 Thinning of an SPE membrane caused by various stressors—resulting in pinhole formation. (a) Membrane thinning, (b) pinhole formation (Reprinted from *Membr. Technol.*, 12, Bessarabov, D. and Kozak, P., Measurement of gas permeability in SPE membranes for use in fuel cells, 6–9, Copyright (2007a), with permission from Elsevier.)

in the fuel, the energy required by the pump is also increased. This energy must be supplied by the FCV that results in a larger stack. In order to keep the nitrogen content at required low levels, the anode loop must be purged, which results in the loss of hydrogen. If nitrogen permeation is reduced sufficiently, the need for recirculation is reduced, while adequate fuel economy and anode stability are still maintained.

Given the magnitude of permeation rates measured in the state-of-the-art (SOA) automotive PFSA membranes, it is unlikely that the PFSA membrane will be able to be significantly thinner in the future, unless the permeability of nitrogen is significantly reduced or adequate hydrocarbon membranes with lower gas permeability are developed (Sexsmith, 2009).

21.2.3 Platinum Precipitation in a PEM

The cathode potential in an automotive fuel cell typically ranges between approximately 0.6 and 0.95 V during normal operation. However, deviations of up to 1.5 V are possible during start-up and shut-down events (Bett et al., 2005). Even though high voltages encourage the formation of a Pt oxide layer, which has been shown to slow the Pt dissolution rate (Uchimura et al., 2008), a voltage cycle that repeatedly forms and removes this layer, such as in a vehicle cycle, may have a very high dissolution rate. From Pourbaix diagrams, it is shown that Pt can dissolve at potentials >1 V (Pourbaix, 1966). Basic mechanisms of Pt dissolution and platinum particles formation and characterization in PFSA membranes have been studied by many groups (Tseung and Dhara, 1975; Honji et al., 1988; Bessarabov et al., 1997, 1998; Antolini, 2003; Mukerjee and Srinivasan, 2003; Xie et al., 2005; Yasuda et al., 2006; de Bruijn et al., 2008; Uchimura et al., 2008; Lee et al., 2009; Zhang et al. 2009).

A common observation from voltage cycling experiments is the migration of platinum into the electrolyte membrane. Figure 21.5 shows a transmission electron microscopy (TEM) image of a CCM after voltage cycle experiments. Platinum particles can be seen in the membrane where it has formed a distinct platinum band approximately 3.8–4.2 μm from the cathode CL edge. Another observation is the presence of a platinum-depleted region in the CL at the catalyst-membrane interface. It indicates that the platinum band originated from the cathode CL.

It has been proposed that the equilibrium location of the Pt band is determined by the oxygen and hydrogen permeability of the ionomer membrane. Bi et al. (2007) assumed that at the platinum band location, hydrogen and oxygen react to produce water. The location is estimated by calculating where all crossover hydrogen and oxygen would be consumed, thus, where concentration of the gases becomes zero.

The Pt band location could be estimated as the point in the membrane where hydrogen crossing over from the anode will completely react with oxygen crossing over from the cathode. Following Bi et al. (2007), the location of the Pt band can be estimated from

$$\frac{1}{x} = 1 + \frac{\alpha_{H_2/O_2}}{2} \frac{p_{H_2}}{p_{O_2}} \tag{21.1}$$

where p is gas partial pressure, x is the dimensionless distance in the membrane with $x = 0$ as the cathode membrane/CL interface, $x = 1$ as the anode membrane/CL interface, α_{H_2/O_2} is the membrane gas selectivity and is given by Equation 21.2.

$$\alpha_{H_2/O_2} = \frac{P_{H_2}}{P_{O_2}} = \frac{D_{H_2} \times S_{H_2}}{D_{O_2} \times S_{O_2}} \tag{21.2}$$

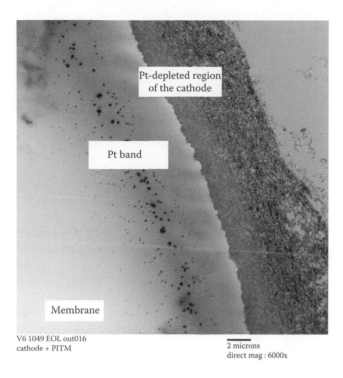

V6 1049 EOL out016
cathode + PITM

2 microns
direct mag : 6000x

FIGURE 21.5 TEM cross-section of a degraded CCM after voltage cycling. (Reprinted from *Membr. Technol.*, 10, Kundu, S. et al. Fingerprint of the automotive fuel cell cathode catalyst degradation: Pt band in the proton-exchange membranes, 7–10, Copyright (2009), with permission from Elsevier.)

where P is gas permeability, D is gas diffusion coefficient in a membrane, and S is Henry's gas solubility. It is seen that the location of the Pt band in a membrane may be influenced not only by operation conditions (humidity and partial pressure), but also by diffusion properties of the ionomer material.

Oxygen and hydrogen membrane gas selectivity in a PFSA membrane was measured *ex situ* at Automotive Fuel Cell Cooperation Corp. (AFCC), as described by Bessarabov and Kozak (2007), and was found to be in the range of 2–2.3 under given relative humidity and temperature conditions (Kundu et al., 2009). Using this selectivity, the Pt band would be expected to form between 3.6 and 4 µm from the cathode, which is in good agreement with the band location as shown in Figure 21.5.

21.2.4 CO_2 and CO Permeation

Permeability of carbon dioxide and oxygen in Nafion 1100 EW membrane has been studied as a function of relative humidity or water content by Ma et al. (2005). The rate of permeation of CO_2 was found to depend strongly on the water content of the membrane. This work mainly targeted direct methanol fuel cells (DMFC) applications, but it has a technical relevance to hydrogen fuel cells as well. It was demonstrated that the permeability coefficient of carbon dioxide increases by two orders of magnitude with the increased water uptake, while the effect is much less pronounced for oxygen. The difference is attributed to the different gas solubilities in water. It is suggested that the CO_2 transport mainly occurred through the hydrophilic region (Ma et al., 2005). It was also further confirmed that carbon dioxide permeation behavior in hydrated Nafion EW1100 is a cofunction of a loss of water content in the membrane, decrease of CO_2 solubility and increase in diffusivity with increasing temperature. It was suggested that hydrophilic domains of PFSA membranes are the major channels for carbon dioxide diffusion. It was confirmed that the permeation behavior in dry Nafion follows the Arrhenius expression. The data were approximated by the following equation:

$$P = 5.43 \times 10^{-11} \exp\left(-\frac{13.18\,\text{kJ}/\text{mol}}{RT}\right) \tag{21.3}$$

The activation energy of carbon dioxide permeation in dry Nafion EW1100 is thus estimated to be 13.18 kJ mol^{-1} by Ma et al. (2007).

The published CO_2 permeation data published are relevant from fundamental and practical points of view for automotive fuel cell applications.

Carbon monoxide (CO) levels as low as 100 ppm in the H_2 feed stream cause a substantial loss in performance in PEMFC operating at 80°C. At this temperature level, the Pt anode CL is deactivated by the adsorption of CO if it is present in the feed gas at such a partial pressure range. The mitigation strategies to improve anode tolerance for CO include special catalysts and/or anode catalyst structure development (Shi et al., 2007).

As membranes in automotive fuel cells become thinner, CO permeation in such membranes may result in critical performance reduction of a fuel cell system due to cathode catalyst poisoning by permeated CO (provided that the fuel feed may contain CO impurities). Some limited data on mass transport properties of CO are available in the literature. CO permeability in a cast Nafion film was measured using a volumetric method under a partial pressure difference (Yasuda and Shimidzu, 1999).

According to Sethuraman et al. (2009), Arrhenius equations for the diffusion coefficient and solubility of CO in Nafion 112, respectively, are

$$D_{CO,\text{Nafion}112} = 4.02 \times 10^{-4} \exp\left(-\frac{2406}{T}\right) \tag{21.4}$$

$$C_{CO,\text{Nafion}112} = 5.43 \times 10^{-6} \exp\left(\frac{449.8}{T}\right) \tag{21.5}$$

More data are required to understand fully CO permeation mechanisms as a function of temperature and humidity in commercial PFSA membranes used in automotive applications as well as in combined heat and power (CHP) applications and potential effect on cathode performance if CO is present in the hydrogen feed and diffused through the PEM. It is envisaged that CO permeation phenomena may play a detrimental role in other fuel cell applications, such as stationary applications, when hydrogen feed is more likely to have higher CO content.

21.2.5 Gas Permeability Requirements for a Cathode in a Pem Fuel Cell

The scope of this chapter does not include detailed quantified analysis of oxygen permeation requirements for the cathode CLs. A brief qualitative discussion is only given. A typical cathode CL includes an ionomer layer and catalytic particles. The rate of oxygen diffusion through such a layer should be as high as possible to increase electrocatalytic efficiency of the fuel cell cathode, provided that other mass-limiting factors associated with a fuel cell system are optimized. It is generally accepted that oxygen transport properties of CLs comprising an ionomer layer and catalyst particles play a detrimental role in providing oxygen to the active sites of a catalyst (Bernardi and Verbrugge, 1992; Buchi et al., 1996; Kulikovsky, 2002).

As reported in many previous publications, for example, by Scott and Mamlouk (2009), the generic polarization curve of the hydrogen fuel cell could be expressed by the terms comprising the Nernst equation, the Butler–Volmer equation, and by an ohmic term as a function of local catalyst conditions. In particular, it was shown that the increase in the oxygen concentration on the catalyst sites results in an increase of a fuel cell performance.

In addition to mass transport effects, an experimental evidence of the effect of oxygen solubility on the kinetics of oxygen reduction was shown by Maruyama et al. (2001). More specifically, it was shown that the kinetics of oxygen reduction on Au electrodes was strongly dependent on the molecular structure of fluorinated additives. Results were discussed in terms of an increase in oxygen solubility in the adsorption layer and availability of the reaction sites on Au electrodes. Experimental methods to measure gas permeation in membranes described further in this chapter can be applied to study gas transport properties of CLs.

21.2.6 Summary

The gas permeability of H_2, O_2, N_2, He, CO, CO_2 seems to have a fundamental influence on all cell phenomena at continuous and transient operation modes (start up, shut down, etc.). The following main effects of gas permeability in membranes were discussed in the literature.

H_2 permeability: affects OCV at start-up, causes a significant fuel loss, mainly at low loads, decreases "real" fuel stoich, decreases cell efficiency, likely influences membrane degradation, plays a role in the formation of the Pt band in the membrane.

O_2 permeability: likely influences membrane degradation; in the fuel starvation case, permeated O_2 may turn cells into a cathode corrosion mode, may decrease "real" fuel stoichiometry and decrease cell efficiency; plays a role in the formation of the Pt band in the membrane. On the other hand, high oxygen permeability and solubility are required for improved cathode performance.

N_2 permeability: significantly affects fuel economy of the system, may lead to low cells and cathode corrosion by building up, causes change of anode gas composition by diluting H_2.

CO permeability: may cause a significant decrease of the cathode performance if permeated across the PEM. It is expected that an increase in the CHP fuel cell applications, where hydrogen is primarily produced from the natural gas, will result in additional fundamental work in the area of CO permeation.

CO_2 and He permeability: may change fuel stoichiometry, however, additional research is required.

Measurements of gas permeation properties in a PEM and establishing related test procedures are important tasks. Section 21.3 provides additional details on the subject.

21.3 Fundamentals of Steady-State Membrane Gas Permeability Measurements

21.3.1 Introduction

Up until recently measurements of gas permeability (steady-state and transient) in polymeric membranes have mostly been a focus of research in the membrane gas-separation community. Industrial applications of gas separation membranes in such areas as air separation, CO_2 purification, hydrogen and hydrocarbon separation and purification as well as a few selected alcohols and water pervaporation processes have successfully driven membrane research from fundamentals to commercial success in a few selected areas (Bessarabov, 1999). Most of the fundamental gas permeability data were obtained for dry single gases or binary gas mixtures across an ambient temperature range to reflect typical operation conditions in gas separation modules and/or packaging films. Extensive experimental methods, procedures and mathematical treatments have also been developed for both steady-state and transient measurements with a focus on conventional nonproton-exchange polymeric gas separation membranes.

PEM fuel cells operate at much wider temperature range (−40–120°C) and various humidity levels. These conditions are not typical for most gas permeability experiments. Lack of data for oxygen, nitrogen, and hydrogen permeability coefficients in various polymeric materials suitable for use as a PEM under elevated temperatures and various humidity levels requires further research. Measuring permeability coefficients at different relative humidity levels also presents an experimental challenge that would be typically avoided in traditional gas permeation experiments. Permeability of gases in membranes under operational conditions existing in a PEM fuel cell has thus become one of the key properties that require comprehensive research and method development.

The existing electrochemical protocols to measure hydrogen and oxygen crossover as a function of electrical current are applicable to MEAs. These experimental procedures may be difficult to use within the wide humidity range and at temperatures above 100°C in a single membrane when fast membrane screening is required.

21.3.2 Definition of Permeability and Measurement Units

The measurable steady-state flux density (J) of a gas permeating through a nonporous (dense) membrane under Fick's law approximations could be presented in terms of a membrane permeability coefficient (P):

$$J/A = \frac{P}{\ell} \Delta p \tag{21.6}$$

where P is the permeability coefficient, ℓ is the membrane thickness, A is the membrane area, and Δp is the differential partial pressure.

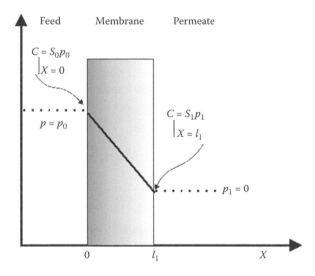

FIGURE 21.6 Schematic diagram of the linear distribution of gas concentration (C) in a membrane. $C_{(X=0)}$ is gas concentration at the membrane inlet. S_0 and p_0 are solubility and partial pressure of the feed gas at the membrane inlet, respectively. $C_{(x=\ell 1)}$ is gas concentration at the membrane permeate side. S_1 and p_1 are solubility and partial pressure of the feed gas at the permeate side, respectively. (Reprinted from *Membr. Technol.*, 12, Bessarabov, D. and Kozak, P., Measurement of gas permeability in SPE membranes for use in fuel cells, 6–9, Copyright (2007a); Bessarabov, D., and Kozak, P., Steady-state gas permeability measurements in proton-exchange membranes for fuel cell application using the GC method. Paper presented at *Annual Meeting of North American Membrane Society (NAMS2007)*, Orlando, FL. Copyright (2007b), with permission from Elsevier.)

For many practical applications it is assumed that there is a linear concentration profile of a gas in a membrane (Ash et al., 1965). The schematic of the linear concentration profile of a gas in a thin dense membrane can be expressed by the diagram in Figure 21.6, where $C_{(X=0)}$ denotes concentration at the inlet of the membrane under partial pressure, $p = p_0$, $C_{(X=\ell 1)}$ denotes concentration at the outlet of the membrane under partial pressure conditions of $p = p_1$. S is the gas solubility. In many cases during gas permeation experiments partial pressure of the permeate is kept at zero (i.e., $p_1 = 0$).

As a widely accepted practical approximation, it is assumed that solubility of gases (S) in membranes under industrial operation conditions follows the Henry's law:

$$C = S \times p \tag{21.7}$$

where C is the gas concentration in membranes (cm^3(gas) cm^{-3}(polymer)) and p is the gas partial pressure. In this case S is the solubility coefficient and has the following units:

$$cm^3(gas) cm^{-3}(polymer) cm^{-1} Hg$$

The investigation of the sorption thermodynamics in a great variety of PFSA polymers used in automotive applications is difficult due to water saturation effects. Fuel cells operate in the presence of water and in the wide temperature range. From the practical point of view measurements of gas permeability in dry PFSA membranes have limited relevance to automotive applications.

It is reasonable to assume that in the presence of absorbed water in the PFSA membrane channels, gas concentration profile could be described by the dual sorption equation, for example,

$$C = Sp + \frac{C_H^* bp}{1 + bp} \tag{21.8}$$

where S is the Henry's law solubility coefficient, b is the Langmuir affinity parameter (adsorption equilibrium constant), and C_H^* is the Langmuir capacity parameter (sorption capacity for the "Langmuir" solute molecular population). The first term in the equation corresponds to the Henry isotherm, the second to the Langmuir isotherm. Other types of sorption isotherms could also be used. Little research has been done in this area and only a few relevant papers are available (Petropoulos, 1992, 1994; Mann et al., 2006; James et al., 2008).

Combining Equations 21.6 and 21.7 results in Equation 21.9 for membrane gas flux density:

$$J/A = \frac{P}{\ell} \Delta p = DS \frac{\Delta p}{\ell} \tag{21.9}$$

where D is the apparent diffusion coefficient, assumed to be a constant, cm^2 s^{-1}.

It is important to note here that the use of only one apparent diffusion coefficient for water-saturated PFSA membranes should be treated as an assumption. Obviously, due to the presence of water in channels of PFSA membranes, multiple diffusion coefficients should be considered.

In the idealized situation, gas permeability coefficient can be defined from Equation 21.9 as a product of diffusion and solubility coefficients:

$$P = D \times S \tag{21.10}$$

It is seen from Equation 21.10 that permeability coefficient P is a function of two variables. In this case, S is a thermodynamic variable (sorption), and D is a kinetics term (diffusion).

Permeability coefficient (constant) is typically expressed in Barrer units.

$$1 \text{ Barrer} = 10^{-10} \text{ cm}^3 \text{ (STP) cm cm}^{-2} \text{ s}^{-1} \text{ cm}^{-1} \text{Hg} = 7.5 \times 10^{-18} \text{ m}^3 \text{(STP) m m}^{-2} \text{ s}^{-1} \text{ Pa}^{-1}$$

The merit of the Barrer composite permeability unit is that it includes practical variables such as membrane area, applied differential pressure and thickness, which allows the calculation of a gas flux, but other units could be used. If the thickness of a membrane (ℓ) is not known, the term permeance is used (Hwang et al., 1974; Yasuda, 1975; Editorial, 2008).

$$\text{Permeance} = P/\ell \tag{21.11}$$

Diffusion coefficients of gases and vapors in polymer membranes depend upon a number of variables which characterize the polymer. These include density of a polymer, presence and frequency of cross-linking, crystallinity degree, nature of the interaction between chains, nature of the substituents, and so on. The diffusion coefficients for gases in different polymers vary in substantially broader ranges than the solubility coefficients.

In the case of PFSA polymers used in fuel cell applications, diffusion coefficients depend strongly on the amount of absorbed water and temperature.

The temperature dependence of the diffusion coefficient and the solubility coefficient can be represented by the following equations:

$$D = D_0 \exp(-E_D/RT) \tag{21.12}$$

$$D = S_0 \exp(-\Delta H_S/RT) \tag{21.13}$$

where E_D is the activation energy of diffusion and ΔH_S is the heat of solution. R is the gas constant.

It is important to note that diffusion and sorption coefficients are fundamental parameters that describe gas mass transport phenomena in dense thin film materials.

The driving force for gas transport in dense polymeric membranes is the concentration gradient across the membrane. The solubility parameter may play a role in determining final permeability. It is generally accepted that the interaction between a permeate (i.e., gas diffusing through the membrane) and nonreacting polymeric membranes is governed by Van Der Waals interactions between a permeate and a polymeric matrix of the membrane. If there is a stronger specific interaction (i.e., electronic donor–acceptor interactions, reversible coordination, hydrogen bond-based interaction, etc.) between the target gas and/or the vapor component and polymeric matrix, then a considerable range in permeability values can be expected. For example, this is the case for NH_3 permeation in PFSA membranes (He and Cussler, 1992).

The permeability coefficient varies with the morphology of polymers and depends on many physical properties such as density, degree of crystallinity, and so on. Structural modification of membranes may result in the change of morphology-related features (e.g., crystallinity, free volume, etc.) that may affect the diffusion coefficient.

As shown by Equations 21.9 and 21.10, steady-state gas flux measurements do not provide information on individual values of the diffusion and sorption coefficients.

For example, Table 21.1 reports permeability data for CO_2 and hydrogen gases in membranes made of a glassy polymer such as polyvinyltrimethylsilane (PVTMS) by Teplyakov and Meares (1990). It is seen

TABLE 21.1 Hydrogen and CO_2 Permeability, Diffusion and Sorption Coefficients in the PVTMS Membrane by Teplyakov and Meares (1990)

Gas	$D \times 10^7$, cm^2 s^{-1}	$S \times 10^3$, cm^3 (gas) cm^{-3} (polymer) cm^{-1} Hg	$P \times 10^8$, cm^3 (STP) cm cm^{-2} s^{-1} cm^{-1}Hg
H$_2$	180	1.1	2.0
CO$_2$	5.0	38.0	1.9

that permeability constants and, therefore, corresponding gas fluxes obtained as a result of steady-state experiments are almost the same. However, diffusion and sorption coefficients measured during separate experiments differ considerably. This indicates different contribution of kinetic and thermodynamic terms to the gas permeation mechanism in the same polymer membrane.

Under steady-state diffusion conditions, permeability coefficient P could also be expressed in terms of steady-state cross-over current density (i_∞):

$$i_\infty = \frac{nFP\Delta p}{\ell} \tag{21.14}$$

where n assumed to be 2 for hydrogen and 4 for oxygen, F is the Faraday constant, P is permeability, Δp is gas partial pressure difference.

21.3.3 Nonelectrochemical Methods

A number of methods and commercial apparatus for measuring gas permeability of polymer films have been developed. Generally, the equipment is rather complex and specific to certain gases. Some equipment and methods generate steady-state permeability data only; some provide diffusion coefficients as well. Nonelectrochemical methods to measure gas permeability typically follow either integral permeation or differential permeation type of experimental procedures.

In integral permeation experiment, a gas permeates through a membrane into a closed chamber and a signal is generated to represent the total mass of the permeating gas. The total amount of gas can be determined experimentally by either a manometric or a volumetric technique.

In the case of the differential method, a flow of a sweep gas continuously removes the permeated gas of interest from one side of the membrane to a detector. The detector provides a concentration signal proportional to the permeation rate.

One of the most versatile methods for gas permeation measurements is based on the differential flow methodology. Although the differential method requires more sensitive analytical equipment, it has some important advantages: high-pressure measurements are possible without imposing stresses on the membrane as the same pressure can be applied on both sides of the membrane (Zanderighi et al., 1979; ASTM F1927-98). This is especially critical when an evaluation of thin PEM membranes (up to 5 μm) is required. Also, the differential flow method allows good control of relative humidity in the feed and sweep gases on both inlet and outlet simultaneously.

This section covers basic principles of steady-state gas permeability measurements using the differential method for fuel cell applications.

21.3.4 Diffusion Cell Design Guidelines

A gas permeation test cell should provide two compartments that are separated by the test membrane. One compartment is the high-feed concentration side, which receives the permeate gas from the feed and vents it out. The second compartment is the lower-permeate concentration side, which receives the carrier (sweep) gas and transports it to a gas detector.

Fundamental principles of a cell design could be found in the following references (Yasuda and Rosengren, 1970; Pye et al., 1976; Zanderighi et al., 1979).

Figure 21.7 provides a schematic of a typical permeation cell design (Bessarabov et al., 1999). A membrane or another thin-film barrier material is sandwiched between flow field plates (feed-side and receiver-side). An O-ring ensures leak-free conditions. A feed gas enters the feed-side compartment, gets distributed evenly along one side of the membrane sample, and comes out from the same cell compartment. The permeate gas stream is collected in the receiver compartment by a sweep-gas and analyzed further using, for example, gas chromatography (GC).

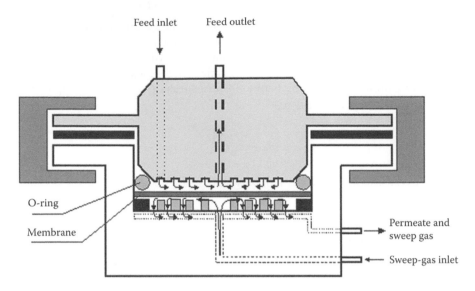

FIGURE 21.7 Schematic of a typical permeation cell and a possible experimental gas flow arrangement. (Reprinted from *Sep. Purif. Technol.*, 16, Bessarabov, D. G. et al., Separation of 1-hexene/*n*-hexane mixtures using a hybrid membrane/extraction system, 167–174, Copyright (1999), with permission from Elsevier.)

A diffusion cell includes the following typical components: flow field plates, Swagelok fittings, a flow plate holding block (the feed-side of the cell), a flow plate holding block (the receiver compartment of the cell), a flow plate holding frame, a flat spacer, O-ring, a through-hole screw a PTFE spacer, and a seal.

A summary of key issues that should be considered when designing a gas permeation cell and a related experiment protocol using GC as a detection method are presented further below, as per Pye et al. (1976), Yasuda and Rosengren (1970), and Zanderighi et al. (1979).

- In general, sensitivity of gas concentration and related gas flux measurements by GC depend on the ratio of the membrane area to the volume of the receiving compartment, in which the total amount of gas transported through the membrane is measured. The sensitivity of the detecting device should be adjusted accordingly.
- A linear dependence of the thermal conductivity on the concentration of the gas carrier with a TCD GC detector is always found in the range of small concentrations. Thus, it is necessary to conduct gas permeation experiments in the low range of concentrations.
- A properly designed experimental method should not depend on the total pressure difference across the membranes.
- A high surface-to-volume ratio allows to maintain boundary conditions of the first order and make permeability calculations with good accuracy.

Figure 21.8 provides a basic schematic of a diffusion cell set-up for steady-state permeability measurements with an option to humidify feed and/or sweep carrier gas streams (Bessarabov and Kozak, 2007b).

A sweep gas, which is the same as a carrier gas in a GC (e.g., He), is fed to the receiver compartment of the cell. A feed gas (or a feed-gas mixture) is fed to the upper compartment of the diffusion cell (the high-permeate concentration side of the cell). The upper compartment of the cell receives the feed gas from the gas cylinder and vents it to the atmosphere. The lower compartment receives the carrier gas from the gas cylinder and also receives the permeated gas through the membrane and sweeps away the resulted gas mixture to the detector for analysis by means of GC. RH sensors provide RH values at the exits of the feed and sweep gases. The flow rate of the sweep gas is typically much lower than that of the

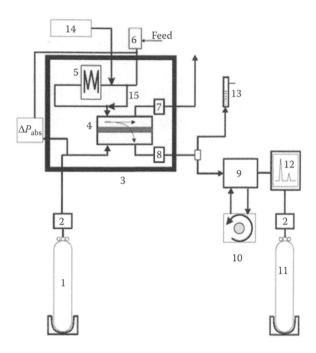

FIGURE 21.8 Basic schematic of a GC and diffusion cell set-up for gas permeability measurements. 1. Sweep gas (same as carrier gas for GC). 2. Pressure regulator. 3. Temperature-controlled chamber. 4. Diffusion cell. 5. Water evaporator (humidifier loop). 6. MF controller. 7 and 8 Outlet RH% sensors. 9. Water vapor condenser. 10. Circulation cooling bath. 11. Gas carrier for GC (same as sweep gas). 12. GC. 13. Soap-bubble flow-meter. 14. Liquid water injection system. 15. Make-up gas loop. (From Bessarabov, D. and Kozak, P. 2007b. Steady-state gas permeability measurements in proton-exchange membranes for fuel cell application using the GC method. Paper presented at *Annual Meeting of North American Membrane Society (NAMS2007)*, Orlando, FL. With permission.)

feed gas, so the RH values are the same. The volume of the cell is minimized to maximize sensitivity. Water is injected into the feed gas using a two-pipe system. In the water evaporator, steam is formed and it is mixed with the feed gas in the make-up feed gas loop.

The permeability constant (P) is calculated using the following equation:

$$P = \frac{C \times F \times \ell}{A \times \Delta p} \quad \left[\frac{cm^3 (STP) cm}{cm^2 \, s \, cm \, Hg} \right] \tag{21.15}$$

where C is the gas concentration measured by GC, F is the sweep gas flow rate at STP, ℓ is the membrane thickness, Δp is the partial pressure difference, A is the membrane sample area. An experimental error is typically 5% due to the nature of the gas-chromatography method.

21.3.5 Electrochemical *ex situ* and *in situ* Methods

Section 21.3.3 describes a classical method for measuring steady-state permeability using the GC detecting method. The advantage of the method includes an option to control humidity of the membrane by adjusting gas humidity levels, which represents real fuel cell operation conditions. For example, the experimental setup used by Broka and Ekdunge (1997) includes equilibration of membrane samples with humidified gases and represents conditions close to fuel cell operations.

The gas permeation data obtained by this method are limited to steady-state permeability data only. Using the steady-state permeation method it is not possible to separate the diffusion coefficient from the gas solubility coefficient. The nonsteady-state measurements using gas chromatography will be discussed further in this chapter. Here a brief discussion of the electrochemical methods to measure gas permeation properties of PFSA membranes is provided.

The challenge is to measure gas permeability properties as a function of temperature, relative humidity, and gas pressure in proton-conducting membranes. Various methods were developed to address these issues. One of the methods is the electrochemical monitoring technique (EMT) (Ogumi et al., 1984, 1985; Tsou et al., 1992; Lehtinen et al., 1998; Haug and White, 2000; Sethuraman et al., 2009). Another well-accepted method includes the use of micro-disk electrodes (Parthasarathy et al., 1991; Buchi et al., 1996; Beattie et al., 1999).

The main difference between these methods is the level of hydration of the membrane sample. In the EMT method, the membrane is in contact with an acidic electrolyte solution. In the micro-electrode method, the membrane sample is exposed to a flow of humidified gas.

The measurements are often time consuming, as special catalyzed membranes are required. All methods yield quite different experimental data (Table 21.2). According to Broka and Ekdunge (1997), neither the EMT nor the micro-disk electrode method gives results corresponding to *in situ* conditions in a fuel cell. In the EMT, the membrane is in direct contact with the liquid electrolyte, whereas the humidity of the membrane in the micro-disk electrode method cannot be controlled precisely. Also, permeability of N_2 is not possible to measure using these electrochemical methods.

One of the methods that is widely used in the automotive fuel cell industry to monitor hydrogen gas permeation includes the use of fuel cell stacks. It can be used as an *ex situ* (single-cell hardware) or an *in situ* (stack level) method. A comprehensive description of the method is given by Kocha et al. (2006).

Figure 21.9 provides a schematic explanation of the *in situ* electrochemical measurement of gas permeation where the gas permeation rate is measured as a mass transfer limited current. Humidified hydrogen is fed on the anode side of the electrochemical cell. The anode side acts as a reference electrode (RE) and a counter electrode (CE). The working electrode (WE) side is fed with a humidified inert gas (Ar, N_2, etc.). The hydrogen crosses over from the anode to the cathode side of the cell and gets oxidized under the applied voltage conditions.

In this electrochemical method, the assumption is that hydrogen diffusing across the membrane is going to be oxidized electrochemically on the opposite side of the membrane. Typical measurements procedure includes potentiostatic steps at a given time at a voltage range between 200 and 600 mV of the cathode potential vs. RHE.

The analysis of the data includes obtaining average current for each potentiostatic step, plotting an *I–V* curve (typically *I* on *y*-axis), fitting a line to the experimental points, calculation of the slope and intercept. The slope is the inverse of the shorting resistance; the intercept is the hydrogen cross-over current. The hydrogen permeability can be calculated from the cross-over current. Hydrogen gas cross-over is typically routinely measured as a part of the MEA evaluation test (Sethuraman et al., 2008). One can also use a very slow potential sweep method. However, when testing various MEAs with different double-layer capacitance, one needs to adjust the scan rate for each MEA.

The gas cross-over current density can be expressed in terms of gas permeability *P* using the following equation:

$$i \left[\frac{A}{cm^2} \right] = \frac{P \times F \times n \times \Delta p}{\ell} \tag{21.16}$$

where $n = 2$ for H_2, *P* is permeability, *F* is the Faraday constant, Δp is the partial pressure difference, and ℓ is the membrane thickness.

It is always important to keep in mind that measurements of gas cross-over current include measurements in an MEA and not in a single membrane.

TABLE 21.2 Updated Oxygen Mass Transport Parameters (Diffusion Coefficient, Solubility, Permeability) for Various PFSA Membranes and PTFE

	"Solubility" as Published		Solubility, S	Diffusion Coefficient, D	Conditions/Assumptions				Permeability, $P(\times10^{10})$	Ref
	Data as Published	Units as Published	$cm^3 \cdot cm^{-3} \cdot cm^{-1}Hg$	Common Units, $cm^2 \cdot sec^{-1}$	T, °C	P, atm	RH%		$P = D \cdot S$	
PTFE	6.00E-01	$cm^3(STP) \cdot cm^{-3}atm^{-1}$	7.89E-03	1.40E-07	20	1	in contact with an electrolyte		11.1	1
PTFE	2.76E-03	$cm^3(STP) \cdot cm^{-3}atm^{-1}$	2.76E-03	1.52E-07	25	1	0		4.2	2
Nafion 117	2.39E-03	$cm^3(STP) \cdot cm^{-3}atm^{-1}$	2.39E-03	4.57E-08	35	various	0		1.1	3
Nafion 117	1.80E-03	$cm^3(STP) \cdot cm^{-3}atm^{-1}$	1.80E-03	8.80E-08	30	various	0		1.6	4
Nafion 117	1.87E-05	$mol \cdot cm^{-3}$	5.51E-03	6.20E-07	25	1	one side soaked in 1N H_2SO_4		34.2	5
Nafion 117	2.60E-02	M	7.66E-03	7.40E-07	25	1	humidified by gas flow		56.7	6
Nafion 120	7.20E-03	M	2.12E-03	2.40E-07	20	1	one side soaked in K_2SO_4		5.1	7
Nafion 120	6.50E-03	M	1.92E-03	2.90E-07	30	1	one side soaked in K_2SO_4		5.6	7
Nafion 120	5.30E-03	M	1.56E-03	4.40E-07	40	1	one side soaked in K_2SO_4		6.9	7
Nafion 120	5.90E-03	M	1.74E-03	5.20E-07	50	1	one side soaked in K_2SO_4		9.0	7
Nafion 117	9.19E-06	$mol \cdot cm^{-3}$	1.35E-03	5.96E-06	30	2	100%		80.7	8
Nafion 117	7.81E-06	$mol \cdot cm^{-3}$	1.15E-03	1.03E-05	70	2	100%		118.7	8
Nafion NRE211	1.60E-05	$mol \cdot cm^{-3}$	2.36E-03	1.04E-06	30	2	100%		24.5	8
Nafion NRE211	1.48E-05	$mol \cdot cm^{-3}$	2.18E-03	2.40E-06	40	2	100%		52.3	8
Nafion NRE211	1.49E-05	$mol \cdot cm^{-3}$	2.20E-03	3.07E-06	50	2	100%		67.4	8
Nafion NRE211	1.47E-05	$mol \cdot cm^{-3}$	2.17E-03	4.08E-06	60	2	100%		88.4	8
Nafion NRE211	1.61E-05	$mol \cdot cm^{-3}$	2.37E-03	2.24E-05	70	2	100%		531.5	8
Nafion 117	1.30E-05	$mol \cdot cm^{-3}$	3.83E-03	7.00E-07	20	1	one side soaked in 0.5N H_2SO_4		26.8	9
Nafion 117	9.30E-06	$mol \cdot cm^{-3}$	2.74E-03	1.90E-06	20	1	100%		52.1	9
Nafion 115	1.00E-05	$mol \cdot cm^{-3}$	5.89E-04	3.00E-06	50	5	100%		17.7	10
Nafion 120	2.70E-05	$mol \cdot cm^{-3}$	1.59E-03	1.80E-06	50	5	100%		28.6	10
Nafion 117	5.20E-06	$mol \cdot cm^{-3}$	1.53E-03	5.60E-07	25	1	100%		8.6	11

(continued)

TABLE 21.2 (continued) Updated Oxygen Mass Transport Parameters (Diffusion Coefficient, Solubility, Permeability) for Various PFSA Membranes and PTFE

	"Solubility" as Published		Solubility, S	Diffusion Coefficient, D	Conditions/Assumptions			Permeability, $P(\times 10^{10})$	Ref
	Data as Published	Units as Published	$cm^3 \cdot cm^{-3} \cdot cm^{-1}Hg$	Common Units, $cm^2 \cdot sec^{-1}$	T, °C	P, atm	RH%	$P = D \cdot S$	
Nafion 117	5.96E−06	mol · cm⁻³	5.86E−04	9.19E−06	30	3	100%	53.8	12
Nafion 117	7.81E−06	mol · cm⁻³	7.67E−04	1.03E−05	70	3	100%	79.1	12
Nafion 117	1.55E−05	mol · cm⁻³	4.57E−03	5.70E−07	25	1	one side is soaked in 1M H_2SO_4	26.0	13
Nafion NRE 211					80	1	85%	46.2	14
Nafion NRE 211					90	1	85%	60.6	14

1   Gibbs, T.K. and Pletcher, D. 1980. *Electrochim. Acta* 25: 1105–1110.
2   Pasternak, R.A. et al. 1970. *Macromolecules* 3: 366–371.
3   Chiou, J.S. and Paul, D.R. 1988. *Ind. Eng. Chem. Res.* 27: 2161–2164.
4   Sakai, T., et al. 1986. *J. Electrochem. Soc.* 133: 88–92.
5   Haug, A.T. and White, R.E. 2000. *J. Electrochem. Soc.* 147: 980–983.
6   Parthasarathy, A. et al. 1991. *J. Electrochem. Soc.* 138: 916–921.
7   Ogumi, Z. et al. 1984. *J. Electrochem.Soc.* 131: 769–773.
8   Peron, J. et al. 2010. *J. Membr. Sci.* 356: 44–51.
9   Lehtinen, T., et al. 1998. *Electrochim. Acta* 43: 1881–1890.
10   Buchi, F.N. et al. 1996. *J. Electrochem. Soc.* 143: 927–932.
11   Gode, P. et al. 2002. *J. Electroanal. Chem.* 518: 115–122.
12   Beattie, P.D. et al. 1999. *J. Electroanal. Chem.* 468: 180–192.
13   Sethuraman, V.A. et al. 2009. *Electrochim. Acta* 54: 6850–6860.
14   Bessarabov, D. and Kozak, P. 2007. *Membr. Techn.* December: 6–9.

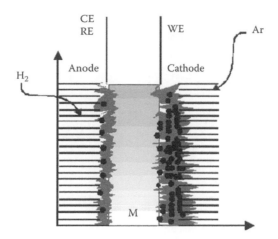

FIGURE 21.9 Schematic explanation of the *in situ* electrochemical measurement of gas permeation of a membrane.

21.3.6 Experimental Results and Discussion

Some of the experimental results measured by the author are presented in this section. Figures 21.10 and 21.11 show the hydrogen and oxygen permeability coefficients of the NRE211 membrane as a function of relative humidity. It is seen that with an increase in water content permeability also increases.

Similar dependencies and values for hydrogen permeation in PFSA membranes were reported by Cleghorn et al. (2003). Also, similar dependencies were reported for nitrogen permeation by Mittelsteadt and Umbrell (2005).

Oxygen mass transport parameters such as the diffusion coefficient and solubility of PFSA materials have been a subject of discussion for the last decade. The issue includes the difficulty of establishing these parameters in the presence of water vapor to address measurements at real operating conditions of fuel cells (Haug and White, 2000).

A critical analysis of the published data on oxygen solubility was conducted by Mann et al. (2006). The issues that were highlighted in the paper included a lack of details on experimental conditions, and a lack of correction of the oxygen partial pressure for the water vapor partial pressure.

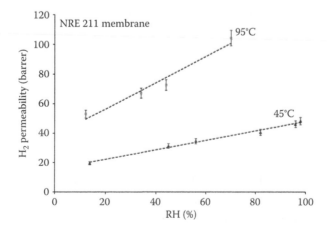

FIGURE 21.10 Hydrogen permeability in a NRE211 membrane as a function of relative humidity. (From Bessarabov, D. and Kozak, P. 2007b. Steady-state gas permeability measurements in proton-exchange membranes for fuel cell application using the GC method. Paper presented at *Annual Meeting of North American Membrane Society (NAMS2007)*, Orlando, FL. With permission.)

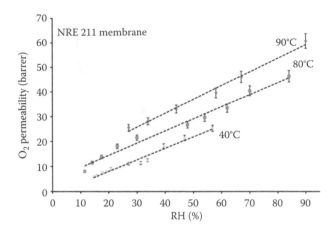

FIGURE 21.11   Oxygen permeability in a NRE211 membrane as a function of relative humidity. (Adapted from Bessarabov, D. and Kozak, P. 2007b. Steady-state gas permeability measurements in proton-exchange membranes for fuel cell application using the GC method. Paper presented at *Annual Meeting of North American Membrane Society (NAMS2007)*, Orlando, FL.)

A review and original experimental data for oxygen mass transport parameters for Nafion 117 membranes were also published by Gode et al. (2002).

More recent updates on gas transport parameters in PFSA membranes including NRE 211 for various gases including oxygen were published by Bessarabov and Kozak (2007a), Sethuraman et al. (2009), and Peron et al. (2010). Bessarabov and Kozak (2007a) also compared the gas permeabilities of PFSA and hydrocarbon membranes.

In this chapter, an updated summary of the oxygen transport parameters for PFSA membranes is given in Table 21.2.

It is seen that the experimental data vary largely. It is proposed in the chapter that a common protocol should be developed and used to measure gas permeability under conditions that are representative of fuel cell operations.

21.4 Nonsteady-State Gas Permeability Measurements

The gas permeability coefficient P is a material property; however, a complete characterization of transport properties of a PFSA membrane requires the separation of P into its components: the diffusion coefficient D and solubility coefficient S. As shown by Equations 21.9 and 21.10, steady-state gas permeation measurements do not provide information on individual values of the diffusion and sorption coefficients. Therefore, nonsteady-state measurement protocols have to be used. This section provides a brief summary of the main principles of designing such experiments and a comparison with electrochemical methods.

21.4.1 Gas Concentration Profile across the Membrane

The one-dimensional distribution of gas concentration (C), when the diffusion coefficient (D) is a constant at a nonsteady state within a nonporous, nonreacting polymeric membrane, is governed by a mathematical solution of the well-known Fick's second law:

$$\frac{\partial C(x,t)}{\partial t} = D \frac{\partial^2 C(x,t)}{\partial x^2} \tag{21.17}$$

where $C(x,t)$ is the concentration of a permeate in the membrane at a position x at a time t.

The solution of Equation 21.17 depends on the boundary conditions. Various solutions have been obtained earlier (Rogers et al., 1954; Crank, 1956) and relevant experimental methods were proposed. It should be noted that a similar mathematical treatment was adopted and used in transient electrochemical measurements, which will be discussed further in the following section.

Typically, a transient experimental method includes simultaneous measurements of the diffusion coefficient, solubility, and permeability for a given gas in a single sample of a film.

It is assumed that the surface concentration of a permeate gas is proportional to its pressure. If sorption of gases in polymeric membranes follows the Henry's law, gas concentration C_0 in a membrane at saturation is presented by Equation 21.7.

21.4.2 Experimental Designs for Gas Diffusion Coefficient Measurements in Membranes

Possible experimental arrangements for measuring the diffusion coefficient in membranes based on the published data are discussed briefly in this section.

1. *Integral method.* It includes measuring the total quantity of a permeate diffused through the membrane sample over a time period (t). In the case of the integral method, the quantity of the permeate $Q(t)$ is calculated as a function of time (t) by means of the following equation:

$$Q(t) = -\int_0^t D \frac{\partial C(x,t)}{\partial x} \partial t \qquad (21.18)$$

The diffusion coefficient can be measured by a so-called time-lag technique associated with the integral method. It involves measuring the cumulative amount of gas that has passed through a membrane. However, the time-lag technique is not efficient when humidification of a membrane sample is required (such as in fuel cell applications) and is not considered in this chapter in details.

2. *Differential method.* It includes measuring a change in the value of the permeate flux at the membrane outlet over time.
3. *Impulse method.* It includes measuring the membrane response to a concentration spike over time.
4. *Sorption method.* It includes measuring the absorption rate of a permeate in the membrane sample.
5. *Transient electrochemical method.* The cross-over current is measured as a function of time. The transient portion of the curve is used to calculate the diffusion coefficient.

This chapter focuses on the differential method as it is the most versatile method for PEM application.

21.4.3 Initial Conditions for the Permeation Experiment

One of the most commonly accepted designs for nonsteady gas permeation experiments includes the following:

1. At the membrane inlet, gas concentration is kept at a constant value (e.g., p_0 is well defined).
2. At the membrane outlet, gas concentration is kept at zero (e.g., $p_0 = 0$).
3. Prior to the experiment, gas concentration throughout the membrane sample is zero (e.g., $p_0 = 0$).

In this case, a simple set of initial and boundary conditions is expressed as follows:

$$C(0,t) = C_0 \tag{21.19a}$$

$$C(\ell,t) = 0 \tag{21.19b}$$

$$C(x,0) = 0 \tag{21.19c}$$

Applications of various analytical solutions of Equation 21.17 for dynamic gas permeation measurements were actively discussed in the past (Barrer, 1941; Ziegel et al., 1969; Pasternak et al., 1970; Yasuda and Rosengren, 1970; Felder, 1978; Marcandalli et al., 1984), and were adopted by groups specializing in membrane physico-chemical research.

An analytical solution of Equation 21.17 for a concentration profile in a membrane sample under boundary and initial conditions (Equations 21.19a, b, and c) is expressed as follows:

$$C(x,t) = \frac{C_0 x}{\ell} + \frac{2C_0}{\pi} \sum_{n=1}^{\infty} \frac{(-1)^n}{n} \sin\frac{n\pi x}{\ell} \exp\left\{-\frac{n^2\pi^2}{\ell^2}Dt\right\} \tag{21.20}$$

It is seen that under steady-state conditions, a linear concentration profile is established in the membrane:

$$C(x,\infty) = \frac{C_0 x}{\ell} \tag{21.21}$$

Owing to relatively small thickness of membrane samples, it is almost impossible to perform experiments to measure a concentration profile of a permeate in membranes unless multiple membrane layers are used. However, it is not practical and, instead, a change in the permeate flux or amount of permeate gas diffused through a membrane is measured.

An increase in the permeate flux over time is measured when using the differential permeability method.

By substitution of the analytical solution (Equation 21.20) of Equation 21.17 into Fick's first law (Equation 21.22):

$$J(t) = -DA\frac{\partial C(x,t)}{\partial x}\bigg|_{x=\ell} \tag{21.22}$$

it is possible to obtain Equation 21.23a, which describes a time-dependent gas flux $J(t)$ exiting from the outlet of the membrane surface. In this case, $J(t)$ is a steady-state flux at the outlet of the membrane with thickness ℓ(e.g., $x = \ell$, $t \to \infty$), A is an active membrane area.

$$J(t) = J_\infty\left[1 + 2\sum_{n=1}^{\infty}(-1)^n \exp\left\{-\left(\frac{n\pi}{\ell}\right)^2 Dt\right\}\right] \tag{21.23a}$$

where

$$J(t) = \frac{DC_0 A}{\ell} = \frac{DSp_0 A}{\ell} = \frac{Pp_0 A}{\ell} \quad \text{at } t \to \infty, \tag{21.23b}$$

The series of Equation 21.23a converge at large values of t. It is also in the suitable experimental measurements range for the transient gas permeation experiments.

If time of diffusion (t) is large, that is, $t \to \infty$ (i.e., there is a so-called steady-state transport through the membrane), then the product of diffusion coefficient (D) and solubility (S) is referred to as permeability (P) (see Equation 21.10).

We are only interested in the value of the permeate flux at the membrane outlet (thickness $x = \ell$). At steady state, a change in flux ΔJ has a limiting value of $\Delta J = DAC_0/\ell$, as C_0 is the initial concentration at the membrane inlet and the concentration of the permeate is kept at zero at the membrane outlet.

21.4.4 Basic Modeling of the Permeability Transient Curve

Equation 21.23a was used to model a gas flux through a membrane as a function of time. The following values for the diffusion coefficient (D) were used:

1. $D = 10^{-6}$ cm^2 s^{-1}
2. $D = 3 \times 10^{-7}$ cm^2 s^{-1}
3. $D = 10^{-7}$ cm^2 s^{-1}
4. $D = 5 \times 10^{-8}$ cm^2 s^{-1}
5. $D = 10^{-8}$ cm^2 s^{-1}

Other parameters were fixed at the following values: the membrane area, $A = 100$ cm^2; partial pressure, $p = 1$ atm; solubility constant, $S = 2 \times 10^{-3}$ cm^3 cm^{-3} atm^{-1}; thickness, $\ell = 0.0025$ cm. The relationship of the flux vs. time is presented in terms of normalized membrane fluxes (J_N): $J_N = J^* \text{max}(J)^{-1}$ in Figure 21.12 Max(J) is the maximum value of each of the fluxes computed for each of the specific diffusion coefficients and Max(J) values are different. The transient time for a kinetic permeation curve was chosen to be 100 s.

In Figure 21.12, curve 1 shows a normalized transient membrane flux as a function of time with the gas diffusion coefficient set at 10^{-6} cm^2 s^{-1}. It is seen that it only takes a few seconds to establish a steady-state flux through the membrane with the thickness 0.0025 cm and solubility 2×10^{-3} cm^3 cm^{-3} atm^{-1}.

Curve 2 shows a normalized transient membrane flux as a function of time with the gas diffusion coefficient set at 3×10^{-7} cm^2 s^{-1}. It is seen that it takes more time to establish a steady-state flux at this value of the diffusion coefficient.

Curve 3 shows a normalized transient membrane flux as a function of time with the gas diffusion coefficient set at 10^{-7} cm^2 s^{-1}. It is seen that a transient period is longer in this case.

Curve 4 shows a normalized transient membrane flux as a function of time with the gas diffusion coefficient set at 5×10^{-8} cm^2 s^{-1}. It is seen that a transient period is even longer in this case.

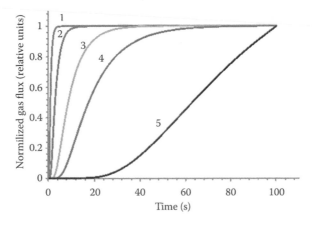

FIGURE 21.12 Shapes of the normalized transient permeability curves for different diffusion coefficients. 1. $D = 10^{-6}$; 2. $D = 3 \times 10^{-7}$; 3. $D = 10^{-7}$; 4. $D = 5 \times 10^{-8}$; 5. $D = 10^{-8}$, cm^2 s^{-1}. (From Bessarabov, D. and Kozak, P. 2007b. Steady-state gas permeability measurements in proton-exchange membranes for fuel cell application using the GC method. Paper presented at *Annual Meeting of North American Membrane Society (NAMS2007)*, Orlando, FL. With permission.)

Curve 5 shows a normalized transient membrane flux as a function of time with the gas diffusion coefficient set at 10^{-8} cm^2 s^{-1}. It is seen that 100 s are not sufficient to establish a steady-state flux under the same conditions.

It is worth to discuss briefly the typical transient permeability curve 3, as shown in Figure 21.12. The curve typically has an asymmetric "S" shape. At the initial time interval, the permeate flux at a membrane outlet increases exponentially. Afterwards, the increase slows down. At a very large time interval, the flux approaches a constant value. The transient permeation curve (flux vs. time) is governed by two variables, that is, P (permeability) and D (diffusion coefficient), which are to be determined experimentally. From the experimental point of view, in certain cases, one needs to consider a third variable, the so-called equipment delay factor, t_e. There is a certain time delay after a fraction of the permeated gas leaves the outlet of the membrane surface until the gas is detected. The equipment delay time is also determined by a time period, which is required for gases to pass through the piping system of an apparatus from the gas cylinder to the diffusion cell. The equipment delay time is especially important in the case of the so-called fast membranes (i.e., when the permeability value is high).

It is seen from the basic modeling results that a change in the values of an effective diffusion coefficient in membranes and/or CLs of FC could significantly change the time required to establish a steady-state gas concentration profile (and, therefore, corresponding permeability values) in MEAs. It may be important for automotive applications when a FC stack frequently operates in transient state.

It is also known from previously published studies that a change in the values of an effective oxygen diffusion coefficient in PFSA membranes and/or CLs of FC could be significant, depending on various water content and temperature (which represent various operation scenarios, such as a wet vs. dry mode, etc.), as well as the nature and quantity of ionomers in a CL and membranes (PFSA vs. hydrocarbons vs. blends vs. composites, etc.).

21.4.5 Interpretation of Transient Permeation Data and Estimation of Diffusion Coefficients

One of the simple methods to determine a diffusion coefficient from a kinetic permeation curve includes the use of special characteristic points of the curve.

Here is an example of how to calculate some of the characteristic points from the following equation see also Equation 21.23a:

$$y = 1 + 2\sum_{n=0}^{\infty}(-1)^n \exp\left\{-\frac{\pi^2 n^2}{\ell^2}Dt\right\} = 1 + 2\sum_{n=0}^{\infty}(-1)^n \exp\left\{-\pi^2 n^2 Bt\right\} \tag{21.24}$$

where $B = D/\ell^2$

As an example, only two components of the series are used in this case:

$$y = 1 - 2e^{-\pi^2 Bt} + 2e^{-\pi^2 4Bt} \tag{21.25}$$

$$y' = 2\pi^2 Be^{-\pi^2 Bt} - 8\pi^2 Be^{-\pi^2 4Bt} \tag{21.26}$$

$$y'' = -2\pi^4 B^2 e^{-\pi^2 Bt} + 32\pi^4 B^2 e^{-\pi^2 4Bt} = 0 \tag{21.27}$$

$$16e^{-4\pi^2 Bt} = e^{-\pi^2 Bt} \tag{21.28}$$

$$\tau_{\text{inf}} = \frac{\ln 16\ell^2}{3\pi^2 D} \quad \text{or} \quad D = \frac{\ell^2}{10.7\tau_{\text{inf}}} \tag{21.29}$$

where τ_{inf} is an inflection characteristic point of the kinetic permeation curve.

The inflection method is based on the determination of the time τ_{inf} at which the transient permeation curve passes through an inflection point that is related to the diffusion coefficient.

A detailed analysis and modeling of Equation 21.24 result in a more accurate value of the inflection coefficient of the transient permeability curve calculated using the inflection time point. It is recommended to use the following equation to calculate diffusion coefficients instead of Equation 21.29:

$$D = \frac{\ell^2}{10.89\tau_{inf}}$$
(21.30)

Another convenient reference point on the differential permeability curve, which can be determined graphically from $J(t)$ versus the time curve, is the time to reach a half of the steady-state permeate flux, $(\tau_{1/2})$. This reference point could be obtained when $\exp\left\{-\pi^2 D\tau_{1/2}/\ell^2\right\} = 0.25$, and, therefore, $D = \ell^2/7.2\tau_{1/2}$, (Ziegel et al., 1969).

In the study of gas permeation in membranes published by DuPont (Pye et al., 1976), this reference point of the dynamic permeation curve was used to measure diffusion coefficients.

Yet, another reference point could be used to calculate a diffusion coefficient. It is a so-called time lag τ_θ derived from the differential permeation curve (Marcandalli et al., 1984). It can be shown that in an integral permeation experiment, the time-lag, θ (i.e., the intercept on the time axis of the asymptotic line obtained by plotting the overall amount of permeated gas against time), is

$$\theta = \frac{\ell^2}{6D}$$
(21.31)

Figure 21.13 represents schematically these three characteristics of the permeability curve, as described above.

The characteristic points could be used as criteria to estimate how homogeneous a diffusion medium (e.g., ionomer film) is. Theoretically, all values of the diffusion coefficients calculated using these reference points should be the same. However, if a diffusion mechanism is mixed, for example, if it includes facilitated and/or multiple diffusion paths, the values of the diffusion coefficients calculated by means of various reference points could be different.

The advantages of using these points include a relatively simple experimental procedure and a straightforward data analysis; however, when only a single point of the transient permeation curve is used to calculate a diffusion coefficient, potentially large experimental errors can be obtained.

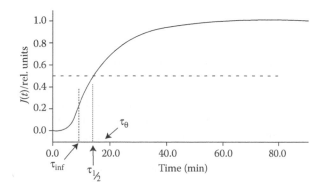

FIGURE 21.13 Three characteristic points of the permeability curve: inflection time at which the transient permeation curve passes through an inflection point, (τ_{inf}); reference point based on the time to reach half of the steady-state permeate flux value, $(\tau_{1/2})$; and a time lag point, (τ_θ).

If all three characteristic points are used to calculate three diffusion coefficients, the confidence interval for the diffusion coefficient D is calculated as follows:

$$D = \bar{D} \pm t(p,f)\bar{s} = \frac{\sum_{i=1}^{n} D_i}{n} \pm t(p,f)\sqrt{\frac{\sum_{i=1}^{n}(\bar{D} - D_i)^2}{n(n-1)}} \qquad (21.32)$$

where $t(p,f)$–Student's t-distribution values, $p = 0.95$–probability value of the given confidence interval, $f = n-1$–degree of freedom, n–number of reference points (in this case $n = 3$, $f = 2$). S is a square root of dispersion of the data point:

$$s = \sqrt{\frac{\sum_{i=1}^{n=3}(\bar{D} - D_i)}{n-1}} \qquad (21.33)$$

And $\bar{s}$ is the standard deviation of the series of diffusion measurements and is calculated as follows:

$$\bar{s} = \frac{s}{\sqrt{n}} \qquad (21.34)$$

21.4.6 Functional Scale Method

The advantage of the functional scale method includes the use of all experimental points of the kinetic permeation curve (Pasternak et al., 1970). The use of all experimental points results in better accuracy of diffusion coefficient calculations. Equation 21.23a could be expressed as follows (Beckman, I. N. 2010. Personal communications, Discussion on statistical moments):

$$\frac{J(t)}{J_\infty} = 1 + 2\sum_{n=1}^{\infty}(-1)^n \exp\left\{-\left(\frac{n\pi}{\ell}\right)^2 Dt\right\} = 1 + 2\sum_{n=1}^{\infty}(-1)^n \exp\left(-n^2\pi^2 u\right) \qquad (21.35)$$

where $u = Dt/\ell^2$.

The following function could be introduced: $F(u) = J(t)/J_\infty$ as a function of time (t) in coordinates (u) and (t). It is expressed in terms of a straight line: $F(u) = D/\ell^2(t - t_0)$, where t_0 is equal to the initial point of the experiment.

A diffusion coefficient is calculated using the following equation:

$$D = \ell^2 \tan(\alpha) \qquad (21.36)$$

21.4.7 Use of Transient Electrochemical Method

Similar mathematical treatment that was developed for transient gas permeation experiments could be used to design transient electrochemical permeation experiments to measure diffusion coefficients. In the papers by Lehtinen et al. (1998), Haug and White (2000), and Sethuraman et al. (2009), transient i–t (current–time) curves were analyzed to measure gaseous transport parameters in ion-conductive membranes.

After combining Equations 21.23a and 21.14 into the following equation:

$$i(t) = nFADc\left[\frac{1}{\ell} + \frac{2}{\ell}\sum_{k=1}^{\infty}(-1)^k \exp\left\{-\left(\frac{k\pi}{\ell}\right)^2 Dt\right\}\right] \qquad (21.37)$$

where c is gas concentration (mol cm^{-3}), the transient limiting current as a function of time could be measured and gas diffusion coefficients could be obtained (Lehtinen et al., 1998).

21.4.8 Method of Statistical Moments

The method of statistical moments is often used for sorption isotherms analyses, but can also be used for the analysis of transient permeation curves (Felder, 1978). A detailed analysis of the permeation curves and data interpretation using statistical moments will be published separately elsewhere, as it could be quite extensive and is beyond the scope of this chapter.

21.5 Conclusions and Future Work

As it has been shown in this chapter, gas permeation in a PFSA significantly affects the performance and durability of fuel cell systems. Permeability of the reactant gases plays a vital role in the degradation of the PFSA membranes and in the formation of the platinum band. Permeation of nitrogen and hydrogen determines fuel economy. The approaches to measure gas diffusion coefficients described in this chapter will contribute to better understanding of gas permeation measurement protocols. Advanced mathematical treatment of the experimental transient permeation curves, such as the method of statistical moments, would allow better understanding of the effect of the membrane morphology on gas permeation in the presence of water. Future work may focus on the improvement of mathematical processing of experimental data to increase the accuracy in measuring diffusion coefficients in highly heterogeneous media and developing adequate mathematical and physical models of gas diffusion.

References

Ahluwalia, R. K. and Wang, X. 2007. Build-up of nitrogen in direct hydrogen polymer-electrolyte fuel cell stacks. *J. Power Sources* 171: 63–71.

Antolini, E. 2003. Formation, microstructural characteristics and stability of carbon supported platinum catalysts for low temperature fuel cells. *J. Mater. Sci.* 38: 2995–3005.

Ash, R., Barrer, R. M., and Palmer, D. G. 1965. Diffusion in multiple laminates. *Br. J. Appl. Phys.* 16: 873–884.

Astrath, N. G. C., Shen, J., Astrath, F. B. G., Zhou, J., Huang, C., and Yuan, X. Z. et al. 2010. Determination of effective gas diffusion coefficients of stainless steel films with differently shaped holes using a Loschmidt diffusion cell. *Rev. Sci. Instrum.* 81: 046104–046104-3.

ASTM International. 2007. Standard test method for determination of oxygen gas transmission rate, permeability and permeance at controlled relative humidity through barrier materials using a coulometric detector. F1927–F1998. DOI: 10.1520/F1927-07. http://www.astm.org

Barrer, R. M. 1941. *Diffusion in and Through Solids.* London: Cambridge University Press.

Beattie, P. D., Basura, V. I., and Holdcroft, S. 1999. Temperature and pressure dependence of O$_2$ reduction at Pt.Nafion 117 and Pt.BAM 407 interfaces. *J. Electroanal. Chem.* 468: 180–192.

Bessarabov, D. and Hitchcock, A. P. 2009. Advances in structural and chemical analysis of fuel cell catalyst-coated membranes (CCM) for hydrogen fuel cell applications. *Membr. Technol.* 12: 6–12.

Bessarabov, D. and Kozak, P. 2007a. Measurement of gas permeability in SPE membranes for use in fuel cells. *Membr. Technol.* 12: 6–9.

Bessarabov, D. and Kozak, P. 2007b. Steady-state gas permeability measurements in proton-exchange membranes for fuel cell application using the GC method. Paper presented at *Annual Meeting of North American Membrane Society* (*NAMS*2007), Orlando, FL.

Bessarabov, D. G., Sanderson, R. D., Popkov, Y. M., and Timashev, S. F. 1998. Characterisation of membranes for electrochemically-aided gas separation: Morphology of platinum deposition. *Sep. Purif. Technol.* 14: 201–208.

Bessarabov, D. G., Sanderson, R. D., Popkov, Y. M., Valuev, V. V., and Timashev, S. F. 1997. New possibilities of electroinduced membrane gas and vapour separation. *Ind. Eng. Chem. Res.* 36: 2487–2489.

Bessarabov, D. G., Theron, J. P., Sanderson, R. D., Schwarz, H.-H., Schossig-Tiedemann, M., and Paul, D. 1999. Separation of 1-hexene/*n*-hexane mixtures using a hybrid membrane/extraction system. *Sep. Purif. Technol.* 16: 167–174.

Bernardi, D. M. and Verbrugge, M. W. 1992. A mathematical model of the solid–polymer–electrolyte fuel cell. *J. Electrochem. Soc.* 139: 2477–2491.

Bessarabov, D. G. 1999. Membrane gas-separation technology in the petrochemical industry. *Membr. Technol.* 107: 9–13.

Bett, J. A. S., Cipollini, N. E., Jarvi, T. D., and Breault, R. D. 2005. Fuel cell having a corrosion resistant and protected cathode catalyst layer. *US Patent*: 6 855 453 B2.

Bi, W., Gray, G. E., and Fuller, T. F. 2007. PEM fuel cell Pt/C dissolution and deposition in Nafion electrolyte. *Electrochem. Solid-State Lett.* 10: B101–B104.

Blunk, R., Zhong, F., and Owens, J. 2006. Automotive composite fuel cell bipolar plates: Hydrogen permeation concerns. *J. Power Sources* 159: 533–542.

Broka, K. and Ekdunge, P. 1997. Oxygen and hydrogen permeation properties and water uptake of Nafion 117 membrane and recast film for PEM fuel cell. *J. Appl. Electrochem.* 27: 117–123.

de Bruijn, F. A., Dam, V. A. T., and Janssen, G. J. M. 2008. Review: Durability and degradation issues of PEM fuel cell components. *Fuel Cells* 8: 3–22.

Buchi, F. N., Wakizoe, M., and Srinivasan, S. 1996. Microelectrode investigation of oxygen permeation in perfluorinated proton exchange membranes with different equivalent weights. *J. Electrochem. Soc.* 143: 927–932.

Chiou, J. S. and Paul, D. R. 1988. Gas permeation in a dry Nafion membrane. *Ind. Eng. Chem. Res.* 27: 2161–2164.

Cleghorn, S., Kolde J., and Liu, W. 2003. Catalyst coated composite membranes. In: *Handbook of Fuel Cells: Fundamentals, Technology and Applications*, eds. W. Vielstich, A. Lamm, and H. A. Gasteiger, Vol. 3, Chapter 44, pp. 566–575. New York: John Wiley and Sons.

Crank, J. 1956. *The Mathematics of Diffusion*. Oxford: Claredon Press.

Curtin, D. E., Lousenberg, R. D., Henry, T. J., Tangeman, P. C., and Tisack, M. E. 2004. Advanced materials for improved PEMFC performance and life. *J. Power Sources* 131: 41–48.

Editorial. 2008. Recommended units for key parameters. *J. Membr. Sci.* 320: III.

Felder, R. M. 1978. Estimation of gas transport coefficients from differential permeation, integral permeation, and sorption rate data. *J. Membr. Sci.* 3: 15–27.

Gibbs, T. K. and Pletcher, D. 1980. The electrochemistry of gases at metallized membrane electrodes. *Electrochim. Acta* 25: 1105–1110.

Gode, P. Lindbergh, G., and Sundholm, G. 2002. *In-situ* measurements of gas permeability in fuel cell membranes using a cylindrical microelectrode. *J. Electroanal. Chem.* 518: 115–122.

Haug, A. T. and White, R. E. 2000. Oxygen diffusion coefficient and solubility in a new proton exchange membrane. *J. Electrochem. Soc.* 147: 980–983.

He, Y. and Cussler, E. L. 1992. Ammonia permeabilities of perfluorosulfonic membranes in various ionic forms. *J. Membr. Sci.* 68: 43–52.

Hiramitsu, Y., Mitsuzawa, N., Okada, K., and Hori, M. 2010. Effects of ionomer content and oxygen permeation of the catalyst layer on proton exchange membrane fuel cell cold start-up. *J. Power Sources* 195: 1038–1045.

Honji, A., Mori, T., Tamura, K., and Hishinuma, Y. 1988. Agglomeration of platinum particles supported on carbon in phosphoric acid. *J. Electrochem. Soc.* 135: 355–359.

Hwang, S. T., Choi, C. K., and Kammermeyer, K. 1974. Gaseous transfer coefficients in membranes. *Sep. Sci. Technol.* 9: 461–478.

Inaba, M., Kinumoto, T., Kiriake, M., Umebayashi, R., Tasaka, A., and Ogumi, Z. 2006. Gas crossover and membrane degradation in polymer electrolyte fuel cells. *Electrochim. Acta* 51: 5746–5753.

James, C. W. Jr., Roy, A., McGrath, J. E., and Marand, E. 2008. Determination of the effect of temperature and humidity on the O_2 sorption in sulfonated poly(arylene ether sulfone) membranes. *J. Membr. Sci.* 309: 141–145.

Kocha, S. S., Yang, J. D., and Yi, J. S. 2006. Characterization of gas crossover and its implications in PEM fuel cells. *AIChE J.* 52: 1916–1925.

Kulikovsky, A. A. 2002. The voltage–current curve of a polymer electrolyte fuel cell: "exact" and fitting equations. *Electrochem. Commun.* 4: 845–852.

Kundu, S., Cimenti, M., Lee, S., and Bessarabov, D. 2009. Fingerprint of the automotive fuel cell cathode catalyst degradation: Pt band in the proton-exchange membranes. *Membr. Technol.* 10: 7–10.

Lehtinen, T., Sundholm, G., Holmberg, S., Sundholm, F., Bjornbom, P., and Burselld, M. 1998. Electrochemical characterization of PVDF-based proton conducting membranes for fuel cells. *Electrochim. Acta* 43: 1881–1890.

Lee, S., Bessarabov, D., and Vohra, R. 2009. Degradation of a cathode catalyst layer in PEM MEAs subjected to automotive-specific test conditions. *Int. J. Green Energy* 6: 594–606.

Liang, D., Shen, Q., Hou, M., Shao, Z., and Yi, B. 2009. Study of the cell reversal process of large area proton exchange membrane fuel cells under fuel starvation. *J. Power Sources* 194: 847–853.

Ma, S., Odgaard, M. and Skou, E. 2005. Carbon dioxide permeability of proton exchange membranes for fuel cells. *Solid State Ionics* 176: 2923–2927.

Majsztrik, P. W., Satterfield, M. B., Bocarsly, A. B., and Benziger, J. B. 2007. Water sorption, desorption and transport in Nafion membranes. *J. Membr. Sci.* 301: 93–106.

Mann, R. F., Amphlett, J. C., Peppley, B. A., and Thurgood, C. P. 2006. Henry's Law and the solubilities of reactant gases in the modelling of PEM fuel cells. *J. Power Sources* 161: 768–774.

Marcandalli, B., Selli, E., Tacchi, R., Bellobono, I. R., and Leidi, G. 1984. Simple device for quality control of polymer films, using measurements of permeation and diffusion coefficients. *Desalin.* 51: 113–122.

Maruyama, J., Inaba, M., Morita, T., and Ogumi, Z. 2001. Effects of the molecular structure of fluorinated additives on the kinetics of cathodic oxygen reduction. *J. Electroanal. Chem.* 504: 208–216.

Mathias, M., Roth, J., Fleming J., and Lehnert, W. 2003. Diffusion media materials and characterization. In: *Handbook of Fuel Cells—Fundamentals, Technology and Applications*, eds. W. Vielstich, A. Lamm, and H. A. Gasteiger, Chapter 46, Vol. 3, pp. 517–537. New York: John Wiley & Sons.

Mittal, V. O., Kunz, H. R., and Fenton, J. M. 2007. Membrane degradation mechanisms in PEMFCs. *J. Electrochem. Soc.* 154: B652–B656.

Mittelsteadt, C. and Umbrell, M. 2005. Gas permeability of perfluorinated sulfonic acid polymer electrolyte membranes. In: *Proceedings of 207th ECS Meeting*, Abstract 770.

Mittelsteadt, C. K. and Liu, H. 2009. Conductivity, permeability, and ohmic shorting of ionomeric membranes. In: *Handbook of Fuel Cells*, eds W. Vielstich, A. Lamm, and H. A. Gasteiger, Chapter 23, Vol. 5, pp. 345–358. New York: John Wiley and Sons.

Mukerjee, S. and Srinivasan, S. 2003. Electrocatalysis. In: *Handbook of Fuel Cells—Fundamentals Technology and Applications*, eds. W. Vielstich, A. Lamm, and H. A. Gasteiger, Vol. 7, pp. 502–519. New York: John Wiley & Sons.

Ogumi, Z., Kuroe, T., and Takehara, Z. 1985. Gas permeation in SPE method II. Oxygen and hydrogen permeation through Nafion. *J. Electrochem. Soc.* 132: 2601–2605.

Ogumi, Z., Takehara, Z., and Yoshizawa, S. 1984. Gas permeation in SPE Method I. Oxygen permeation through Nafion and Neosepta. *J. Electrochem. Soc.* 131: 769–773.

Parthasarathy, A., Martin, C. R., and Srinivasan, S. 1991. Investigations of the oxygen reduction reaction at the Platinum/Nafion interface using a solid-state electrochemical cell. *J. Electrochem. Soc.* 138: 916–921.

Pasternak, R. A., Schimscheimer, J. F., and Heller, J. 1970. A dynamic approach to diffusion and permeation measurements. *J. Polym. Sci.* Part A-2 8: 467–479.

Peron, J., Mani, A., Zhao, X., Edwards, D., Adachi, M., Soboleva, T. et al. 2010. Properties of Nafion® NR-211 membranes for PEMFCs. *J. Membr. Sci.* 356: 44–51.

Petropoulos, J. H. 1992. Plasticization effects on the gas permeability and permselectivity of polymer membranes. *J. Membr. Sci.* 75: 47–59.

Petropoulos, J. H. 1994. Mechanisms and theories for sorption and diffusion of gases in polymers. In: *Polymeric Gas Separation Membranes*, eds. D. R. Paul, Yu. P. Yampolskii, Chapter 2, pp. 17–81. Boca Raton, FL: CRC Press, Inc.

Pianca, M., Barchiesi, E., Esposto, G., and Radice, S. 1999. End groups in fluoropolymers. *J. Fluorine Chem.* 95: 71–84.

Pharoah, J. G. 2005. On the permeability of gas diffusion media used in PEM fuel cells. *J. Power Sources* 144: 77–82.

Pourbaix, M. 1966. *Atlas of Electrochemical Equilibria in Aqueous Solutions.* Oxford: Pergamon Press.

Pye, D. G., Hoehn, H. H., and Panar, M. 1976. Measurement of gas permeability of polymers. Apparatus for determination of permeabilities of mixed gases and vapours. *J. Appl. Polym. Sci.* 20: 287–301.

Qiao, J., Saito, M., Hayamizu, K., and Okada, T. 2006. Degradation of perfluorinated ionomer membranes for PEM fuel cells during processing with H_2O_2. *J. Electrochem. Soc.* 153: A967–A974.

Rogers, W. A., Buritz, R. S., and Alpert, D. 1954. Diffusion coefficient, solubility, and permeability for helium in glass. *J. Appl. Phys.* 25: 868–875.

Sakai, T., Takenaka, H., and Torikai, E. 1986. Gas diffusion in the dried and hydrated Nafions. *J. Electrochem. Soc.* 133: 88–92.

Scott, K. and Mamlouk, M. 2009. A cell voltage equation for an intermediate temperature proton exchange membrane fuel cell. *Int. J. Hydrogen Energy* 34: 9195–9202.

Sethuraman, V. A., Khan, S., Jur, J. S., Haug, A. T., and Weidner, J. W. 2009. Measuring oxygen, carbon monoxide and hydrogen sulfide diffusion coefficient and solubility in Nafion membranes. *Electrochim. Acta* 54: 6850–6860.

Sethuraman, V. A., Weidner, J. W., Haug, A. T., and Protsailo, L. V. 2008. Durability of perfluorosulfonic acid and hydrocarbon membranes: Effect of humidity and temperature. *J. Electrochem. Soc.* 155: B119–B124.

Sexsmith, M. P. 2009. Key remaining challenges in automotive PEM fuel cell stack design. *Paper presented at the 238th ACS National Meeting*, Washington, DC, August 16–20.

Shen, J., Zhou, J., Astrath, N. G. C., Navessin, T., Liu, Z.-S. et al. 2011. Measurement of effective gas diffusion coefficients of catalyst layers of PEM fuel cells with a Loschmidt diffusion cell. *J. Power Sources* 196: 674–678.

Shi, W., Hou, M., Shao, Z., Hu, J., Hou, Z., Ming P., and Yi, B. 2007. A novel proton exchange membrane fuel cell anode for enhancing CO tolerance. *J. Power Sources* 174: 164–169.

Stanic V. and Hoberecht, M. 2004. Mechanism of pinhole formation in membrane electrode assemblies for PEM fuel cells. NASA report (Web ref (b)): http://ntrs.nasa.gov/archive/nasa/casi.ntrs.nasa.gov/20050198939_2005199186.pdf.

Teplyakov, V. and Meares, P. 1990. Correlation aspects of the selective gas permeabilities of polymeric materials and membranes. *Gas Sep. Purif.* 4: 66–74.

Tseung, A. C. C. and Dhara, S. C. 1975. Loss of surface area by platinum and supported platinum black electrocatalyst. *Electrochim. Acta* 20: 681–683.

Tsou, Y., Kimble, M. C., and White, R. E. 1992. Hydrogen diffusion, solubility, and water uptake in Dow's short-side-chain perfluorocarbon membranes. *J. Electrochem. Soc.* 139: 1913–1917.

Uchimura, M., Sugawara, S., Suzuki, Y., Zhang, J., and Kocha, S. 2008. Electrocatalyst durability under simulated automotive drive cycles. *ECS Trans.* 16: 225–234.

Web (a): http://www1.eere.energy.gov/hydrogenandfuelcells/mypp/pdfs/fuel_cells.pdf.

Weber, A. Z. 2008. Gas-crossover and membrane-pinhole effects in polymer-electrolyte fuel cells. *J. Electrochem. Soc.* 155: B521–B531.

Xie, J., Wood D. L. III, More, K. L., Atanassov, P., and Borup, R. L. 2005. Microstructural changes of membrane electrode assemblies during PEFC durability testing at high humidity conditions. *J. Electrochem. Soc.* 152: A1011–A1020.

Yamanaka, I. and Otsuka, K. 1991. The partial oxidations of cyclohexene and benzene on the $FeCl_3$-embedded cathode during the O_2–H_2 fuel cell reactions. *J. Electrochem. Soc.* 138: 1033–1038.

Yasuda, H. 1975. Units of gas permeability constants. *J. Appl. Polym. Sci.* 19: 2529–2536.

Yasuda, H. and Rosengren, K. J. 1970. Isobaric measurement of gas permeability of polymers. *J. Appl. Polym. Sci.* 14: 2839–2877.

Yasuda, A. and Shimidzu, T. 1999. Electrochemical carbon monoxide sensor with a Nafion film. *React. Funct. Polym.* 41: 235–243.

Yasuda, H., Taniguchi, A., Akita, T., Ioroi, T., and Siroma, Z. 2006. Platinum dissolution and deposition in the polymer electrolyte membrane of a PEM fuel cell as studied by potential cycling. *Phys. Chem. Chem. Phys.* 8: 746–752.

Yu, Z. and Carter, R. N. 2010. Measurement of effective oxygen diffusivity in electrodes for proton exchange membrane fuel cells. *J. Power Sources* 195: 1079–1084.

Zanderighi, L., Marcandalli, B., Faedda, M. P., Bellobono, I. R., and Riva, M. 1979. The transport of gases in polymeric membranes. A simple cell for the determination of diffusion and permeability coefficients. *Gazetta Chim. Ital.* 109: 505–509.

Zhang, J., Tang, Y., Song, C., Zhang, J., and Wang, H., 2006. PEM fuel cell open circuit voltage (OCV) in the temperature range of 23°C to 120°C. *J. Power Sources* 163: 532–537.

Zhang, S., Yuan, X.-Z., Hin, J. N. C., Wang, H., Friedrich, K. A., and Schulze, M. 2009. A review of platinum-based catalyst layer degradation in proton exchange membrane fuel cells. *J. Power Sources* 194: 588–600.

Ziegel, K. D., Frensdorff, H. K., and Blair, D. E. 1969. Measurement of hydrogen isotope transport in poly-(vinyl fluoride) films by the permeation-rate method. *J. Polym. Sci.* 7: 809–819.

22

Species Detection

Khalid Fatih
Institute for Fuel Cell Innovation

22.1 Introduction

The proton exchange membrane fuel cell (PEMFC) is seen to be one of the most promising power devices with a wide range of applications including transportation, stationary, portable, and micro-systems. Transportation seems to be the most demanding application because of the system constraints and the aggressive cost targets. Transportation is, by all measures, the driver of the R&D activities for fuel cell commercialization, which ultimately benefits other applications (Wilkinson, 2001). PEMFC technology has reached a significant degree of maturity but there are still a number of stringent requirements for commercialization, which mostly concern the performance, cost, and reliability of the system. Fuel cell reliability and durability limitations are impeding large-scale adoption even for applications that may support a larger $/kW cost. Studying and improving the durability of a PEMFC stack have recently become a major topic in fuel cells R&D. Durability is a complex topic because there are a strong but not yet fully resolved correlations between the fuel cell's variable operating conditions and the interdependent performance degradation mechanisms and failure modes, which in turn are related to the different components involved in a fuel cell stack (Borup et al., 2007). One of the most vulnerable components of a fuel cell stack is the proton exchange membrane (PEM). The degradation of the PEM has been studied at length in recent years. Many review papers addressed the extensive work done on this topic (Borup et al., 2007; de Bruijn et al., 2008; Collier et al., 2006; Mauritz and Moore, 2004). In general, typical polymer electrolytes used in durability studies of fuel cells (single cell or stack) are based on perfluorinated sulfonic acid (PFSA) membranes because of their unique chemical and thermal stability as well as their high proton conductivity. The lifetime of a PFSA-based PEM had earlier been related to its chemical stability. Baldwin et al. (1989, 1990) established a relationship between the ultimate life of a PEM in an operating cell (SPE® fuel cell or electrolyzer) and the fluoride emission rate (FER) as shown in Figure 22.1. It was reported that this relationship was established based on millions of hours of cells operation for over 30 years. However, the relationship was also based on the assumption that the cell, stack, and system designs were configured to prevent premature mechanical failures. It has been found that the FER represented an excellent parameter to measure the health and life expectancy of the PEM (Baldwin et al., 1990).

The correlation between FER and the degradation of PFSA is currently considered a standard approach for studying the chemical and electrochemical degradation of PFSA ionomer in the membranes and

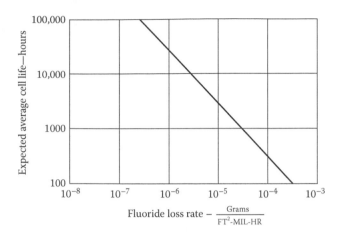

FIGURE 22.1 Expected lifetime of SPE cells with perfluorocarbon proton exchange membranes. (Reprinted from *J. Power Sources*, 29, Baldwin, R. et al. Hydrogen–oxygen proton-exchange membrane fuel cells and electrolyzers. 399–412. Copyright (1990), with permission from Elsevier.)

catalyst layers. This chapter addresses the speciation of fluoride as a diagnostic tool in fuel cell degradation studies. Special focus will be on methods and techniques used to measure the FER such as ion-selective electrode (ISE) and ion chromatography (IC). ISE is discussed in more details in terms of basic theory, testing protocols, and experimental limitations. IC and nuclear magnetic resonance (NMR) spectroscopy of ^{19}F (^{19}F NMR), although, are very relevant techniques to this topic and are desired methods for detecting low molecular weight degradation products of PFSA, are briefly discussed here. Studies using the Fenton's reagent test, which is a widely used method for *ex situ* study of the chemical stability of membranes and membrane electrode assemblies (MEAs), will not be covered in this chapter. The focus will mostly be on FER used as a tool in induced degradation tests during an operating fuel cell, which combines the assessment of both the chemical and the electrochemical stability of membranes and MEAs.

22.2 Fluoride Emission Rate

22.2.1 Chemical Structure and Degradation Mechanism of PFSA Membrane

E. I. DuPont developed PFSA membrane in the mid-1960s under the brand name Nafion® (Connolly and Gresham, 1966). The original form of Nafion consists of copolymer of tetrafluoroethylene (TFE) and perfluoro(4-methyl-3,6-dioxa-7-octene-sulfonyl fluoride) (Curtin et al., 2004). The form sulfonyl fluoride (–SO$_2$F) is a thermoplastic resin that was originally used to make tick Nafion membrane (>125 μm). This was done using an established extrusion-cast manufacturing process followed by a conversion to the sulfonic acid form (–SO$_3$H) through a chemical process (Curtin et al., 2004; Smith and Withers, 1984). E. I. DuPont later developed a new manufacturing process based on PFSA solution casting technology intended to produce low defect and thin membranes and MEAs (Preischl et al., 2001). Figure 22.2 shows some chemical variations of commercial PFSA membranes that have been developed for PEMFC applications (Borup et al., 2007).

The manufacturing process of the PFSA ionomers has a significant effect on their thermal and chemical stability. The process has seen extensive development since first introduced by DuPont. It is generally believed that the weakest link in a PFSA chemical structure is the presence of low concentration of end groups particularly H-containing ones (Zhou et al., 2007). Pianca et al. (1999) have identified some end groups in TFE polymers, which could be present in all TFE-containing copolymers. The nature of the

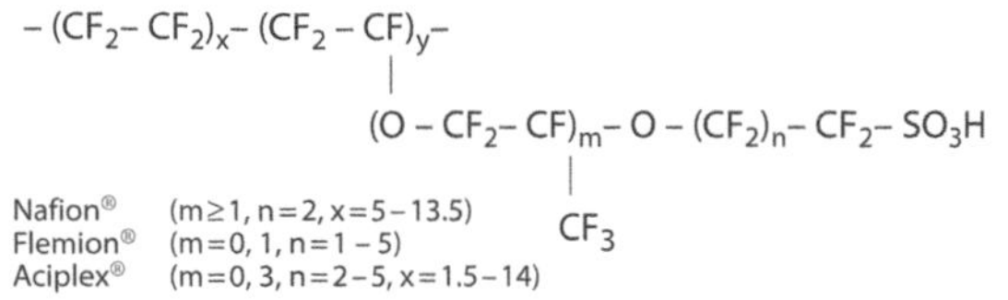

$$- (CF_2- CF_2)_x- (CF_2 - CF)_y-$$
$$|$$
$$(O - CF_2- CF)_m- O - (CF_2)_n- CF_2- SO_3H$$
$$|$$

Nafion®   (m≥1, n=2, x=5–13.5)
Flemion® (m=0, 1, n=1–5) CF_3
Aciplex® (m=0, 3, n=2–5, x=1.5–14)

FIGURE 22.2   Chemical structure of PFSA membranes.

end groups depends on the synthesis and manufacturing process. The authors identified the following possible end groups:

- Carboxylic acid ($-CF_2-COOH$)
- Amide ($-CF_2-CONH_2$)
- Perfluorovinyl ($-CF_2-CF=CF_2$)
- Acyl fluoride ($-CF_2-COF$)
- Difluoromethyl ($-CF_2-CF_2H$) and
- Ethyl ($-CF_2-CH_2-CH_3$)

In fuel cell applications, the observed mass loss or thinning of the membrane and detected fluoride ions in product water as well as low molecular weight PFSA and CO_2 are strong indications of a chemical degradation mechanism (Curtin et al., 2004; Healy et al., 2005; LaConti et al., 2003; Takasaki et al., 2009). It is now widely accepted that peroxide radicals attack on residual H-containing terminal bonds at the end groups of the polymer is the primary mechanism of degradation (Curtin et al., 2004). In this proposed mechanism, hydrogen peroxide (H_2O_2), which is generated *in situ* during fuel cell operation (LaConti et al., 2003), decompose to form hydroxy ($^\bullet$OH) and/or hydroperoxy ($^\bullet$OOH) radicals. Hydroxyl radicals could also be directly generated during oxygen reduction reaction at the cathode side (Panchenko et al., 2004). These radicals attack residual H-containing terminal bonds and initiate the membrane decomposition through an unzipping reaction of the main chain. Curtin et al. (2004) suggested the following steps in the case of a carboxylic acid end groups:

$$R_f -CF_2-COOH + {}^\bullet OH \rightarrow R_f -CF_2{}^\bullet + CO_2 + H_2O \tag{22.1}$$

$$R_f -CF_2{}^\bullet + {}^\bullet OH \rightarrow R_f -CF_2OH \rightarrow R_f -COF + HF \tag{22.2}$$

$$R_f -COF + H_2O \rightarrow R_f -COOH + HF \tag{22.3}$$

The chemical degradation, which is based on H_2O_2 generation and its decomposition to hydroxyl ($^\bullet$OH) and/or hydroperoxyl ($^\bullet$OOH) radicals are still not a fully understood mechanism in the context of an operating fuel cell. However, it is accepted that H_2O_2 is generated through gas crossover and catalytic reaction at the anode particularly at open-circuit conditions (Inaba et al., 2006; LaConti et al., 2003).

H_2O_2 is also generated electrochemically through incomplete O_2 reduction reaction at high-current density at the cathode (LaConti et al., 2003). Aoki et al. (2005) found that the PEM decomposes when H_2, O_2, and Pt catalyst coexist. LaConti et al. (2003) demonstrated using SEM/EDX that the majority of membrane degradation occurs on the anode side in the case of poly(styrene sulfonic acid). When trace amounts of impurities such as transition metal (Fe^{2+}, Cu^{2+}), chloride, and carbon monoxide (Markovic et al., 2001; Schmidt et al., 2001) are present, the degradation rate increases substantially. Under open-circuit conditions, Inaba et al. (2006) also observed that the ionomer degradation take place mainly at the anode side due to oxygen crossover initiating formation of H_2O_2 and subsequent radicals. Conversely, Kundu et al. (2008a) had recently demonstrated, under open-circuit conditions, that the FER detected at the cathode outlet was much higher than the anode side as shown in Figure 22.3a,b.

The authors observed a severe thinning of the ionomer in the cathode side using SEM micrograph, which indicated that the degradation rather occurs in the cathode side of the MEA. Although under

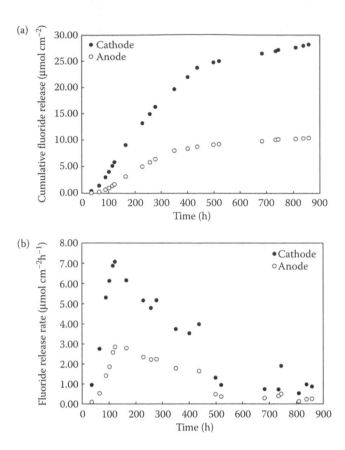

FIGURE 22.3 Anode and cathode (a) cumulative fluoride emission, and (b) FER during the duration of testing. (Reprinted from *J. Power Sources*, 183(2), Kundu, S. et al. Degradation analysis and modeling of reinforced catalyst coated membranes operated under OCV conditions. 619–628. Copyright (2008), with permission from Elsevier.)

open-circuit conditions, the scattered results and conclusions obtained from these studies on the degradation mechanism of PFSA ionomer, shows further investigations are still needed to understand this degradation. Takasaki et al. (2009) detected large amount of side-chain fragments in the catalyst layer based on fuel cell OCV hold and load-cycling tests as well as H_2O_2 vapor-exposure test. The authors suggested that FER alone is not always a sufficient parameter to evaluate PFSA polymers degradation. Teranishi et al. (2006) used direct gas mass spectroscopy to analyze the cathode outlet of a PEMFC at open-circuit condition. The authors assigned the identified molecular weight to HF, H_2O_2, CO_2, SO, SO_2, H_2SO_2, and H_2SO_3 and concluded that the degradation of the ionomer occurred mainly in the catalyst layer and was initiated by H_2O_2 generated at the cathode due to hydrogen crossover.

22.2.2 Application of FER as a Diagnostic Tool

The importance of measuring the rate of fluoride emission can be seen in the number of recently published works using this parameter when performing fuel cell durability studies. FER is currently used as one of the main parameters to help elucidate degradation mechanisms and failure modes of the membrane and the ionomer in the catalyst layer. In general, experimental FER values are dependent on fuel cell operating conditions and testing protocols; type of fuel cell components particularly PFSA membranes; and the FER analysis technique. While reported values of FER in the literature are scattered over a wide range (10^{-6}–10^{-9} gF $cm^{-2}h^{-1}$) and may not be directly used to compare different studies, the variation of FER is however an excellent and useful tool to gain insight on the mechanism of the ionomer degradation in the membrane and the catalyst layer. Indeed, FER has been used as a function of time to investigate the effect of fuel cell operating parameters, PFSA membrane chemical structure, catalyst layer structure, catalyst nature, and some mitigation strategies against contamination such as air bleeding. Mittal et al. (2007) used ISE to quantify the FER of model fuel cell, under accelerated decay conditions, to investigate the membrane degradation mechanism. The authors analyzed the FER for three-cell configurations including anode-only mode, cathode-only mode, and bilayer-membrane mode. The authors proposed that molecular H_2 and O_2 resulting from crossover react on the surface of Pt catalyst to form a membrane's degrading species, X, according to the following reaction (Mittal et al., 2007):

$$H_2 + O_2 \xrightarrow{\text{Pt}} X \tag{22.4}$$

The reaction is thought to be independent of the catalyst potential and chemical in nature. Madden et al. (2009) studied the effect of O_2 concentration, relative humidity (RH), temperature, and membrane thickness, under open-circuit conditions, on the fluoride released from a PFSA-based MEAs, which was measured by IC. FER was found to increase with temperature but decrease with increasing RH. FER was also found to increase with increasing the membrane thickness. The FER at the cathode was found to decrease with increasing O_2 concentration. Based on this trend and TEM imaging of Pt particles in the bulk of the membrane, the authors concluded on a peroxide and radicals formation mechanism within the bulk of the membrane where dissolved Pt precipitates and react with gaseous H_2 and O_2 available from crossover.

Sethuraman et al. (2009) showed that FER could be used to effectively assess the effect of the catalyst composition on the ionomer degradation as shown in Figure 22.4 (Sethuraman et al., 2009). The authors pointed out that despite that studied Pt-alloys have a higher selectivity toward H_2O_2, they showed lower FER compared to Pt. The authors suggested a higher solubility of Pt compared to Pt-alloys and hence migration of Pt^{2+} into the membrane reacting with available gaseous H_2 and O_2.

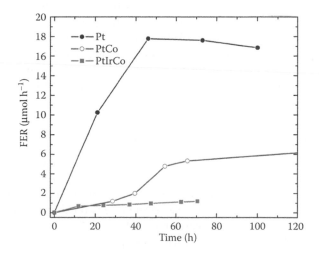

FIGURE 22.4 Fluorine emission rates per μmol h^{-1} measured as a function of time from single-sided MEAs. The electrode with Pt (-●-), PtCO (-○-), and PtIrCo (-■-) is held at 600 mV versus Au reference electrode throughout the test. (Reprinted from *Electrochim. Acta*, 54, Sethuraman, V. A. et al. Importance of catalyst stability vis-a-vis hydrogen peroxide formation rates in PEM fuel cell electrodes. 5571–5582. Copyright (2009), with permission from Elsevier.)

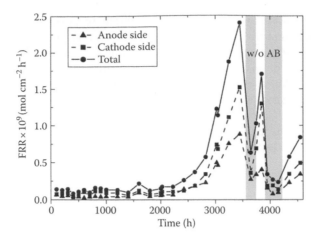

FIGURE 22.5 Variation of FER, in effluent water at the anode and cathode, during an accelerated air-bleeding test. (Reprinted from *J. Power Sources*, 178, Inaba, M. et al. Impacts of air bleeding on membrane degradation in polymer electrolyte fuel cells. 699–705. Copyright (2008), with permission from Elsevier.)

Inaba et al. (2008) studied the effect of air bleeding on PEMFC degradation by measuring FER in a long-term accelerated test of 4600 h. The authors were able to detect the onset of a severe degradation after 2000 h because of the diagnostic opportunity offered by FER (Figure 22.5).

22.3 Fluoride Analysis Methods

22.3.1 Ion-Selective Electrode

Ion-selective electrodes (ISEs) are electrochemical sensors commonly used when a fast, nondestructive, reproducible, sensitive, and accurate analysis of certain ionic species is required. ISEs are based on ion-selective membranes, as the sensing component of the electrochemical cell. One of the most attractive aspects of an ISE is its ability to sense the change in the activity of some particular ionic species in a complex media containing several electrolytes. Among the most successful applications of an ISE beside the pH measurement is the fluoride speciation. Fluoride ISE is applied to measure fluoride in drinking water, environmental monitoring, food industry, and physiological fluids. In 1966, Frant and Ross Jr (1966) developed the fluoride ISE. This method is now widely used for FER measurement in fuel cell durability studies to monitor PFSA membrane and ionomer degradation.

22.3.1.1 Basic Principle of an Ion-Selective Electrode

ISE is based on a potentiometric measurement of the activity of certain ions in the presence of other ions. Conventional ISE is a galvanic half-cell, consisting of an ion-selective membrane, an internal contacting solution or a solid contact and an internal reference electrode. The other half-cell is an external reference electrode in contact with a reference electrolyte. The two half-cells are connected through a salt bridge. The basic schematic diagram is represented in Figure 22.6.

ISE electrodes are already manufactured either as a single-electrode configuration separated from the external reference electrode or in a combined configuration. When an ion-selective membrane contacts a test solution containing the target ions, an electrode potential develops across the membrane. This potential is linearly dependent on the logarithm of the activity of that particular ion in the test solution. The dominant component of the potential response is the free-energy change associated with the phase-boundary processes. Nicolsky equation generally describes the potential of an

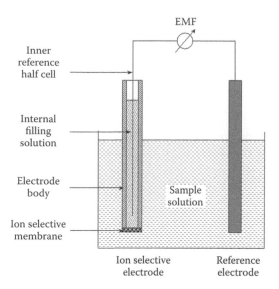

FIGURE 22.6 Schematic diagram of a membrane electrode measuring circuit and cell assembly.

electrochemical cell containing an ion-selective membrane. The extended form of this equation is as follows (Morf, 1981; Skoog, 1985):

$$E = C + \left(RT/z_i F \right) \ln \left[a_i + \sum_{j \neq i} k_{ij} \left(a_j \right)^{z_i/z_j} \right]$$ (22.5)

where E is the experimentally determined galvanic potential difference of ISE cell (in V) when the only variables are activities in the test solution; R is the gas constant; T is the absolute temperature; F is the Faraday constant; a_i is the activity of the primary ion; a_j are activities of the interfering ions; k_{ij} is the selectivity coefficient for the interfering ion with respect to the primary ion; z_i is the charge of the primary ion, and z_j are the charge of interfering ions. The constant C includes the standard potential of the ISE, the reference electrode potential and the junction potential. The ISE usually exhibits what is known as a Nernstian behavior when the plot of the potential versus the logarithm of the activity is linear over a particular activity range with a slop value of $2.303RT/z_i F$ ($59.16/z_i$ mV decade^{-1} at 25°C).

Ion-selective membranes are classified into crystalline and noncrystalline. The crystalline membranes can be either a single crystal such as LaF_3 for sensing fluoride ions or polycrystalline (homogenous/mixed) such as Ag_2S for sensing S^{2-} and Ag^+. For noncrystalline membranes, there are glass-based types for sensing ions such as Na^+ and H^+; liquid type used for sensing Ca^{2+} and also liquid immobilized in inert matrix. Depending on the membrane type, different mechanisms have been proposed to explain the development of an ion-sensitive potential across the membrane and the research is still progressing to understand the membrane behavior. Not only the mechanism but also the theory to describe the relationship between the potential and the activity seem to be unsatisfactory and may not explain many of the experimental behavior when performing the analysis. For this reason, fundamental aspects of the ISEs will not be discussed further.

22.3.1.2 Fluoride ISE

The ISE for fluoride detection is often made of lanthanum trifluoride single crystal, LaF_3. This membrane is an ideal material to manufacture an ion-selective membrane for fluoride analysis. However, only doped LaF_3 seems appropriate for ISE membrane application. Doping LaF_3 with europium (II) fluoride, EuF_2, further enhances the membrane conductivity by creating more vacancies in the lattice and hence more mobile fluoride ions (Koryta, 1986). The charge transfer mechanism can be presented by the equation Ure, 1957:

$$LaF_3 + Vacancy \rightarrow LaF_2^+ + F^-$$ (22.6)

Van den Winkel et al. (1977) proposed a more detailed mechanism for the membrane conductivity for EuF_2-doped LaF_3. The authors suggested a mixed ionic and electronic conduction mechanism:

$$(EuF_2)_L + F_{aq}^- \rightarrow (EuF_3)_L + e$$ (22.7)

$$(LaF_3)_L + Vacancy \rightarrow (LaF_2^+)_L + (F^-)_L$$ (22.8)

$$(LaF_2^+)_L + F_{aq}^- \rightarrow (LaF_3)_L$$ (22.9)

The electrode is often prepared by cutting disks from a EuF_2-doped LaF_3 single crystal. The detection limit of the fluoride ISE correspond to the solubility of LaF_3, 7×10^{-7} M for single crystal.

The solubility limit of a freshly precipitated salt is 3.3×10^{-5} M (Koryta, 1986). In practice, when the appropriate experimental precautions are taken, the potential developed at the ISE electrode usually shows Nernstian response to fluoride concentration in the test solution according to the expression:

$$E = C - 2.303RT/F \log\left[F^-\right]$$ (22.10)

In simple electrolytes this response takes place over a range of $1-10^{-6}$ M of fluoride as shown in Figure 22.7. It is important to mention that fluoride ISE electrode measures only the free fluoride ions (hydrated ions) in an aqueous test solution. This means that any fluoride, that reacts to form complexes,

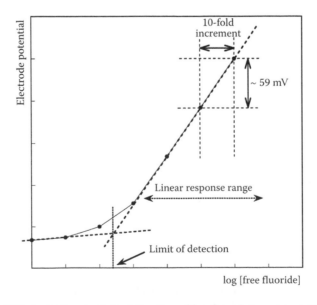

FIGURE 22.7 Typical ISE electrode potential as a function of free fluoride ion concentration (calibration curve).

is not accounted for by ISE. The fluoride ISE is sensitive to changes in temperature. Most electrodes appear to reach phase-boundary equilibrium and stabilize rapidly at higher temperatures (Campbell, 1987) but the lifetime of the electrode and measurements reproducibility may suffer serious decreases.

22.3.1.3 Experimental Aspects Affecting Fluoride Ion-Selective Electrode

Several analytical approaches could be used for species' quantification. The approach depends on the technique, the sample constraint, detection limits, presence of interfering species, and so on. For fluoride ISE, direct calibration is a widely used method for routine aqueous solution analysis. It is also one of the easiest for measuring a large number of samples. The method is based on calibration using series of standards. It is important to stress that the precision of the direct calibration method is directly dependent on the amount and distribution of measurements used to plot the calibration curve. The characteristics of a typical calibration curve for a fluoride ISE are shown in Figure 22.7. Many experimental factors may affect the response of the electrode potential and the Nersntian behavior shown in the calibration curve (Figure 22.7).

22.3.1.3.1 Effect of Interfering Species

Two types of interferences are possible with fluoride ISE namely electrode (affect the electrode response) and chemical interferences (affect the true concentration of fluoride). Most anions such as Cl^-, Br^-, I^-, SO_4^{2-}, HCO_3^-, PO_4^{3-} and acetate do not interfere with fluoride electrode response. The only significant electrode interference that induces errors when using fluoride ISE is the presence of the hydroxide ion, OH^-, since it has the same charge and similar ionic radii. Some anions such as CO_3^{2-} or PO_4^{3-} increase the pH of the solution, which increase the concentration of OH^-, but are not direct electrode interferences. Many mechanisms have been proposed to understand the OH^- interference. Frant and Ross Jr (1966) considered a replacement mechanism of fluoride with hydroxide in the crystal membrane. Vesely and Stulik (1974) suggested a competitive adsorption of fluoride and hydroxide over sites at the membrane surface. Cammann and Rechnitz (1976) proposed a surface equilibrium reaction of the type:

$$LaF_3 + 3OH^- \rightarrow La(OH)_3 + 3F^- \tag{22.11}$$

De Marco et al. (1989) studied the surface of LaF_3 using XPS technique suggested the formation of mixed insoluble hydroxo-complexes on the surface (15–20 nm in depth) according to the equation:

$$LaF_3(s) + nOH^- (aq) \rightarrow LaF_{3-n}(OH)_n (s) + nF^- (aq) \tag{22.12}$$

Chemical interferences are dissolved specics in the test solution that react with fluoride and form stable complexes. Fluoride is known to complex with Al^{3+}, Fe^{3+}, and other polyvalent cations as well as with hydrogen. These interferences are the ones that present more challenges. These species interact with the fluoride ion, consequently decreasing the measured fluoride content since the electrode senses only free fluoride ions. For an accurate measurement, fluoride must be decomplexed prior to analysis.

22.3.1.3.2 Effect of Temperature

Nernst Equation 22.10 shows the dependence of the potential response of the electrode on temperature. The temperature of both the standard and the test solution need to be controlled and set at the same value. Any change in the electrode potential induces an error in the fluoride concentration. Commercial fluoride ion selective electrode indicates that samples and standard solutions should be within ±1°C (±2°F) of each other. A 1°C difference in temperature was shown to induce a 2% error in the concentration level of fluoride of 10^{-3} M (Nicholson, 1983). The absolute potential of the reference electrode also shifts slowly with temperature because of the solubility equilibrium on which the electrode potential is based. When working at a significantly different temperature compared to room temperature, the time to reach the equilibrium could be significantly higher. Nicholson and Duff found that the reproducibility of the electrode is much lower at 50°C compared to 20°C (Nicholson and Duff, 1981).

22.3.1.3.3 Effects of pH

The operating pH range for a fluoride ISE is constrained at pH below 5 by the formation of stable complexes of fluoride and hydrogen namely, the undissociated acid HF and the ion HF_2^- (Campbell, 1987). These fluoride species are not detected by the fluoride electrode (Srinivasan and Rechnitz, 1968) and hence lead to an underestimated content of the total fluoride ions. For higher pH, hydroxide ion interferes with fluoride when its concentration is higher than 10% of the fluoride concentration (Nicholson, 1983). A pH of 5.5 appears to be an optimum for the analysis of fluoride. At this pH there is a little formation of HF and HF_2^- complexes (Vesely and Stulik, 1974).

22.3.1.3.4 Effects of the Ionic Strength and Liquid-Junction Potential

The potential of a fluoride selective electrode is sensitive to its activity and not to concentration. The interactions with other ions become important and hence the ionic strength of the test solution and the standard need to be controlled and kept constant in order to obtain a valid calibration curve. To keep the activity proportional to the concentration, the activity coefficient needs to be kept constant. Up to a certain point, this is possible by keeping the ionic strength constant according to the expression of Debye–Huckel, which relates the activity coefficient and ionic strength. ISE manufacturers recommended a dilution of the test solution before buffer addition when the ionic strength is greater than 0.1 M (ThermoFisher Scientific, 2007). The liquid-junction potential, which develops due to concentration difference at the boundary between the ISE and the reference electrode salt bridge, contributes to the measured potential of the cell. If the potential at the liquid junction is not minimized, it can lead to potential drift or unstable potential (Nicholson, 1983).

22.3.1.3.5 Other Possible Effects

Other experimental factors could also affect the analysis of fluoride content in a test solution. Factor such as memory effect requires measurement to be performed from low to high concentration. The adsorption effect, which originates from the possibility of fluoride to adsorb on the container's wall, was also demonstrated in a study involving silver analysis with silver selective electrode (Durst and Duhart, 1970). For fluoride, this requires the use of plastic containers as a precaution when performing the analysis (ThermoFisher Scientific, 2007). The effect of stirring also needs to be addressed as it affects the liquid-junction potential and the response time of the electrode. The stirring rate parameter needs to be kept similar for both the standard and test solution.

22.3.1.4 Ion-Selective Electrode Test Protocol for Fuel Cell Applications

As mentioned in previous sections, the rate of fluoride emission in the product water of an operating fuel cell correlates well with the degradation rate of PFSA ionomers in the membrane and catalyst layer of an MEA. The measure of FER using a fluoride ISE needs to be carefully conducted to avoid errors in fluoride content. To reduce to the effect of temperature, interference and fluoride complexes, pH, liquid-junction potential, and to keep the fluoride activity proportional to its concentration, buffer solutions are usually applied to both the standards and the test solution at the same temperature. These buffer solutions are commonly called total ionic strength adjustment buffer (TISAB) solutions. TISAB solution allows the following functions:

1. Adjusts the pH, usually to 5–5.5, which reduces the electrode interference from OH^- and avoids formation of fluoride complexes at lower pH
2. Allows relatively high and constant ionic strength of the solution to minimize the effect of liquid-junction potential as well as to keep the proportionality between the activity and the concentration of fluoride
3. Allows the breakdown of complexes and free fluoride for more accurate content determination

Various TISAB solutions are commercially available and consist of pH buffer, ionic strength adjusting salts, and complexing ligand(s). The TISAB solution could be tailored to meet the specificity of the test solution. For fuel cell application, this means that the stack and system components such as catalyst, membrane, bipolar plates, gaskets, tubing and generally most of their related corrosion products has to be taken into account when choosing or formulating TISAB solution for an accurate determination of the fluoride content. Generally, fluoride ISE detailed operation protocols are provided by the ISE supplier, which takes into account the specificity of the electrode design and its tolerance to operational parameters.

22.3.2 Other Methods for Fluoride Analysis

22.3.2.1 Ion Exchange Chromatography

Ion exchange chromatography (commonly known as ion chromatography, IC) is another technique frequently used to measure FER in fuel cell durability studies (Akiyama et al., 2010; Aoki et al., 2005; Healy et al., 2005; Inaba et al., 2006, 2008; Kundu et al., 2008a,b; Xie et al., 2005). A more detailed description of principles and applications of this versatile method could be found in the literature (Haddad and Jackson, 1990). The IC technique belongs to liquid–solid phases chromatography family. The method was developed four decades ago and saw a rapid growth lead by Dionex Corp. In general, IC methods consist of a separation step followed by a detection step of ions or polar molecules belonging to the following groups: (i) inorganic anions and cations; (ii) low molecular weight water-soluble organic acids and bases; and (iii) ionic organometallic compounds (Haddad and Jackson, 1990). The overall separation mechanism of these methods is based on ion retention (ion exchange) on the chromatographic column due to ionic interactions with the stationary phase, which includes appropriate functional group(s) of opposite charge in its surface. For anion exchange process such as for fluoride analysis, the exchange equilibrium can be described as follows (Haddad and Jackson, 1990):

$$M^+C^- + A^- \rightarrow M^+A^- + C^- \tag{22.13}$$

where M^+ is the functional group on the stationary phase, which could be a polymer-coated silica; functionalized silica; synthetic polymeric resin; or a hydrous oxide, C^- is the counter ion, and A^- is the analyte in the mobile phase. The mobile phase also known as eluent in liquid–solid chromatography techniques often consists of an aqueous solution of a competing ion, which percolates through the stationary phase and acts mainly to elute the analytes in an optimal time interval. Because the mobile phase's elution characteristics depend on its pH as well as the nature and the concentration of the competing ion, its properties can be tailored by appropriate composition of salt(s) and buffer to efficiently elute the target analyte.

After the separation process, the quantification process involves the detection of the analyte. Chromatography techniques use variety of detection methods such as conductivity; electroanalytical (amperometric or coulometric); potentiometric; spectroscopic; postcolumn reaction or a combination of these methods to accurately measure the concentration of the analyte (Haddad and Jackson, 1990). For the analysis of inorganic anions and cations, the conductivity detection offers an advantage due to its high sensitivity, particularly when a low background conductance mobile phase is used. This requires a diluted mobile phase, which in turn requires a low ion-exchange capacity in order for the elution to process in an optimal time interval. To reduce the background conductance of the mobile phase, a second column, called the suppressor, is usually introduced between the chromatographic column and the detector. The role of the suppressor is to transform both the mobile phase and the analyte for an enhanced sensitivity of the conductivity detector.

The advantage of IC technique over ISE for fuel cells durability studies is the ability to eliminate most of the interferences potentially existing with ISE and simultaneously measure FER and other anionic degradation products release rate such as for SO_4^{2-}. However, while IC is highly automated method and based on physical separation of the ion before detection for enhanced accuracy, ISE remains a simple approach, more convenient and an elegant method for FER measurement for fuel cell durability studies. Both methods have the potential to be automated for an online monitoring of fluoride in fuel cell's effluent water.

22.3.2.2 ^{19}F NMR

Nuclear magnetic resonance spectroscopy (NMR) is a well-known dominant technique for the structure determination of organic compounds. The technique is based on the spin properties of charged nucleus with an odd atomic mass such as ^{1}H, ^{13}C, ^{19}F, and ^{31}P. These nucleus have all a spin number $I = 1/2$. By applying an external magnetic field, the spin energy states of the nucleus split into two states corresponding to the two possible orientations of the spin axis with respect to the magnetic field (Skoog, 1985). The energy difference between the two states is proportional to the strength of the magnetic field for a particular nucleus and is usually in the radio-frequency region of the electromagnetic spectrum. NMR is based upon the measurement of absorption in the radio-frequency region, which involves a flipping of the magnetic moment from a low-energy state (spin 1/2) to high-energy state (spin −1/2) (Skoog, 1985). NMR technique was tentatively used for fuel cells durability study. Contrary to ISE or IC, which are used to quantitatively determine traces of fluoride ions and hence fluoride release rate in the effluent water of an operating fuel cell, ^{19}F NMR, requires concentrated samples and was used mainly with objective to better understand the chemical and electrochemical degradation of PFSA ionomer in the PEM membrane and catalyst layer. ^{19}F NMR is often combined with other analytical technique such as mass spectrometry (MS) (Healy et al., 2005) or Fourier transform infrared spectroscopy (FT-IR) (Kinumoto et al., 2006) in order to be able to characterize complex fluorinated degradation products. The technique could be classified as *ex situ* and destructive for durability studies since the MEA is often analyzed at the end of durability tests where the degradation products need to be extracted. This extraction is performed either by dissolving a piece of the MEA or membrane in an organic solvent (Patil et al., 2010); immersion of the membrane in water for few days (Healy et al., 2005); or by using Soxhlet extraction method (Iojoiu et al., 2007). The later allows an increased amount of water-soluble degradation products while maintaining a low volume of water. ^{19}F NMR and MS demonstrated similarities in PFSA degradation mechanism in both *ex situ* MEA degradation using Fenton's test and *in situ* where the MEA is degraded during a fuel cell operation (Healy et al., 2005). ^{19}F NMR combined to FT-IR spectroscopy revealed that PFSA decomposition, using Fenton's test, occur both in the main and side chains through radical de-polymerization (Kinumoto et al., 2006). Gebel group studied the hygrothermal aging of Nafion, using ^{19}F NMR, ^{1}H NMR, and FT-IR and suggested a sulfonic anhydrides formation resulting in crosslinked side chains (Collette et al., 2009).

22.4 Concluding Remarks

Fluorine analysis represents a convenient tool to predict chemical structure degradation occurring in PFSA-based components of a PEMFC during its operation. By measuring FER, it is possible to gain insight on cell failure modes and to predict the relative lifetime of a fuel cell stack. Fluoride ions can be easily analyzed in effluent water, which make it a widely adopted *ex situ* approach in fuel cell degradation studies. Many failure modes and mitigation approaches have been proposed based on fluoride ion monitoring. However, a better understanding of components degradation in fuel cell may requires the monitoring of other membrane's and/or catalyst layer's degradation products. Some of these products are water-soluble and could be detected in effluent water but others are water insoluble and require cell opening and extraction from the MEA. Even larger water-soluble molecules that are products of degradation may require cell opening and extraction from the MEA. Hence, it is difficult to correlate between degradation products that cannot be detected under similar conditions.

Fluoride ISE and ion exchange chromatography methods were both shown to provide a fairly good sensitivity and accuracy for fluoride ion detection. For the purpose of fluoride monitoring during fuel cell durability evaluation, ISE and IC are comparable analytical methods but the former appears to be relatively simple and more straightforward. Online monitoring of fuel cell degradation is also possible through fluoride ion analysis but it requires further development of ISE and IC methods to support continuous measurements of fluoride concentration without additional analytical interferences and/or increases in fuel cell system's complexity.

References

Akiyama, Y., Sodaye, H., Shibahara, Y. S. et al. 2010. Study on degradation process of polymer electrolyte by solution analysis. *J. Power Sources* 195: 5915–5921.

Aoki, M., Uchida, H. and Watanabe, M. 2005. Novel evaluation method for degradation rate of polymer electrolytes in fuel cells. *Electrochem. Commun.* 7: 1434–1438.

Baldwin, R., Pham, M., Leonida, A. et al. 1989. Hydrogen–oxygen proton-exchange membrane fuel cells and electrolyzers. NASA Glenn Research Center, Report N: 89N22996.

Baldwin, R., Pham, M., Leonida, A. et al. 1990. Hydrogen–oxygen proton-exchange membrane fuel cells and electrolyzers. *J. Power Sources* 29: 399–412.

Borup, R., Meyers, J., Pivovar, B. et al. 2007. Scientific aspects of polymer electrolyte fuel cell durability and degradation. *Chem. Rev.* 107: 3904–3951.

Cammann, K. and Rechnitz, G. A. 1976. Exchange kinetics at ion-selective membrane electrodes. *Anal. Chem.* 48: 856–862.

Campbell, A. D. 1987. Determination of fluoride in various matrices. *Pure Appl. Chem.* 59: 695–702.

Collette, F. M., Lorentz, C., Gebel, G. et al. 2009. Hygrothermal aging of Nafion. *J. Membr. Sci.* 330: 21–29.

Collier, A., Wang, H., Yuan, X. et al. 2006. Degradation of polymer electrolyte membranes. *Int. J. Hydrogen Energy* 31: 1838–1854.

Connolly, D. J. and Gresham, W. F. 1966. Fluorocarbon vinyl ether polymers. US Patent 3,282,875.

Curtin, D. E., Lousenberg, R. D., Henry, T. J. et al. 2004. Advanced materials for improved PEMFC performance and life. *J. Power Sources* 131: 41–48.

De Bruijn, F. A., Dam, V. A., and Janssen, G. J. M. 2008. Review: Durability and degradation issues of PEM fuel cell components. *Fuel Cells* 8: 3–22.

De Marco, R., Cattrall, R. W., Liesegang, J. et al. 1989. XPS studies of the fluoride ion-selective electrode membrane LaF$_3$: Ion interferences. *Surf. Interface Anal.* 14: 457–462.

Durst, R. A. and Duhart, B. T. 1970. Ion-selective electrode study of trace silver ion adsorption on selected surfaces. *Anal. Chem.* 42: 1002–1004.

Frant, M. S. and Ross, J. W. Jr. 1966. Electrode for sensing fluoride ion activity in solution. *Science* 154: 1553–1555.

Haddad, P. R. and Jackson, P. E. 1990. *Ion Chromatography: Principles and Applications*. Netherlands: Elsevier Science B. V.

Healy, J., Hayden, C., Xie, T. et al. 2005. Aspects of the chemical degradation of PFSA ionomers used in PEM fuel cells. *Fuel Cells* 5: 302–308.

Inaba, M., Kinumoto, T., Kiriake, M. et al. 2006. Gas crossover and membrane degradation in polymer electrolyte fuel cells. *Electrochim. Acta* 51: 5746–5753.

Inaba, M., Sugishita, M., Wada, J. et al. 2008. Impacts of air bleeding on membrane degradation in polymer electrolyte fuel cells. *J. Power Sources* 178: 699–705.

Iojoiu, C., Guilminot, E., Maillard, F. et al. 2007. Membrane and active layer degradation following PEMFC steady-state operation. *J. Electrochem. Soc.* 154: B1115–B1120.

Kinumoto, T., Inaba, M., Nakayama, Y. et al. 2006. Durability of perfluorinated ionomer membrane against hydrogen peroxide. *J. Power Sources* 158: 1222–1228.

Koryta, J. 1986. Ion selective electrodes. *Ann. Rev. Mat. Res.* 16: 13–27.

Kundu, S., Fowler, M. W., Simon, L. C. et al. 2008a. Degradation analysis and modeling of reinforced catalyst coated membranes operated under OCV conditions. *J. Power Sources* 183(2): 619–628.

Kundu, S., Karan, K., Fowler, M. et al. 2008b. Influence of micro-porous layer and operating conditions on the fluoride release rate and degradation of PEMFC membrane electrode assemblies. *J. Power Sources* 179: 693–699.

LaConti, A. B., Hamdan, M., and McDonald, R. C. 2003. Mechanisms of membrane degradation. In *Handbook of Fuel Cells—Fundamentals, Technology and Applications*, eds W. Vielstich, H. Gasteiger, A. Lamm, 49: 647–662. Chichester, England: John Wiley & Sons, Ltd.

Madden, T., Weiss, D., Cipollini, N. et al. 2009. Degradation of polymer-electrolyte membranes in fuel cells. *J. Electrochem. Soc.* 156: B657–B662.

Markovic, N. M., Schmidt, T. J., Stamenkovic, V. et al. 2001. Oxygen reduction reaction on Pt and Pt bimetallic surfaces: A selective review. *Fuel Cells* 1: 105–116.

Mauritz, K. A. and Moore, R. B. 2004. State of understanding of Nafion. *Chem. Rev.* 104: 4535–4586.

Mittal, V. O., Kunz, H. R., and Fenton, J. M. 2007. Membrane degradation mechanisms in PEMFCs. *J. Electrochem. Soc.* 154: B652–B656.

Morf, W. E. 1981. *The Principles of Ion-Selective Electrodes and of Membrane Transport.* Amsterdam: Elsevier Scientific Publishing Company.

Nicholson, K. 1983. Fluorine determination in geochemistry: Errors in the electrode method of analysis. *Chem. Geo.* 38: 1–22.

Nicholson, K. and Duff, E. J. 1981. Errors in the direct potentiometric electrode method of fluoride determination: Adsorption, illumination and temperature effects. *Analyst* 106: 985–991.

Panchenko, A., Dilger, H., Kerres, J. et al. 2004. *In-situ* spin trap electron paramagnetic resonance study of fuel cell processes. *Phys. Chem. Chem. Phys.* 6(11): 2891–2894.

Patil, Y. P., Jarrett, W. L., and Mauritz, K. A. 2010. Deterioration of mechanical properties: A cause for fuel cell membrane failure. *J. Membr. Sci.* 356: 7–13.

Pianca, M., Barchiesi, E., Esposto, G. et al. 1999. End groups in fluoropolymers. *J. Fluor. Chem.* 95: 71–84.

Preischl, C., Hedrich, P., and Hahn, A. 2001. Continuous method for manufacturing a laminated electrolyte and electrode assembly. US Patent 6,291,091.

Schmidt, T. J., Paulus, U. A., Gasteiger, H. et al. 2001. The oxygen reduction reaction on a Pt/carbon fuel cell catalyst in the presence of chloride anions. *J. Electroanal. Chem.* 508: 41–47.

Sethuraman, V. A., Weidner, J. W., Haug, A. T. et al. 2009. Importance of catalyst stability vis-á-vis hydrogen peroxide formation rates in PEM fuel cell electrodes. *Electrochim. Acta* 54: 5571–5582.

Skoog, D. A. 1985. *Principles of Instrumental Analysis*, 3rd Edition. Philadelphia: Saunders College Publishing.

Smith, R. A. and Withers, M. S. 1984. Coextruded multilayer cation exchange membranes. US Patent 4,437,952.

Srinivasan, K. and Rechnitz, G. A. 1968. Activity measurements with a fluoride-selective membrane electrode. *Anal. Chem.* 40: 509–512.

Takasaki, M., Nakagawa, Y., Sakiyama, Y. et al. 2009. Degradation study of PFSA polymer electrolytes: Approach from decomposition product analysis. *ECS Trans.* 17: 439–447.

Teranishi, K., Kawata, K., Tsushima, S. et al. 2006. Degradation mechanism of PEMFC under open circuit operation. *Electrochem. Solid-State Lett.* 9: A475–A477.

ThermoFisherScientific. 2007. *User Guide: Fluoride Ion Selective Electrode.* ISE User Guide: Thermo Fisher Scientific Inc.

Ure, J. 1957. Ionic conductivity of calcium fluoride crystals. *J. Chem. Phys.* 26: 1363–1373.

Van den Winkel, P., Mertens, J., Boel, T. et al. 1977. The application of concentration jump and impedance measurements to the mechanistic study of the fluoride solid-state electrode. *J. Electrochem. Soc.* 124: 1338–1342.

Vesely, J. and Stulik, K. 1974. The effect of solution acidity on the response of the lanthanum trifluoride single-crystal electrode. *Anal. Chim. Acta* 73: 157–166.

Wilkinson, D. P. 2001. Fuel cells. *Elecrochem. Soc. Interface* 10: 22–25.

Xie, J., Wood III, D. L., Wayne, D. M. et al. 2005. Durability of PEFCs at high humidity conditions. *J. Electrochem. Soc.*152: A104–A113.

Zhou, C., Guerra, M. A., Qiu, Z. M. et al. 2007. Chemical durability studies of perfluorinated sulfonic acid polymers and model compounds under mimic fuel cell conditions. *Macromolecules* 40: 8695–8707.

23

Rotating Disk Electrode/ Rotating Ring-Disk Electrode

Xuan Cheng
Xiamen University

Hengyi Li
Xiamen University

Qiaoming Zheng
Xiamen University

23.1 Introduction

The membrane electrode assembly (MEA), consisting primarily of an anode catalyst, a cathode catalyst, and a proton exchange membrane (PEM), is the key component of PEM fuel cells. The electrocatalytic activity, selectivity, and stability of catalysts, as well as the electrochemical reactions taking place at the catalyst–PEM interface, play a crucial role in determining the performance and durability of fuel cells. The electrochemical hydrogen oxidation reaction (HOR) and oxygen reduction reaction (ORR) on platinum (Pt) in acidic media have been extensively studied because of their great importance in fuel cells, and various electrochemical techniques have been widely used as diagnostic tools in PEM fuel cell research (Wu et al., 2008).

Generally, electrochemical techniques involve the application of potential (current) to an electrode and the measurement of the resulting current (potential) through an electrochemical cell for a period of time. Thus, electrochemical techniques can be described as a function of potential, current, and time. The analytical advantages of various electrochemical techniques include excellent sensitivity, a large number of useful solvents and electrolytes, a wide range of temperatures, rapid analysis time, simultaneous determination of several analytes, the ability to determine kinetic and mechanistic parameters, a well-developed theoretical framework, and hence the ability to reasonably estimate the values of unknown parameters.

The rotating disk electrode (RDE) or rotating ring-disk electrode (RRDE) is one of the most commonly used analytical tools in many areas, including fundamental studies of oxidation and reduction processes in various media, adsorption processes on surfaces, electron transfer and reaction mechanisms, and the kinetics of electron transfer processes. For instance, RDE/RRDE has been applied to investigate the corrosion inhibition behaviors of copper in O_2-saturated solutions (Amin and Khaled, 2010), copper–nickel alloy in seawaters (Kear et al., 2007a,b), carbon steel in oil-in-water emulsions (Becerra et al., 2000), as well as the electrodeposition mechanisms of manganese dioxide (Clarke et al., 2006), copper (Ibl and Schadegg, 1967; Low and Walsh, 2008; Vazquez-Arenas, 2010), tin and tin–copper alloys (Low and Walsh, 2008), and samarium-based films (Ruiz et al., 2006). This chapter will provide a brief review of the theory and applications of RDE/RRDE technology, with an emphasis on recent progress in RDE/RRDE studies for different catalysts used in PEM fuel cells.

23.2 Fundamentals of RDE and RRDE

23.2.1 Principles of RDE and RRDE

An RDE is a small metal disk inlaid into an insulating cylinder having a large base. The disk is situated in the center of the base. The cylinder is mounted on a metallic axle connected to an electric motor. The axle is perpendicular to the base, lies in the axis of the cylinder, and is connected to a potentiostat by graphite brushes and to the metal disk by a wire. Alternatively, it may bear a metallic bell that rotates in a mercury pool to obtain noiseless electrical contact with the potentiostat.

As shown in Figure 23.1, a disk electrode is set in an insulating rod, which is rotated at a constant frequency in a solution. The solution near its base moves angularly, both radially and axially. Because of friction, a thin layer of solution in contact with the base follows the rotation of the cylinder. Due to the solution's viscosity this angular movement extends into a thicker layer of the solution, but its rate decreases exponentially with the distance from the cylinder's surface. Rotation of the solution induces a centrifugal force that causes radial movement. To preserve the pressure, the movement is offset by axial flux from the bulk of the solution.

The components of fluid velocity depend on the angular velocity of the disk (ω), that is,

$$\omega = 2f\pi \tag{23.1}$$

where f is the rotation frequency in Hertz, as well as on the radial distance from the center of the disk (r), the coefficient of the fluid's kinematic viscosity (ν), and the axial distance from the surface of the disk (y), as given below

$$v_r = r\omega F(\lambda) \tag{23.2}$$

$$v_\varphi = r\omega G(\lambda) \tag{23.3}$$

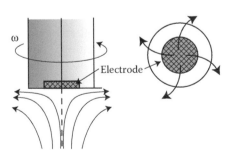

FIGURE 23.1   Movement of solution at an RDE.

$$v_y = (\omega v)^{1/2} H(\lambda) \tag{23.4}$$

where

$$\lambda = y(\omega/v)^{1/2} \tag{23.5}$$

Around the center of the cylinder's base, where the metal disk lies, the axial component of the solution velocity is most important, since the electroactive material is transported toward the surface in this direction only. Under chronoamperometric conditions, a diffusion layer develops at the electrode surface and extends far into the solution, as the flux at the surface is not equal to the mass transport rate in the bulk of the solution. Under steady-state conditions and the solution's laminar flow, the distance δ depends on the electrode rotation rate:

$$\delta = 1.61 D^{1/3} v^{1/6} \omega^{-1/2} \tag{23.6}$$

where D is the diffusion coefficient. In cyclic voltammetry (CV), the RDE may operate under either steady-state or transient conditions. If the scan rate is very slow, the response is a wave with a limiting current proportional to the square root of the rotation rate. At fast scan rates, a voltammetric peak appears during the forward scan but not during the reverse.

An RRDE is a double-working electrode (WE) used in hydrodynamic voltammetry, very similar to an RDE. The electrode actually rotates during experiments, inducing a flux of the analyte to the electrode. The difference between an RRDE and an RDE is the addition of a second WE in the form of a ring around the central disk of the first WE. The two electrodes are separated by a nonconductive barrier and connected to the potentiostat through different leads. To operate such an electrode it is necessary to use a bipotentiostat or a potentiostat capable of controlling a four-electrode system. This rotating hydrodynamic electrode motif can be extended to rotating double-ring electrodes, rotating double-ring-disk electrodes, and other even more esoteric constructions to suit a given experiment.

Figure 23.2 shows a typical structure of an RRDE, consisting of a thin metallic ring inlaid around the metallic disk situated in the center of the base of the insulating cylinder. Because of the radial component of the solution's movement, which is caused by the cylinder's rotation, the products of the electrode reaction formed on the disk electrode are carried over the insulating gap toward the ring electrode, where they can be detected and analyzed. The electrode rotates around its own axis, the solution rotates as it does in an RDE, and the ring and the disk are related via electrolyte transmission.

When the reductants in the solution transmit to the disk, the generalized reaction taking place on the disk electrode is written as

$$\text{Red} - ne^- \rightarrow \text{Ox} \tag{23.7}$$

The reaction product Ox transmits to the ring electrode, so the production at a certain potential of Ox can be detected thus

$$\text{Ox} + n'e^- \rightarrow \text{Red}' \tag{23.8}$$

When Reactions (Equations 23.7 and 23.8) occur on the disk and the ring electrode, the current generated from the disk (i_D) and the ring (i_R) can be obtained simultaneously. According to the Levich equation (Bard and Faulkner, 2001), the limiting diffusion current of an RDE ($i_{d,l}$) is expressed as follows:

$$i_{d,l} = 0.62 \pi r^2 n F D^{2/3} v^{-1/6} \omega^{1/2} C \tag{23.9}$$

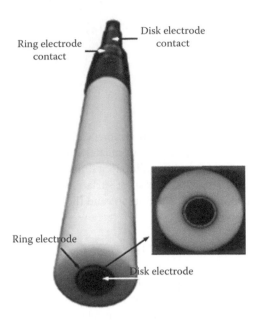

FIGURE 23.2 An RRDE.

where r is the radius of the disk electrode, F is the Faraday constant, and C is the concentration of the active species in the disk reaction. The Levich equation takes into account both the rate of diffusion across the stagnant layer and the complex solution flow pattern. Thus, a plot of the current versus the square root of the rotation yields a straight line. Equation 23.9 can be simplified to

$$i_{d,l} = B\omega^{1/2} \tag{23.9}$$

where B is the Levich slope and equals $0.62\pi r^2\, nFCD^{2/3}\, v^{-1/6}$. The Levich equation (23.9) is used for reversible systems. In more general cases, the Koutecky–Levich equation (Bard and Faulkner, 2001) can be used to describe the total current measured in RDE/RRDE:

$$\frac{1}{i} = \frac{1}{i_k} + \frac{1}{i_{d,l}} \tag{23.10}$$

where i_k represents the current in the absence of any mass-transfer effects—that is, the current that would flow under the kinetic limitations if the mass transfer were efficient enough to keep the concentration at the electrode surface equal to the bulk value, regardless of the electrode reaction. The first term on the right-hand side of Equation 23.10 is the activity-determining current, which is not dependent on the rotation rate. The second term is a reciprocal of Equation 23.9 and means the mass-transfer resistance, which depends on the rotation rate.

23.2.2 Measurements and Data Treatments

A potentiostat is fundamental for modern electrochemical studies using three electrode systems to investigate reaction mechanisms related to redox chemistry and other chemical phenomenon. As illustrated in Figure 23.3, the system functions either by controlling the potential of the WE at a constant level with respect to the reference (Ref) electrode while measuring the current at an auxiliary electrode

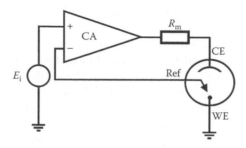

FIGURE 23.3 Scheme of a potentiostat.

(the so-called counterelectrode (CE)) or by adjusting the current flowing through the CE and WE while measuring the potential between the WE and Ref.

Most early potentiostats could function independently, providing data output through a physical data trace. Modern potentiostats are designed to connect compatibly with a personal computer and operate through a dedicated software package. The automated software allows the user to rapidly shift between experiments and experimental conditions. Just as important, the computer allows data to be stored and analyzed more effectively, rapidly, and accurately than historic methods.

Electrochemical measurements obtained using hydrodynamic voltammetry (RDE/RRDE) are carried out while the WE is rotating. To determine the mass-transfer rate with good reproducibility, it is very important to keep the electrode rotating at a stable speed during measurements. The WE potential is slowly swept back and forth across the formal potential of the analyte. The WE itself is rotated at a very high speed, and this rotational motion sets up a well-defined flow of solution toward the RDE surface. The flow pattern is akin to a vortex that literally sucks the solution (and the analyte) toward the electrode. Experimental results are generally plotted as a graph of current versus potential, and a typical rotated disk voltammogram is shown in Figure 23.4. The voltammogram exhibits a sigmoidal wave, the height of which provides the analytical signal. It is important to note that the layer of solution immediately adjacent to the surface of the electrode behaves as if it were stuck to the electrode. While the bulk of the solution is being stirred vigorously by the rotating electrode, this thin layer of solution manages to cling to the surface of the electrode and appears (from the perspective of the rotating electrode) to be motionless. This layer is called the stagnant layer to distinguish it from the remaining bulk of the solution.

The analyte is conveyed to the electrode surface by a combination of two types of transport. First, via convection, the vortex flow in the bulk solution continuously brings the fresh analyte to the outer edge of the stagnant layer. Then, the analyte moves across the stagnant layer via simple molecular diffusion. The thinner the stagnant layer, the sooner the analyte can diffuse across it and reach the electrode surface. Faster electrode rotation makes the stagnant layer thinner. Thus, greater rotation rates allow the

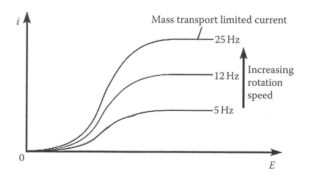

FIGURE 23.4 Typical rotating disk voltammograms.

analyte to reach the electrode more quickly, resulting in a higher current being measured at the electrode.

In the case of electrochemical measurements using an RRDE, the potentials from the ring and the disk electrode (E_D and E_R) are independently controlled by a common Ref and CE. The currents of the ring and disk electrode (i_D and i_R) are recorded by a dual potentiostat. When the RRDE is rotated, convection occurs near the electrode surface; therefore, the diffusion layer thickness is constant and a diffusion-limited current is observed. This feature is the same for an RDE. The advantage of an RRDE is that the electrolytic product in the disk electrode is transported to the ring electrode by centrifugal force and is detected in the ring electrode. The electrochemical reaction mechanism in the disk electrode can be analyzed in more detail accordingly.

Equations 23.9 and 23.10 can be used to determine the kinetic parameters, which are necessary for elucidating the mechanism of the electrode reaction. The information obtained from RDE measurements for an irreversible system include the kinetically controlled current (i_k) and the electron transfer number (n), which in the case of a catalyst electrode can be calculated from the slope of a Koutecky–Levich plot. The RDE/RRDE system has been widely applied in mechanistic/kinetic studies because it is thereby possible to detect the unstable products formed by electrode reactions in the disk region. An example of a typical slow electron transfer process (usually an irreversible system) is the reduction of dissolved oxygen in acidic solution, which is a multielectron reaction often associated with the formation of reaction intermediates (Damjnovic et al., 1966):

23.2.2.1 4-electron Reduction in Two Steps

$$O_2 + 2e^- + 2H^+ \rightarrow H_2O_2 \qquad (23.11)$$

$$H_2O_2 + 2e^- + 2H^+ \rightarrow 2H_2O \qquad (23.12)$$

23.2.2.2 Direct 4-Electron Reduction to Water

$$O_2 + 4e^- + 4H^+ \rightarrow 2H_2O \qquad (23.13)$$

The amount of the reaction intermediate, hydrogen peroxide (H_2O_2), formed from one mole of oxygen molecule at the disk electrode can be calculated using the following equation (Markolvic et al., 1995):

$$x(H_2O_2) = \frac{2i_R/N}{i_D + i_R/N} \times 100 \qquad (23.14)$$

or

$$x(H_2O_2) = \frac{2i_R/i_D}{N + i_R/i_D} \times 100 \qquad (23.14)$$

where $x(H_2O_2)$ is the percentage of H_2O_2 formation and N is the collection efficiency. The electron transfer number for the overall ORR can be calculated by

$$n = \frac{4i_D}{i_D + i_R/N} \quad (4 \geq n > 0) \qquad (23.15)$$

More detailed information about the considerations and applications for RDE/RRDE in the diagnoses of charge-transfer mechanisms can be found elsewhere (Bard and Faulkner, 2001; Treimer et al., 2002).

23.3 Advantages and Limitations of RDE/RRDE

The RDE and RRDE techniques can be applied as a diagnosis tool in probing reaction mechanisms and evaluating the performance/stability of catalysts used in PEM fuel cells. The major advantages of RDE or RRDE include simple construction, rigorous theoretical treatment, and hydrodynamic and thermal modulation.

23.3.1 Simple Construction

The RDE or RRDE is rather simple to construct, consisting of a disk of the electrode material imbedded in a rod of insulating material. For example, a common RRDE form uses a Pt wire sealed in glass tubing, with the sealed end ground smooth and situated perpendicularly to the rod axis. More frequently, the metal is imbedded in Teflon, epoxy resin, or other plastics. The catalysts are usually coated or electrodeposited onto the electrodes; sometimes Nafion is coated onto the electrodes.

23.3.2 Rigorous Theoretical Treatment

The RDE or RRDE is one of the few convective electrode systems for which the hydrodynamic equations and the convective–diffusion equation have been solved rigorously for the steady state. Initially, the simplest treatments of convective systems are based on a diffusion layer approach. Research on the reaction mechanism begins with the convective–diffusion equation and the velocity profiles in the solution, which are solved either analytically or, more frequently, numerically. In most cases, only the steady-state solution is desired. The observation of current transients at the disk or the ring electrode following a potential step can sometimes be of use in understanding an electrochemical system.

23.3.3 Hydrodynamic and Thermal Modulation

The RDE or RRDE can change the rotation rate of the electrode to obtain different steady-state values of ω when the current measurements are carried out. The RDE or RRDE can also be modulated thermally by irradiating the back of the disk electrode with a laser or jumping temperature while heat is carried away from the electrode surface by thermal diffusion and convection into the solution. Eventually, a steady-state disk temperature is attained. However, the temperature change at the electrode surface will cause changes in a number of parameters. While RDE or RRDE is capable of extracting thermodynamic information about a reaction, both the theory and the experimental setup are sufficiently complex that it has not yet found widespread use.

RDE/RRDE measurements have a few drawbacks. The current density values obtained from RDE measurements are usually much lower than those estimated from fuel cells in actual use. Because of the fast kinetics, mass transport in the usual version of an RDE is not sufficient for reliable determination of the current density of hydrogen electrode reactions on Pt in acidic media. As a result, the current density values obtained from Pt RDE measurements are usually lower than they should be, with reported values obtained at room temperature spanning two orders of magnitude.

Construction of hydrodynamic electrodes that provide known, reproducible mass-transfer conditions is more difficult than construction of stationary electrodes. The theoretical treatments involved in these methods are also more difficult and involve solving a hydrodynamic problem (e.g., determining solution flow velocity profiles as functions of rotation rates, solution viscosities, and densities) before the electrochemical one can be tackled. Rarely, closed-form or exact solutions can be obtained. Even though the number of possible electrode configurations and flow patterns in these methods are limited only by the imagination and resources of the experimenter, the most convenient and widely used system is the RDE.

23.4 Applications and Recent Advances of RDE/RRDE in PEM Fuel Cells

It is extremely valuable to develop a better understanding of the kinetic mechanisms of the HOR and the ORR on Pt-based electrodes, as both are important reactions occurring in PEM fuel cells at the anode and cathode catalysts, respectively. Furthermore, it is also urgently necessary to establish a scientific basis for a more comprehensive understanding of the HOR and ORR on other types of catalysts, including non-Pt catalysts, which have recently been receiving increased attention.

Theoretical studies on the HOR kinetics of Pt electrodes are based on the Tafel–Heyrovsky–Volmer mechanism (Gennero de Chialvo and Chialvo, 2004).

$$H_2 + 2Pt \rightleftharpoons 2Pt-H_{ads} \left(\text{Tafel reaction}\right) \tag{23.16}$$

$$H_2 + Pt \rightleftharpoons Pt\text{-}H_{ads} + H^+ + e^- (\text{Heyrovsky reaction}) \tag{23.17}$$

$$Pt\text{-}H_{ads} \rightleftharpoons Pt + H^+ + e^- (\text{Volmer reaction}) \tag{23.18}$$

$$H_2 \rightleftharpoons 2H^+ + 2e^- (\text{overall reaction}) \tag{23.19}$$

The HOR on Pt is much faster than the ORR on Pt. The overall HOR occurring in fuel cells may be expressed by considering Reactions (Equations 23.13 and 23.19), whereby the catalysts are kinetically required in fuel cell applications:

$$2H_2 + O_2 \rightleftharpoons 2H_2O \ E_0 = 1.229 \text{ V} \tag{23.20}$$

The total fuel cell voltage (E_{cell}) is the sum of the following potentials:

$$E_{cell} = E_0 - \eta_a - \eta_c - iR_\Omega \tag{23.21}$$

where E_0, η_a η_c i, and R_Ω are the reversible potential, the anodic overpotential, the cathodic overpotential, the current passing through the cell, and the internal resistance, respectively. The overpotentials are usually determined by the electrocatalytic activity and the mass-transfer property, which can generally be divided into activation and diffusion (or concentration) overpotentials. For a generalized redox reaction shown below,

$$Ox + ne^- \rightleftharpoons Red \quad (\text{cathode reaction}) \tag{23.22}$$

$$Red - ne^- \rightleftharpoons Ox \quad (\text{anode reaction}) \tag{23.23}$$

the relationship between the activation overpotential (η) and the current density (j) is described by the Butler–Volmer equation:

$$j = j_0 \left[\exp\frac{\alpha_a F\eta}{RT} - \exp\left(-\frac{\alpha_c F\eta}{RT}\right) \right] \tag{23.24}$$

where j_0 is the exchange current density, R is the gas constant, T is the temperature, while α_a and α_c are the charge-transfer coefficients of the anode and cathode, respectively. The total charge-transfer coefficient (α) is determined by the total electron transfer number (n) and the number of rate-determining steps involved in the reactions (υ):

$$\alpha = \alpha_a + \alpha_c = n / \upsilon \tag{23.25}$$

The performance and the stability of catalysts greatly affect the reaction rate, and therefore, significantly influence the cell performance and durability. The following section presents an overview and highlights of recent progress in the study of the HOR and ORR on different catalysts using RDE and RRDE techniques, as well as RDE/RRDE applications in PEM fuel cells.

23.4.1 Activity and Stability of Catalysts for Electrode Reactions

23.4.1.1 Analysis of the Hydrogen Oxidation Reaction

Pt is an important catalyst in PEM fuel cells due to its high electrocatalytic activity for the HOR and ORR, and its high stability in acidic environments. However, Pt is precious, expensive, and commonly used as metal or alloy nanoparticles supported on conductive carbon black (e.g., commercially available Vulcan XC72 carbon (VC)). The most commonly used commercial Pt-based catalysts are Pt/C and PtRu/C. Recent studies have focused on improving catalytic efficiency by increasing the electrochemical surface area (ECSA) and decreasing the Pt loading.

Figure 23.5 presents Tafel plots obtained for high-performance E-TEK 20% Pt/C at different rotation speeds (Esparbe et al., 2009). Because a well-defined Tafel region is absent, j_0 cannot be obtained by extrapolating from the Tafel line. However, after correcting for diffusion, j_0 can be estimated from the slope of the linear polarization region near the reversible potential, according to

$$\frac{\Delta E}{\Delta j} = \left(\frac{RT}{nF}\right)\left(\frac{1}{j_0} + \frac{1}{j_L}\right) \tag{23.26}$$

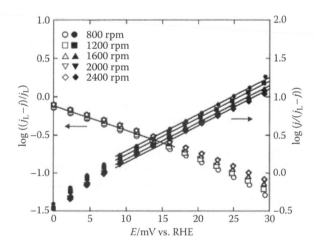

FIGURE 23.5 Tafel slopes assuming reversible (open symbols) or irreversible (filled symbols) HOR at different RDE rotation speeds. (Reprinted from *J. Power Sources*, 190, Esparbe, I. et al., Structure and electrocatalytic performance of carbon-supported platinum nanoparticles, 201–209, Copyright (2009), with permission from Elsevier.)

where j_L is the limiting current density. The HOR appears to be electrochemically reversible, with an estimated j_0 value of 0.27 mA cm^{-2}. More j_0 values, obtained mainly by RDE from various Pt electrodes in different electrolytes, are available from the literature (Sun et al., 2010).

Based on the Tafel–Heyrovsky–Volmer mechanism, a dual-pathway kinetic equation has been developed to describe the HOR on Pt over the entire overpotential region pertinent to the reaction, in the absence of any noticeable site-blocking effect (Wang et al., 2006):

$$j_k = S\left[j_{0T}\left(1 - e^{-2F\eta/\gamma RT}\right) + j_{0H}\left(e^{F\eta/2RT} - e^{-F\eta/\gamma RT}e^{-F\eta/2RT}\right)\right] \tag{23.27}$$

where γ is the potential range constant for the adsorption isotherm which is determined by the exchange rates for the three elemental reactions (Equations 23.16 through 23.18), S is the dimensionless scaling factor, defined as

$$S = \left(C_{H_2}^0 / C_{H_2}^{Ref}\right) A^r = P^r A^r \tag{23.28}$$

$C_{H_2}^0$ denotes the concentration of H_2 at E_0, $C_{H_2}^{Ref}$ is the saturated concentration of H_2 under 101,325 Pa of pressure, while P^r and A^r are the actual ratio of pressure with respect to 101,325 Pa and the ratio of the real Pt surface area to the geometric area of the electrode, respectively. The values of three essential kinetic parameters for a Pt microelectrode were found to be $j_{0T} = 470$ mA cm^{-2}, $j_{0H} = 10$ mA cm^{-2}, $\gamma = 1.2$ in H_2-saturated 0.1 M H_2SO_4 solutions at 25°C (Wang et al., 2006), while the values for a Pt nanocatalyst dispersed on an interactive, homemade titanium suboxide support, $Ti_nO_{2n-1}/Pt(5\ \mu g)$, in H_2-saturated 0.5 M $HClO_4$ solutions at 25°C were $j_{0T} = 50$ mA, $j_{0H} = 3$ mA, $\gamma = 1.38$ (Babic et al., 2009). Although the specific surface area was relatively low, the $Ti_nO_{2n-1}/Pt(5\ \mu g)$ catalyst exhibited a specific activity for the HOR similar to that of a conventional carbon-supported Pt catalyst.

The commercial catalysts used in fuel cells are usually prepared by adding Nafion solutions. The HOR on high-performance E-TEK Pt/C catalyst has been electrochemically characterized by a one-step deposition of inks, made with the catalyst and varying amounts of Nafion, on glassy carbon (GC) electrodes using CV and RDE (Esparbe et al., 2009). Equation 23.10 can be modified thus

$$\frac{1}{j_L} = \frac{1}{j_d} + \frac{1}{j_k} + \frac{1}{j_f} \tag{23.29}$$

where j_d is the current density limited by H_2 diffusion through the electrochemical double layer and is identical to that given in Equation 23.9, j_k is the kinetic current density, and j_f is the current density limited by H_2 diffusion into Nafion. The inverse of $(1/j_k + 1/j_f)$ can be calculated by extrapolation from a Koutecky–Levich diagram and then plotted against the inverse of the Nafion content in Figure 23.6. An exponential rise of $(1/j_k + 1/j_f)^{-1}$ with respect to the inverse of the Nafion content was found, and the values of $(1/j_k + 1/j_f)^{-1}$ approached 15 mA cm^{-2} when the Nafion content in the thin-film catalyst layer was <15%.

A series of carbon-supported IrM (M = V, Mn, Fe, Co, Ni) binary alloys have been prepared and investigated as non-Pt catalyst candidates for the HOR in PEM fuel cells (Qiao et al., 2009). The results revealed that the electrocatalytic activities of heat-treated 50% IrCo/C and IrV/C at 800°C in an argon atmosphere were comparable to that of 20% Pt/C at 25°C in terms of both mass and specific activities; however, the stability of these catalysts was still lower than that of Pt/C.

23.4.1.2 Analysis of the Oxygen Reduction Reaction

One of the major problems in PEM fuel cells is the slow reaction rate of the ORR. The mechanism of the ORR on Pt is not fully understood. It is commonly accepted that the ORR is a multielectron reaction

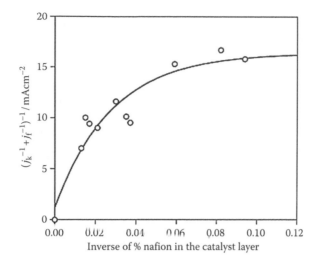

FIGURE 23.6 The plot of $(1/j_k + 1/j_f)^{-1}$ versus the inverse of wt% Nafion content for the HOR. (Reprinted from *J. Power Sources*, 190, Esparbe, I. et al., Structure and electrocatalytic performance of carbon-supported platinum nanoparticles, 201–209, Copyright (2009), with permission from Elsevier.)

that occurs via two main pathways: one involving the transfer of two electrons to give peroxide (Equation 23.11), and the other the so-called direct 4-electron pathway to give water (Equation 23.12). The latter involves rupturing of the O—O bond. The generalized ORR mechanism in acidic media (Damjonovic and Brusic, 1967; Yeager, 1984) is schematically presented in Figure 23.7.

The nature of the electrode strongly influences the preferred pathway. RDE and RRDE techniques are powerful tools in the study of the ORR. The disk electrode, which is surrounded by a ring set to a constant potential and acts as an amperometric sensor of peroxide, is the WE. The ring-disk electrode can be calibrated with the Koutecky–Levich equation to determine the amount of reactant detected on the ring. It is possible to determine the amount of H_2O_2 generated in the reaction (Equation 23.14) while studying the reduction of O_2 on a ring-disk electrode. The apparent electron transfer number during the ORR can also be obtained through RDE/RRDE measurements by using Equation 23.15.

23.4.1.2.1 Activity and Stability of Pt-based Catalysts

Pt-based alloys may exhibit better electrocatalytic activity than pure Pt toward the ORR. Sode et al. (2006) found that the ORR on PtZn alloy took place 30 mV higher than on pure Pt. The activity of PtZn was comparable to the activities of other Pt alloys, which show a similar decrease in the ORR overpotential (30 mV). Furthermore, a 30–70 mV higher cell voltage over the entire range of explored superficial current densities up to 800 mA cm^{-2} could be achieved using MEAs with Zn-treated Pt cathodes. Using

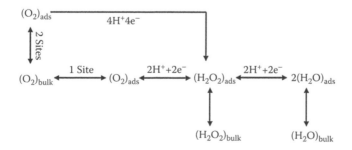

FIGURE 23.7 A sketch of pathways for the electrochemical reduction of molecular oxygen in acidic media.

sputter deposition, Pt films and $Pt_{1-x}Ir_x$ ($0 < x < 0.3$) layers have been deposited onto both 0.05 μm mirror-polished and nanostructured thin-film (NSTF)-coated GC disks, with the ORR activity of the disks being determined by RDE measurements (Liu et al., 2010). The onset potentials and the half-wave potentials for the catalysts deposited on flat surfaces, including polycrystalline Pt and sputtered $Pt/Pt_{1-x}Ir_x$ materials on mirror-polished GC disks, were 50–100 mV lower than those for the catalysts deposited onto NSTF-coated GC disks. Extended potential cycling and prolonged exposure to liquid electrolyte had no apparent effect on the catalytic activity of the catalysts deposited onto the NSTF-coated GC disks. After the data were normalized for active Pt surface area, the specific current densities were all the same and showed identical Tafel slopes, regardless of support. In addition, the specific current densities for catalyst films on the mirror-polished and NSTF-coated GC disks were identical to those measured on the polycrystalline Pt disk, suggesting that the active component of the ORR catalyst in all four types of catalyst/ film combinations was Pt. Using an RDE, the kinetics of the HOR on PtSb and PtSn have also been studied (Grgur et al., 1998; Colmenares et al., 2006; Innocente and Angelo, 2006). It was shown that the higher HOR activities obtained with PtSb and PtSn were attributable to alteration of the electron density of the surface adsorption sites, which markedly improved the surface adsorption of hydrogen molecules.

The stabilities of Ketjen black carbon-supported Pt, PtX (X = Co, Ru, Rh, V, Ni) and PtCoX (X = Ir, Rh) catalysts have been examined in detail in fuel cells by performing RRDE measurements and using potential cycling (Sethuraman et al., 2009). As is evident in Figure 23.8, all Pt-based alloy catalysts

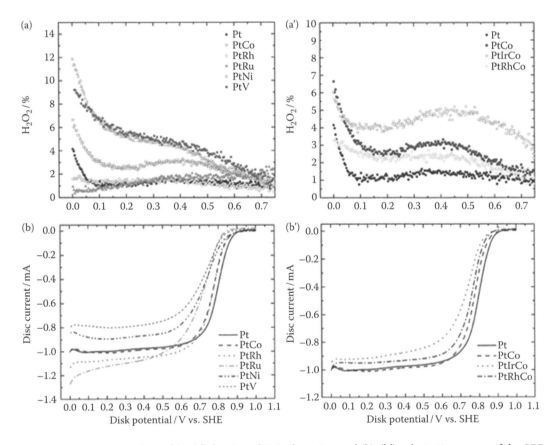

FIGURE 23.8 A comparison of (a), (a') fraction of H_2O_2 formation and (b), (b') polarization curves of the ORR for various catalysts in 1 M $HClO_4$ solutions at 25°C. (Reprinted from *Electrochim. Acta*, 54, Sethuraman, V. A. et al., Importance of catalyst stability vis-à-vis hydrogen peroxide formation rates in PEM fuel cell electrodes, 5571–5582, Copyright (2009), with permission from Elsevier.)

showed better selectivity toward H_2O_2 formation during the ORR than did unalloyed Pt. Among the binary alloys studied, PtNi exhibited the highest selectivity toward H_2O_2 formation, followed by PtV and PtCo. The authors who studied these alloys concluded that PtCo, PtRhCo, and PtIrCo displayed good activity and stability, and that MEAs with alloyed catalysts were more durable.

Lim et al. synthesized PdPt bimetallic nanodendrites, consisting of a dense array of Pt branches on a Pd core, by reducing K_2PtCl_4 with l-ascorbic acid in the presence of uniform Pd nanocrystal seeds in an aqueous solution (Lim et al., 2009). They found that the specific ECSA (i.e., the ECSA per unit weight of metal) of the PdPt nanodendrites (48.5 m^2 g_{Pd+Pt}^{-1}) was 66% of the Pt/C catalyst's (74.0 m^2 g_{Pt}^{-1}). At room temperature, the PdPt nanodendrites exhibited a mass activity of 0.204 mA μg_{Pd+Pt}^{-1} on the basis of the total mass of Pd and Pt at 0.9 V, which were 2.1 and 4.3 times greater, respectively, than those of Pt/C catalyst (0.095 mA μg_{Pt}^{-1}) and Pt black (0.048 g_{Pt}^{-1}). At 60°C, the Pt mass activity of PdPt nanodendrites (0.433 g_{Pt}^{-1}) was still greater than that of Pt/C catalyst (0.204 g_{Pt}^{-1}) and Pt black (0.078 g_{Pt}^{-1}). After 4000 cycles, an ECSA loss of 30% was observed for PdPt nanodendrites, of 36% for Pt/C catalyst, and of 33% for Pt black. However, PdPt nanodendrites showed a loss of 50% after 10,000 cycles.

23.4.1.2.2 *Activity and Stability of Non-Pt Catalysts*

Non-Pt catalysts have been actively investigated as alternatives for cathode catalysts (Zhang et al., 2006; Bezerra et al., 2008b). On a graphite electrode, the catalytic activities of adsorbed cobalt tetramethoxyphenyl porphyrin (CoTMPP) toward O_2 and H_2O_2 reduction have been studied using CV and RDE techniques (Liu et al., 2006). It was found that in acidic solutions the ORR occurred by a 2-electron process. The adsorption of anions on CoTMPP resulted in a negative shift of the onset potential for the ORR. In a study comparing the pyrolyzing of poly-o-phenylenediamine (PPDA) on carbon black (PPDA/C) with iron-PPDA and cobalt-PPDA deposited on carbon black (FePPDA/C and CoPPDA/C), Wang et al. reported an n value of 3.7 for FePPDA/C at 0.5 and 0 V, and of 2.7 and 3.4 for CoPPDA/C at 0.5 and 0 V (Wang et al., 2007). They suggested that the oxygen bound in the carbon support, as well as the transition metals Fe and Co added as salts, could be helpful in the formation of active sites on the catalysts during the high-temperature heat treatment processes applied to the precursors. The ORR mechanism on the Fe-based catalyst favors the 4-electron pathway, as the Fe-based catalyst is more active than the Co-based one or pyrolyzed PPDA/C alone. This has been further confirmed by the ORR/RRDE studies of five types of Fe-based catalysts, which found that the values of $n > 3.9$ and $xH_2O_2 < 5\%$ at the maximum catalytic activities of those catalysts were comparable to those of 2% Pt/C (Lefèvre and Dodelet, 2003).

23.4.1.2.3 *Formation of Hydrogen Peroxide*

RRDE technology can quantitatively detect the possible intermediate species, H_2O_2, involved in the ORR mechanism. The ORR takes place on the center of the disk electrode, and the produced H_2O_2 is either oxidized or reduced on the concentric ring electrode, depending on the electrode potential. RRDE studies allow evaluation of the molar proportion of H_2O_2 formation on bulk Pt in acidic media. Equation 23.14 can be used to calculate the fraction of H_2O_2 formation during the ORR.

The Pt mass loading has been shown to influence H_2O_2 formation. When different supporting materials have been compared, such as Vulcan carbon, NSTF, and TiO_2, H_2O_2 production at 0.4 V dropped significantly with increases in Pt loadings (Timperman et al., 2009). The n values for the ORR on GC RDE and RRDE (Kothandaraman et al., 2009) were estimated to be 3.57 ± 0.07 from the slope of a Koutecky–Levich plot, indicating a nearly complete reduction of oxygen to water. By applying an RRDE, the fractional H_2O_2 generation remained near 6% in the range of 0.4–0.9 V with $n \approx 3.88$, demonstrating high selectivity toward the 4-electron reduction of oxygen to water. The observed maximum H_2O_2 production is about 6.5% with a catalyst loading of 200 μg cm^{-2} at 0.60 V. Sarkar et al. (2010) synthesized Pt-encapsulated Pd_xCo_{100-x} nanoalloy electrocatalysts supported on carbon, through a rapid, microwave-assisted, solvothermal method. Their RDE data, including hydrodynamic polarization curves, mass-transfer-corrected Tafel plots, and mass-specific current density, showed that the 75%

$Pd_{80}Co_{20}$ + 25% Pt sample with a Pt loading of 5% exhibited higher catalytic activity for the ORR compared to either $Pd_{80}Co_{20}$ or Pt synthesized by the same method. Furthermore, the 75% $Pd_{80}Co_{20}$ + 25% Pt sample showed an almost threefold higher activity per unit of ECSA compared to commercial Pt. Single-cell PEMFC data confirmed the RDE data and the enhancement of catalytic activity per unit mass of Pt by encapsulating PdCo with Pt.

Heat treatment appears to be an effective way to enhance the ORR activity and stability of non-Pt catalysts (Bezerra et al., 2007). The fraction of H_2O_2 formation can be reduced from 88% with unpyrolyzed cobalt polypyrrole, CoPPy/C, to 61% with pyrolyzed CoPPy/C at 1000°C, suggesting that a 2-electron peroxide pathway becomes the dominant mechanism for unpyrolyzed CoPPy/C (Lee et al., 2009). A comparison between the values of n and xH_2O_2 in the ORR for various Fe-based catalysts with different heat-treatment temperatures provided strong evidence for the obvious effect of heat treatment (Lefèvre and Dodelet, 2003). In particular, without carbon support, FeTPP/CoTPP catalysts heat-treated at 600°C (Jiang and Chu, 2000) tended to prefer a 4-electron transfer pathway. Table 23.1 lists more values, obtained by RDE/RRDE measurements, for n and xH_2O_2 during the ORR for different types of catalysts.

Figure 23.9 compares the effects of catalyst loading on H_2O_2 production for RuSe/C (Bonakdarpour et al., 2008) and $CoSe_2$/C (Feng et al., 2009). The maximum H_2O_2 production at the ring electrode decreased from 30% to 15% when the loading rate of $CoSe_2$/C increased from 20% to 50%. An increase in the number of catalytic sites (or mass surface density) with higher catalyst loading is necessary to produce the higher reaction rate, and hence to reduce H_2O_2 production.

Suarez-Alcantara et al. evaluated the ORR activity of $Ru_xM_ySe_z$ (M = Cr, Mo, W)-type electrocatalysts in fuel cells (Suarez-Alcantara et al., 2006; Suarez-Alcantara and Solorza-Feria, 2008, 2009). Figure 23.10a gives a set of RDE polarization curves obtained on $Ru_xW_ySe_z$ catalysts in O_2-saturated 0.5 M H_2SO_4 at 25°C (Suarez-Alcantara and Solorza-Feria, 2009, 2010). Well-defined charge-transfer kinetic control, mixed kinetic-diffusion, and mass transport-limited currents were observed. If the Nafion film resistance is small, the film's diffusion-limited current density can be neglected. The mass-transfer-corrected Tafel plots at several temperatures for $Ru_xW_ySe_z$ electrocatalyst are compared in Figure 23.10b. Increases in the catalytic current and shifts in the curves to more positive potentials are observable with increasing temperature. The Arrhenius equation was used to calculate an average value of the apparent activation energy, ΔE_a, found to be 52.1 ± 0.4 kJ mol⁻¹ in the temperature range of 25–60°C.

$$\Delta E_a = -2.3R \left[\frac{d\log j_o}{d(1/T)} \right] \tag{23.30}$$

The kinetic data obtained with $Ru_xMo_ySe_z$ catalyst (Suarez-Alcantara and Solorza-Feria, 2008) indicated that incorporating molybdenum with ruthenium chalcogenide improved the catalytic activity and selectivity of molecular oxygen toward the 4-electron reduction, leading to water formation with a maximum of 2.5% H_2O_2 as a by-product. The ΔE_a value was 45.6 ± 0.5 kJ mol⁻¹ at 25–60°C. At 0.30 V, maximum power densities of 240 and 180 mW cm⁻² were achieved for $Ru_xMo_ySe_z$ and $Ru_xW_ySe_z$, respectively, with cathode catalyst loadings of 1.0 mg cm⁻² for 20% $Ru_xMo_ySe_z$ and 1.4 mg cm⁻² for 40% $Ru_xW_ySe_z$, dispersed on VC with a 10% Pt/C anodic loading of 0.8 mg cm⁻².

23.4.2 Effects of Contaminants

The effects of various contaminants on PEM fuel cell performance and durability have previously been reviewed (Schmidt et al., 1999a,b; Kaiser et al., 2006; Cheng et al., 2007). Minor impurities from fuels or oxidants can have significant impacts on the electrocatalytic activity and stability of catalysts. The most well-known contaminant effect is carbon monoxide (CO) poisoning of Pt catalysts. Schmidt et al. (1998) quantitatively evaluated the CO electrooxidation activity of colloid-based PtRu catalyst toward CO/H_2

TABLE 23.1 Values of n and xH_2O_2 on Various Catalysts, Determined by RDE/RRDE

	Catalyst/Electrolyte		Parameter		References
			n	$xH_2O_2(\%)$	
Pt	Pt	0.1 M HClO$_4$	≈4	<3	Sarapuu et al. (2008)
	Pt	0.05 M H$_2$SO$_4$	≈4	<3	Sarapuu et al. (2008)
	Pt/C	0.1 M H$_2$SO$_4$	3.95		Li et al. (2010a)
	Pt/C	0.1 M HClO$_4$	2.7–3.9	0.22	Maruyama et al. (2003)
	Pt/C	0.1 M HClO$_4$	≈4		Wang et al. (2007)
	Pt/C	0.1 M NaOH	3.9		Jiang et al. (2009)
Au	Au	0.5 M H$_2$SO$_4$	≈2		Sarapuu et al. (2008)
	Au/BP	0.5 M H$_2$SO$_4$	2.5–3.2		Bron (2008)
		0.5 M H$_2$SO$_4$			
	Au/C	0.5 M H$_2$SO$_4$	2.1–2.6		Bron (2008)
	AuNP/C		2.0 3.5		Vazquez-Huerta et al. (2010)
Pt-based alloy	PtAu	0.1 M HClO$_4$	3.9 ± 0.1	<5	Sarapuu et al. (2006)
	PtAu	0.05 M H$_2$SO$_4$	3.9 ± 0.1	<5	Sarapuu et al. (2006)
	PtRu/C	0.1 M NaOH	3.8	<4	Jiang et al.(2009)
	PtCo/C	0.5 M H$_2$SO$_4$	~4	<4.5	Garsany et al. (2007)
	Pt$_3$Co/C	0.1 M HClO$_4$	2.2		Garsany et al. (2007)
	PtBi/C	0.2 M NaOH	3.7–4.0	<15	Demarconnay (2008)
	PtFe/C	0.5 M H$_2$SO$_4$	3.8		Zeng et al. (2010)
Pd$_2$Co/C		0.1 M HClO$_4$	≈4	0.84	Wang et al. (2007)
IrV/C		0.5 M H$_2$SO$_4$	3.868		Qiao et al. (2010)
Ru$_x$Se$_y$		0.5 M H$_2$SO$_4$	~4		Gago et al. (2010)
Ru$_x$Mo$_y$Se$_z$		0.5 M H$_2$SO$_4$	≈4	<2.5	Suarez-Alcantara and Solorza-Feria (2008)
CoSe$_2$/C		0.5 M H$_2$SO$_4$	3.5	50–10	Feng et al. (2009)
CoPc/C		0.1 M NaOH	1.4–2.4	20–50	Chen et al. 2009
FePc/C		0.1 M NaOH	3.8	few	Chen et al. (2009)
FeTPTZ/C		0.5 M H$_2$SO$_4$	3.5–3.8	10–30	Bezerra et al. (2008b)
CoTPTZ/C		0.5 M H$_2$SO$_4$	3.1–3.7	5–20	Li et al. (2010b)
Mo$_2$N/C		0.5 M H$_2$SO$_4$	3.8		Zhong et al. (2006)
FeCoN/C		0.5 M H$_2$SO$_4$	3.80–3.95	3–10	Li et al. (2010c)
CoTETA/C		0.5 M H$_2$SO$_4$	3.6		Zhang et al. (2010b)
Fe phenantroline/C		0.5 M H$_2$SO$_4$	3.7	15	Bron (2002)
ClFeTMPP		0.1 M H$_2$SO$_4$	3.45–4	28–0	Gojkovic et al. (1999)
FeTPP/CoTPP		0.5 M H$_2$SO$_4$	4.0	0	Jiang and Chu (2000)
ZrO$_x$N$_y$/C		0.5 M H$_2$SO$_4$	3.8		Liu et al. (2007)

gas mixtures in a true RDE configuration, with well-defined mass transport properties and negligible film diffusion resistances through Nafion film. Comparison with other fuel cell catalysts demonstrated that the former has activity similar to or slightly higher than commercially available PtRu catalyst. The CO tolerance assessed from RDE measurements with supported catalysts in dilute CO/H$_2$ mixtures is consistent with the reported CO tolerance of PEM fuel cell anodes.

Kaiser et al. (2006) evaluated the performance of three different carbon-supported PtRu catalysts: (1) carbon-supported PtRu alloy particles (PtRu/C), (2) Pt and Ru particles codeposited on the same carbon support (Pt + Ru)/C, and (3) a mixture of carbon-supported Pt and carbon-supported Ru (Pt/C + Ru/C) used as anode catalysts by feeding a H$_2$/CO (2%) gas mixture (simulated reformate)

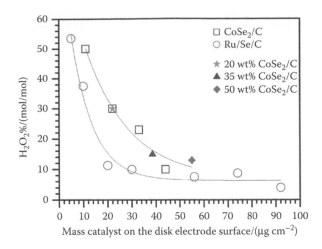

FIGURE 23.9 Hydrogen peroxide formation as a function of the mass of catalyst per unit surface for the ORR on Ru/Se/C and CoSe₂/C[15] systems. (Reprinted from *Electrochem. Commun.*, 10, Bonakdarpour, A. et al., Loading of Se/Ru/C electrocatalyst on a rotating ring-disk electrode and the loading impact on a H2O2 release during oxygen reduction reaction, 611–615, Copyright (2008), with permission from Elsevier.)

under relevant fuel cell conditions (elevated temperature, continuous reaction, and controlled reactant transport) in an RDE setup. The polarization curves in Figure 23.11 show a distinctly different behavior for the bimetallic catalyst electrodes and the monometallic Pt/C and Ru/C catalyst electrodes, as would be expected from CO_{ads} stripping behavior. The oxidation current of the Pt/C catalyst only increased at potentials >0.6 V, while the oxidation current of the Ru/C increased slowly at potentials >0.3 V and then increased rapidly at >0.37 V. These values coincide with CO_{ads} stripping behavior, which starts at about 0.3 V and becomes significant at potentials >0.37 V. However, in contrast with the other electrodes, the oxidation of H₂/CO (2%) reached its maximum current density near 0.42 V,

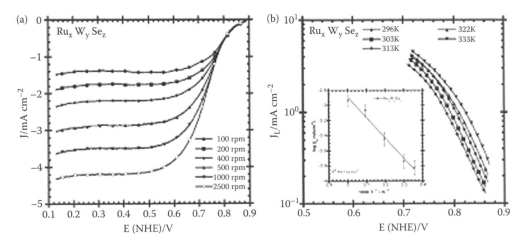

FIGURE 23.10 (a) RDE polarization curves for the ORR on $Ru_xW_ySe_z$; (b) mass-transfer-corrected Tafel plots for the ORR on $Ru_xW_ySe_z$ at different temperatures. The inset shows the electrochemical Arrhenius plot. (Suarez-Alcantara, K. and Solorza-Feria, O.: Evaluation of RuxWySez catalyst as a cathode electrode in a polymer electrolyte membrane fuel cell. *Fuel Cells*. 2010. 10. 84–92. Copyright Wiley-VCH Verlag GmbH & Co. KgaA. Reproduced with permission.)

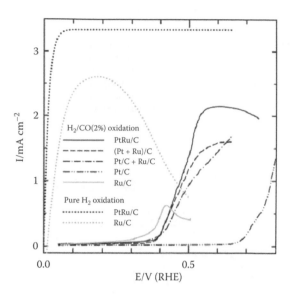

FIGURE 23.11 Potentiodynamic oxidation (positive-going scan) of a H$_2$/CO (2%) mixture on alloyed PtRu/C catalyst (solid line), codeposited (Pt + Ru)/C catalyst (long-dashed line), and mixed (Pt/C + Ru/C) catalyst (short-dashed line). For comparison, pure H$_2$ oxidation on PtRu/C catalyst is included (dotted line). (Kaiser, J. et al., On the role of reactant transport and (surface) alloy formation for the CO tolerance of carbon supported PtRu polymer electrolyte fuel cell catalysts. *Fuel Cells*. 2006. 6. 190–202. Copyright Wiley-VCH Verlag GmbH & Co. KgaA. Reproduced with permission.)

then decreased. For the bimetallic catalysts, pronounced current increases were found at potentials around 0.4 V, while at lower potentials the catalyst surface was largely blocked by CO$_{ads}$. The currents on the bimetallic catalysts were below the transport-limited current obtained for H$_2$ oxidation, indicating that the oxidation currents were kinetically determined. Due to the fast HOR kinetics, the currents observed above 0.39 V were dominated by contributions from the HOR. At these potentials this was only possible on Pt sites, which required the removal of CO$_{ads}$ to create the surface sites needed for the adsorption and oxidation of H$_2$.

In addition to PtRu alloy catalysts, which are well known to have excellent tolerance to CO poisoning, RDE/RRDE has also been employed to study other Pt-based alloys for the HOR in the presence of CO contaminants. By examining the electrochemical oxidations of H$_2$, CO, and H$_2$/CO mixtures on PtMo bulk alloy using RDEs in 0.5 M H$_2$SO$_4$ solutions at 60°C (Grgur et al., 1998; Colmenares et al., 2006; Innocente and Angelo, 2006), researchers have reported that the Mo mechanism on the Pt surface that enhances the HOR in the presence of CO seemed to be very similar to that of Ru, that is, a reduction in the steady-state coverage of CO$_{ads}$ by oxidative removal, freeing Pt sites for H$_2$ oxidation.

More H$_2$O$_2$ can be produced when Co^{2+} is present, according to Li et al. (2010a). At a disk potential of 0.5 V, the presence of 0.05 M Co^{2+} led to an approximate 9% increase in H$_2$O$_2$ formation, indicating an increase in the portion of the 2-electron transfer pathway producing H$_2$O$_2$. In addition, the overall n values decreased from 3.95 in the absence of Co^{2+} to 3.42 in the presence of 0.2 M Co^{2+}. The effects of sulfur poisoning on Pt$_3$Co/VC catalyst activity for the ORR, and the efficiency of water versus H$_2$O$_2$ production, have been investigated using RRDE and density function theory (Garsany et al., 2007, 2009; Pillay et al., 2010). The RRDE results are shown in Figure 23.12. A clean Pt$_3$Co/VC electrode exhibited a well-defined, diffusion-limiting disk current density of –6 mA cm^{-2} at potentials (E) of 0.10–0.80 V, followed by a region under mixed kinetic-diffusion control at 0.82 < E < 1.00 V. In the kinetic-diffusion control region the ring currents were a small portion. The fraction of H$_2$O$_2$ produced on Pt$_3$Co/VC was

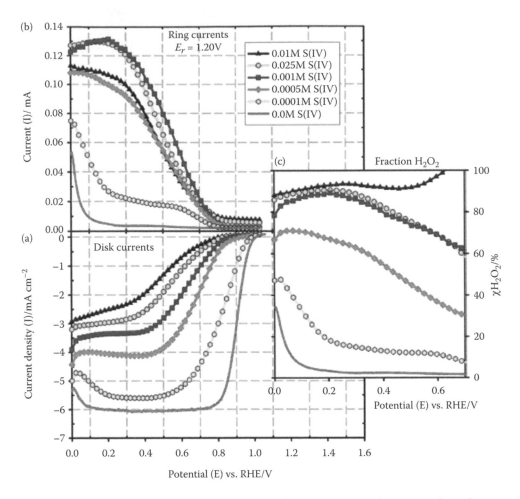

FIGURE 23.12 First anodic scan from RDE voltammetry of clean Pt_3Co/VC and Pt_3Co/VC electrodes poisoned with various sulfur concentrations. (a) Disk currents for the ORR, (b) ring currents for H_2O_2 oxidation, and (c) fraction of H_2O_2 formation. (From Garsany, Y., Baturina, O. A., and Swider-Lyons, K. E. 2009. *J. Electrochem. Soc.*, 156: B848–B855. With permission.)

negligible at $E = 0.20$ V, indicating that the ORR proceeded almost entirely by means of a 4-electron pathway to water.

23.4.3 Effects of Heat Treatment

The effects of heat treatment on the catalytic activity and stability of different types of catalysts used in PEM fuel cells have been reviewed by Bezerra et al. (2007). Heat treatment plays an important role in improving ORR activity and stability (Faubert et al., 1996; Gojkovic et al., 1999; Zhang et al., 2007, 2009; Bezerra et al., 2008a; Charreteur et al., 2009; Li et al., 2010b). The influence of pyrolysis gas on the activity and stability of iron porphyrin-based catalysts has also been studied; Charreteur et al. (2009) showed that pyrolysis in ammonia gas resulted in a low (5%) H_2O_2 yield but unstable catalysts, while pyrolysis in argon gas led to less activity with a high (26%) H_2O_2 yield, but was more stable for at least 15 h at high catalyst loadings. In the case of PtSn bimetallic nanoparticles, the particle sizes of heat-treated catalysts tended to increase with rising temperature, resulting in a slight reduction in ORR current densities at 0.55 V from

2.74 mA cm^{-2} without heat treatment to 2.04 mA cm^{-2} with 800°C treatment, which corresponded to an increase in H$_2$O$_2$ yields from 0.76% to 2.30% (Jeyabharathi et al., 2008).

23.4.4 Analysis of Nafion–Catalyst Interfaces

Proton exchange membranes are used exclusively as the electrolyte in PEM fuel cells because of their high chemical and thermal stability, and fairly high ionic conductivity. Nafion is the most commonly used commercially available perfluorosulfonate membrane. The kinetics of the ORR at Nafion–catalyst interfaces has been studied to optimize catalytic activity, electrode morphology, and operating conditions. Using the design of a minicell, the measured ORR activity at a Pt-recast Nafion interface under conditions identical to those of operating fuel cells has been used to evaluate the degree of Pt catalyst utilization in an MEA (Uribe et al., 1992). The kinetic currents of the ORR at a Pt-Nafion electrode (i_k (Pt-Nafion)) have been found to be higher than those at a bare Pt electrode (i_k (Pt)) in 0.7 M H$_3$PO$_4$ solutions; however, the enhancement factor (i_k (Pt-Nafion))/(i_k (Pt)) was lower based on O$_2$ solubility, mainly due to the effects of morphology and the micelle-like microstructure of Nafion (Lawson et al., 1988).

23.5 Conclusion

The instruments for RDE/RRDE measurements are commercially available, and RDE/RRDE is the most convenient and widely used technology to investigate the performance of catalysts and fuel cells. In particular, RDE/RRDE combines ease of construction and use with the ability to achieve well-controlled hydrodynamics. It also has significant advantages in the detection of short-lived intermediate species, and can be used for quantitative analysis. By providing voltammograms (current–potential curves) of the disk and ring, this technique permits detailed study of the kinetics and mechanisms of electrode reactions. RDE/RRDE technology is commonly applied as a diagnostic tool in PEM fuel cells to study the electrocatalytic activity, selectivity, and stability of various catalysts in different environments, whereby the performance and durability of fuel cells are also evaluated. With developments in hydrodynamic methods and theory, and the optimization of experimental setups, RDE/RRDE will be capable of extracting more useful parameters to facilitate a better understanding of the complex phenomena involved in PEM fuel cell functioning.

Acknowledgment

The authors acknowledge the National Natural Science Foundation of China for its support under Grant no. 10472098.

References

Amin, M. A. and Khaled, K. F. 2010. Copper corrosion inhibition in O$_2$-saturated H$_2$SO$_4$ solutions. *Corros. Sci.* 52: 1194–1204.

Babic, B., Gulicovski, J., Gajic-Krstajic, L., Elezovic, N., Radmilovic, V. R., Krstajic, N. V., and Vracar, L. M. 2009. Kinetic study of the hydrogen oxidation reaction on sub-stoichiometric titanium oxide-supported platinum electrocatalyst in acid solution. *J. Power Sources* 193: 99–106.

Bard, A. J. and Faulkner, L. R. 2001. *Electrochemical Methods: Fundamentals and Applications (second ed.)*, New York: John Wiley & Sons.

Becerra, H. Q., Retamoso, C., and Macdonald, D. D. 2000. The corrosion of carbon steel in oil-in-water emulsions under controlled hydrodynamic conditions. *Corros. Sci.* 42: 561–575.

Bezerra, C. W. B., Zhang, L., Lee, K., Liu, H., Zhang, J., Shi, Z., Marques, A. L. B., Marques, E. P., Wu, S., and Zhang, J. J. 2008a. Novel carbon-supported Fe-N electrocatalysts synthesized through heat treatment of iron tripyridyl triazine complexes for the PEM fuel cell oxygen reduction reaction. *Electrochim. Acta* 53: 7703–7710.

Bezerra, C. W. B., Zhang, L., Lee, K. C., Liu, H. S., Marques, A. L. B., Marques, E. P., Wang, H. J., and Zhang, J. J. 2008b. A review of Fe-N/C and Co-N/C catalysts for the oxygen reduction reaction. *Electrochim. Acta* 53: 4937–4951.

Bezerra, C. W. B., Zhang, L., Liu, H., Lee, K., Marques, A. L. B., Marques, E. P., Wang, H., and Zhang, J. 2007. A review of heat-treatment effects on activity and stability of PEM fuel cell catalysts for oxygen reduction reaction. *J. Power Sources* 173: 891–908.

Bonakdarpour, A., Delacote, C., Yang, R., Wieckowski, A., and Dahn, J. R. 2008. Loading of Se/Ru/C electrocatalyst on a rotating ring-disk electrode and the loading impact on a H_2O_2 release during oxygen reduction reaction. *Electrochem. Commun.* 10: 611–615.

Bron, M. 2008. Carbon black supported gold nanoparticles for oxygen electroreduction in acidic electrolyte solution. *J.Electroanal.Chem.* 624: 64–68.

Bron, M., Fiechter, S., Hilgendorff, M., and Bogdanoff, P. 2002. Catalysts for oxygen reduction from heat-treated carbon-supported iron phenantroline complexes. *J. Appl. Electrochem.* 32: 211–216.

Charreteur, F., Jaouen, F., and Dodelet, J.-P. 2009. Iron porphyrin-based cathode catalysts for PEM fuel cells: Influence of pyrolysis gas on activity and stability. *Electrochim. Acta* 54: 6622–6630.

Chen, R., Li, H., Chu, D., and Wang, G. 2009. Unraveling oxygen reduction reaction mechanisms on carbon-supported Fe-phthalocyanine and Co-phthalocyanine catalysts in alkaline solutions. *J. Phys. Chem. C* 113: 20689–20697.

Cheng, X., Shi, Z., Glass, N., Zhang, L., Zhang, J., Song, D., Liu, Z.-S., Wang, H., and Shen, J. 2007. A review of PEM hydrogen fuel cell contamination: Impacts, mechanisms, and mitigation. *J. Power Sources* 165: 739–756.

Clarke, C. J., Browning, G. J., and Donne, S. W. 2006. An RDE and RRDE study into the electrodeposition of manganese dioxide. *Electrochim. Acta* 51: 5773–5784.

Colmenares, L., Wang, H., Jusys, Z., Jiang, L., Yan, S., Sun, G. Q., and Behm, R. J. 2006. Ethanol oxidation on novel, carbon supported Pt alloy catalysts—Model studies under defined diffusion conditions. *Electrochim. Acta* 52: 221–233.

Damjnovic, A., Genshaw, M., and Bockris, J. 1966. The role of hydrogen peroxide in the reduction of oxygen at platinum electrodes. *J. Phys. Chem.* 45: 3761.

Damjonovic, A. and Brusic, V. 1967. Electrode kinetics of oxygen reduction on oxide-free platinum electrodes. *Electrochim. Acta* 12: 625.

Demarconnay, L., Coutanceau, C., and Léger, J. M. 2008. Study of the oxygen electroreduction at nanostructured PtBi catalysts in alkaline medium. *Electrochim. Acta* 53: 3232–3241.

Esparbe, I., Brillas, E., Centellas, F., Garrido, J. A., Rodriguez, R. M., Arias, C., and Cabot, P. L. 2009. Structure and electrocatalytic performance of carbon-supported platinum nanoparticles. *J. Power Sources* 190: 201–209.

Faubert, G., Lalande, G., Côté, R., Guay, D., Dodelet, J. P., Weng, L. T., Bertrand, P., and Dénès, G. 1996. Heat-treated iron and cobalt tetraphenylporphyrins adsorbed on carbon black: Physical characterization and catalytic properties of these materials for the reduction of oxygen in polymer electrolyte fuel cells. *Electrochim. Acta* 41: 1689–1701.

Feng, Y., He, T., and Alonso-Vante, N. 2009. Oxygen reduction reaction on carbon-supported CoSe2 nanoparticles in an acidic medium. *Electrochim. Acta* 54: 5252–5256.

Gago, A. S., Arriaga, L. G., Gochi-Ponce, Y., Feng, Y. J., and Alonso-Vante, N. 2010. Oxygen reduction reaction selectivity of RuxSey in formic acid solutions. *J. Electroanal. Chem.* 648: 78–84.

Garsany, Y., Baturina, O., and Swider-Lyons, K. 2007. Impact of SO_2 on the kinetics of Pt_3Co/vulcan carbon electrocatalysts for oxygen reduction. *ECS Transactions* 11: 863–875.

Garsany, Y., Baturina, O. A., and Swider-Lyons, K. E. 2009. Oxygen reduction reaction kinetics of SO_2-contaminated Pt_3Co and Pt/Vulcan carbon Electrocatalysts. *J. Electrochem. Soc.* 156: B848–B855.

Gennero De Chialvo, M. R. and Chialvo, A. C. 2004. Hydrogen diffusion effects on the kinetics of the hydrogen electrode reaction. Part I. Theoretical aspects. *Physical Chemistry Chemical Physics* 6: 4009–4017.

Gojkovic, S. L., Gupta, S., and Savinell, R. F. 1999. Heat-treated iron(III) tetramethoxyphenyl porphyrin chloride supported on high-area carbon as an electrocatalyst for oxygen reduction: Part III. Detection of hydrogen-peroxide during oxygen reduction. *Electrochim. Acta* 45: 889–897.

Grgur, B. N., Markovic, N. M., and Ross, P. N. 1998. Electrooxidation of H_2, CO, and H_2/CO Mixtures on a Well-Characterized $Pt_{70}Mo_{30}$ Bulk Alloy Electrode. *J. Phys. Chem. B* 102: 2494–2501.

Ibl, N. and Schadegg, K. 1967. Surface roughness effects in the electrodeposition of copper in the limiting current range. *J. Electrochem. Soc.* 114: 54–58.

Innocente, A. F. and Angelo, A. C. D. 2006. Electrocatalysis of oxidation of hydrogen on platinum ordered intermetallic phases: Kinetic and mechanistic studies. *J. Power Sources* 162: 151–159.

Jeyabharathi, C., Venkateshkumar, P., Mathiyarasu, J., and Phani, K. L. N. 2008. Platinum-tin bimetallic nanoparticles for methanol tolerant oxygen-reduction activity. *Electrochim. Acta* 54: 448–454.

Jiang, L., Hsu, A., Chu, D., and Chen, R. 2009. Oxygen reduction on carbon supported Pt and PtRu catalysts in alkaline solutions. *J. Electroanal. Chem.* 629: 87–93.

Jiang, R. and Chu, D. 2000. Remarkably active catalysts for the electroreduction of O_2 to H_2O for Use in an acidic electrolyte containing concentrated methanol. *J. Electrochem. Soc.* 147: 4605–4609.

Kaiser, J., Colmenares, L., Jusys, Z., Mortel, R., Bonnemann, H., Kohl, G., Modrow, H., Hormes, J. and Behm, R. J. 2006. On the role of reactant transport and (surface) alloy formation for the CO tolerance of carbon supported PtRu polymer electrolyte fuel cell catalysts. *Fuel Cells* 6: 190–202.

Kear, G., Barker, B. D., Stokes, K. R., and Walsh, F. C. 2007a. Electrochemistry of non-aged 90–10 copper-nickel alloy (UNS C70610) as a function of fluid flow: Part 1: Cathodic and anodic characteristics. *Electrochim. Acta* 52: 1889–1898.

Kear, G., Barker, B. D., Stokes, K. R., and Walsh, F. C. 2007b. Electrochemistry of non-aged 90–10 copper-nickel alloy (UNS C70610) as a function of fluid flow: Part 2: Cyclic voltammetry and characterisation of the corrosion mechanism. *Electrochim. Acta* 52: 2343–2351.

Kothandaraman, R., Nallathambi, V., Artyushkova, K., and Barton, S. C. 2009. Non-precious oxygen reduction catalysts prepared by high-pressure pyrolysis for low-temperature fuel cells. *Appl. Catal., B: Environmental* 92: 209–216.

Lawson, D. R., Whiteley, L. D., Martin, C. R., Szentirmay, M. N., and Song, J. I. 1988. Oxygen reduction at nafion film-coated platinum electrodes: Transport and kinetics. *J. Electrochem. Soc.* 135: 2247–2253.

Lee, K., Zhang, L., Lui, H., Hui, R., Shi, Z., and Zhang, J. 2009. Oxygen reduction reaction (ORR) catalyzed by carbon-supported cobalt polypyrrole (Co-PPy/C) electrocatalysts. *Electrochim. Acta* 54: 4704–4711.

Lefèvre, M. and Dodelet, J.-P. 2003. Fe based catalysts for the reduction of oxygen in polymer electrolyte membrane fuel cell conditions: Determination of the amount of peroxide released during electroreduction and its influence on the stability of the catalysts. *Electrochim. Acta* 48: 2749–2760.

Li, H., Tsay, K., Wang, H., Wu, S., Zhang, J., Jia, N., Wessel, S., Abouatallah, R., Joos, N., and Schrooten, J. 2010a. Effect of Co^{2+} on oxygen reduction reaction catalyzed by Pt catalyst, and its implications for fuel cell contamination. *Electrochim. Acta* 55: 2622–2628.

Li, S., Zhang, L., Liu, H., Pan, M., Zan, L., and Zhang, J. 2010b. Heat-treated cobalt-tripyridyl triazine (Co-TPTZ) electrocatalysts for oxygen reduction reaction in acidic medium. *Electrochim. Acta* 55: 4403–4411.

Li, S., Zhang, L., Kim, J., Pan, M., Shi, Zh., and Zhang, J. 2010c. Synthesis of carbon-supported binary FeCo-N non-noble metal electrocatalysts for the oxygen reduction reaction. *Electrochimi. Acta.* 55: 7346–7353.

Lim, B., Jiang, M., Camargo, P. H. C., Cho, E. C., Tao, J., Lu, X., Zhu, Y., and Xia, Y. 2009. Pd-Pt bimetallic nanodendrites with high activity for oxygen reduction. *Science* 324: 1302–1305.

Liu, G., Zhang, H. M., Wang, M. R., Zhong, H. X., and Chen, J. 2007. Preparation, characterization of ZrOxNy/C and its application in PEMFC as an electrocatalyst for oxygen reduction. *J. Power Sources* 172: 503–510.

Liu, G. C. K., Sanderson, R. J., Vernstrom, G., Stevens, D. A., Atanasoski, R. T., Debe, M. K., and Dahn, J. R. 2010. RDE measurements of ORR activity of $Pt_{1-x}Ir_x$ (0 < x < 0.3) on high surface area NSTF-coated glassy carbon disks. *J. Electrochem. Soc.* 157: B207–B214.

Liu, H. S., Zhang, L., Zhang, J. J., Ghosh, D., Jung, J., Downing, B. W., and Whittemore, E. 2006. Electrocatalytic reduction of O_2 and H_2O_2 by adsorbed cobalt tetramethoxyphenyl porphyrin and its application for fuel cell cathodes. *J. Power Sources* 161: 743–752.

Low, C. T. J. and Walsh, F. C. 2008. Electrodeposition of tin, copper and tin-copper alloys from a methane-sulfonic acid electrolyte containing a perfluorinated cationic surfactant. *Surf. Coat. Technol.* 202: 1339–1349.

Maruyama, J. and Abe, I. 2003. Cathodic oxygen reduction at the catalyst layer formed from Pt/carbon with adsorbed water. *J. Electroanal. Chem.* 545: 109–115.

Pillay, D., Johannes, M. D., Garsany, Y., and Swider-Lyons, K. E. 2010. Poisoning of Pt3Co electrodes: A combined experimental and DFT study. *J. Phys. Chem. C* 114: 7822–7830.

Qiao, J., Li, B., and Ma, J. 2009. Carbon-supported IrM (M = V, Mn, Fe, Co, and Ni) binary alloys as anode catalysts for polymer electrolyte fuel cells. *J. Electrochem. Soc.* 156: B436–B440.

Qiao, J., Lin, R., Li, B., Ma, J., and Liu, J . 2010. Kinetics and electrocatalytic activity of nanostructured Ir-V/C for oxygen reduction reaction. *Electrochimi. Acta.* 55: 8490–8497.

Ruiz, E. J., Ortega-Borges, R., Godinez, L. A., Chapman, T. W., and Meas-Vong, Y. 2006. Mechanism of the electrochemical deposition of samarium-based coatings. *Electrochim. Acta* 52: 914–920.

Sarapuu, A., Kallip, S., Kasikov, A., Matisen, L., and Tammeveski, K. 2008. Electroreduction of oxygen on gold-supported thin Pt films in acid solutions. *J. Electroanal. Chem.* 624: 144–150.

Sarapuu, A., Nurmik, M., Mändar, H., Rosental, A., Laaksonen, T., Kontturi, K., Schiffrin, D. J., and Tammeveski, K. 2008. Electrochemical reduction of oxygen on nanostructured gold electrodes. *J. Electroanal. Chem.* 612: 78–86.

Sarkar, A., Murugan, A. V., and Manthiram, A. 2010. Pt-Encapsulated Pd-Co Nanoalloy Electrocatalysts for Oxygen Reduction Reaction in Fuel Cells. *Langmuir* 26: 2894–2903.

Schmidt, T. J., Gasteiger, H. A., and Behm, R. J. 1999a. Electro-oxidation of H_2 and CO/H_2-mixtures on a carbon-supported Pt_3Sn catalyst. *J. New Mater. Electrochem. Syst.* 2: 27–32.

Schmidt, T. J., Gasteiger, H. A., and Behm, R. J. 1999b. Rotating disk electrode measurements on the CO tolerance of a high-surface area Pt/Vulcan carbon fuel cell catalyst. *J. Electrochem. Soc.* 146: 1296–1304.

Schmidt, T. J., Noeske, M., Gasteiger, H. A. et al. 1998. PtRu alloy colloids as precursors for fuel cell catalysts—A combined XPS, AFM, HRTEM, and RDE study. *J. Electrochem. Soc.* 145: 925–931.

Sethuraman, V. A., Weidner, J. W., Haug, A. T., Pemberton, M., and Protsailo, L. V. 2009. Importance of catalyst stability vis-à-vis hydrogen peroxide formation rates in PEM fuel cell electrodes. *Electrochim. Acta* 54: 5571–5582.

Sode, A., Li, W., Yang, Y. G., Wong, P. C., Gyenge, E., Mitchell, K. A. R., and Bizzotto, D. 2006. Electrochemical formation of a Pt/Zn alloy and its use as a catalyst for oxygen reduction reaction in fuel cells. *J. Phys. Chem. B* 110: 8715–8722.

Suarez-Alcantara, K., Rodriguez-Castellanos, A., Dante, R., and Solorza-Feria, O. 2006. RuxCrySez electrocatalyst for oxygen reduction in a polymer electrolyte membrane fuel cell. *J. Power Sources* 157: 114–120.

Suarez-Alcantara, K. and Solorza-Feria, O. 2008. Kinetics and PEMFC performance of RuxMoySez nanoparticles as a cathode catalyst. *Electrochim. Acta* 53: 4981–4989.

Suarez-Alcantara, K. and Solorza-Feria, O. 2009. Comparative study of oxygen reduction reaction on RuxMySez (M = Cr, Mo, W) electrocatalysts for polymer exchange membrane fuel cell. *J. Power Sources* 192: 165–169.

Suarez-Alcantara, K. and Solorza-Feria, O. 2010. Evaluation of RuxWySez Catalyst as a Cathode Electrode in a Polymer Electrolyte Membrane Fuel Cell. *Fuel Cells* 10: 84–92.

Sun, Y. B., Lu, J. T., and Zhuang, L. 2010. Rational determination of exchange current density for hydrogen electrode reactions at carbon-supported Pt catalysts. *Electrochim. Acta* 55: 844–850.

Timperman, L., Feng, Y. J., Vogel, W., and Alonso-Vante, N. 2009. Substrate effect on oxygen reduction electrocatalysis. *Electrochimi. Acta* 55: 7558–7563.

Treimer, S., Tang, A., and Johnson, D. 2002. A consideration of the application of Koutecký-Levich plots in the diagnoses of charge-transfer mechanisms at rotated disk electrodes. *Electroanalysis* 14: 165–171.

Uribe, F. A., Springer, T. E., and Gottesfeld, S. 1992. A Microelectrode Study of Oxygen Reduction at the Platinum/Recast-Nafion Film Interface. *J. Electrochem. Soc.* 139: 765–773.

Vazquez-Arenas, J. 2010. Experimental and modeling analysis of the formation of cuprous intermediate species formed during the copper deposition on a rotating disk electrode. *Electrochim. Acta* 55: 3550–3559.

Vázquez-Huerta, G., Ramos-Sánchez, G., Rodríguez-Castellanos, A., Meza-Calderón, D., Antaño-López, R., and Solorza-Feria, O. 2010. Electrochemical analysis of the kinetics and mechanism of the oxygen reduction reaction on Au nanoparticles. *J. Electroanal. Chem.* 645: 35–40.

Wang, J. X., Springer, T. E., and Adzic, R. R. 2006. Dual-pathway kinetic equation for the hydrogen oxidation reaction on Pt electrodes. *J. Electrochem. Soc.* 153: A1732–A1740.

Wang, P., Ma, Z. Y., Zhao, Z. C., and Ha, L. X. 2007. Oxygen reduction on the electrocatalysts based on pyrolyzed non-noble metal/poly-o-phenylenediamine/carbon black composites: New insight into the active sites. *J. Electroanal. Chem.* 611: 87–95.

Wu, J., Yuan, X. Z., Wang, H., Blanco, M., Martin, J. J., and Zhang, J. 2008. Diagnostic tools in PEM fuel cell research: Part I Electrochemical techniques. *Int. J. Hydrogen Energy* 33: 1735–1746.

Yeager, E. 1984. Electrocatalysts for O_2 reduction. *Electrochim. Acta* 29: 1527.

Zeng, J., Liao, S., Lee, J. Y., and Liang, Z. 2010. Oxygen reduction reaction operated on magnetically-modified PtFe/C electrocatalyst. *Int. J. Hydrogen Energy* 35: 942–948.

Zhang, H.-J., Yuan, X., Sun, L., Zeng, X., Jiang, Q.-Z., Shao, Z., and Ma, Z.-F. 2010. Pyrolyzed CoN4-chelate as an electrocatalyst for oxygen reduction reaction in acid media. *Int. J. Hydrogen Energy* 35: 2900–2903.

Zhang, H.-J., Yuan, X., Wen, W., Zhang, D.-Y., Sun, L., Jiang, Q.-Z., and Ma, Z.-F. 2009. Electrochemical performance of a novel CoTETA/C catalyst for the oxygen reduction reaction. *Electrochem. Commun.* 11: 206–208.

Zhang, L., Lee, K., and Zhang, J. 2007. The effect of heat treatment on nanoparticle size and ORR activity for carbon-supported Pd-Co alloy electrocatalysts. *Electrochim. Acta* 52: 3088–3094.

Zhang, L., Zhang, J., Wilkinson, D. P., and Wang, H. 2006. Progress in preparation of non-noble electrocatalysts for PEM fuel cell reactions. *J. Power Sources* 156: 171–182.

Zhong, H., Zhang, H., Liu, G., Liang, Y., Hu, J., and Yi, B. 2006. A novel non-noble electrocatalyst for PEM fuel cell based on molybdenum nitride. *Electrochem. Commun.* 8: 707–712.

24

Porosimetry and Characterization of the Capillary Properties of Gas Diffusion Media

Jeff T. Gostick
McGill University

Michael W. Fowler
University of Waterloo

Mark D. Pritzker
University of Waterloo

Marios A. Ioannidis
University of Waterloo

24.1 Introduction

Polymer electrolyte membrane fuel cells (PEMFCs) use a porous electrode structure usually referred to as a membrane-electrode assembly (MEA). A schematic cross-section of a standard MEA with typical dimensions listed is shown in Figure 24.1. PEMFC operation involves multiphase transport of gaseous species and liquid water through the porous layers of the MEA. Because of the multiphase aspects involved, a full mechanistic understanding of transport in fuel cell electrodes requires knowledge of capillary properties. This chapter will outline some important and relevant concepts of capillarity and wettability, describe the relevant experimental techniques that have been developed to study wettability and capillary behavior in electrode materials and discuss the most recent findings concerning the effect of degradation and aging on wettability.

The polymer electrolyte membrane (PEM) material is typically a fluoro-polymer with side chains possessing strong acidic sites. When the membrane is humidified, the polymer matrix becomes highly protonated with a high ionic conductivity; this type of material is usually called an ionomer. The requirement that the ionomer be humidified is perhaps one of the most important limitations of the PEMFC and is discussed further below. The catalyst layer is a porous three-dimensional reaction zone consisting of a catalyst (usually platinum) supported on carbon particles and held together with an ionomer binder. The ionomeric binder allows protons to travel into the 3D reaction zone while the carbon-supported catalyst particles provide electrically conductive pathways for electrons to reach the reactive sites. The porosity of the reaction zone allows gaseous reaction species to be transported to the reactive sites.

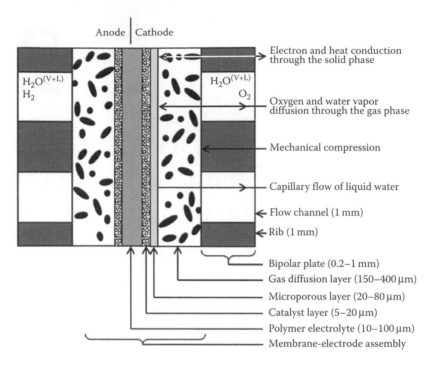

FIGURE 24.1 Schematic diagram of membrane electrode assembly typically used in PEM fuel cells.

The cathodic reaction occurs at the interfaces between catalyst sites and ionomer, where electrons, protons and oxygen dissolved in the ionomeric phase meet. The gas diffusion layer (GDL) is typically an electronically conductive carbon fiber paper. It acts as a spacer to allow gaseous reactants to diffuse laterally from the channel to the areas of the catalyst layer under the channel. Similarly, it allows electrons produced over the channels easier access to the ribs where current is collected. The GDL also provides mechanical support for the membrane and supports thermal conduction of heat away from the electrode.

One of the major challenges when designing, modeling and/or operating a cell with a porous electrode structure is the problem of multiphase flow. The requirement that that the membrane be as highly humidified as possible means cells must be operated close to the dew point, leading to the inevitable formation of liquid water in the porous electrode. The presence of liquid water in the pore network of the MEA hinders the transport of gaseous reactants to the catalyst sites, reduces the maximum power density of the cell, negatively affects efficiency and vehicle range, creates degradation problems (Meyers and Darling, 2006) and complicates operation, startup and shutdown at subzero temperatures. Since water is also produced by the cathode reaction, its rate of generation inside the cell varies with current density. Heat is also produced by the reaction and so the cell temperature and the dew point in the cell vary with time during the course of operation and whenever the operating conditions change. Prevention of the formation of liquid water in the electrode is virtually impossible over the entire range of operating conditions. To address the problem of liquid water in the electrodes, interest has been focused on developing high-temperature ionomer materials to enable operation above 100°C (Li et al., 2003; Zhang et al., 2006). Until such materials become available, however, it is necessary to minimize the negative effects due to the presence of liquid water in the electrodes. A great deal of effort has been directed at understanding the properties related to multiphase flow in the porous electrode and most importantly the air–water capillary pressure behavior (Ji and Wei, 2009; Sinha et al., 2007). Accordingly, this chapter focuses on the capillary pressure properties of fuel cell electrode components and closely examines the conceptual challenges associated with such measurements, the various measurement methods that have been proposed and used and interpretation of the experimental data so obtained.

24.2 Capillarity and Wettability

Capillary pressure curves are one of the most basic and essential pieces of information regarding a porous material. These curves describe how fluid pressure alters the saturation of the fluid when multiple phases are present. Since many transport parameters such as effective diffusivity and relative permeability depend on phase saturation, accurate capillary pressure data are essential to understand multiphase flow and transport in porous electrodes.

24.2.1 Capillary Pressure

Capillary pressure is the pressure difference across a static interface formed between two immiscible phases. This pressure difference is fundamentally related to the mean curvature H of the interface through the Young–Laplace equation (Dullien, 1992):

$$P_C \equiv P_1 - P_2 = 2\sigma H \tag{24.1}$$

where P_C is the capillary pressure defined as the difference between the pressure in phase 1 (P_1) and phase 2 (P_2), σ is the surface tension of the liquid–gas interface and H is the mean radius of curvature of the interface. At equilibrium the phase pressure difference is balanced by the tension of the interface. Smaller and more highly curved surfaces (higher values of H) are able to withstand larger pressure differences while still maintaining equilibrium and a mechanically stable (i.e., static) interface. Equation 24.1 applies to any interface between immiscible fluid phases, such as a lens on a flat surface, a spherical droplet suspended in another fluid or a finger of liquid confined in a capillary tube. This latter case pertains directly to porous media where each pore can be approximated as a cylindrical tube. To illustrate this, consider the situation shown in Figure 24.2 (top) where phase 1 is advancing into a capillary tube and displacing fluid 2. As can be deduced from Equation 24.1, the pressure in fluid 1 must be higher than fluid 2 for this displacement to occur. Since H has a positive value when measured from phase 1, the phase on the positively curved or convex side of an interface must be the phase with the higher pressure. Phase 1 is conventionally called the nonwetting phase and phase 2 the wetting phase. The wettability of each phase can be identified by either its relative pressure or position relative to the meniscus. This general and straightforward definition of phase wettability leads to some difficulties as will be discussed below.

24.2.2 Capillary Pressure Curves

A porous medium can be simplistically viewed as a 3D network of capillary tubes of varying size. The capillary pressure curve is the cumulative effect of the capillary behavior of each pore in the network. A capillary pressure curve is typically obtained by injecting a nonwetting fluid and monitoring the fluid uptake as the pressure is increased. In this way, pores of progressively smaller size can be penetrated by the invading fluid until the medium is completely saturated. A capillary pressure curve is the resulting relationship between the applied capillary pressure and the pore volume filled at each pressure. One important feature of capillary pressure measurement that is often overlooked is that the accessibility of a pore is not determined solely by its geometric dimensions but also by network limitations. For example, increasing pressure during injection enables smaller pores to be penetrated but also allows larger pores previously shielded by smaller pores to be filled as the smaller pores themselves are penetrated as

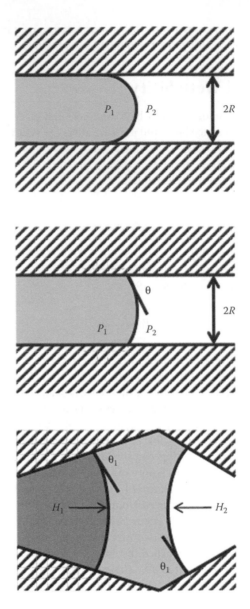

FIGURE 24.2 Invasion of a nonwetting phase into idealized capillary tubes. Top: A perfectly nonwetting fluid penetrates a capillary of radius R. Middle: A nonwetting fluid with a contact angle of θ penetrates a cylindrical tube of radius R. Bottom: A nonwetting fluid with a contact angle θ_1 invades a diverging–converging pore in two steps. At step 1, the interface has a positive curvature of H_1 when measured from the invading phase, while at step 2 the interface has a negative curvature of H_2, despite identical contact angles.

shown in Figure 24.3. Thus, the measurement actually being described in a capillary pressure curve is the *accessible* pore volume as a function of applied capillary pressure.

The measurement of capillary pressure curves can be performed by either controlling the pressure applied to the fluid and tracking the volume of fluid injected or controlling the fluid injection and monitoring the pressure response. Either approach can be done in a stepwise or scanning fashion. Understanding the differences between these four approaches is necessary since each can produce different results. At each increment in a pressure-step experiment, the pressure is held fixed until the invading fluid fills

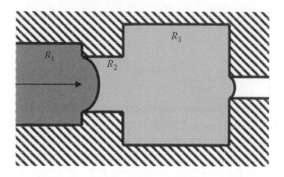

FIGURE 24.3 Schematic diagram showing the filling of a shielded pore of size R_3 upon penetration of a smaller constriction R_2. The use of porosimetry to obtain pore size distributions from capillary pressure data can falsely attribute the volume of pore R_3 to pores of size R_2.

all the accessible pore volume before proceeding to the next step. Pressure–scanning capillary curves are often used in commercial mercury intrusion porosimetry (MIP) devices to speed up data collection, with only minimal deviation from equilibrium (Leon, 1998). Volume-controlled methods are much more difficult to interpret reliably because they are essentially transient (Cerepi et al., 2002; Sygouni et al., 2006; Yuan and Swanson, 1989). The main difficulty with volume-scanning methods (also called rate-controlled methods) is that the results depend on the injection rate applied. Because viscous pressure losses occur in the fluid, it becomes easier for the fluid to invade smaller pores near the injection source rather than larger pores farther away. This means that the volume injected at a given pressure is always lower than the equilibrium saturation. The deviation is negligible at sufficiently low rates, but grows as the rate increases. At low rates, however, the system pressure response exhibits large oscillations due to the networked nature of porous materials. The pressure rises as the fluid invades small constrictions or throats and then drops as the larger pores behind the constriction are accessed and filled. Very useful information concerning the pore structure is embedded within the spiked pressure response, but detailed pore scale modeling is required to extract it (Knackstedt et al., 1998; Toledo et al., 1994). In the volume-step-controlled method the system response depends on the size of the volume increments as well as the injection rate. It is highly unlikely that the total accessible pore volume will be filled after each fluid addition. Inevitably, this will be followed by some redistribution of the fluid through the available pore space which causes an additional pressure decay that further complicates the interpretation of the response considerably. Methods to measure air–water capillary pressure curves of GDL materials discussed by Gostick et al. (2010a) have employed three of the four possible experimental approaches. Gostick et al. (2008) employed a pressure-step-controlled approach, Harkness et al. (2009) used a volume-scanning method and Fairweather et al. (2007) used volume-step control.

24.2.3 Contact Angle

A drop of fluid placed on a solid surface will either spread into a flat pool or form a lens. In the former case, the fluid is said to wet the solid perfectly, while in the latter the fluid ability to wet the solid can be quantified by the angle formed between the plane of the solid and the tangent to the lens surface at the liquid–solid contact line measured through the droplet. A contact angle near 0° indicates that the fluid preferentially wets the solid, while a contact angle near 180° is obtained for a fluid that is highly nonwetting. There is no specific value of contact angle where a fluid transitions from wetting to nonwetting, but rather a gradual transition from wetting to nonwetting. The widespread use of 90° as the transition from wetting and nonwetting behavior is based on the conceptual situation shown in

Figure 24.2 (middle). In this special case of a cylindrical tube it can be shown that the mean radius of curvature is described by

$$H = \frac{-\cos(\theta)}{R} \tag{24.2}$$

where θ is the contact angle formed between the liquid and the solid surface and R is the radius of the capillary tube (Dullien, 1992; Washburn, 1921a). According to Equation 24.2 the capillary pressure changes sign at $\theta = 90°$ and therefore the phases switch wettabilities. Although this may be true in straight cylindrical capillaries, it may no longer be valid, for example, when the pore has only the slightly more complicated geometry of the diverging–converging tube in Figure 24.2 (bottom). An invading fluid (grey) is shown at two different positions in a capillary with nonparallel walls. The contact angle is identical for both menisci although their curvatures H have different signs and magnitudes despite being poised at positions with identical radii. This apparent switch in wettability is more likely to occur for contact angles closer to 90° for obvious geometrical reasons.

24.2.4 Wettability

The description of a fluid as wetting or nonwetting implies certain associated behaviors. For instance, one might expect that a wetting fluid will spontaneously imbibe or wick into a porous material, as in the case of a water droplet imbibing into a tissue paper. (The term imbibition is usually reserved in the porous media literature for the spontaneous uptake of wetting fluid, while drainage refers to the opposite process of applying some force to displace a wetting phase with a nonwetting phase). In practice, many systems do not display clear and distinctive wetting behavior; water does not imbibe into a dry specimen or air does not imbibe into a wet one. Such systems are said to exhibit intermediate or neutral wettability. An example of the counterintuitive behavior in intermediate wetting materials is forced imbibition where the nonwetting phase must be removed by forcefully injecting the wetting phase. In this situation, the nonwetting phase is at a lower pressure than the wetting phase as well as being on the convex side of the interface. This would seem to contradict the definitions of wetting and nonwetting phases given above according to their relative pressures or positions relative to the meniscus. On the other hand, neutrally wet materials can exhibit some of the features expected of typical wetting and nonwetting systems. For instance, the wetting phase saturation can be reduced to zero during injection of a nonwetting fluid (drainage), while the nonwetting phase saturation may not be decreased to zero during withdrawal (imbibition and/or forced imbibition). When discussing displacements in neutrally wet materials, it may be beneficial to avoid the use of terms such as imbibition, drainage, wetting, and nonwetting. Unambiguous terms such as water injection and water withdrawal are preferable and are used throughout this chapter.

Anderson (1986, 1987) has produced a series of informative review articles discussing many practical aspects and implications of neutrally wettable systems, which he classified as those displaying a contact angle between 60° and 120°. Neutral wettability has a direct bearing on fuel cell operation and behavior since water forms neutral contact angles on carbon/graphite and poly-tetra-fluoro-ethylene (PTFE) which are major constituents of PEMFC electrodes.

24.2.5 Porosimetry

The simple inverse relationship between capillary pressure and interface size described by Equation 24.1 is the basis of the widely used porous media characterization technique of porosimetry. With the assumption that all pores can be approximated as cylinders, porosimetry can be used to study the structure of porous materials using the relationship between the pressure required to forcefully inject a nonwetting fluid and pore size given by the Washburn equation obtained by combining Equations 24.1 and

24.2 (Washburn, 1921a,b). In principle, measurement of the volume of nonwetting fluid injected into a porous sample at a given capillary pressure P_C gives the volume of the pores in the medium with a size of $2\sigma \cos \theta/P$. In reality, however, a porosimetry experiment only provides the volume of pore space accessible at a given pressure (i.e., a capillary pressure curve). Use of the Washburn equation to determine pore sizes is limited by the fact that pores are neither circular nor straight and regions of large pores are shielded by small pores in actual 3D porous networks. An implication of the first limitation is that the relationship between entry pressure and pore size is not as simple as that given by Equation 24.2 (Lindquist, 2006), while the second factor undermines the use of the intruded volume to estimate the fraction of pores of a given size. Pore size distributions obtained via porosimetry serve as useful estimates, but one must be mindful of the above limitations.

24.3 Experimental Techniques

24.3.1 Mercury Intrusion Porosimetry

The most commonly used experimental method for studying the capillary pressure characteristics of porous materials is MIP (Leon, 1998). In this method, the volume of mercury injected into a porous sample is tracked as a function of applied pressure, yielding a capillary pressure curve. The use of mercury as the invading fluid gives MIP a crucial advantage: because mercury is highly nonwetting on almost all solid surfaces, MIP data are affected only by pore geometry (i.e., pore sizes) and not by factors associated with wettability or surface chemistry in the media. This is advantageous when structural information is sought. For instance, the aging of a GDL may lead to erosion and loss of PTFE leading to a larger average pore size. This may be indicated in an MIP experiment by a shift of the capillary pressure curve to lower intrusion pressures. On the other hand, if aging of a GDL only involves a chemical process such as oxidation of the graphite fibers that alters their wettability, this would not be detected by MIP.

In cases such as quartz- and silica-based materials typical of oil reservoirs and hydro-geologic formations, it is possible to obtain an equivalent air–water capillary pressure curve from MIP capillary pressure data. Since these surfaces are strongly hydrophilic, water is a highly wetting fluid and air nonwetting. The invasion of mercury into such a porous material filled with air is analogous to the invasion of air into a water-filled structure. The capillary pressure need only be scaled by the different surface tensions in each case by rearranging Equation 24.1 (solving for H) and equating them to yield $P_{C,Air-Water}/\sigma_{Air-Water} = P_{C,Hg-Air}/\sigma_{Hg-Air}$. This is possible because the wettability of the nonwetting phase in each system is similar so that their curvatures H are equal at a given saturation. On the other hand, water is less wetting and air less nonwetting on GDL surfaces and so the invasion of mercury is not completely analogous to the invasion of air into a water-filled porous material, the conversion between capillary pressures is not as simple. It is common to use Equation 24.2 to include the effect of wettability via the contact angle when converting from one type of capillary pressure curve to the other. This yields the relation $P_{C,Air-Water}/(\sigma_{Air-Water} \cdot \cos\theta_{Air-Water}) = P_{C,Hg-Air}/(\sigma_{Hg-Air} \cdot \cos\theta_{Hg-Air})$. The failure of contact angle to reliably account for wettability differences in noncylindrical pores has already been discussed in Section 24.2.3. Any attempts in this direction are further undermined by the fact the surfaces of many real materials are heterogeneous with regard to structure and chemical composition. This is the case for GDLs, which are comprised of carbon fiber substrates partially coated with PTFE to enhance hydrophobicity. Due to this compound or dual wettability, at least two contact angles operate over the surface of this material and so any attempts to convert capillary data from one fluid system to that of another (i.e., mercury–air to water–air) require knowledge of the fluid contact angles on the different types of surfaces, the relative proportion of each type of surface and perhaps even their breakdown with respect to pore size, which is virtually impossible to determine. Consequently, it is necessary to use fluid phases of direct interest to PEMFCs and not rely on MIP data in order to properly study wettability effects in GDLs.

24.3.2 Air–Water Capillary Pressure

A number of techniques for measuring air–water capillary pressure curves in GDLs have been presented in recent years (Fairweather et al., 2007; Gallagher et al., 2008; Gostick et al., 2006a; Gostick et al., 2008; Harkness et al., 2009; Koido et al., 2008; Nguyen et al., 2008; Rensink et al., 2008; Sole and Ellis, 2008). Measurement of these curves for GDLs requires an approach that is significantly modified from that used for the more traditional measurements on rock cores, soil samples and sand packs due to the following three factors. First, because the system of interest is air–water–GDL and GDLs typically receive various wettability altering treatments, it is necessary that air and water are used as the working fluids. The use of mercury or another substitute fluid with the intention of converting the data so obtained to an air–water capillary pressure is not useful and is essentially futile for the reasons discussed above. Second, GDLs are very thin (200–400 μm) and therefore have a very low pore volume which demands precise measurement of the fluid saturation. Third, due to the neutral wettability of GDLs to water, it is necessary to apply positive capillary pressure to force water into a GDL, but a negative capillary pressure to draw it out. A successful method must also be able to scan the capillary pressure in both increasing and decreasing directions over both positive and negative ranges. A recent review by Gostick et al. (2010a) describes and critically analyzes the many techniques that have been proposed in the literature. In this review, it was concluded that only two methods fully meet all of the necessary criteria: the pressure-step-controlled approach of Gostick et al. (2008, 2009b) and the volume-scanning method of Harkness et al. (2009). To avoid duplication, the present discussion will focus only on these two methods.

Both the methods of Gostick et al. (2008) and Harkness et al. (2009) effectively use the same sample holder, which is based on the classic porous plate technique. A schematic diagram of this type of sample holder is shown in Figure 24.4. Below the sample is a porous hydrophilic membrane through which liquid water flows in and out of the sample. The purpose of this membrane is to prevent air from leaving the system during water withdrawal. This membrane must therefore be chosen so that its breakthrough or bubble point pressure is lower than the most negative capillary pressure reached during the experiment. The hydrophobic membrane above the sample serves the reverse purpose by allowing air but not water to leave the system. Typically the hydrophobic barrier is a PTFE membrane with pore size smaller than 500 nm, while a number of materials can be used as the hydrophilic membrane (e.g., PVDF with 220 nm pores was used by Gostick et al. (2008, 2009b)). With this setup,

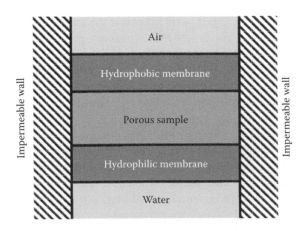

FIGURE 24.4 Schematic diagram of porous-plate-type sample holder typically used to measure GDL capillary pressure curves. Water flow through the hydrophilic membrane is unimpeded but cannot breach the hydrophobic membrane at pressures reached in the experiment. The hydrophobic membrane is a capillary barrier. Displaced air escapes through the upper hydrophobic membrane unimpeded, but cannot breach the hydrophilic membrane below.

the sample is completely isolated so that a capillary pressure higher than that required to penetrate the sample can be applied.

The method developed by Gostick et al. (2008) controls the capillary pressure in the sample and tracks the water uptake over time. The main limitation of this approach is that high positive capillary pressures are generated by creating a gas vacuum above the sample ($P_C = P_L - P_G$). Thus, it is limited to capillary pressures less than the vapor pressure of water ($P_C \approx 95{,}000$ Pa); in practice, it is limited to even lower pressures to reduce vaporization of the water into the gas space above the sample. On the other hand, there is no theoretical limit to the negative capillary pressures that can be reached. The rate-controlled or volume-controlled method of Harkness et al. (2009) utilizes a syringe pump to inject and withdraw water at a controlled rate while monitoring the liquid pressure response. Harkness et al. (2009) were aware of the potential problems of volume-controlled methods and demonstrated that the flow rates used did not introduce viscous pressure loss effects or nonequilibrium effects. This method has no limitations on the positive capillary pressure that can be reached, making it the only option for studying components such as the microporous layer or catalyst layer that have much smaller pores than the GDL. On the other hand, it is limited with regard to negative capillary pressures since water will cavitate once the vapor pressure of water is reached. Depending on the expected behavior of the materials being tested, the respective capillary pressure limits of each method may become a factor. The volume-controlled approach is better suited for testing microporous layers with small pores and high breakthrough pressure, while the pressure-controlled approach is more suitable for GDLs with enhanced hydrophilic properties (Gallagher et al., 2008; Litster et al., 2007).

The extremes of pressure reached in these experiments are not expected to occur during fuel cell operation. For characterization purposes, however, it is necessary to reach full saturation on injection and residual saturation on withdrawal. Also, since condensation may lead to very high electrode saturations, it is necessary to study the extremes of saturation in order to more fully investigate of all processes that may occur during operation. One very important aspect that has not yet been explored is the effect of temperature on fuel cell capillary properties due to its influence, in particular, on the surface tension of water in GDLs. Surface tension strongly affects wettability of the GDL material so it is expected that temperature will alter the capillary pressure behavior in fuel cells which are operated at temperatures as high as 95°C.

24.3.2.1 Capillary Pressure Hysteresis

The existence of capillary pressure hysteresis is ubiquitously observed in measurements on all types of porous materials (Hammervold et al., 1998). There are numerous factors contributing to the hysteresis ranging from contact angle hysteresis (Leon, 1998) to thermodynamic irreversibility of interfacial movements (Melrose, 1968; Morrow, 1970) to the fact that invasion is controlled by pore throat sizes while withdrawal by pore body sizes (Lenormand et al., 1983). The hysteresis in GDLs may appear to be a special extreme case since water injection occurs at positive capillary pressures and withdrawal at negative capillary pressures. Typical capillary pressure curves for a GDL with and without PTFE shown in Figure 24.5 clearly reveal negative withdrawal pressures. As mentioned above, the hysteresis is occurring for usual reasons and the fact that the hysteresis switches sign is a product of the GDL's neutral wettability. Numerous factors can contribute to this switch and cause the interface to switch direction of curvature, including contact angle hysteresis (Cheung et al., 2009), the converging–diverging nature of constrictions in fibrous media (Harkness et al., 2009) and cooperative pore filling during water withdrawal (Gostick et al., 2008). Each of these effects has the ability to alter the sign of the mean curvature of the interface, particularly when contact angles are in the range of intermediate wettability.

The switch in capillary pressure to negative values suggests that the phase wettability has reversed and water has become the wetting phase. This is the problem associated with the definition of phase wettability that was discussed in Section 24.2.4. Since there is no reason that water should suddenly gain an affinity for the GDL simply because it is flowing in the opposite direction, the definition appears to break down for materials with neutral wettability. Alternatively, one can look at the behavior of the

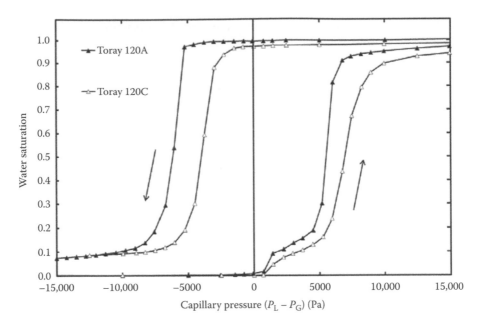

FIGURE 24.5 Capillary pressure curves for Toray 120C with 10 wt% PTFE coating and Toray 120A with no PTFE coating. Arrows denote direction of pressure scan. The addition of PTFE to the GDL results in a shift to higher water injection pressures indicating increased hydrophobicity, but water withdrawal still requires negative capillary pressures. (Reproduced from *J. Power Sources*, 194, Gostick, J. T. et al. Wettability and capillary behavior of fibrous gas diffusion media for polymer electrolyte membrane fuel cells. 433–444, Copyright (2009), with permission from Elsevier.)

fluids to deduce their wettability. For instance, the observation that some residual water remains in the pores after withdrawal indicates that it is behaving as a nonwetting fluid since wetting fluids can be reduced to zero saturation. Furthermore, water withdrawal shows no finite size effects (Gostick et al., 2010a) indicating that a forced imbibition is occurring, thus water is nonwetting. This uncertainty has led to significant confusion about the wettability of GDLs. The early data of Gostick et al. (2006a) showed that water removal from GDLs required negative capillary pressure which was attributed to the existence of a "hydrophilic pore network." Subsequent work has clearly shown that no such network exists but the concept of a hydrophilic network persists. Recently, Gostick et al. (2009b) attempted to quantify the GDL wettability using the US Bureau of Mines wettability index. This approach accounts for the amount of work required for injection versus withdrawal. It was found that untreated GDLs required slightly more work for withdrawal than injection, while the opposite was observed for GDLs treated with PTFE, indicating that the samples switched from slightly hydrophilic to slightly hydrophobic upon application of PTFE (see Figure 24.5). This index has also been used by Fairweather et al. (2010) with similar results.

The hysteresis in GDLs has important implications for fuel cell operation since it indicates that not all water will be removed from GDL pores under normal operating conditions in PEM fuel cells. The inability to remove water from the cell prior to shutdown in subzero conditions could cause damage during freezing, for instance. Some elaborate schemes have been employed to suck water from the GDL (Bett et al., 1996; Litster et al., 2007). An alternative is to promote evaporation using subsaturated air, but this creates unwanted humidity cycles in the membrane. The use of a GDL that is highly hydrophobic to the extent that water is spontaneously ejected would seem beneficial, but the increased capillary pressure necessary for water to flow through it may increase catalyst layer flooding. Clearly a need exists for innovation in the area of MEA water management.

24.3.2.2 Breakthrough Conditions

Liquid water generated in the cathode catalyst layer by oxygen reduction must flow out of the cell through the GDL via an invasion percolation process. In this situation, liquid water will penetrate many GDL pores to find a pathway across the GDL resulting in many dead-end liquid pathways. Since this process is so essential to fuel cell operation, it is important to characterize the pressure and liquid saturation at the breakthrough point. The measurement of breakthrough pressure can be very straightforward as described by Benziger et al. (2005). They essentially mounted a small GDL sample on the end of a tube and increased the water head in the tube until breakthrough occurred. Liquid saturation was determined by measuring the change in sample weight before and after the test, but it is difficult with this approach to remove loose water from the sample without altering the internal saturation. More elaborate approaches have been described by Gao et al. (2009) and Djilali and co-workers (Bazylak et al., 2007; Litster et al., 2006) using a pressure sensor to track the liquid pressure with time as water was injected with a syringe pump, but the breakthrough saturation was not reported in these cases. Gostick et al. (2010b) used the syringe pump method but also calculated the water saturation from knowledge of the pump rate and experimental time.

The breakthrough pressure and saturation obtained for GDLs with and without PTFE treatment were reported by Gostick et al. (2010b). They found no increase in breakthrough pressure with PTFE application, but the saturation was reduced by nearly 50%. Benziger et al. (2005) reported an increase in breakthrough pressure with the addition of PTFE content. It is unclear why the two groups obtained different results in this regard. Interestingly, Benziger et al. (2005) found that although the breakthrough pressure was affected by the presence of PTFE, it was not affected by the amount actually present, similar to the effect observed by Gostick et al., (2009b) that the capillary pressure curves were not affected by PTFE loading (Gostick et al., 2009b). The use of breakthrough conditions as a means of estimating GDL wettability has several advantages. It is significantly faster to determine the breakthrough point since only one measurement is required. It should also be more practical to measure the breakthrough point at elevated temperature since no high precision equipment is required. The breakthrough pressure and/or saturation have been shown to be sensitive to wettability so such measurements may provide an excellent means of determining changes in GDL properties due to degradation, for example.

24.3.2.3 Microporous Layers

Most MEA's include a microporous layer (shown on the cathode side in Figure 24.1). This layer is usually applied to the GDL as a paste of carbon black and PTFE particles and then sintered. This creates a consolidated layer of very small pores less than 500 nm in size (Ostadi et al., 2010). This layer is known to improve fuel cell performance but its exact function is still being debated. Weber and Newman (2005) suggested that the MPL is essentially impermeable to water and therefore forces water generated in the cathode catalyst layer to flow across the membrane to the anode, but this was not observed in subsequent water balance experiments (Atiyeh et al., 2007). Gostick et al. (2009a) and Nam et al. (2009) proposed that water flows into the MPL only from a limited number of injection points which thereby reduces the GDL saturation. It has also been proposed that the presence of the MPL increases the temperature gradient across the MEA which leads to vaporization of water at the catalyst layer and condensation at the channels, the so-called phase change-induced flow (Kim and Mench, 2009). Regardless of which mechanism operates, the capillary properties of the MPL play a very important role.

Because the pores of the MPL are much smaller than those in the GDL, it is expected that significantly higher pressure is required for water invasion. This higher pressure should not be confused with higher hydrophobicity since the constituent materials are identical (carbon and PTFE). To date no capillary pressure curves on MPL materials have been published. The high entry pressures of the MPL pores require that the volume-controlled method of Harkness et al. (2009) be utilized. Gostick et al. (2009a) performed tests on GDLs with MPL using the pressure-controlled approach and observed essentially no water injection into the MPL at pressures up to 25,000 Pa, although breakthrough was observed at lower pressures likely through cracks.

24.3.3 Contact Angle Measurement

Despite the difficulties with interpreting wettability in terms of contact angle discussed in Section 24.2.3, it can still be a useful qualitative tool for making direct comparisons between GDLs, such as between fresh and degraded samples. Contact angle is often used as a parameter in multiphase flow models that use the Leverett normalization function to scale capillary pressure curves obtained for material X to those for model material Y. This was originally used by Wang et al. (1999, 2001) to adjust capillary pressure data from sand packs to model GDLs since GDL-specific capillary data were not available at that time. Now that extensive capillary pressure data for actual GDLs are available and readily measured there is no need to use the Leverett function or contact angle in multiphase flow models.

The most direct technique for contact angle measurement is the sessile drop method, where a drop of liquid is placed on a solid surface and the resulting angle formed between the plane of the solid and the tangent to the drop at the solid–liquid contact is measured. This approach becomes complicated when the solid of interest is a porous media. Due to the roughness and porosity of the surface, a droplet only contacts a fraction of the solid which causes the observed contact angle to deviate from the actual value. Using the sessile drop approach, Gostick et al. (2006a) attempted to determine the actual contact angle by correcting the observed value through a combination of the Wenzel equation (1936) to account for roughness and the Cassie–Baxter equation (1944) to account for porosity. Although reasonable values were obtained, the number of assumptions made to obtain the correction raise questions concerning the approach that require further validation in the future. An equivalent but alternative method is the Wilhelmy plate technique, which involves dipping a strip of the GDL into a reservoir of water and measuring the meniscus height. Mathias et al. (2003) utilized this approach to measure the water contact angle by dipping GDLs into water. Since water does not imbibe into GDLs, the porous material acts as a solid. (*Note:* In the Washburn method, a similar measurement is made, but it is applied in situations where the fluid wicks into the dipped sample, as discussed below.) Mathias et al. (2003) inserted and retracted the sample and observed different advancing and receding contact angles and even found different values for PTFE-treated and virgin GDL samples, showing the qualitative sensitivity of the approach. They did not correct for the roughness or porosity to obtain an actual contact angle. Determination of an actual contact angle from either approach is probably not feasible due to the corrections necessary to account for the porous nature of the sample; however, observation of variations in the contact angle due to changes of some other variable (e.g., degradation) is quite useful for qualitative comparisons. For instance, Lim and Wang (2004) used the Wilhelmy plate technique. Lim found that the contact angle decreased as temperature was increased.

An alternative means of measuring contact angle is the Washburn technique, which is similar to the Wilhelmy technique, but applies when the liquid wicks into the sample. By measuring the height and rate of imbibition into the porous sample, the contact angle can be calculated. Since this obviously requires that the liquid imbibe, such tests must use wetting fluids such as octane, dodecane, and so on. Gurau et al. (2006) were the first to apply this approach to GDLs. They measured the capillary rise of several different wetting liquids and then applied Owens–Wendt theory to infer the water contact angle. This approach has recently been applied by several workers (Friess and Hoorfar, 2010; Parry et al., 2010). Wood et al. (2010) have also employed the Washburn method, but analyzed the results according to the Zisman theory as well as the Owens–Wendt theory. They also used the Wilhelmy technique to measure the contact angle on single fibers extracted from GDL samples to avoid the difficulties associated with measurements on a porous surface.

24.4 Applications and Degradation

The effects of aging and degradation on GDL capillary and wettability properties have not been well studied. This is largely because experimental methods for testing capillary pressure and wettability have only recently become available. A number of recent review articles have focused on degradation in PEMFCs and a few have included discussion of GDL degradation (Borup et al., 2007; Wu et al., 2008; Zhang et al., 2009), but minimal coverage of GDL degradation indicates that very little work has been done in this area. The following section is divided according to the mechanism considered to be responsible for GDL degradation and discusses the use of each characterization method described above.

24.4.1 Aging

Perhaps the most obvious concern about studying GDL degradation is the instability of its wetting characteristics with time. Presumably, a GDL is chosen for its desirable beginning-of-life (BOL) capillary properties (among other qualities) and any deviation from these properties toward the end-of-life (EOL) of the GDL is undesirable. The evolution of wetting properties from BOL to EOL can be very broadly referred to as aging. In similarly broad terms, aging can be said to occur by processes causing either chemical and/or structural changes. An example of chemical alterations would be oxidation of the carbon fiber surfaces in the GDL. Since un-Teflonated GDLs are rather hydrophobic when they are new, any chemical changes to the carbon fiber might lead to more hydrophilic behavior. Structural changes to the GDL could involve deterioration of the PTFE coating caused by loss of adhesion and subsequent erosion. Also, plastic materials like PTFE could undergo sintering at cell operating temperatures causing agglomeration and loss of hydrophobic surface area. The classification of aging into structural and chemical effects suggests that MIP would be ideally suited to study the former while some form of water–air capillary pressure measurement could be used to study the later. Unfortunately, no comprehensive efforts have been made to study aged GDLs with either technique. It has been shown, however, that MIP experiments on GDLs with 0%, 10%, and 20% PTFE loading yield capillary pressure curves with no significant differences (Gostick et al., 2009b), which is somewhat discouraging since it casts doubt on the possibility of differentiating between structural and chemical sources of wettability changes. Ide and Ikeda (2010) performed MIP on gas diffusion electrodes aged up to 12,000 h of operation and observed only small changes over the entire range of pore sizes. Air–water capillary pressure curves, on the other hand, should be able to resolve any wettability changes that occur from aging.

It is clear that the environment to which a GDL is exposed inside a fuel cell will alter its wettability. For example, Borup et al. (2007) reported a simple experiment where a GDL was submerged in high temperature, oxygenated water. They found that the contact angle decreased with exposure time. This situation is very comparable to the conditions on the cathode side of an operating cell. Conditions on the anode side are also likely to affect a GDL since low hydrogen partial pressures lead to carbon corrosion. Chen et al. (2009) performed an accelerated aging test on GDL by electrochemically oxidizing them in H_2SO_4 solution and found that the contact angle decreased as the applied potential was raised. Imaging analysis showed significant loss of material from the MPL due to carbon corrosion. The loss of carbon was accompanied by loss of PTFE due to the deterioration of the MPL structure. Loss of PTFE from the GDL could also be caused by oxidation of the underlying supporting carbon fibers, although these fibers are more highly graphitized than carbon black powder in the MPL and therefore more resistant to oxidation (Borup et al., 2007).

The techniques that have been developed to study the capillary properties of the GDL are fairly new and have yet to receive wide adoption. Efforts to relate BOL GDL capillary properties to PEMFC performance have barely been undertaken (Harkness et al., 2009). Investigations of EOL changes are equally critical, particularly since cell durability, reliability, and longevity are key steps in PEMFC commercialization.

24.4.2 Compression

Fuel cell assembly requires that the stack be compressed to ensure tight gas seals and good electrical conductivity at the GDM–bipolar plate interface. The membrane material also swells as much as 20% depending on the humidity levels in the cell. Since the GDL is the most compliant component in the fuel cell stack, it will absorb all the compression due to assembly, humidity cycles, and thermal cycles. The effect of compression on the transport and capillary properties of the GDL is of interest and has been studied fairly extensively (Flückiger et al., 2008; Gostick et al., 2006b, 2009b). The extent of the permanent damage caused by compression and the effect on GDL properties has not been studied as extensively.

Compression leads to reduced porosity, smaller average pore size, and possibly other microstructure changes in the GDL. Since capillary properties are controlled by pore size, it is expected that the capillary behavior of GDLs should differ depending on whether or not they are compressed. It has been shown (Gostick et al., 2009b; Harkness et al., 2009) that GDL capillary pressure curves are altered by compression as expected, with higher water entry pressure required in compressed samples. Figure 24.6 shows capillary pressure curves for Teflon-treated Toray 120 under compression.

Neither study focused on the permanent damage that was likely caused by the compression. Escribano et al. (2006) conducted repeated compressive cycling experiments on GDLs which clearly showed permanent deformation. The effect of permanent damage to the GDL on its capillary properties has not received much attention to date. Bazylak et al. (2007) studied the injection of water into and through GDLs and found that water tended to breakout near compressed areas. This was contrary to expectation since compressed areas should exhibit smaller pores that are harder to penetrate. Subsequent SEM analysis showed breakage of the fibers in the compressed region which could have created preferential flow paths. Very few other investigations into the compressive damage of the GDL have been reported.

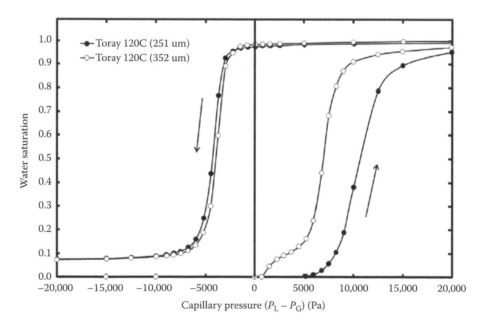

FIGURE 24.6 Capillary pressure curves for Toray 120C (10 wt% PTFE coating) in uncompressed (white markers) and compressed state (black markers). Arrows denote direction of pressure scan. The thickness of the samples are given in the legend. (Reproduced from *J. Power Sources*, 162, Gostick, J. T. et al. In-plane and through-plane gas permeability of carbon fiber electrode backing layers. 228–238, Copyright (2006), with permission from Elsevier.)

24.4.3 Freezing

The main concern of freezing in a porous electrode is damage caused by the volume expansion of ice. If water freezes within the GDL, it can lead to carbon fiber breakage, weaker PTFE adhesion to the fibers and therefore loss of hydrophobicity, as well as increased cracking of the MPL. The vast majority of freeze studies simply use SEM images to visually inspect the damage which can be considerable (Kim and Mench, 2007). As with most problems related to GDL degradation and capillary properties, relatively few *ex situ* studies of freeze- and ice-damaged GDLs have been reported.

The presence of liquid water freezing in the GDL results in expansion forces that presumably have an equivalent effect to the damage caused by compressive forces described in the previous section. Guo and Qi (2006) presented qualitative images showing the extent of the damage that ice formation can cause in a GDL. In controlled and less destructive tests, the in-plane and through-plane gas permeability have both been shown to increase after repeated freezing cycles (Lee and Merida, 2007). Since permeability is closely linked to pore size, one would expect the capillary behavior of liquid water to also be altered. Lee and Merida (2007) observed no change in water contact angle after the freeze cycles, which is expected since contact angle is independent of pore size.

Understanding the effect of freeze damage on GDL properties is critical since some amount of ice formation is probably unavoidable in actual operation. It is also preferable that the GDL deform to accommodate the expansion of ice rather than undergo a catastrophic frost heave of the entire stack. Determination of the amount and location of ice formation that a GDL can tolerate without damage and altered wetting behavior will require the application of the techniques described previously.

24.5 Conclusion and Outlook

Methods to measure the air–water capillary pressure properties of GDLs have only recently become available. Consequently, the efforts to fully understand capillary behavior in these materials are just now beginning. At this point, it is not clear what capillary pressure properties an ideal GDL should have since only relatively little work to correlate such data to fuel cell performance has been done. It is known that GDL wettability can change with time during operation, generally moving toward a more hydrophilic state (Borup et al., 2007), but clearly this topic needs further investigation. Application of air–water capillary pressure techniques to aged and degraded GDL materials has still been limited. The use of contact angle to gauge wettability has been proposed and shown to be sensitive to age-related wettability changes. Contact angle is only an indicator of wettability, however, and does not give detailed information regarding the actual capillary behavior in GDL and its link to the nature of the material. Although this chapter focused exclusively on the capillary properties of the GDL, the methods discussed can feasibly be adapted to study MPL and catalyst layers materials in future investigations. The effect of elevated temperature on capillary properties is also a matter of pressing interest.

References

Anderson, W. G. 1986. Wettability literature survey—Part 2: Wettability measurement. *J. Petrol. Tech.* 38: 1246–1262.

Anderson, W. G. 1987. Wettability literature survey—Part 4: Effects of wettability on capillary-pressure. *J. Petrol. Tech.* 39: 1283–1300.

Atiyeh, H. K., Karan, K., Peppley, B., Phoenix, A., Halliop, E., and Pharoah, J. 2007. Experimental investigation of the role of a microporous layer on the water transport and performance of a PEM fuel cell. *J. Power Sources* 170: 111–121.

Bazylak, A., Sinton, D., Liu, Z. S., and Djilali, N. 2007. Effect of compression on liquid water transport and microstructure of PEMFC gas diffusion layers. *J. Power Sources* 163: 784–792.

Benziger, J., Nehlsen, J., Blackwell, D., Brennan, T., and Itescu, J. 2005. Water flow in the gas diffusion layer of PEM fuel cells. *J. Membr. Sci.* 261: 98–106.

Bett, J. A. S., Wheeler, D. J., and Bushnell, C. 1996. Porous Carbon Body with Increased Wettability by Water. Patent number: US 5840414. S. Winsor, CT: International Fuel Cells, Inc.

Borup, R., Meyers, J., Pivovar, B. et al. 2007. Scientific aspects of polymer electrolyte fuel cell durability and degradation. *Chem. Rev.* 107: 3904–3951.

Cassie, A. B. D. and Baxter, S. 1944. Wettability of porous surfaces. *Faraday Soc. Trans.* 40: 546–551.

Cerepi, A., Humbert, L., and Burlot, R. 2002. Dynamics of capillary flow and transport properties in porous media by time-controlled porosimetry. *Colloids Surf., A* 206: 425–444.

Chen, G., Zhang, H., Ma, H., and Zhong, H. 2009. Electrochemical durability of gas diffusion layer under simulated proton exchange membrane fuel cell conditions. *Int. J. Hydrogen Energy* 34: 8185–8192.

Cheung, P., Fairweather, J. D., and Schwartz, D. T. 2009. Characterization of internal wetting in polymer electrolyte membrane gas diffusion layers. *J. Power Sources* 187: 487–492.

Dullien, F. A. L. 1992. *Porous Media: Fluid Transport and Pore Structure*. New York, NY: Academic Press.

Escribano, S., Blachot, J. F., Etheve, J., Morin, A., and Mosdale, R. 2006. Characterization of PEMFCs gas diffusion layers properties. *J. Power Sources* 156: 8–13.

Fairweather, J. D., Cheung, P., and Schwartz, D. T. 2010. The effects of wetproofing on the capillary properties of PEMFC gas diffusion layers. *J. Power Sources* 195: 787–792.

Fairweather, J. D., Cheung, P., St Pierre, J., and Schwartz, D. T. 2007. A microfluidic approach for measuring capillary pressure in PEMFC gas diffusion layers. *Electrochem. Commun.* 9: 2340–2345.

Flückiger, R., Freunberger, S. A., Kramer, D., Wokaun, A., Scherer, G. G., and Büchi, F. N. 2008. Anisotropic, effective diffusivity of porous gas diffusion layer materials for PEFC. *Electrochim. Acta* 54: 551–559.

Friess, B. R. and Hoorfar, M. 2010. Measurement of internal wettability of gas diffusion porous media of proton exchange membrane fuel cells. *J. Power Sources* 195: 4736–4742.

Gallagher, K. G., Darling, R. M., Patterson, T. W., and Perry, M. L. 2008. Capillary pressure saturation relations for PEM fuel cell gas diffusion layers. *J. Electrochem. Soc.* 155: B1225–B1231.

Gao, B., Steenhuis, T. S., Zevi, Y., Parlange, J. Y., Carter, R. N., and Trabold, T. A. 2009. Visualization of unstable water flow in a fuel cell gas diffusion layer. *J. Power Sources* 190: 493–498.

Gostick, J. T., Fowler, M. W., Ioannidis, M. A., Pritzker, M. D., Volfkovich, Y. M., and Sakars, A. 2006a. Capillary pressure and hydrophilic porosity in gas diffusion layers for polymer electrolyte fuel cells. *J. Power Sources* 156: 375–387.

Gostick, J. T., Fowler, M. W., Pritzker, M. D., Ioannidis, M. A., and Behra, L. M. 2006b. In-plane and through-plane gas permeability of carbon fiber electrode backing layers. *J. Power Sources* 162: 228–238.

Gostick, J. T., Ioannidis, M. A., Fowler, M. W., and Pritzker, M. D. 2008. Direct measurement of the capillary pressure characteristics of water–air–gas diffusion layer systems for PEM fuel cells. *Electrochem. Commun.* 10: 1520–1523.

Gostick, J. T., Ioannidis, M. A., Fowler, M. W., and Pritzker, M. D. 2009a. On the role of the microporous layer in PEMFC operation. *Electrochem. Commun.* 11: 576–579.

Gostick, J. T., Ioannidis, M. A., Fowler, M. W., and Pritzker, M. D. 2009b. Wettability and capillary behavior of fibrous gas diffusion media for polymer electrolyte membrane fuel cells. *J. Power Sources* 194: 433–444.

Gostick, J. T., Ioannidis, M. A., Fowler, M. W., and Pritzker, M. D. 2010a. Characterization of the capillary properties of gas diffusion media. In *Modern Aspects of Electrochemistry*, Eds. C. Y. Wang and U. Pasaogullari. Berlin: Springer.

Gostick, J. T., Ioannidis, M. A., Pritzker, M. D., and Fowler, M. W. 2010b. Impact of liquid water on reactant mass transfer in PEM fuel cell electrodes. *J. Electrochem. Soc.* 57: B563–B571.

Guo, Q. and Qi, Z. 2006. Effect of freeze–thaw cycles on the properties and performance of membrane-electrode assemblies. *J. Power Sources* 160: 1269–1274.

Gurau, V., Bluemle, M. J., De Castro, E. S., Tsou, Y. M., Mann, J. A., and Zawodzinski, T. A. 2006. Characterization of transport properties in gas diffusion layers for proton exchange membrane fuel cells: 1. Wettability (internal contact angle to water and surface energy of GDL fibers). *J. Power Sources* 160: 1156–1162.

Hammervold, W. L., Knutsen, Ø., Iversen, J. E., and Skjæveland, S. M. 1998. Capillary pressure scanning curves by the micropore membrane technique. *J. Petrol. Sci. Eng.* 20: 253–258.

Harkness, I. R., Hussain, N., Smith, L., and Sharman, J. D. B. 2009. The use of a novel water porosimeter to predict the water handling behaviour of gas diffusion media used in polymer electrolyte fuel cells. *J. Power Sources* 193: 122–129.

Ide, M. and Ikeda, H. 2010. Investigation into the gas diffusion electrodes of polymer electrolyte membrane fuel cell under long-term durability test. *J. Renewable Sustainable Energy* 2 (013110):1–11.

Ji, M. B. and Wei, Z. D. 2009 A review of water management in polymer electrolyte membrane fuel cells. *Energies* 2: 1057–1106.

Kim, S. and Mench, M. M. 2007. Physical degradation of membrane electrode assemblies undergoing freeze/thaw cycling: Micro-structure effects. *J. Power Sources* 174: 206–220.

Kim, S. and Mench, M. M. 2009. Investigation of temperature-driven water transport in polymer electrolyte fuel cell: Phase-change-induced flow. *J. Electrochem. Soc.* 156: B353–B362.

Knackstedt, M. A., Sheppard, A. P., and Pinczewski, W. V. 1998. Simulation of mercury porosimetry on correlated grids: Evidence for extended correlated heterogeneity at the pore scale in rocks. *Phys. Rev. E: Stat., Nonlinear, Soft Matter Phys.* 58: R6923–R6926.

Koido, T., Furusawa, T., and Moriyama, K. 2008. An approach to modeling two-phase transport in the gas diffusion layer of a proton exchange membrane fuel cell. *J. Power Sources* 175: 127–136.

Lee, C. and Merida, W. 2007. Gas diffusion layer durability under steady-state and freezing conditions. *J. Power Sources* 164: 141–153.

Lenormand, R., Zarcone, C., and Sarr, A. 1983. Mechanisms of the displacement of one fluid by another in a network of capillary ducts. *J. Fluid Mech.* 135: 337–353.

Leon, C. A. 1998. New perspectives in mercury porosimetry. *Adv. Colloid Interface Sci.* 76–77: 341–372.

Li, Q. F., He, R. H., Jensen, J. O., and Bjerrum, N. J. 2003. Approaches and recent development of polymer electrolyte membranes for fuel cells operating above 100 degrees C. *Chem. Mater* 15: 4896–4915.

Lim, C. and Wang, C. Y. 2004. Effects of hydrophobic polymer content in GDL on power performance of a PEM fuel cell. *Electrochim. Acta* 49: 4149–4156.

Lindquist, W. B. 2006. The geometry of primary drainage. *J. Colloid Interface Sci.* 296: 655–668.

Litster, S., Buie, C. R., Fabian, T., Eaton, J. K., and Santiago, J. G. 2007. Active water management for PEM fuel cells. *J. Electrochem. Soc.* 154: B1049–B1058.

Litster, S., Sinton, D., and Djilali, N. 2006. *Ex situ* visualization of liquid water transport in PEM fuel cell gas diffusion layers. *J. Power Sources* 154: 95–105.

Mathias, M. F., Roth, J., Fleming, J., and Lehnert, W. 2003. Diffusion media materials and characterization. In *Handbook of Fuel Cells—Fundamentals, Technology and Applications*, eds. W. Vielstich, H. A. Gasteiger, and A. Lamm (Vol. 3, Part 1). New York, NY: John Wily & Sons.

Melrose, J. C. 1968. Thermodynamic aspects of capillarity. *Ind. Eng. Chem.* 60: 53–70.

Meyers, J. P. and Darling, R. M. 2006. Model of carbon corrosion in PEM fuel cells. *J. Electrochem. Soc.* 153: A1432–A1442.

Morrow, N. R. 1970. Physics and thermodynamics of capillary action in porous media. *Ind. Eng. Chem.* 62: 32–56.

Nam, J. H., Lee, K.-J., Hwang, G.-S., Kim, C.-J., and Kaviany, M. 2009. Microporous layer for water morphology control in PEMFC. *In. J. Heat Mass Transfer* 52: 2779–2791.

Nguyen, T. V., Lin, G., Ohn, H., and Wang, X. 2008. Measurement of capillary pressure property of gas diffusion media used in proton exchange membrane fuel cells. *Electrochem. Solid-State Lett.* 11: B127–B131.

Ostadi, H., Rama, P., Liu, Y., Chen, R., Zhang, X. X., and Jiang, K. 2010. 3D reconstruction of a gas diffusion layer and a microporous layer. *J. Membr. Sci.* 351: 69–74.

Parry, V., Appert, E., and Joud, J. C. 2010. Characterisation of wettability in gas diffusion layer in proton exchange membrane fuel cells. *Appl. Surf. Sci.* 256: 2474–2478.

Rensink, D., Fell, S., and Roth, J. 2008. Liquid water transport and distribution in fibrous porous media and gas channels. *Proceedings of the Sixth International Conference on Nanochannels, Microchannels and Minichannels.* Darmstadt, Germany.

Sinha, P. K., Mukherjee, P. P., and Wang, C. Y. 2007. Impact of GDL structure and wettability on water management in polymer electrolyte fuel cells. *J. Mater. Chem.* 17: 3089–3103.

Sole, J. and Ellis, M. W. 2008. Determination of the relationship between capillary pressure and saturation in PEMFC gas diffusion media. *Proceedings of the 6th International Conference on Fuel Cell Science, Engineering and Technology.* Denver, CO.

Sygouni, V., Tsakiroglou, C. D., and Payatakes, A. C. 2006. Capillary pressure spectrometry: Toward a new method for the measurement of the fractional wettability of porous media. *Phys. Fluids* 18(053302): 1–15.

Toledo, P. G., Scriven, L. E., and Davis, H. T. 1994. Pore-space statistics and capillary–pressure curves from volume-controlled porosimetry. *Soc. Petrol. Eng.: Formation Evaluation* 9: 46–54.

Wang, C. Y., Wang, Z. H., and Pan, Y. 1999. Two-phase transport in proton exchange membrane fuel cells. *ASME HTD* 364: 351–357.

Wang, Z. H., Wang, C. Y., and Chen, K. S. 2001. Two-phase flow and transport in the air cathode of proton exchange membrane fuel cells. *J. Power Sources* 94: 40–50.

Washburn, E. W. 1921a. The dynamics of capillary flow. *Phys. Rev.* 17: 273–283.

Washburn, E. W. 1921b. Note on a method of determining the distribution of pore sizes in a porous material. *PNAS* 7: 115–116.

Weber, A. Z. and Newman, J. 2005. Effects of microporous layers in polymer electrolyte fuel cells. *J. Electrochem. Soc.* 152: A677–A688.

Wenzel, R. N. 1936. Resistance of solid surfaces to wetting by water. *Ind. Eng. Chem.* 28: 988–994.

Wood, D. L., Iii, Rulison, C., and Borup, R. L. 2010. Surface properties of PEMFC gas diffusion layers. *J. Electrochem. Soc.* 157: B195–B206.

Wu, J., Yuan, X. Z., Martin, J. J., Wang, H., Zhang, J., Shen, J. et al. 2008. A review of PEM fuel cell durability: Degradation mechanisms and mitigation strategies. *J. Power Sources* 184: 104–119.

Yuan, H. H. and Swanson, B. F. 1989. Resolving pore-space characteristics by rate-controlled porosimetry. *Soc. Petrol. Eng.: Form. Eval.* 4: 17–24.

Zhang, J. L., Xie, Z., Zhang, J. J., Tang, Y. H., Song, C. J., Navessin, T. et al. 2006. High temperature PEM fuel cells. *J. Power Sources* 160: 872–891.

Zhang, S., Yuan, X., Wang, H., Mérida, W., Zhu, H., Shen, J. et al. 2009. A review of accelerated stress tests of MEA durability in PEM fuel cells. *Int. J. Hydrogen Energy* 34: 388–404.

Index

G